Multicomponent Reactions

Synthesis of Bioactive Heterocycles

Multicomponent Reactions

Synthesis of Bioactive Heterocycles

Edited by

K. L. Ameta and Anshu Dandia

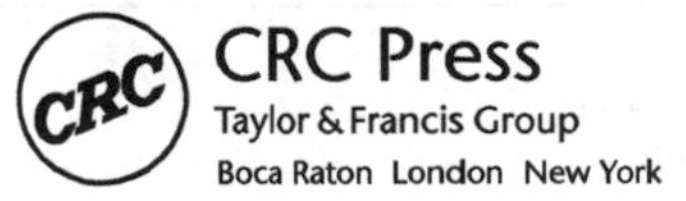

CRC Press
Taylor & Francis Group
Boca Raton London New York

CRC Press is an imprint of the
Taylor & Francis Group, an informa business

CRC Press
Taylor & Francis Group
6000 Broken Sound Parkway NW, Suite 300
Boca Raton, FL 33487-2742

Printed on acid-free paper

ISBN: 9781498734127 (Hardback)
ISBN: 9780367573126 (Paperback)

Library of Congress Cataloging-in-Publication Data

Names: Ameta, K. L. | Dandia, Anshu.
Title: Multicomponent reactions : synthesis of bioactive heterocycles / [edited by] K.L. Ameta and Anshu Dandia.
Description: Boca Raton : CRC Press, 2017. | Includes bibliographical references and index.
Identifiers: LCCN 2016049529 | ISBN 9781498734127 (alk. paper)
Subjects: LCSH: Heterogeneous catalysis. | Heterocyclic compounds--Synthesis. | Bioactive compounds. | Heterocyclic compounds.
Classification: LCC QD505 .M8475 2017 | DDC 547/.59--dc23
LC record available at https://lccn.loc.gov/2016049529

Visit the Taylor & Francis Web site at
http://www.taylorandfrancis.com

and the CRC Press Web site at
http://www.crcpress.com

Contents

Preface

For many decades, biologically active heterocycles have been one of the main topics of interest for medicinal chemists because they display a number of pharmacological activities. Nitrogen, sulfur, and oxygen containing five- and six-membered heterocyclic compounds have been of enormous significance in the field of medicinal chemistry. The majority of pharmaceuticals and biologically active agrochemicals are heterocycles; countless additives and modifiers used in industrial applications, such as cosmetics, reprography, information storage, and plastics, are heterocyclic in nature.

A multicomponent reaction (MCR) is a process in which three or more easily accessible components are combined in a single reaction vessel to produce a final product. An MCR is a domino process, a sequence of elementary steps according to a program in which subsequent transformations are determined by the functionalities produced in the previous step. MCRs constitute an especially attractive synthetic strategy since they provide easy and rapid access to large libraries of organic compounds with diverse substitution patterns. As MCRs are one-pot reactions, they are easier to carry out than multistep syntheses. In spite of the significant useful attributes of MCRs for modern organic chemistry and their suitability for building up large compound libraries, these reactions were of limited interest in the past 50 years. However, in the last decade, with the introduction of high-throughput biological evaluation, the importance of MCRs for drug discovery has been recognized, and considerable efforts from both academic and industrial researchers have been focused especially on the design and development of multicomponent procedures for the generation of various bioactive heterocycles. This growing interest is stimulated by the significant therapeutic potential that is associated with many heterocycles.

Many of these techniques are reported in the reviews that you can read in the chapters of this book. Thus, the purpose of the book is to provide a succinct summary of protocols for the synthesis of various differently sized bioactive heterocycles using multicomponent reaction approaches.

We would like to thank all the authors who contributed so promptly to this book, providing extraordinary reviews on the syntheses of different and highly valuable bioactive heterocyclic compounds through the use of MCR. We trust that both undergraduate and graduate students, as well as researchers and faculties, will find in these pages an interesting guide and an updated state-of-the-art report on the enormous versatility of MCR in the synthesis of bioactive heterocycles. We think that even an expert in the research field will receive inspiration and motivation for future projects.

K. L. Ameta
Anshu Dandia

Editors

Dr. K. L. Ameta is an associate professor of chemistry at the Department of Chemistry, College of Arts, Science and Humanities, Mody University of Science and Technology, Lakshmangarh, Rajasthan, India. He received his doctorate degree in organic chemistry from M. L. Sukhadia University, Udaipur, India, in 2002. He has published more than 50 research papers, 25 book chapters, and 6 edited books with publishers of international repute. He has vast experience in teaching both graduate and postgraduate level students. His research area involves synthesis, characterization, and biological evaluation of different sized bioactive heterocyclic systems. In addition to this, he has keen interest in heterogeneous catalyzed organic synthesis and photocatalysis.

Dr. Anshu Dandia is a professor and head of the Department of Chemistry, University of Rajasthan, Jaipur, India. She is also the coordinator of the University Grants Commission–Special Assistance Programme (UGC-SAP) at the University of Rajasthan and vice president of the Indian Chemical Society, Kolkata. Inducted as a fellow of the Royal Society of Chemistry (FRSC), London, in 2012, Prof. Dandia has published over 170 research papers in the field of synthetic organic chemistry in journals of international repute. Her fields of research interest are green and nanotechnology in organic synthesis; heterocyclic, medicinal chemistry; and organo-fluorine chemistry. Her awards include the Science Legend Award from the World Science Congress Association in 2015; the bronze medal awarded by the Chemical Research Society of India, the Best Chemistry Teacher Award in India from Tata Chemicals Ltd., Confederation of Indian Industries, and the Association of Chemistry Teachers, all in 2012; International Distinguished Woman Scientist Award at the international conference, "Chemistry for Mankind," organized by the joint venture of American Chemical Society (ACS), Royal Society of Chemistry (RSC), and Indian Society of Chemists and Biologists (ISCB) at Nagpur in 2011; and the top researcher award from the vice chancellor, University of Rajasthan, based on highest *H*-index in a survey done by DST New Delhi in 2009. Apart from this, she is a dedicated teacher and strives to keep abreast of new developments in teaching methodologies.

Contributors

Somayeh Ahadi
Peptide Chemistry Research Center
K. N. Toosi University of Technology
Tehran, Iran

Nahid S. Alavijeh
Peptide Chemistry Research Center
K. N. Toosi University of Technology
Tehran, Iran

K. L. Ameta
Department of Chemistry
College of Arts, Science and Humanities
Mody University of Science and Technology
Lakshmangarh, Rajasthan, India

Mohammad Imran Ansari
Department of Pharmaceutical Sciences
University of Maryland
Baltimore, Maryland

Saeed Balalaie
Peptide Chemistry Research Center
K. N. Toosi University of Technology
Tehran, Iran

and

Medical Biology Research Center
Kermanshah University of Medical Sciences
Kermanshah, Iran

Arghya Banerjee
Department of Chemistry
Indian Institute of Technology
Guwahati, Assam, India

Anshu Dandia
Center of Advanced Studies
Department of Chemistry
University of Rajasthan
Jaipur, India

Esmail Doustkhah
Organic and Nano Group (ONG)
Department of Chemistry
Faculty of Science
University of Maragheh
Maragheh, Iran

Mohammad Kamil Hussain
Department of Chemistry
Government Raza Post Graduate College
Rampur, India

Irfan Khan
Department of Chemistry
Mohan Lal Sukhadia University
Udaipur, Rajasthan, India

Mohammad Faheem Khan
Department of Chemistry
University of Lucknow
Lucknow, India

Shahnawaz Khan
Center of Advanced Studies
Department of Chemistry
University of Rajasthan
Jaipur, India

Begraj Kumawat
Center of Advanced Studies
Department of Chemistry
University of Rajasthan
Jaipur, India

Thomas J. Laughlin
Laboratory for Environmentally Friendly Organic Synthesis
Department of Chemistry
Illinois Wesleyan University
Bloomington, Illinois

Susruta Majumdar
Department of Neurology
Memorial Sloan Kettering Cancer Center
New York, New York

Sandeep B. Mane
Institute of Chemistry
Academia Sinica
Taipei, Taiwan

J. Carlos Menéndez
Department of Organic and Pharmaceutical Chemistry
Faculty of Pharmacy
Complutense University
Madrid, Spain

Ram S. Mohan
Laboratory for Environmentally Friendly Organic Synthesis
Department of Chemistry
Illinois Wesleyan University
Bloomington, Illinois

Sandeep More
Department of Fibers and Textile Processing
Institute of Chemical Technology
Mumbai, India

and

Department of Chemistry
Mahatma Gandhi Mahavidyalaya
Ahemadpur, India

Vijaykumar Paike
Department of Research and Development
Florentis Pharmaceuticals Pvt. Ltd.
Tathawade, Pune, India

Travis C. Palmer
Department of Neurology
Memorial Sloan Kettering Cancer Center
New York, New York

Vijay Parewa
Center of Advanced Studies
Department of Chemistry
University of Rajasthan
Jaipur, India

Bhisma K. Patel
Department of Chemistry
Indian Institute of Technology
Guwahati, Assam, India

Sunil N. Patil
Organic Chemistry Division
CSIR-National Chemical Laboratory
Pune, India

Rajendra P. Pawar
Department of Chemistry
Deogiri College
Aurangabad, India

M. Teresa Ramos
Department of Organic and Pharmaceutical Chemistry
Faculty of Pharmacy
Complutense University
Madrid, Spain

Sadegh Rostamnia
Organic and Nano Group (ONG)
Department of Chemistry
Faculty of Science
University of Maragheh
Maragheh, Iran

Mohammad Saquib
Department of Chemistry
S. S. Khanna Girls Degree College
Allahabad, India

Satavisha Sarkar
Department of Chemistry
Indian Institute of Technology
Guwahati, Assam, India

Amit Sharma
Center of Advanced Studies
Department of Chemistry
University of Rajasthan
Jaipur, India

Sandip S. Shinde
Organic Chemistry Division
CSIR-National Chemical Laboratory
Pune, India

Jagdamba Singh
Department of Chemistry
University of Allahabad
Allahabad, India

Ruby Singh
Department of Chemistry
School of Basic Sciences
Jaipur National University
Jaipur, Rajasthan, India

Yu Sun
Key Laboratory of Pesticide and Chemical Biology
Ministry of Education
College of Chemistry
Central China Normal University
Wuhan, People's Republic of China

Padmakar Suryavanshi
Division of Chemistry and Biological Chemistry
School of Physical and Mathematical Science
Nanyang Technological University
Singapore

Giammarco Tenti
Department of Organic and Pharmaceutical Chemistry
Faculty of Pharmacy
Complutense University
Madrid, Spain

S. N. Thore
Department of Chemistry
Deogiri College
Aurangabad, India

András Váradi
Department of Neurology
Memorial Sloan Kettering Cancer Center
New York, New York

Miao Wang
Key Laboratory of Pesticide and Chemical Biology
Ministry of Education
College of Chemistry
Central China Normal University
Wuhan, People's Republic of China

Anxin Wu
Key Laboratory of Pesticide and Chemical Biology
Ministry of Education
College of Chemistry
Central China Normal University
Wuhan, People's Republic of China

Jiachen Xiang
Key Laboratory of Pesticide and Chemical Biology
Ministry of Education
College of Chemistry
Central China Normal University
Wuhan, People's Republic of China

Joshua S. Yoo
Laboratory for Environmentally Friendly Organic Synthesis
Department of Chemistry
Illinois Wesleyan University
Bloomington, Illinois

1 Synthesis of Pyridines by Multicomponent Reactions

*Giammarco Tenti, M. Teresa Ramos, and J. Carlos Menéndez**

CONTENTS

1.1 INTRODUCTION

The pyridine nucleus was described for the first time by the Scottish chemist T. Anderson in 1846 during his studies of bone pyrolysis. In the course of these experiments, he could separate "a colorless liquid with unpleasant odor" from which he initially isolated picoline (the first known pyridine base) and, at a later stage, pyridine itself. Owing to its flammability, Anderson gave to this substance the name we still use, derived from the Greek πυρ (pyr), which means fire; the suffix -idine was used for all aromatic bases.[1] Some years later, Körner in 1869 and Dewar in 1871 independently proposed the correct structure of pyridine, describing it as the mono-aza analog of benzene. Since its identification, this nucleus became one of the most investigated aromatic compounds and nowadays it has widespread applications in various fields of chemistry, ranging from supramolecular[2] to coordination chemistry,[3] as well as in polymer science.[4] It is also of relevance in organic synthesis, being part

* Corresponding author: josecm@farm.ucm.es.

of various organic[5] and organometallic catalysts.[6] However, the most attractive aspect of this nucleus is related to its biological importance. The first pyridine-based natural compound to come to the fore was niacin (already known as vitamin B_3 or nicotinic acid), when, in the 1930s, it was identified as an efficient cure for pellagra, one of the most common life-threatening diseases known at that time, that was characterized by dementia and dermatitis.[7] At the turn of twentieth century, beyond niacin, other pyridines such as nicotine, nicotinamide and the oxidoreductive NAD–NADH coenzymes had already been identified and described. To date many other natural compounds containing a pyridine nucleus are known and have been studied for their interesting biological activity and potential pharmacological applications (Figure 1.1).[8]

In the fields of medicinal chemistry and drug discovery, various compounds containing the pyridine ring show numerous interesting pharmacological activities such as anti-inflammatory,[9] antiasthmatic,[10] antiretroviral,[11] antihistaminic,[12] anticancer,[13] antiulcer,[14] and antidiabetic[15] ones (Figure 1.2). Indeed, due to its pharmacological versatility, pyridine can be defined as a "privileged scaffold"—that is, as defined by Evans,[16] a molecular structure with the ability to interact with different groups of pharmacological receptors. Owing to this characteristic, the use of this kind of structure for drug discovery is very attractive.[17]

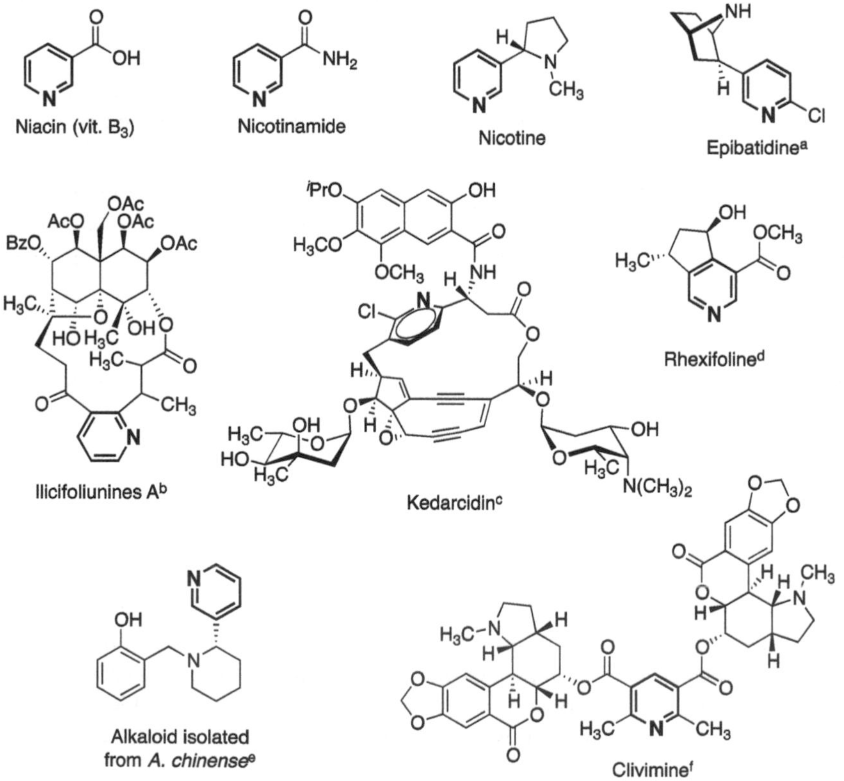

FIGURE 1.1 Some natural compounds containing a pyridine structural fragment.

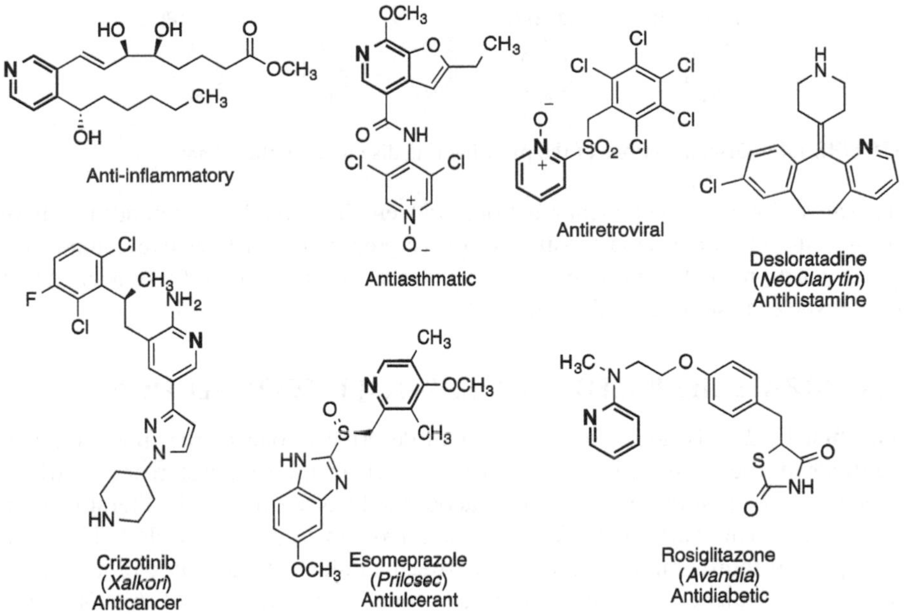

FIGURE 1.2 Some pharmacologically active molecules containing the pyridine nucleus.

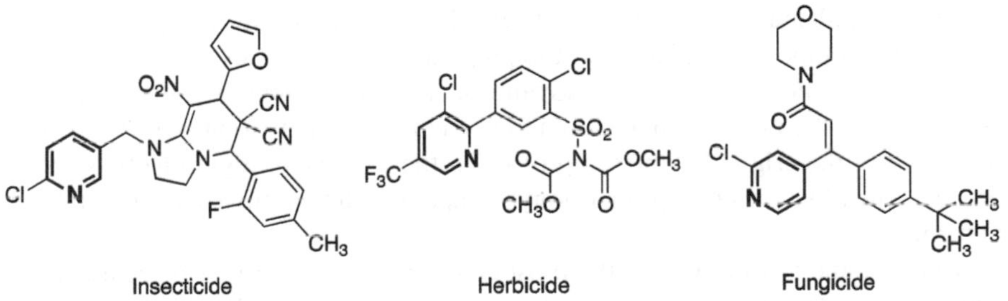

FIGURE 1.3 Some agrochemical compounds containing the pyridine nucleus.

Finally, it is worth mentioning that pyridine compounds are also important in the agrochemical field, showing different activities such as insecticidal,[18] herbicidal,[19] and fungicidal[20] (Figure 1.3).

1.2 SYNTHETIC ACCESS TO THE PYRIDINE NUCLEUS BASED ON MULTICOMPONENT REACTIONS: GENERAL ASPECTS

The biological interest in the pyridine scaffold and its synthetic utility has led to the development of a large number of methods for its synthesis. These methods have been reviewed several times in the last decade,[21,22] with one of these reviews specifically devoted to the use of metal-free multicomponent procedures.[23] In the reminder of this chapter, we will provide a general overview of the synthesis of pyridine

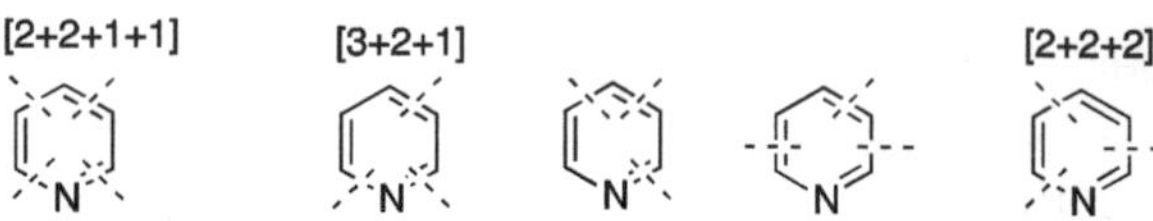

FIGURE 1.4 Disconnections of the pyridine ring discussed in this chapter.

derivatives by multicomponent reactions, with emphasis on those methods that have been found to be particularly suitable for the preparation of bioactive compounds. These reactions will be presented taking into account the type of disconnection that they involve, as summarized in Figure 1.4.

1.3 STRATEGIES BASED ON A [2+2+1+1] DISCONNECTION

The first [2+2+1+1] approach that we will describe is one of the most common methods for the construction of the pyridine nucleus: the Hantzsch reaction, which was one of the first multicomponent reactions to be described in the literature.[24] In its original form, it allows the formation of a 1,4-dihydropyridine (DHP) derivative by a pseudo-four-component reaction between an aldehyde, a source of ammonia, and two equivalents of a β-dicarbonyl compound. The oxidation of this DHP into pyridine has been thoroughly investigated and it is usually performed with well-known oxidizing agents,[25] including some environmentally friendly protocols that involve O_2 as oxidant.[26] Although in most cases this oxidation is usually performed in a sequential manner,[27] in the literature there are also multicomponent procedures where emphasis is placed on the preparation of pyridines in a one-pot fashion.[28] Many of these methods take advantage of the use of microwave irradiation to enhance both the multicomponent reaction and the aromatization step.[29] The main limitation of the original Hantzsch protocol is the absence of regioselectivity of the process when two different β-dicarbonyl compounds are employed, as illustrated in Scheme 1.1.

To overcome this drawback, the most efficient strategy is to form separately the intermediates that take part in the Hantzsch reaction: namely, the Knoevenagel adduct between one of the β-dicarbonyl compounds and the aldehyde, and the

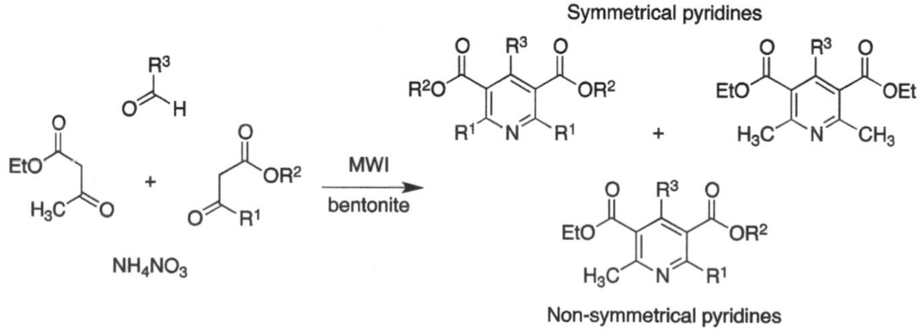

SCHEME 1.1 The combinatorial study developed by Khmelnitsky employing the Hantzsch pyridine synthesis.[28a]

β-enaminone intermediate derived from the reaction between ammonia and the second β-dicarbonyl compound (Scheme 1.2). This methodology, known as the "modified three-component Hantzsch approach," allows the control of the regioselectivity of the reaction and is usually employed in the generation of nonsymmetrical 1,4-DHPs (where "nonsymmetrical" indicates the presence of two different ester groups at the DHP C-3 and C-5 positions).[30] We will focus on the modified three-component Hantzsch strategy in Section 1.4.1, taking into account that it leads to the formation of the pyridine ring by a [3+2+1] disconnection.

In recent years, some variations to the original [2+2+1+1] Hantzsch pyridine synthesis based on the use of substrates bearing at least one nitrile functionality to replace one or both of β-dicarbonyl compounds have been described.[31] In this context, the group of Zhou reported in 2008 a very interesting example of the synthesis of 2-hydroxypyridines employing a β-dicarbonyl compound, an α-cyanoester, an aldehyde, and ammonium acetate. In spite of the nonsymmetrical structure of the final product, the reaction was fully regioselective, affording only one pyridine derivative, even though in low yield (Scheme 1.3).[32]

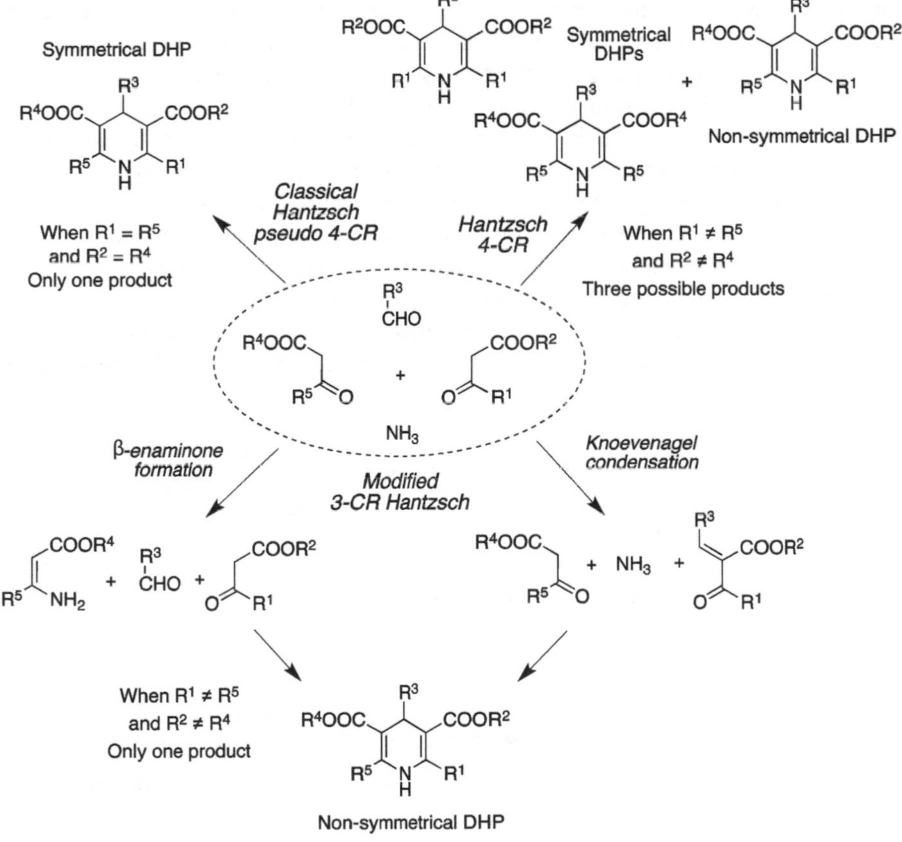

SCHEME 1.2 Regioselectivity issues in the Hantzsch pyridine synthesis.

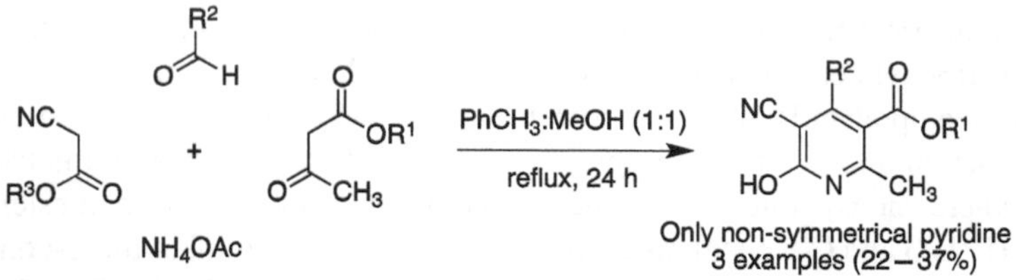

SCHEME 1.3 The four-component synthesis of 2-hydroxypyridines developed by Zhou.

At the beginning of the twentieth century, the Russian chemist Aleksei Chichibabin described a pseudo-four-component synthesis of the pyridine ring based on the [2+2+1+1] approach, involving the reaction between three equivalents of an enolizable aldehyde and one equivalent of ammonia (Scheme 1.4).[33]

The low yield, harsh conditions, and numerous by-products combined to leave this protocol in oblivion until the end of the 1940s, when Frank and Seven carried out several experiments in an effort to better understand the reaction. These mechanistic studies suggested that an α,β-unsaturated carbonyl compound, generated from the condensation of two equivalents of the enolizable aldehyde, was a plausible intermediate. In order to confirm this hypothesis, they decided to study the condensation of α,β-unsaturated compounds and ammonia employing the Chichibabin reaction conditions, observing the formation of an unexpected pattern of substitution in the pyridine ring (2,4,6-trisubstitution instead of the 2,3,5-trisubstitution described by Chichibabin). Pointing at the reversible nature of the aldol condensation, they envisioned a reaction pathway going through the formation of a 1,5-dicarbonyl compound with a final ring-closure step by intervention of the ammonia source (Scheme 1.5).[34] It must be remarked at this point that, confusingly, 2,4,6-trisubstituted pyridines are known as Kröhnke pyridines, irrespectively of the method used for their preparation, because they were first obtained by application of the Kröhnke synthesis, as will be discussed in Section 1.4.2.

These mechanistic studies allowed overcoming some of the limitations of the original protocol and, from that moment, most of the variations of this pyridine

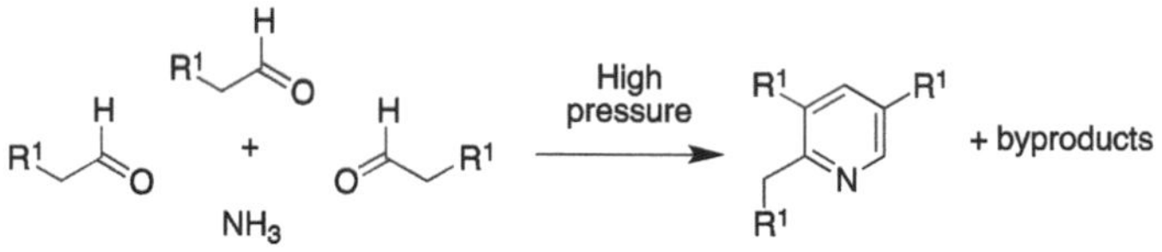

SCHEME 1.4 The original [2+2+1+1] Chichibabin pyridine synthesis.

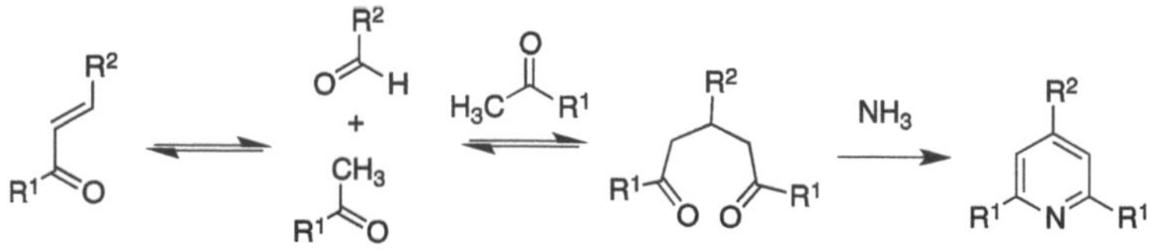

SCHEME 1.5 Mechanistic studies of the Chichibabin reaction performed by Frank and Seven.

synthesis have been directed to the formation of 2,4,6-trisubstituted pyridines by reaction between an ammonia source, a nonenolizable aldehyde, one equivalent of an enolizable ketone, and a second equivalent of the same ketone or a 1,3-dicarbonyl compound. It is worth emphasizing that, beyond yield improvement, one of the most important benefits achieved with the development of this variation to the original Chichibabin pyridine synthesis is the complete regioselectivity of the reaction.

Acetophenone derivatives are usually the enolizable ketones of choice for this multicomponent protocol. In the last decade, several successful variations of the Chichibabin pyridine synthesis, based on the use of a wide range of these derivatives in many different reaction conditions, have been reported.[35] In some cases, acetophenone analogs were also employed—for example, in the report by Penta and Vedula in 2013 of the reaction between aromatic aldehydes, ammonium acetate, and acetylpyran-2-one as an acetophenone equivalent, in refluxing water and catalyzed by cerium(IV) ammonium nitrate (CAN) (Scheme 1.6).[36]

In 2009, Wu and co-workers reported a new modified Chichibabin methodology showcasing a fascinating mode of reactivity. This pseudo-five-component pyridine synthesis was based on the use not only of two equivalents of the acetophenone component and one equivalent of ammonium acetate, but also of two equivalents of the aldehyde partner. This reaction, carried out under microwave irradiation (MWI) in the presence of an imidazolium-type ionic liquid as the catalyst, allows the formation of six new bonds (three C–N bonds, one C–O bond, three C–C bonds), giving the access to the chromenopyridine nucleus in short times and good yields. As shown in Scheme 1.7, the authors proposed a mechanism that proceeds through a tandem aldol condensation–Michael addition process between two equivalents of acetophenone and one equivalent of aldehyde, leading to the formation of a 1,5-dicarbonyl intermediate. This derivative first undergoes another aldol condensation with the second equivalent of aldehyde and then reacts with the ammonium acetate, leading to the final cyclization.[37]

Considering the chemical and pharmacological importance of fused heterocyclic compounds, cyclic enolizable ketones are among the most employed partners in the modified Chichibabin pyridine synthesis. The partners of choice for this multicomponent pyridine synthesis are usually ketones bearing only one enolizable α-position, such as 1-indanone or α-tetralone, in order to ensure the formation of a single reaction product.[38] In 2007, Tu's group attempted to develop a variation of the Chichibabin pyridine synthesis employing one equivalent of 1-indanone, an acetophenone derivative,

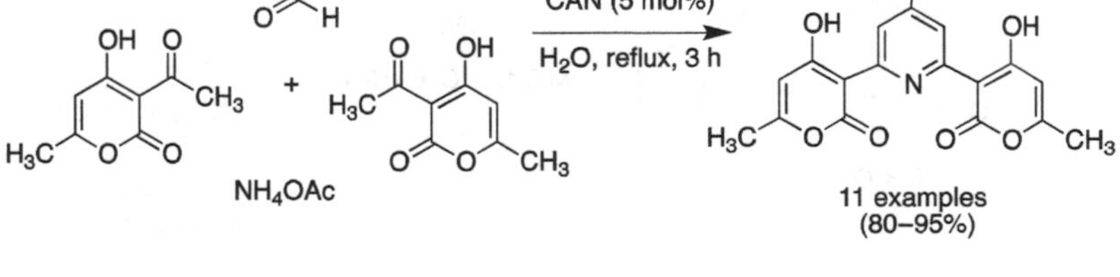

SCHEME 1.6 The [2+2+1+1] Chichibabin-related pseudo-four-component strategy for pyridine synthesis developed by Penta and Vedula.

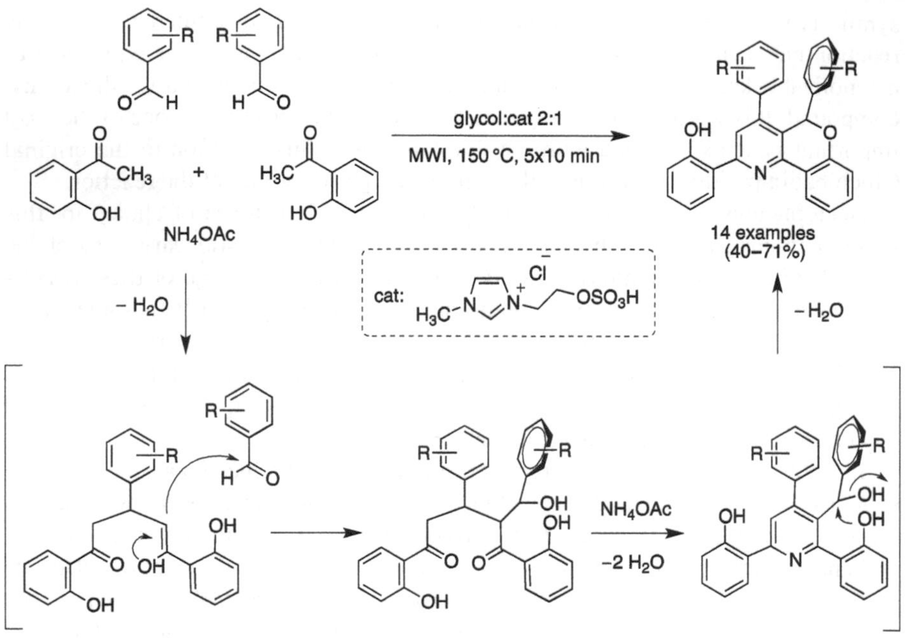

SCHEME 1.7 The pseudo-five-component variation of the Chichibabin synthesis developed by Wu.

an aromatic aldehyde, and ammonium acetate; unfortunately, in this case the reaction lost its product selectivity, affording a mixture of pentacyclic, tricyclic, and monocyclic pyridines. This limitation could be avoided by the prior formation of a Knoevenagel adduct between the 1-indanone and the aromatic aldehyde that was employed at a later stage in a multicomponent strategy in combination with the acetophenone derivative and the ammonium acetate, leading to the formation of only one product. Remarkably, in the same publication, the authors also described a four-component methodology for the formation of the pyridine ring based on the use of 1,3-indanedione as a cyclic enolizable ketone, in combination with one equivalent of ammonium acetate, aromatic aldehyde and an acetophenone derivative. In this case the reaction was carried out in DMF, under microwave irradiation and without any catalyst, and was completely regioselective, affording only the expected tricyclic products (Scheme 1.8).[39]

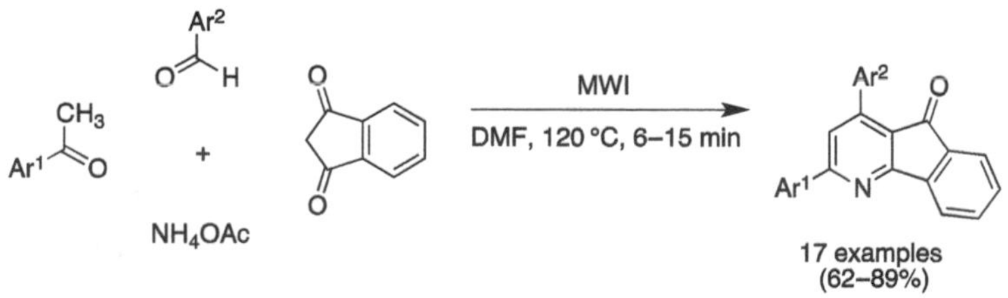

SCHEME 1.8 The four-component strategy for pyridine synthesis developed by Tu, involving the use of 1,3-indanedione.

In 1980 Kambe and Saito described an interesting variation of the Chichibabin reaction by replacing one of the enolizable ketones by malononitrile; this was the first variation of the Chichibabin pyridine synthesis involving a real four-component reaction, as opposed to a pseudo-four-component reaction.[40] This approach to 2-amino-3-cyanopyridines was thoroughly investigated in the last decade and many groups reported several variations focused on improving yield and scope of the methodology.[41] Murata's group exploited this strategy to carry out the synthesis of various 2-amino-3-cyano-4,6-diarylpyridines, designed as analogs of a lead compound that emerged from high-throughput screening as a potent inhibitor of IκB kinase β (IKK-β). This serine-threonine protein kinase is involved in the activation of the nuclear transcription factor NF-κB in response to various inflammatory stimuli, and its inhibition could be therapeutically useful in the treatment of inflammatory diseases and cancer. After an extensive structure–activity relationship study, they identified a highly potent inhibitor of IKK-β with good oral bioavailability in mice and rats and with a significant *in vivo* activity in anti-inflammatory models. The synthesis of the pyridine core of this drug candidate was achieved by a four-component reaction between malononitrile, ammonium acetate, an aliphatic aldehyde, and an acetophenone derivative in 1,4-dioxane at 110 °C (Scheme 1.9).[42]

In 2014, Reddy and collaborators reported an interesting application of this strategy, combining it with Pd-mediated Sonogashira/Heck coupling reactions to achieve the synthesis of alkynyl/alkenyl substituted pyridine derivatives. The entire process is a sequential reaction starting with the initial heterocyclization between malononitrile, ammonium acetate, bromobenzaldehyde, and acetophenone in the presence of pyrrolidine as catalyst, followed by Pd-mediated coupling reactions in a mixture of water and dimethoxyethane under reflux conditions (Scheme 1.10).[43]

Similarly to malononitrile, β-ketonitriles and β-cyanoesters[44] are also commonly employed as partners in this modification of the Chichibabin reaction. Owing to the importance of the pyridine and indole nuclei among the pharmacologically active heterocycles, 3-(cyanoacetyl)indoles are among the most employed building blocks in this methodology. Moreover, as published by Zeng and Cai in 2013, when they are combined with acetyl heteroaromatic components, together with ammonium acetate and aromatic aldehyde, they allow the formation of interesting polyheterocyclic skeletons. The reaction was carried out in the presence of a catalytic amount of molecular iodine in acetic acid or under neat conditions and afforded the expected indolopyridines in moderate to good yields (Scheme 1.11).[45]

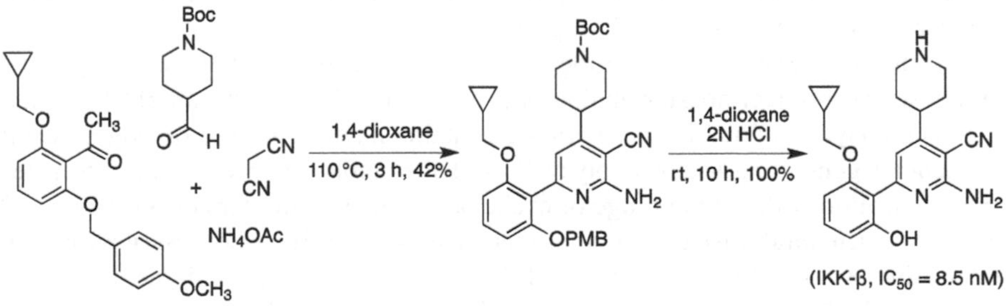

Boc N CH3 O O H CN + CN NH4OAc O OCH3 — 1,4-dioxane, 110 °C, 3 h, 42% → Boc N CN N NH2 OPMB — 1,4-dioxane, 2N HCl, rt, 10 h, 100% → H N CN N NH2 OH (IKK-β, IC_{50} = 8.5 nM)

SCHEME 1.9 The Chichibabin reaction-based strategy for the synthesis of IKK-β inhibitors.

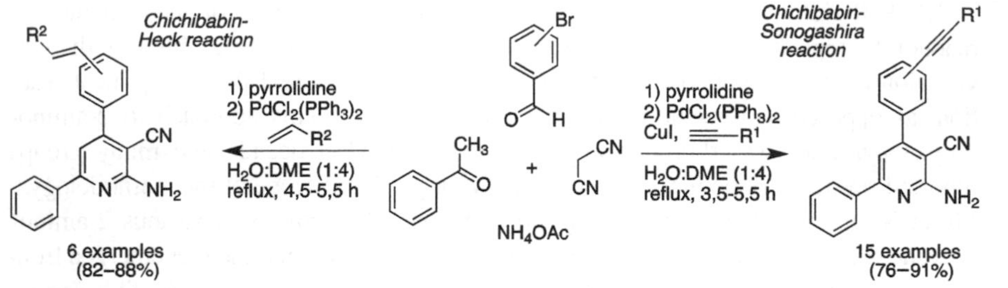

SCHEME 1.10 The sequential Chichibabin–Sonogashira/Heck process developed by Reddy.

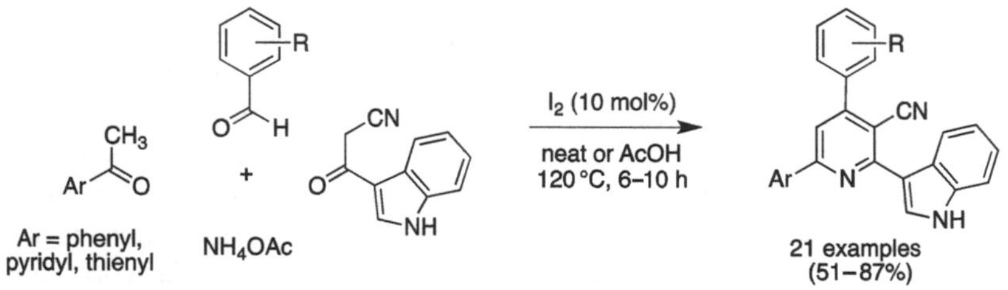

SCHEME 1.11 The indolylpyridine synthesis described by Zeng and Cai.

1.4 STRATEGIES BASED ON A [3+2+1] DISCONNECTION

1.4.1 Hantzsch Type [3+2+1] Strategies

As already mentioned in the previous section, the most efficient method to synthesize nonsymmetrical Hantzsch pyridines is to employ the modified three-component Hantzsch approach, leading to the pyridine ring via a [3+2+1] disconnection. In this context, the most commonly employed tactic involves the use of synthetic equivalents of the β-enaminoester component, generally represented by amino-aromatic or -heteroaromatic substrates such as pyrazole, isoxazole, pyrimidine, or aniline. This strategy is especially fruitful in medicinal chemistry, since it allows a straightforward access to variously fused nitrogen-containing heterocyclic systems that exhibit diverse and important biological properties.

As shown in Scheme 1.12, in the modified three-component Hantzsch pyridine synthesis the β-enaminoester analog is generally combined with an aldehyde and a β-dicarbonyl compound. The literature contains many examples of experimental procedures generating various polycyclic systems employing β-enamine surrogates in combination with a wide range of cyclic or acyclic β-ketoesters or β-diketones.[46] This vast structural variation was further increased by Tu, who described a very interesting protocol for the synthesis of fused pyrazolopyridopyrimidine derivatives employing barbituric acids as 1,3-dicarbonyl partners. By combining these derivatives with pyrazolamines and aldehydes in an acid-catalyzed modified three-component Hantzsch reaction, under microwave irradiation, the synthesis of this important

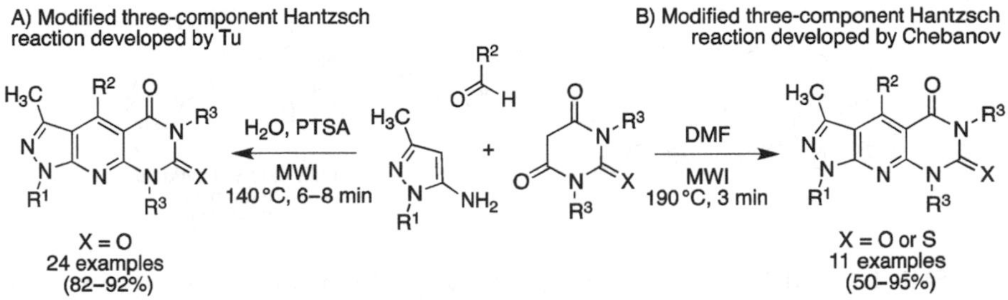

A) Modified three-component Hantzsch reaction developed by Tu

B) Modified three-component Hantzsch reaction developed by Chebanov

H_2O, PTSA, MWI, 140 °C, 6–8 min

DMF, MWI, 190 °C, 3 min

X = O
24 examples
(82–92%)

X = O or S
11 examples
(50–95%)

SCHEME 1.12 The modified three-component Hantzsch reaction involving (thio)barbituric acids.

nitrogen-containing heterocyclic scaffold was achieved in good yields and short times. One year later, Chebanov's group extended the scope of this methodology by tuning the substitution pattern of the pyrazolamine component, including also the use of thiobarbituric acids (Scheme 1.12).[47]

Malononitrile, β-ketonitriles, and β-cyanoesters have also been extensively used as 1,3-dicarbonyl partners in the modified three-component Hantzsch pyridine synthesis.[48] In 2002, Nasr and co-workers reported the multicomponent synthesis of pyridopyrimidines employing malononitrile or ethyl cyanoacetate, aromatic aldehydes, and 6-aminouracil as a β-enamine surrogate; the expected products were obtained under reflux condition in good yields and short times. These scaffolds were further functionalized, leading to new products that exhibited interesting antiviral activities against herpes simplex virus and also cytotoxic properties (Scheme 1.13).[49]

In 2011, Magedov's group described further pharmaceutically relevant work based on the use of 6-aminouracil as the β-enamine; the modified three-component Hantzsch was achieved in this case with aldehydes and indane-1,3-dione as the β-dicarbonyl partner. The camptothecin-inspired tetracyclic heterocycles thus synthesized exhibited antiproliferative properties against human cancer cells by inhibiting the activity of topoisomerase II. Unfortunately, the multicomponent protocol developed by Magedov and co-workers, based on heating the reagents at 120 °C in a mixture of acetic acid/ethylene glycol (2:1), generated the fully oxidized pyridine ring only when formaldehyde was used; otherwise, the reaction afforded 1,4-DHPs,

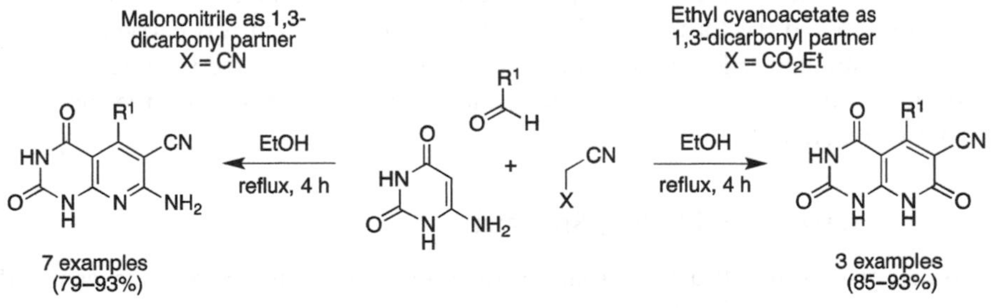

Malononitrile as 1,3-dicarbonyl partner
X = CN

Ethyl cyanoacetate as 1,3-dicarbonyl partner
X = CO_2Et

EtOH, reflux, 4 h

EtOH, reflux, 4 h

7 examples
(79–93%)

3 examples
(85–93%)

SCHEME 1.13 The multicomponent pyridopyrimidine synthesis described by Nasr.

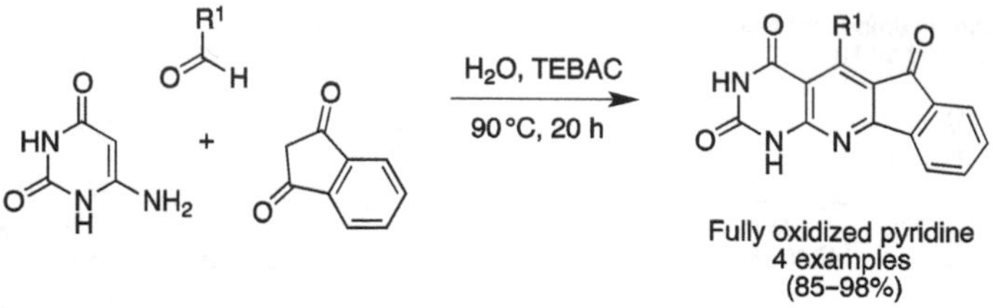

SCHEME 1.14 Multicomponent synthesis of camptothecin-inspired tetracyclic systems described by Shi's group.

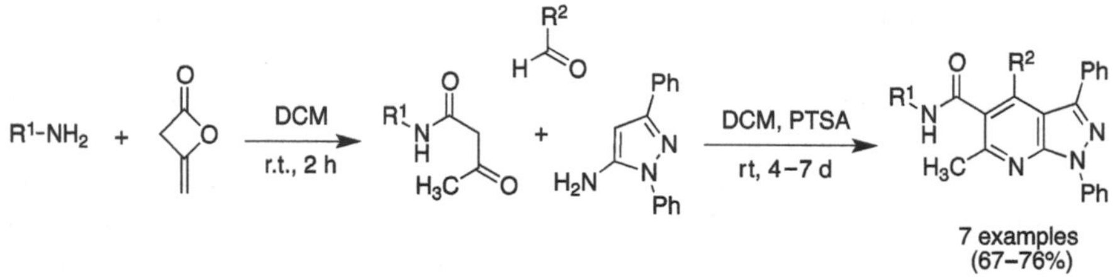

SCHEME 1.15 The modified three-component Hantzsch protocol developed by Shabani's group.

requiring therefore an extra oxidation step to get the desired pyridines.[50] One year later, Shi's group refined the methodology by heating the same reaction partners at 90 °C in water and in the presence of benzyltriethylammonium chloride (TEBAC) as catalyst, obtaining directly the expected tetracyclic heterocycles bearing the aromatic pyridine ring in good yields (Scheme 1.14).[51]

Unlike their ester counterparts, β-ketoamides have received relatively little attention in multicomponent processes. In 2009, Shabani's group described an interesting modified three-component Hantzsch protocol centered on the use of aminopyrazoles as β-enamine analogs, aromatic aldehydes, and β-ketoamides that were generated *in situ* by a reaction between a diketene and a primary amine. The reaction, carried out in dichloromethane at room temperature, was catalyzed by *p*-toluenesulphonic acid and afforded the desired pyrazolopyridine derivatives in moderate to good yields. In order to avoid the formation of side products, the reaction was carried out in a sequential manner, allowing initially the formation *in situ* of the β-ketoamide and adding at a later stage, in a one-pot fashion, the other partners for the modified Hantzsch reaction (Scheme 1.15).[52]

1.4.2 Povarov Type [3+2+1] Strategies

In recent years some attractive variations to the modified Hantzsch reaction, centered on the replacement of the aldehyde or the 1,3-dicarbonyl components, have been described. In this context some groups have focused their efforts on the use of simple carbonyl compounds in combination with aldehydes and a β-enamine surrogate

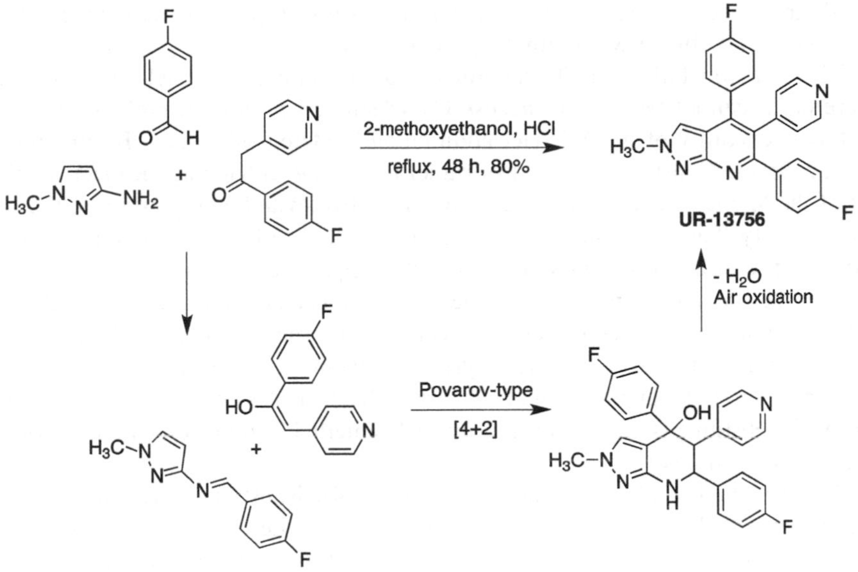

SCHEME 1.16 The multicomponent synthesis of UR-13756 developed by Backley.

to build the pyridine nucleus. As an example, in 2010, Backley and co-workers applied this multicomponent strategy to the synthesis of UR-13756, a highly selective inhibitor of p38 mitogen-activated protein kinases (MAPKs) that exhibited good bioavailability and pharmacokinetic properties and could be potentially used in the treatment of Werner syndrome, a rare genetic disorder characterized by the appearance of premature aging. The pyrazolopyridine UR-13756 was synthesized by an acid catalyzed three-component reaction between 1-methyl-3-aminopyrazole, 4-fluorobenzaldehyde, and a ketone component under reflux conditions (Scheme 1.16).[53] Although the authors did not carry out a mechanistic study, the pseudosymmetric [6+5] bicyclic core of UR-13756 is formed via a Povarov-type [4+2] cycloaddition between the enolic form of the ketone and the imine deriving from the condensation of the aminopyrazole with the aldehyde, as confirmed some years later by other groups.[54]

1.4.3 Michael-Initiated [3+2+1] Strategies

A simple method allowing the metal-free synthesis of pyridines from 1,3-dicarbonyl compounds was reported by Constantieux and Rodriguez, and involves the reaction between β-dicarbonyl compounds (1,3-diketones, β-ketoesters, or β-ketoamides), β-unsubstituted α,β-unsaturated aldehydes or ketones and ammonium acetate, in refluxing toluene under heterogeneous catalysis by 4Å molecular sieves.[55] A mechanistic study revealed that the initial intermediate of this three-component process is a 1,5-dicarbonyl derivative, arising from the molecular sieve-promoted Michael addition of the β-dicarbonyl component to the unsaturated carbonyl derivative, and discarded the initial formation of a β-enaminone. The 1,5-dicarbonyl compound would then react with ammonia to furnish a 1,4-dihydropyridine derivative, whose *in situ* oxidation by air would finally yield the observed pyridine (Scheme 1.17).

This reaction fails with β-substituted α,β-unsaturated carbonyl substrates and therefore it cannot be used to access C4-substituted pyridines, probably because of the reversibility of the Michael addition on hindered substrates. To circumvent this drawback, Constantieux and Rodriguez resorted to enhancing the electrophilicity of the Michael acceptor by attaching an electron-withdrawing group to the carbonyl moiety. Thus, the use of α-keto-β,γ-unsaturated carbonyl compounds allowed the presence of alkyl or aryl substituents at the pyridine C-4 position. In this case, besides the previously mentioned molecular sieve catalyst, the reactions had to be performed in the presence of activated charcoal in order to complete the air oxidation of the intermediate 1,4-dihydropyridine (Scheme 1.18).[56,57] This chemistry was also extended to the preparation of biheterocyclic scaffolds bearing a variety of heterocyclic substituents (e.g., 2-furyl, 2-pyridyl, 1-methyl-2-imidazolyl) at the pyridine C-2 or C-6 positions.[58]

A related method allowing the synthesis of 2,3,4-trisubstituted pyridines involves the proline catalyzed reaction between cinnamaldehyde derivatives, symmetrical dibenzyl ketones, and ammonium acetate. This transformation was proposed to take place by iminium organocatalysis and thus be initiated by the reaction between

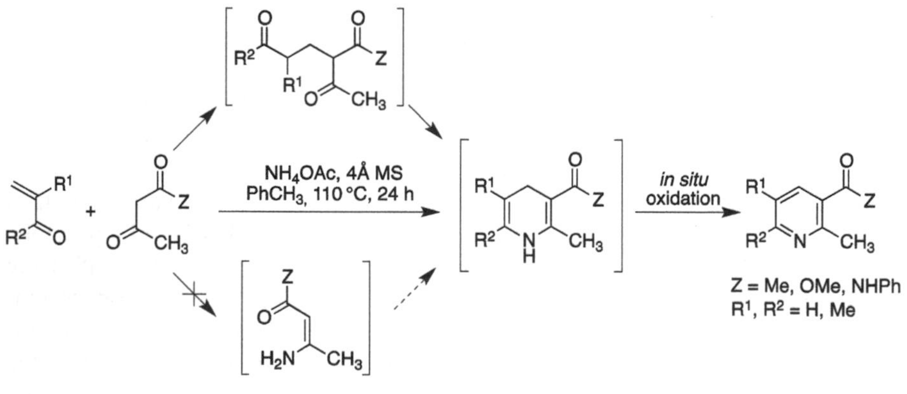

SCHEME 1.17 The [3+2+1] Michael-initiated multicomponent strategy for pyridine synthesis developed by Constantieux and Rodriguez.

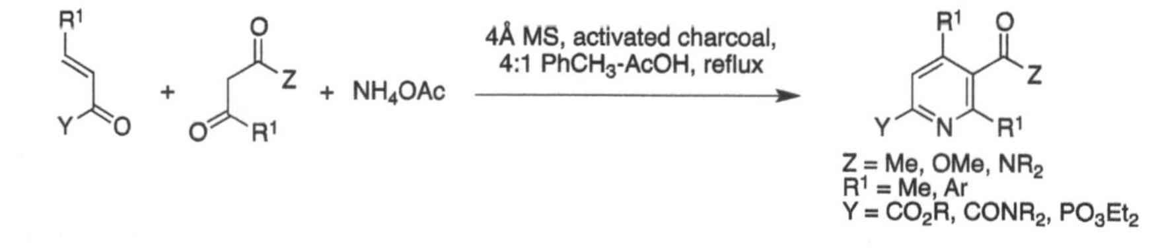

SCHEME 1.18 A modified [3+2+1] Michael-initiated pyridine synthesis leading to 4 substituted pyridines.

the unsaturated aldehyde and proline to give an iminium species that would then be attacked by the enol tautomer of the starting ketone. Reaction of the resulting intermediate with ammonia to yield a β-enaminone, followed by cyclization with concomitant loss of the proline catalyst and a final air oxidation of a nonisolated dihydropyridine intermediate, explains the isolation of the observed products (Scheme 1.19).[59]

A less studied approach to the pyridine nucleus that exploits the Michael-initiated [3+2+1] disconnection involves the use of Mannich bases as starting materials. The Mannich reaction is a dialkylaminomethylation of an enolizable aldehyde or ketone with an amine, usually secondary, and formaldehyde under acid catalysis, via an intermediate iminium species.[60] Mannich bases can be easily converted into α,β-unsaturated carbonyl compounds by elimination, and thus participate in Michael addition reactions. Due to this reactivity, they have been used to prepare polycyclic pyridines from β-aminoketone hydrochlorides, cyclic ketones, and ammonium acetate in boiling ethanol in moderate yields[61] (Scheme 1.20) and for the synthesis of fused oligopyridines with different molecular shapes.[62]

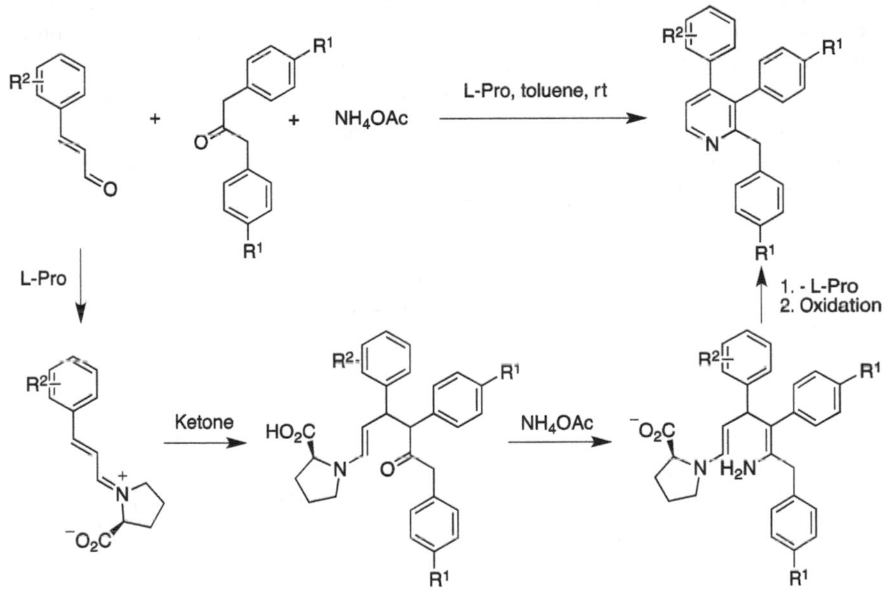

SCHEME 1.19 Iminium organocatalysis in the proline-catalyzed synthesis of 2,3,4-trisubstituted pyridines.

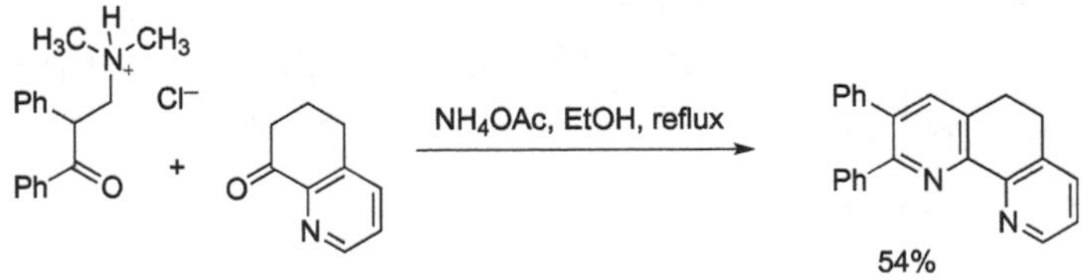

SCHEME 1.20 Synthesis of pyridines from Mannich bases.

Recently, this strategy has been studied in water as an environmentally friendly approach to trisubstituted pyridines and tetrahydroquinolines from carbonyl compounds and different Mannich bases.[63] The authors propose the formation of an enone from the Mannich base, and an enamine from ammonium acetate and the active methylene compound; a subsequent Michael addition gives an intermediate that forms the heterocyclic ring in a final cyclocondensation reaction (Scheme 1.21).

The classical Kröhnke synthesis represents another Michael addition-initiated process leading to pyridines. This methodology is based on the use of an α-pyridinium methyl ketone salt as a surrogate of the 1,3-dicarbonyl compound to react with an α,β-unsaturated carbonyl compound, and ammonium acetate as the source of the nitrogen atom to give 2,4,6-triarylsubstituted pyridines (Kröhnke pyridines) simply heating in glacial acetic acid or ethanol and without the need for an oxidation step (Scheme 1.22).[64] The reaction provides access to pyridines bearing up to four substituents, and since substituents on both fragments can be aromatic, heteroaromatic, or aliphatic, it constitutes a very useful method for the synthesis of pyridine derivatives like oligopyridines or polyaryl substituted pyridines.

The starting α-pyridinium methyl ketone salts are easy to prepare by treatment of α-bromomethyl ketones with pyridine, or directly by treatment of methyl ketones with iodine in pyridine (Ortoleva–King reaction) (Scheme 1.23).

The Kröhnke reaction has been proposed to proceed via the generation of a 1,5-dicarbonyl intermediate by the Michael addition of a pyridinium ylide onto the α,β-unsaturated ketone, followed by ring formation by reaction with ammonia and final aromatization by loss of pyridine and water (Scheme 1.24).

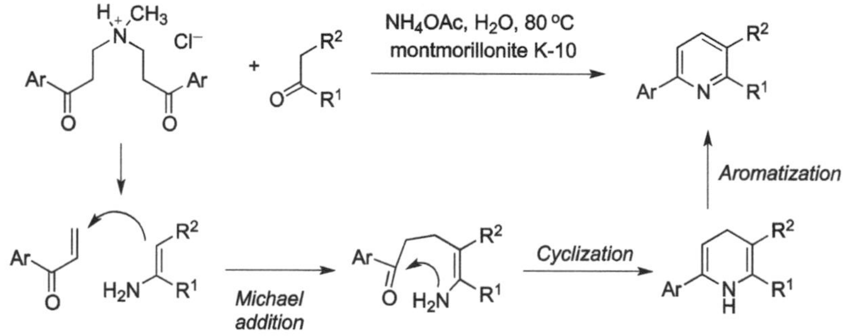

SCHEME 1.21 Synthesis of pyridines from Mannich bases.

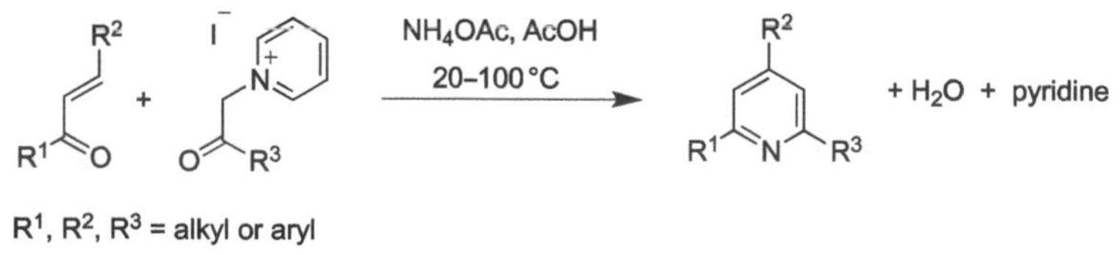

SCHEME 1.22 The [3+2+1] one-pot Kröhnke pyridine synthesis.

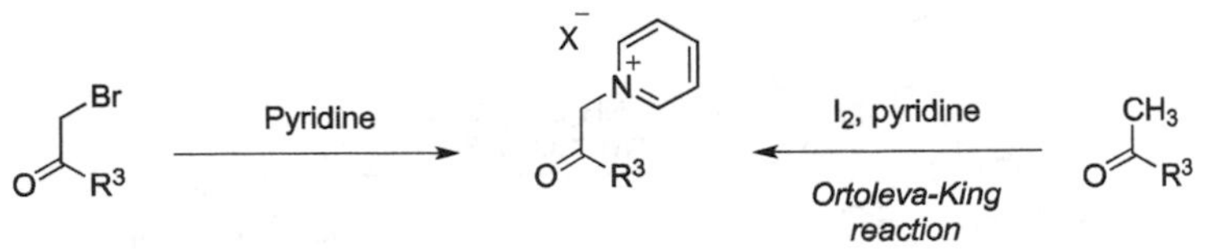

SCHEME 1.23 Preparation of α-pyridinium methyl ketone salts, the starting materials for the Krönke pyridine synthesis.

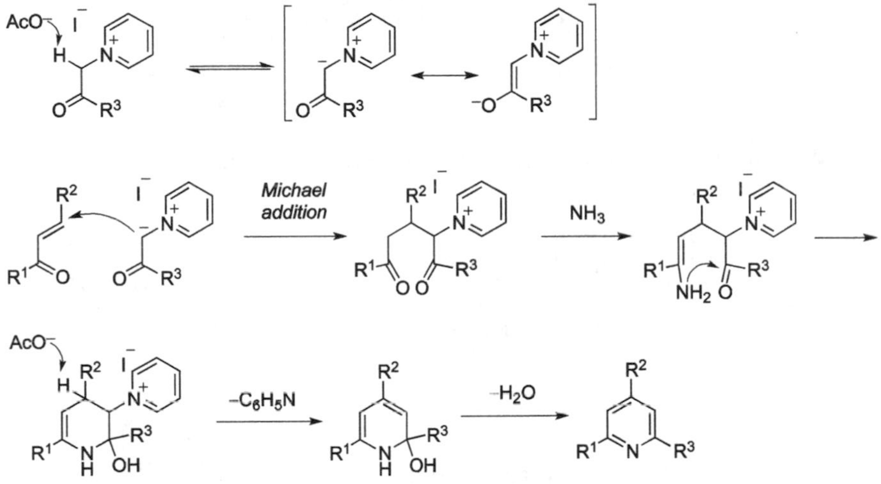

SCHEME 1.24 Proposed mechanism for the Kröhnke pyridine synthesis.

Some variations of the Kröhnke reaction have been described, including one methodology based on a [2+2+1+1] disconnection characterized by the *in situ* formation of an α,β-unsaturated carbonyl derivative (Scheme 1.25). This four-component reaction of *N*-phenacylpyridinium salts, aromatic aldehydes, acetophenones,[65] or cyclic ketones[66] in the presence of ammonium acetate and acetic acid was assisted by microwave irradiation.

The synthesis of the antimalarial agent enpiroline, structurally related to quinine, involves a Kröhnke reaction to build the central pyridine ring.[67] The reaction of the appropriate aroyl acrylic acid, as chalcone component with an α-pyridinium methyl ketone salt, under Kröhnke conditions gives a derivative of the isonicotinic acid, which reacts with excess of 2-lithiopyridine to afford a dipyridyl ketone. Catalytic hydrogenation under acidic conditions affects the carbonyl group and the more basic heterocycle—that is, the one bearing fewer electron-withdrawing substituents. The resulting mixture of diastereoisomers was separated by fractional crystallization to afford enpiroline (Scheme 1.26).

Alpha-beta unsaturated aldehydes can also be used as a C–C fragment, and in this case the products are C6-unsubstituted pyridines. An optimized, scalable route to the SMO receptor antagonist SEN794 begins with the synthesis of a key pyridine intermediate by the Kröhnke reaction of methacrolein and the appropriate

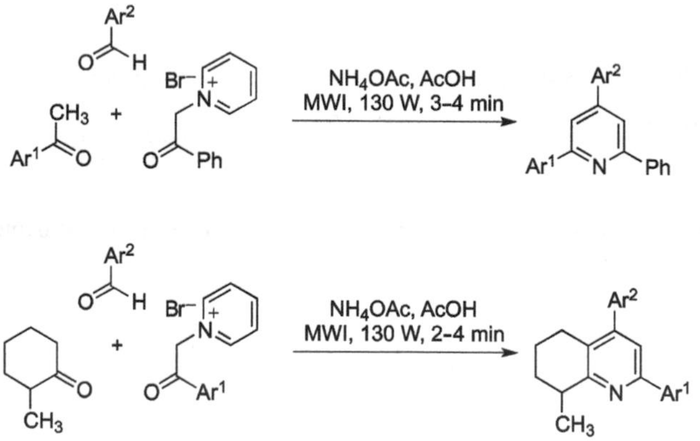

SCHEME 1.25 Four-component modification of the Kröhnke reaction.

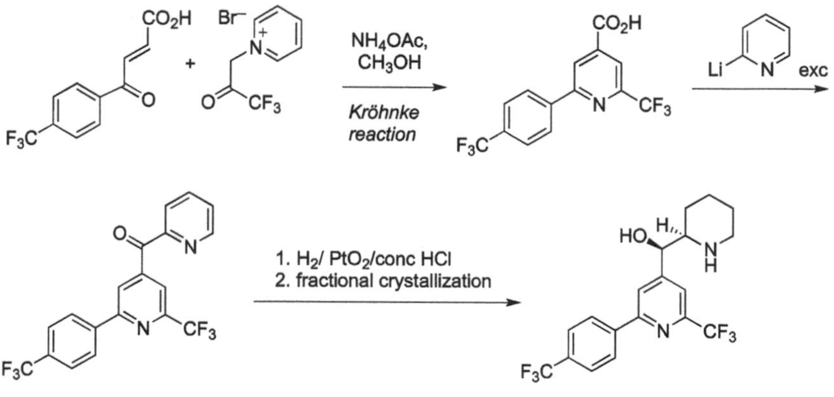

SCHEME 1.26 Synthesis of the antimalarial enpiroline based on a Krönke pyridine synthesis.

pyridinium salt, obtained in a two-step sequence that avoids the use of excess pyridine.[68] The haloarylpyridine thus obtained is then coupled with ethyl isonipecotate via a Pd catalyzed *N*-arylation; finally, hydrolysis of the ester group followed by amide bond formation gives the four-ring structure of SEN794 (Scheme 1.27). This compound is a potent antagonist of the SMO (smoothened) receptor that mediates the hedgehog signaling pathway, which is critical to cell development, differentiation, growth, and migration.

Some analogs of these compounds have been prepared and evaluated as topoisomerase I and II inhibitors and for their cytotoxicity, including 2,6-dithienyl-4-furyl pyridines and 2,4-di(hydroxyphenyl)-6-(2-furyl) pyridines.[69] The structures of the most promising compounds are shown in Scheme 1.28. Related compounds with a coumarin ring have been screened as antimicrobial agents.[70]

Other heterocycles have been used to replace pyridine as the substituent α to the ketone, acting equally as stabilizers of the intermediate ylide and then as leaving groups.

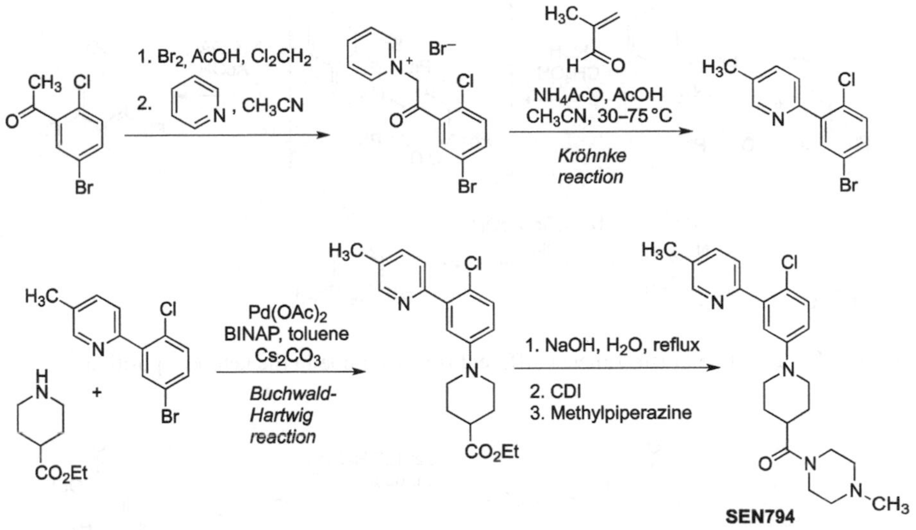

1. Br_2, AcOH, Cl_2CH_2
2. pyridine, CH_3CN

NH_4AcO, AcOH
CH_3CN, 30–75 °C
Kröhnke reaction

$Pd(OAc)_2$
BINAP, toluene
Cs_2CO_3
Buchwald-Hartwig reaction

1. NaOH, H_2O, reflux
2. CDI
3. Methylpiperazine

SEN794

SCHEME 1.27 Synthesis of the SEN794, an antagonist of the SMO receptor.

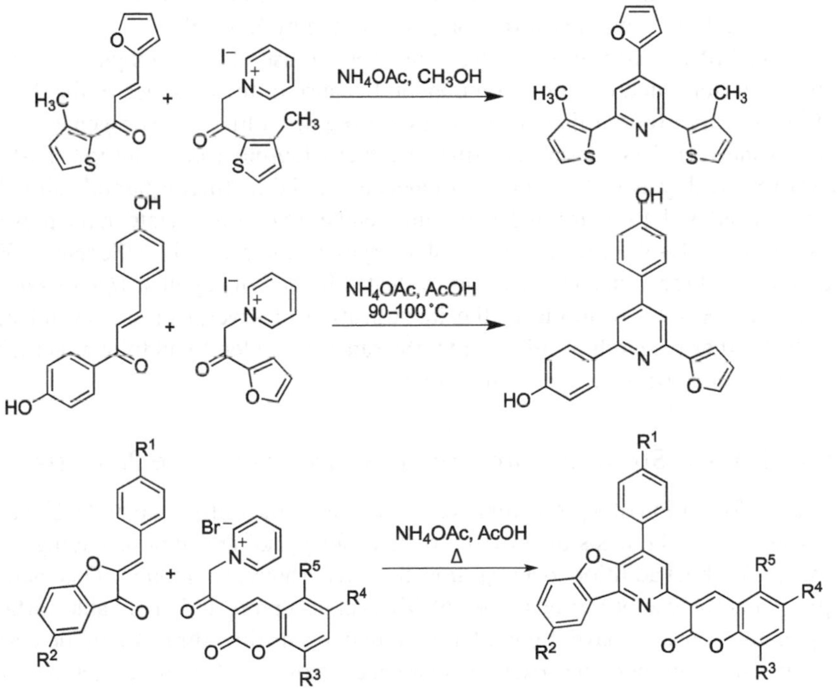

NH_4OAc, CH_3OH

NH_4OAc, AcOH
90–100 °C

NH_4OAc, AcOH
Δ

SCHEME 1.28 Krönke synthesis of some biologically active pyridines.

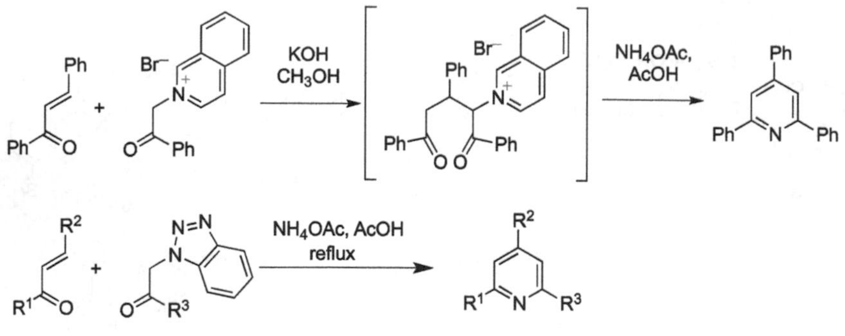

SCHEME 1.29 Use of substituents different from pyridine in the ketone α position.

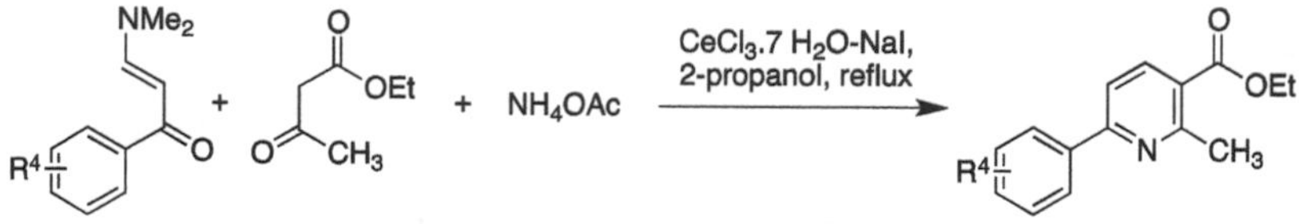

SCHEME 1.30 Synthesis of 6-arylpyridines from β-dimethylaminoenones.

They include isoquinoline, which was used in the very first reported Kröhnke reaction (Scheme 1.29),[64] and benzotriazole, as reported by Katritzky.[71]

One final Michael-initiated strategy involves the use of β-dimethylaminoenones as the α,β-unsaturated carbonyl component. Because of the presence of the dimethylamino leaving group, this reaction does not require a final dehydrogenation step. Thus, Kantevari has described that 3-dimethylamino-1-phenylprop-2-en-1-one derivatives, readily available from acetophenones and dimethylformamide dimethylacetal, reacted with β-dicarbonyl compounds and ammonium acetate in the presence of the $CeCl_3.7H_2O$-NaI catalyst to furnish 6-arylpyridine derivatives (Scheme 1.30).[72] The pyridine library thus obtained was evaluated for activity against *Mycobacterium tuberculosis* H37Rv, resulting in the identification of several promising hit compounds. Further research in this area by the same group led to additional antitubercular compounds bearing thienyl substituents.[73]

1.4.4 [3+2+1] Strategies Involving β-Enaminone-Initiated Processes

The *in situ* formation of a β-enaminone, besides being the initial step of the Hantzsch synthesis, is also the basis of several strategies for pyridine synthesis having as the key step a Michael addition to an α,β-unsaturated carbonyl compound. This methodology allows the use of various types of Michael acceptors and enolizable carbonyl compounds. Thus, we have reported a route to polysubstituted nicotinamide derivatives via a three-component reaction between chalcones, β-ketoamides, and ammonium acetate in the presence of Ce(IV) ammonium nitrate (CAN) as a Lewis acid catalyst.[74] This reaction was shown to start by the formation of a β-aminocrotonamide intermediate, which would then give a Michael addition to the starting chalcone, followed by a cyclocondensation with loss of a molecule of water to furnish a dihydropyridine derivative and the final air oxidation of the latter intermediate (Scheme 1.31). It is interesting to note that a similar reaction using β-ketoesters as the dicarbonyl component gave 1,4-dihydropyridines exclusively.[75]

Our group has also reported the synthesis of 6-alkoxy-1,4,5,6-tetrahydropyridines by a four-component reaction between primary amines, β-dicarbonyl compounds, α,β-unsaturated aldehydes, and alcohols in the presence of CAN via a formal aza [3+3] process that involved the generation of two C–N bonds and one C–C and one C–O bond.[76a] Unfortunately, an attempt to adapt this method to pyridine synthesis by replacing the primary amine by ammonia was unsuccessful. Instead, we resorted to the use of dimethylhydrazine as the primary amine, finding that, as expected, its reaction with β-dicarbonyl compounds, acrolein and ethanol, using indium trichloride as catalyst in this case, allowed the formation of 6-ethoxy-1-dimethylamino-1,4,5,6-tetrahydropyridine derivatives. These compounds were subsequently converted into 2,3-disubstituted pyridines by a double elimination reaction induced by heating in the presence of palladium on charcoal. Interestingly, we found that the whole sequence could be performed without the need for purifying any intermediate, although workup was required after the initial multicomponent step (Scheme 1.32).[76b]

In a subsequent development, we discovered that the use of 2-furylmethylamine as the primary amine component for the multicomponent reaction, followed by

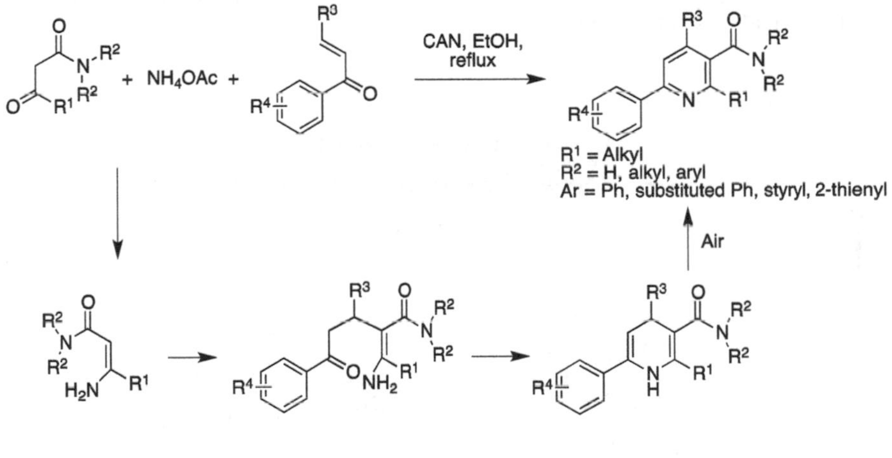

+ NH_4OAc + CAN, EtOH, reflux

R^1 = Alkyl
R^2 = H, alkyl, aryl
Ar = Ph, substituted Ph, styryl, 2-thienyl

Air

SCHEME 1.31 Three-component synthesis of nicotinamides.

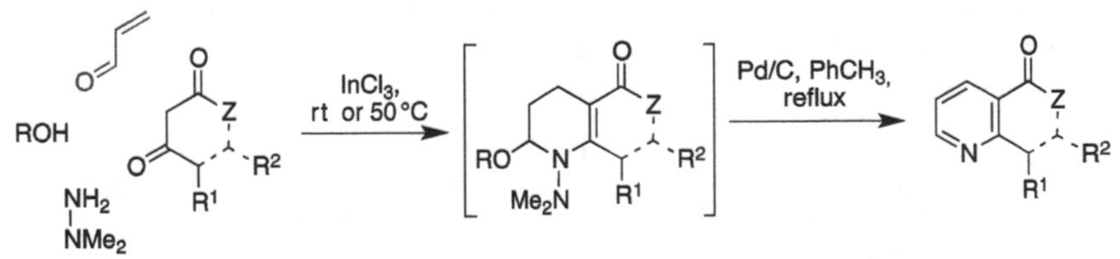

ROH
NH_2–NMe_2
$InCl_3$, rt or 50 °C
Pd/C, $PhCH_3$, reflux

SCHEME 1.32 One-pot synthesis of pyridines based on a four-component reaction using dimethylhydrazine as an ammonia equivalent.

microwave irradiation of the crude tetrahydropyridine in the absence of solvent, afforded good to excellent yields of highly substituted pyridines, with loss of the 2-furylmethyl side chain as the corresponding ethyl ether (Scheme 1.33). This protocol was quite general in terms of the pyridine substitution pattern and functionalization, and it was also applied to the synthesis of fused pyridine systems including quinolines, isoquinolines, phenanthridines, and benzo[*a*]phenanthridines.[77]

The previously mentioned β-enaminone-initiated methods for pyridine synthesis require a final *in situ* dehydrogenation step. In order to avoid it, one possible strategy would involve replacement of the double bond in the starting α,β-unsaturated carbonyl derivative by a triple bond. The Bohlmann–Rahtz synthesis of pyridines was first described in 1957 as a two-step procedure to obtain 2,3,6-trisubstituted pyridines from β-aminocrotonates and ethynyl ketones with an aminodienone as isolable intermediate[78] (Scheme 1.34). The use of the alkyne leads directly to the aromatic ring, and no final oxidation step is necessary. Alkynones can be synthesized by addition of ethynylmagnesium bromide to an aldehyde, followed by oxidation of the resulting secondary alcohol.

This transformation has been proposed to proceed by an initial Michael addition of the enamine onto the unsaturated ketone to afford isolable aminodieneketones with a geometry that precludes their spontaneous cyclization. However, upon heating, these intermediates undergo an efficient *Z*–*E* isomerization of the double bond followed by a cyclocondensation with loss of water to give pyridine derivatives in excellent yields (Scheme 1.35).[79]

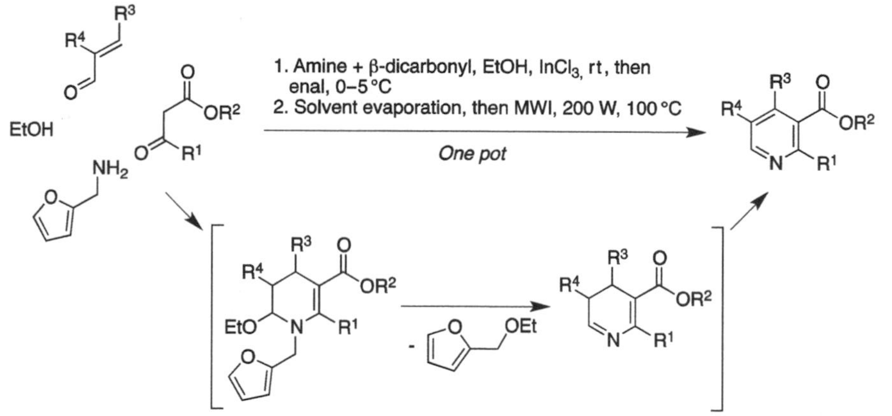

SCHEME 1.33 One-pot synthesis of pyridines based on a four-component reaction using 2-furylmethylamine as an ammonia equivalent.

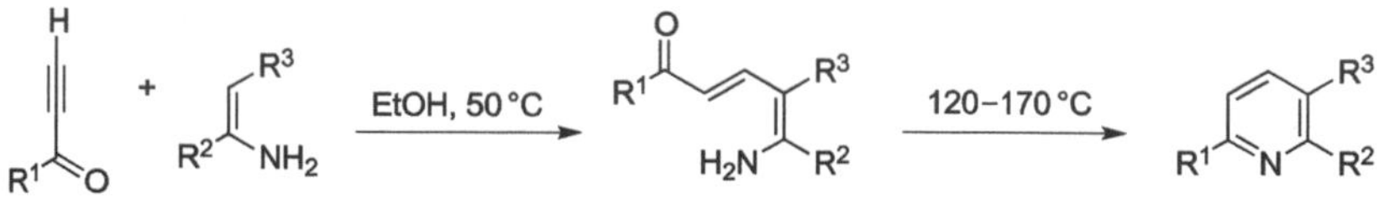

SCHEME 1.34 The original Bohlmann–Rahtz synthesis of pyridines.

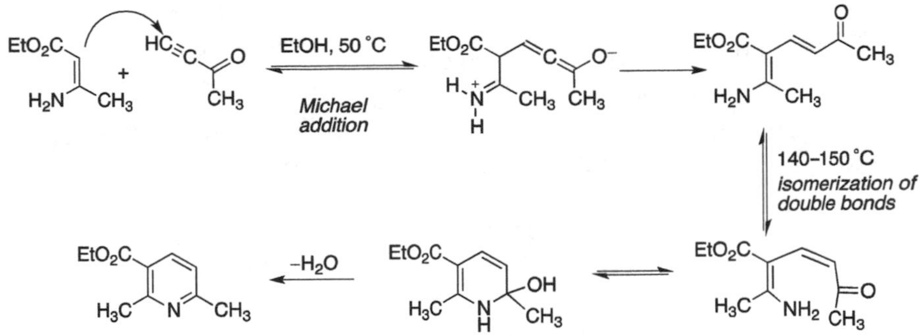

SCHEME 1.35 Proposed mechanism for the Bohlmann–Rahtz reaction.

Following a comprehensive study of the reaction, Bagley developed a three-component version of the Bohlmann–Rahtz synthesis, by refluxing in toluene a mixture of a β-ketoester, an alkynone and a source of ammonia in the presence of an acid catalyst like acetic acid, Amberlist 15, or zinc bromide.[80] The enamine is generated *in situ* from the β-ketoester and ammonium acetate, and then it reacts with the alkynone present in the reaction mixture (Scheme 1.36).

The reaction also succeeds under milder conditions and has been used in natural product synthesis. Dimethyl sulfomycinamate is one of the products identified after the acidic methanolysis of sulfomycins, a family of the thiopeptide antibiotics characterized by a common oxazole-thiazole-pyridine domain in a macrocyclic peptidic structure. The synthesis of the central pyridine core of this compound was initially accomplished in full regioselectivity by a two-step Bohlmann–Rahtz reaction, and then it was improved with a three-component version of the same transformation by refluxing in methanol the appropriate β-oxoamide, methyl oxobutynoate, and ammonium acetate in excess.[81] Reaching the final target still required six additional steps, and the overall yield of the whole sequence was 9% over 12 steps (Scheme 1.37).

The Bagley variation of the Bohlmann–Rahtz reaction has been applied to the synthesis of other thiazole-pyridine cores of thiopeptide antibiotics,[82] sometimes being carried out in a sequential one-pot, multicomponent manner. Cyclothiazomycin is another thiopeptide natural product, which produces, upon hydrolysis, a chiral γ-lactam on a thiazole-pyridine structure. The synthesis of the pyridine ring of this γ-lactam was planned using the Bohlmann–Rahtz strategy, and after several attempts, the best results in terms of chemical yield and chiral integrity of the starting material were obtained when a sequential one-pot procedure was followed[83] (Scheme 1.38).

Bagley has also developed a domino process involving a secondary alcohol-oxidation Bohlmann–Rahtz sequence to obtain pyridines from propargylic alcohols,

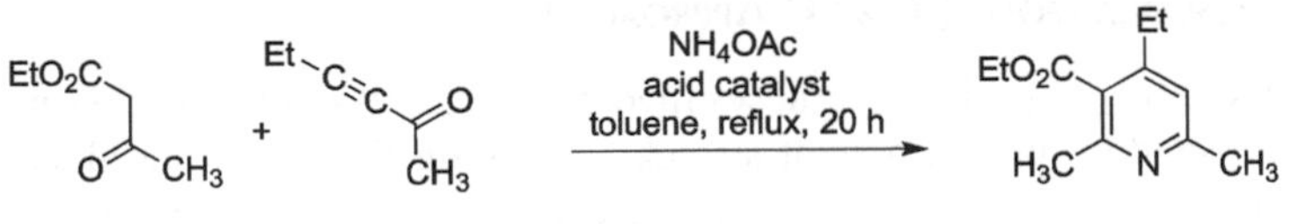

SCHEME 1.36 Three-component version of the Bohlmann–Rahtz reaction.

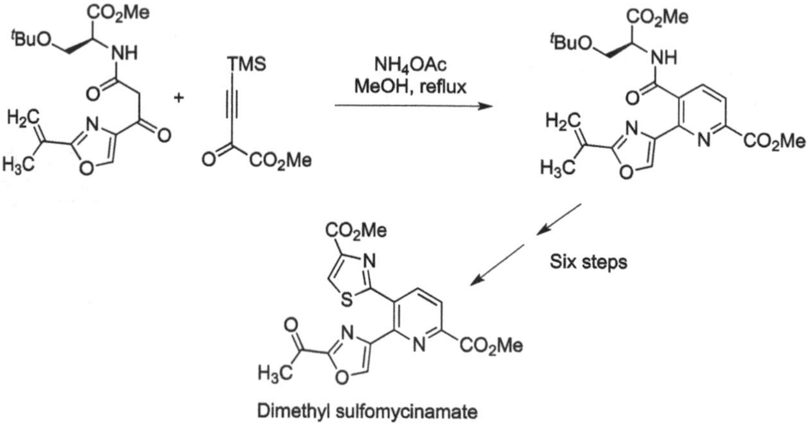

SCHEME 1.37 Synthesis of dimethyl sulfomycinamate based on a multicomponent reaction.

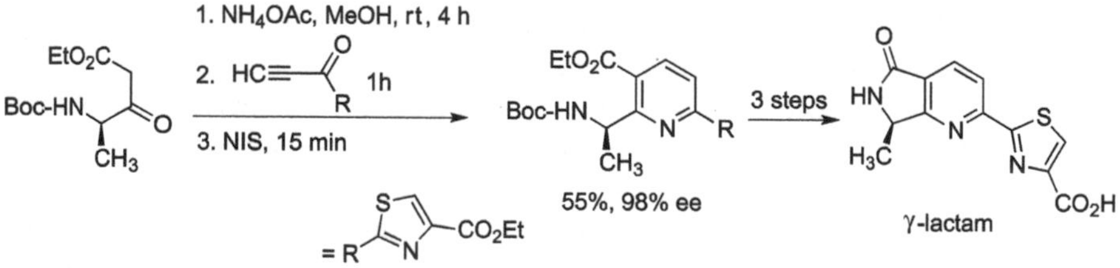

SCHEME 1.38 Synthesis of the γ-lactam produced by hydrolysis of cyclothiazomycin.

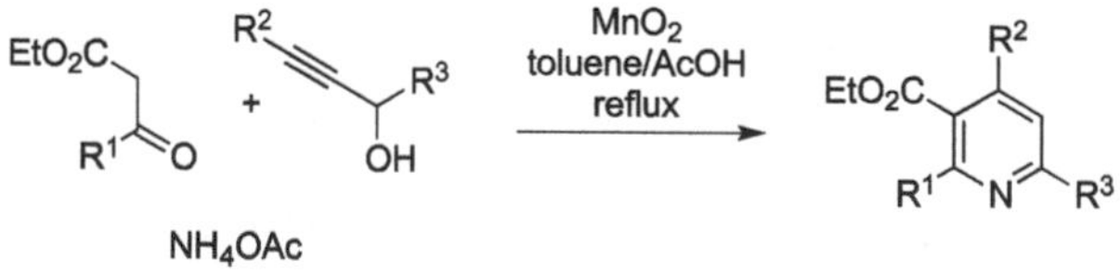

SCHEME 1.39 A domino alcohol oxidation Bohlmann–Rahtz process.

which are oxidized *in situ* with manganese dioxide to afford alkynones.[84] This reaction is carried out by refluxing the β-ketoester, propargylic alcohol, and ammonium acetate in toluene-acetic acid (5:1) in the presence of manganese dioxide, and it affords pyridines in good yields (Scheme 1.39).

1.4.5 Miscellaneous [3+2+1] Approaches

A number of recent pyridine syntheses involve the construction of the six-membered ring by a pseudo-four-component strategy involving the use of an aldehyde, two equivalents of malononitrile, and a nucleophile as starting materials. Several nucleophiles have been identified as suitable for this transformation, including thiols,[85] selenols,[86] alkoxides,[87] ammonia,[88] and aromatic[89] and aliphatic[90] amines. In the latter case, it is relevant to mention that some types of primary amines (e.g., benzylamine derivatives) react differently, giving Hantzsch products. The common mechanism for these reactions involves an initial Knoevenagel condensation between the aldehyde and one molecule of malononitrile, followed by the Michael addition of the second malononitrile to give an intermediate tetranitrile species. Attack of the nucleophile onto one of the nitrile groups induces an anionic domino process leading to the generation of the six-membered ring, and a final air oxidation completes the pyridine synthesis (Scheme 1.40).

Another [3+2+1] disconnection with an unusual three-carbon component was employed for the synthesis of the central pyridine ring of a COX-2 specific inhibitor with structure of phenylbipyridine[91] (Scheme 1.41). Several three-carbon electrophiles were assayed in order to complete the synthesis of this compound, and the best results

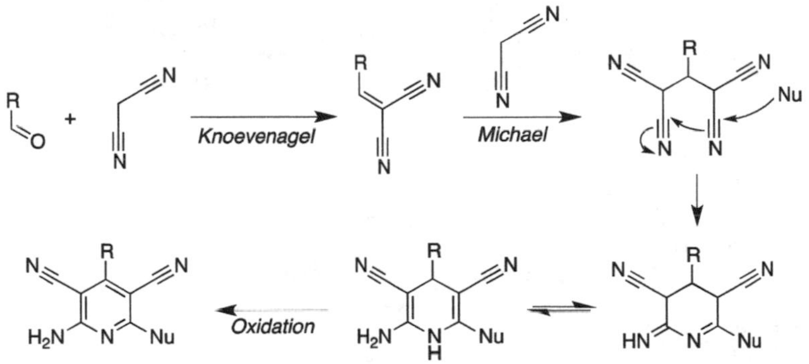

SCHEME 1.40 Pseudo-four-component synthesis of pentasubstituted pyridines.

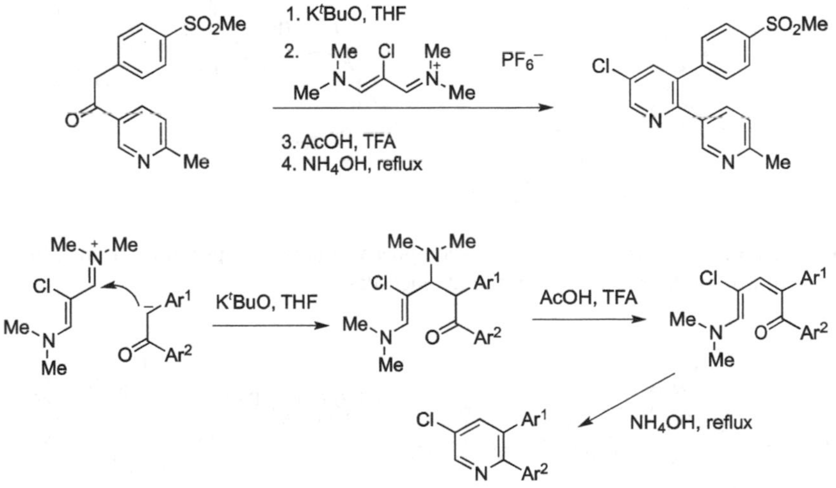

SCHEME 1.41 Sequential multicomponent synthesis of a specific COX-2 inhibitor.

were obtained when vinamidinium salts in basic media were used. In this sequential procedure, the enolate anion of the ketone reacted with the vinamidinium salt to form a diamino-enone intermediate; quenching with acid gives presumably a dienone that, upon heating at reflux with aqueous ammonium hydroxide, forms the heterocyclic ring.

A copper catalyzed domino process involving the reaction of oxime acetates, activated methylene compounds, and a wide range of aldehydes efficiently affords tri- or tetrasubstituted pyridines.[92] This reaction involves cleavage of one N–O bond and the formation of one C–C and one C–N bond, and was proposed to take place by the mechanism summarized in Scheme 1.42.

Holzer has described a three-component domino reaction that is initiated by a palladium-promoted Sonogashira cross coupling between 5-chloropyrazole-4-carbaldehydes and alkynyl-(hetero)arenes, followed by ring formation in the presence of *tert*-butylamine as a source of the pyridine nitrogen (Scheme 1.43).[93]

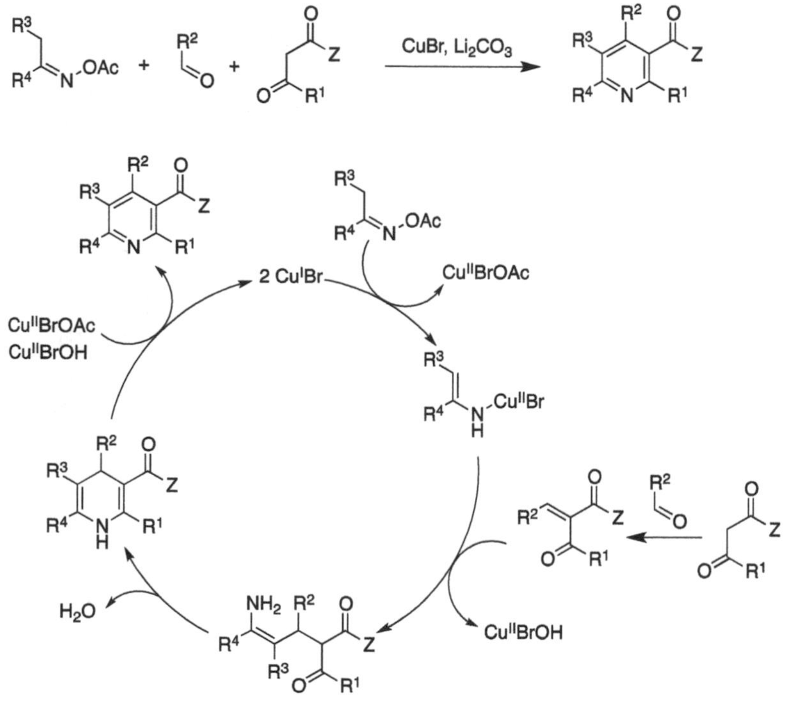

SCHEME 1.42 Copper catalyzed three-component synthesis of pyridines from oxime acetates, activated methylene compounds, and aldehydes.

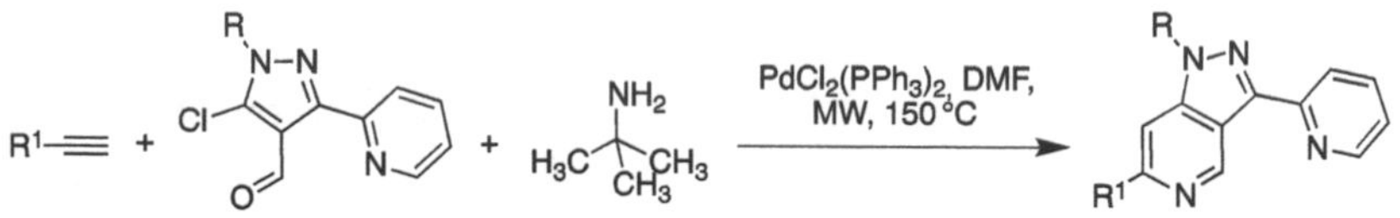

SCHEME 1.43 Synthesis of pyrazolo[4,3-*c*]pyridines initiated by a Sonogashira reaction.

1.5 STRATEGIES BASED ON A [2+2+2] DISCONNECTION

Formal [2+2+2] cycloadditions are a very atom-efficient way to construct six-membered rings.[94] The application of this strategy to the construction of the pyridine nucleus involves the use of one nitrile and two electron-rich alkyne units, and is known as the Bönnemann reaction.[95] Although this transformation was first performed in the presence of a cobalt catalyst, more recent work shows that it can be promoted by several additional transition metal catalysts.[96] Three-component versions of this reaction are plagued by regioselectivity problems when nonsymmetric alkynes are employed, as shown in Scheme 1.44.

Barluenga and Valdés have reported the regioselective synthesis of 2,3,5-triarylpyridines via a pseudo-three-component reaction between styryl bromides (two equivalents) and *N*-(trimethylsilyl) arylimines in the presence of morpholine, a Pd catalyst together with a phosphine ligand and ytterbium triflate as a Lewis acid. The mechanism of this process involves the formation of an enamine by Pd catalyzed amination of the alkenyl bromide with morpholine, the concomitant formation of a 2-aza-1,3-butadiene by Pd catalyzed cross coupling of the trimethylsilylimine with a second molecule of the alkenyl bromide and, finally, a Lewis acid catalyzed cycloaddition between the enamine and the azadiene (Scheme 1.45). The authors also developed conditions allowing the use of two different styryl bromides, leading to nonsymmetrical pyridines.[97]

R—≡N + 2 ≡ —CpCo(COD), 120–130 °C→

R—≡N + 2 R¹—≡ —CpCo(COD), 120–130 °C→ +

SCHEME 1.44 The Bönnemann pyridine synthesis.

Pd$_2$(dba)$_3$, DavePhos, NaO^{t}Bu, Yb(OTf)$_3$, morpholine, toluene, 90 °C

Pd

Yb(OTf)$_3$ [4+2]

- morpholine

Air oxidation

SCHEME 1.45 Pseudo-three-component [2+2+2] synthesis of 2,3,5-triarylpyridines.

1.6 CONCLUSIONS

The synthesis of pyridine derivatives is a vital task in many ambits of chemistry, especially in the pharmaceutical and agrochemical industries. The high synthetic efficiency of multicomponent reactions has encouraged their use in this field, although there is still much room for the improvement of known methods and for the challenging task of designing new ones. We hope that this chapter encourages synthetic chemists to enter this fascinating area.

REFERENCES

1. Anderson, T. *Liebigs Ann. Chem.* **1846**, 60, 86–103.
2. (a) Kozhevnikov, V. N.; Kozhevnikov, D. N.; Nikitina, T. V.; Rusinov, V. L.; Chupakhin, O. L.; Zabel, M.; König, B. *J. Org. Chem.* **2003**, 68, 2882–2888; (b) Hofmeier, H.; Schubert, U. S. *Chem. Commun.* **2005**, 2423–2432; (c) Šmejkal, T.; Breit, B. *Angew. Chem. Int. Ed.* **2008**, 47, 311–315.
3. (a) Halcrow, M. A. *Coord. Chem. Rev.* **2005**, 249, 2880–2908; (b) de Ruiter, G.; Lahav, M; van der Boom, M. E. *Acc. Chem. Res.* **2014**, 47, 3407–3416.
4. Raje, V. P.; Bhat, R. P.; Samant, S. D. *Synlett* **2006**, 2676–2678.
5. (a) Fu, G. C. *Acc. Chem. Res.* **2004**, 37, 542–547; (b) De Rycke, N.; Couty, F.; David, O. R. P. *Chem. Eur. J.* **2011**, 17, 12852–12871.
6. (a) Shibatomi, K.; Muto, T.; Sumikawa, Y.; Narayama, A.; Iwasa, S. *Synlett* **2009**, 241–244; (b) Lin, S.; Lu, X. *Org. Lett.* **2010**, 12, 2536–2539.
7. Elvehjem, C. A.; Madden, R. J.; Strong, F. M.; Woolley, D. W. *J. Biol. Chem.* **1938**, 123, 137–149.
8. (a) Epibatidine: Daly, J. W.; Garraffo, H. M.; Spande, T. F.; Decker, M. W.; Sullivan J. P.; Williams, M. *Nat. Prod. Rep.* **2000**, 17, 131–135; (b) Ilicifoliunines A: Santos, V. A. F. F. M.; Regasini, L. O.; Nogueira, C. R.; Passerini, G. D.; Martinez, I.; Bolzani, V. S.; Graminha, M. A. S.; Cicarelli, R. M. B.; Furlan, M. *J. Nat. Prod.* **2012**, 75, 991–995; (c) Kedarcidin: Leet, J. E.; Schroeder, D. R.; Langley, D. R.; Colson, K. L.; Huang, S.; Klohr, S. E.; Lee, M. S.; Golik, J.; Hofstead, S. J.; Doyle, T. W.; Matson, J. A. *J. Am. Chem. Soc.* **1993**, 115, 8432–8443; (d) Rhexifoline: Roby, M. R.; Stermitz, F. R. *J. Nat. Prod.* **1984**, 47, 846–853; (e) alkaloid isolated from *A. chinense*: Zhang, Y.; Liu, Y.-B.; Li, Y.; Ma, S.-G.; Li, L.; Qu, J.; Zhang, D.; Chen, X.-G.; Jiang, J.-D.; Yu, S.-S. *J. Nat. Prod.* **2013**, 76, 1058–1063; (f) clivimine: Ieven, M.; Vlietinck, A. J.; Vanden Berghe, D. A.; Totte, J. J. *Nat. Prod.* **1982**, 45, 564–573.
9. Duffy, C. D.; Maderna, P.; McCarthy, C.; Loscher, C. E.; Godson, C.; Guiry, P. J. *Chem. Med. Chem.* **2010**, 5, 517–522.
10. Buckley, G. M.; Cooper, N.; Davenport, R. J.; Dyke, H. J.; Galleway, F. P.; Gowers, L.; Haughan, A. F.; Kendall, H. J.; Lowe, C.; Montana, J. G.; Oxford, J.; Peake, J. C.; Picken, C. L.; Richard, M. D.; Sabin, V.; Sharpe, A.; Warneck, J. B. H. *Bioorg. Med. Chem. Lett.* **2002**, 12, 509–512.
11. Balzarini, J.; Stevens, M.; De Clercq, E.; Schols, D.; Pannecouque, C. *J. Antimicrob. Chemother.* **2005**, 55, 135–138.
12. Bachert, C. *Clin. Ther.* **2009**, 31, 921–944.
13. Ou, S.-H. I. *Drug Des. Dev. Ther.* **2011**, 5, 471–485.
14. Saccar, C. L. *Expert Opin. Drug Metab. Toxicol.* **2009**, 5, 1113–1124.
15. Canmtello, B. C. C.; Cawthorne, M. A.; Cottam, G. P.; Duff, P. T.; Haigh, D.; Hindley, R. M.; Lister, C. A.; Smith, S. A.; Thurlby, P. L. *J. Med. Chem.* **1994**, 37, 3977–3985.

16. Evans, B. E.; Rittle, K. E.; Bock, M. G.; DiPardo, R. M.; Freidinger, R. M.; Whitter, W. L.; Lundell, G. F.; Veber, D. F.; Anderson, P. S.; Chang, R. S. L.; Lotti, V. J.; Cerino, D. J.; Chen, T. B.; Kling, P. J.; Kunkel, K. A.; Springer, J. P.; Hirshfield, J. *J. Med. Chem.* **1988**, 31, 2235–2246.
17. (a) Costantino, L.; Barlocco, D. *Curr. Med. Chem.* **2006**, 13, 65–85; (b) Welsch, M. E.; Snyder, S. A.; Stockwell, B. R. *Curr. Opin. Chem. Biol.* **2010**, 14, 347–361.
18. Zhang, W.; Chen, Y.; Chen, W.; Liu, Z.; Li, Z. *J. Agric. Food Chem.* **2010**, 58, 6296–6299.
19. Xie, Y.; Chi, H.-W.; Guan, A.-Y.; Liu, C.-L.; Ma, H.-J.; Cui, D.-L. *J. Agric. Food Chem.* **2014**, 62, 12491–12496.
20. Yan, X.; Qin, W.; Sun, L.; Qi, S.; Yang, D.; Qin, Z.; Yuan, H. *J. Agric. Food Chem.* **2010**, 58, 2720–2725.
21. Henry, G. D. *Tetrahedron* **2004**, 60, 6043–6061.
22. Hill, M. D. *Chem. Eur. J.* **2010**, 16, 12052–12062.
23. Allais, C.; Grassot, J.-M.; Rodriguez, J.; Constantieux, T. *Chem. Rev.* **2014**, 114, 10829–10868.
24. Hantzsch, A. *Chem. Ber.* **1881**, 14, 1637–1638.
25. For examples of dihydropyridine oxidation into pyridine, see (a) Liao, X.; Lin, W.; Lu, J.; Wang, C. *Tetrahedron Lett.* **2010**, 51, 3859–3861; (b) Jia, X.; Yu, L.; Huo, C.; Wang, Y.; Liu, J.; Wang, X. *Tetrahedron Lett.* **2014**, 55, 264–266; (c) Sánchez, L. M.; Sathicq, A. G.; Baronetti, G. T.; Thomas, H. J.; Romanelli, G. P. *Catal. Lett.* **2014**, 44, 172–180; (d) Saikh, F.; De, R.; Ghosh, S. *Tetrahedron Lett.* **2014**, 55, 6171–6174.
26. For examples of dihydropyridine oxidation into pyridine employing O_2 as oxidant, see (a) Shen, L.; Cao, S.; Wu, J.; Zhang, J.; Li, H.; Liu, N.; Qian, X. *Green Chem.* **2009**, 11, 1414–1420; (b) Abdel-Mohsen, H. T.; Conrad, J.; Beifuss, U. *Green Chem.* **2012**, 14, 2686–2690; (c) Wei, X.; Wang, L.; Jia, W.; Du, S.; Wu, L.; Liu, Q. *Chin. J. Chem.* **2014**, 32, 1245–1250.
27. (a) Xia, J. J.; Wang, G. W. *Synthesis* **2005**, 2379–2383; (b) Das Sharma, S.; Hazarika, P.; Konwar, D. *Catal. Commun.* **2008**, 9, 709–714.
28. (a) Cotterill, I. C.; Usyatinsky, A. Y.; Arnold, J. M.; Clark, D. S.; Dordick, J. S.; Michels, P. C.; Khmelnitsky, Y. L. *Tetrahedron Lett.* **1998**, 39, 1117–1120; (b) De Paolis, O.; Baffoe, J.; Landge, S.; Török, B. *Synthesis* **2008**, 3423–3428; (c) Shen, L.; Cao, S.; Wu, J.J.; Zhang, J.; Li, H.; Liu, N. J.; Qian, X. H. *Green Chem.* **2009**, 11, 1414–1420; (d) Mirza-Aghayan, M.; Asadi, F.; Boukherroub, R. *Monatsh. Chem.* **2014**, 145, 1919–1924.
29. Vanden Eyde, J. J.; Mayence, A. *Molecules* **2003**, 8, 381–391.
30. (a) Andersen, K. H.; Nordlander, M.; Westerlund, R. C. U.S. Patent 5856346, 1999; (b) Dondoni A, Massi A, Aldhoun M *J. Org. Chem.* **2007**, 72, 7677–7687; (c) Vishnu, N. R.; Puruschottam, J. A.; Kumar, S. S. Eur. Patent EP 2386544, **2011**.
31. (a) Zhu, S.-L.; Ji, S.-J.; Su, X.-M.; Sun, C.; Liu, Y. *Tetrahedron Lett.* **2008**, 49, 1777–1781; (b) Biradar, J. S.; Sharanbasappa, B. *Green Chem. Lett. Rev.* **2009**, 2, 237–241.
32. Zhou, Y.; Kijima, T.; Kuwahara, S.; Watanabe, M.; Izumi, T. *Tetrahedron Lett.* **2008**, 49, 3757–3761.
33. Chichibabin, A. E. *J. Russ. Phys. Chem. Soc.* **1906**, 37, 1229–1231.
34. Frank, R. L.; Seven, R. P. *J. Am. Chem. Soc.* **1949**, 71, 2629–2635.
35. For examples of pyridine synthesis based on variations to the Chichibabin synthesis involving acetophenone derivatives, see: (a) Davoodnia, A.; Bakavoli, M.; Moloudi, R.; Takavoli-Hoseini, N.; Khashi, M. *Monatsh. Chem.* **2010**, 141, 867–870; (b) Li, J.; He, P.; Yu, C. *Tetrahedron* **2012**, 68, 4138–4144; (c) Moosavi-Zare, A. R.; Zolfigol, M. A.; Farahmand, S.; Zare, A.; Pourali, A. R.; Ayazi-Nasrabadi, R. *Synlett* **2014**, 25, 193–196.
36. Penta, S.; Vedula, R. R. *J. Heterocycl. Chem.* **2013**, 50, 859–862.

37. Wu, H.; Wan, Y.; Chen, X.-M.; Chen, C.-F.; Lu, L.-L.; Xin, H.-Q.; Xu, H.-H.; Pang, L.-L.; Ma, R.; Yue, C.-H. *J. Heterocycl. Chem.* **2009**, 46, 702–707.
38. For examples of pyridine synthesis based on variations to the Chichibabin synthesis involving cyclic ketones, see: (a) Tu, S.; Jia, R.; Jiang, B.; Zhang, J.; Zhang, Y.; Yao, C.; Ji, S. *Tetrahedron* **2007**, 63, 381–388; (b) Rong, L.; Han, H.; Wang, S.; Zhuang, Q. *Synth. Commun.* **2008**, 38, 1808–1814; (c) Wu, P.; Cai, X.-M.; Wang, Q.-F.; Yan, C.-G. *Synth. Commun.* **2011**, 41, 841–850.
39. Tu, S.; Jiang, B.; Jia, R.; Zhang, J.; Zhang, Y. *Tetrahedron Lett.* **2007**, 48, 1369–1374.
40. Kambe, S.; Saito, K.; Sakurai, A.; Midorikawa, H. *Synthesis* **1980**, 366–368.
41. For examples of pyridine synthesis based on variations of the Chichibabin synthesis involving malononitrile, see (a) Jiang, B.; Wang, X.; Shi, F.; Tu, S.-J.; Li, G. *Org. Biomol. Chem.* **2011**, 9, 4025–4028; (b) Pagadala, R.; Maddila, S.; Moodley, V.; van Zyl, W. E.; Jonnalagadda, S. B. *Tetrahedron Lett.* **2014**, 55, 4006–4010; (c) Pagadala, R.; Kommidi, D. R.; Rana, S.; Maddila, S.; Moodley, B.; Koorbanally, N. A.; Jonnalagadda, S. B. *RCS Adv.* **2015**, 5, 5627–5632.
42. (a) Murata, T.; Shimada, M.; Sakakibara, S.; Yoshino, T.; Kadono, H.; Masuda, T.; Shimazaki, M.; Shintani, T.; Fuchikami, K.; Sakai, K.; Inbe, H.; Takeshita, K.; Niki, T.; Umeda, M.; Bacon, K. B.; Ziegelbauer, K. B.; Lowinger, T. B. *Bioorg. Med. Chem. Lett.* **2003**, 13, 913–918; (b) Murata, T.; Shimada, M.; Kadono, H.; Sakakibara, S.; Yoshino, T.; Masuda, T.; Shimazaki, M.; Shintani, T.; Fuchikami, K.; Bacon, K. B.; Ziegelbauer, K. B.; Lowinger, T. B. *Bioorg. Med. Chem. Lett.* **2004**, 14, 4013–4017; (c) Murata, T.; Shimada, M.; Sakakibara, S.; Yoshino, T.; Masuda, T.; Shintani, T.; Sato, H.; Koriyama, Y.; Fukushima, K.; Nunami, N.; Yamauchi, M.; Fuchikami, K.; Komura, H.; Watanabe, A.; Ziegelbauer, K. B.; Bacon, K. B.; Lowinger, T. B. *Bioorg. Med. Chem. Lett.* **2004**, 14, 4019–4022.
43. Bodireddy, M. R.; Reddy, N. C. G.; Kumar, S. D. *RCS Adv.* **2014**, 4, 17196–17205.
44. For examples of pyridine synthesis based on variations to the Chichibabin synthesis involving β-ketonitriles and β-cyanoesters, see (a) Thirumurugan, P.; Perumal, P. T. *Tetrahedron Lett.* **2009**, 50, 4145–4150; (b) Thirumurugan, P.; Nandakumar, A.; Muralidharan, D.; Perumal, P. T. *J. Comb. Chem.* **2010**, 12, 161–167; (c) El-Sayed, N. S.; Shirazi, A. N.; El-Meligy, M. G.; El-Ziaty, A. K.; Rowley, D; Sun, J.; Nagib, Z. A.; Parang, K. *Tetrahedron Lett.* **2014**, 55, 1154–1158.
45. Zeng, L.-Y.; Cai, C. *Synth. Commun.* **2013**, 43, 705–718.
46. For examples of pyridine synthesis based on modified three-component Hantzsch approach involving β-ketoesters and β-diketones, see (a) Shi, D.-Q.; Ni, S.-N.; Yang, F.; Shi, J.-W.; Dou, G.-L.; Li, X.-Y.; Wang, X.-S.; Ji, S.-J. *J. Heterocycl. Chem.* **2008**, 45, 693–702; (b) Shi, D.-Q.; Yang, F.; Ni, S.-N. *J. Heterocycl. Chem.* **2009**, 46, 469–476; (c) ang, H.-Y.; Shi, D.-Q. *J. Heterocycl. Chem.* **2012**, 49, 212–216.
47. (a) Shi, F.; Zhou, D.; Tu, S.; Li, C.; Cao, L.; Shao, Q. *J. Heterocycl. Chem.* **2008**, 45, 1305–1310; (b) Muravyova, E.; Shishkina, S.; Musatov, V.; Knyazeva, I.; Shishkin, O.; Desenko, S.; Chebanov, V. *Synthesis* **2009**, 1375–1385.
48. For examples of pyridine synthesis based on modified three-component Hantzsch approach involving malononitrile and β-ketonitriles, see (a) Zhu, S.-L.; Ji, S.-J.; Zhao, K.; Liu, Y. *Tetrahedron Lett.* **2008**, 49, 2578–2582; (b) Zare, L.; Mahmoodi, N. O.; Yahyazadeh, A.; Mamaghani, M. *Synth. Commun.* **2011**, 41, 2323–2330; (c) Huang, Z.; Hu, Y.; Zhou, Y.; Shi, D. *ACS Comb. Sci.* **2011**, 13, 45–49.
49. Nasr, M. A.; Gineinah, M. M. *Arch. Pharm.* **2002**, 335, 289–295.
50. Evdokimov, N. M.; Van Slambrouck, S.; Heffeter, P.; Tu, L.; Le Calvé, B.; Lamoral-Theys, D.; Hooten, C. J.; Uglinskii, P. Y.; Rogelj, S.; Kiss, R.; Steelant, W. F. A.; Berger, W.; Yang, J. J.; Bologa, C. G.; Kornienko, A.; Magedov, I. V. *J. Med. Chem.* **2011**, 54, 2012–2021.
51. Shi, D.-Q.; Li, Y.; Wang, H.-Y. *J. Heterocycl. Chem.* **2012**, 49, 1086–1090.

52. Shaabani, A.; Seyyedhamzeh, M.; Maleki, A.; Behnam, M.; Rezazadeh, F. *Tetrahedron Lett.* **2009**, 50, 2911–2913.
53. Bagley, M. C.; Davis, T.; Rokicki, M. J.; Widdowson, C. S.; Kipling, D. *Future Med. Chem.* **2010**, 2, 193–201.
54. (a) Jiang, B.; Liu, Y.-P.; Tu, S.-J. *Eur. J. Org. Chem.* **2011**, 3026–3035; (b) Xu, B.-H.; Tu, M.-S.; Liu, Y.-P.; Jiang, B.; Wang, X.-H; Tu, S.-J. *J. Heterocycl. Chem.* **2014**, 51, 1591–1594; (c) Anvar, S.; Mohammadpoor-Baltork, I; Tangestaninejad, S.; Moghadam, M.; Mirkhani, V.; Khosropour, A. R.; Isfahani, A. L.; Kia, R. *ACS Comb. Sci.* **2014**, 16, 93–100.
55. Liéby-Muller, F.; Allais, C.; Constantieux, T.; Rodriguez, J. *Chem. Commun.* **2008**, 4207–4209.
56. Allais, C.; Constantieux, T.; Rodriguez, J. *Chem. Eur. J.* **2009**, 15, 12945–12948.
57. Allais, C.; Liéby-Muller, F.; Constantieux, T.; Rodriguez, J. *Eur. J. Org. Chem.* **2013**, 4131–4145.
58. Allais, C.; Liéby-Muller, F.; Constantieux, T.; Rodriguez, *J. Adv. Synth. Catal.* **2012**, 354, 2537–2544.
59. Khanal, H. D.; Lee, Y. R. *Chem. Commun.* **2015**, 51, 9467–9470.
60. Mannich, C.; Krösche, W. *Arch. Pharm. (Weinheim)* **1912**, 250, 647–666.
61. Keuper, R.; Risch, N. *Liebigs Ann.* **1996**, 717–723.
62. (a) Westerwelle, U.; Risch, N. *Tetrahedron Lett.* **1993**, 34, 1775–1778. (b) Keuper, R.; Risch, N.; Flörke, U.; Haupt, H. *J. Liebigs Ann.* **1996**, 705–715. (c) Kelly, T. R.; Lebedev, R. L. *J. Org. Chem.* **2002**, 67, 2197–2205.
63. Hanashalshahaby, E. H. A.; Unaleroglu, C. *ACS Comb. Sci.* **2015**, 17, 374–380.
64. Zecher, W.; Kröhnke, F. *Chem. Ber.* **1961**, 94, 690–697.
65. Yan, C.-G.; Cai, X.-M.; Wang, Q.-F.; Wang, T.-Y.; Zheng, M. *Org. Biomol. Chem.* **2007**, 5, 945–951.
66. Yan, C.-G.; Wang, Q.-F.; Cai, X.-M.; Sun, J. *Cent. Eur. J. Chem.* **2008**, 6, 188–198.
67. Ash, A. B.; LaMontagne, M. P.; Markovac, A. US Patent 3,886,167, **1975**.
68. Betti, M.; Castagnoli, G.; Panico, A.; Sanna Coccone, S.; Wiedenau, P. *Org. Process Res. Dev.* **2012**, 16, 1739–1745.
69. Karki, R.; Park, C.; Jun, K.-Y.; Kadayat, T.-M.; Lee, E.-S.; Kwon, Y. *Eur. J. Med. Chem.* **2015**, 90, 360–378.
70. Patel, A. K.; Patel, N. H.; Patel, M. A.; Brahmbhant, D. I. *J. Heterocycl. Chem.* **2012**, 49, 504–510.
71. Katritzky, A. R.; Abdel-Fattah, A. A. A.; Tymoshenko, D. O.; Essawy, S. A. *Synthesis* **1999**, 2114–2118.
72. Kantevari, S.; Patpi, S. R.; Addla, D.; Putapatri, S. R.; Sridhar, B.; Yogeeswari, P.; Sriram, D. *ACS Comb. Sci.* **2011**, 13, 427–435.
73. Kantevari, S.; Patpi, S. R.; Sridhar, B.; Yogeeswari, P.; Sriram, D. *Bioorg. Med. Chem. Lett.* **2011**, 21, 1214–1217.
74. Tenti, G.; Ramos, M. T.; Menéndez, J. C. *ACS Comb. Sci.* **2012**, 14, 551–557.
75. Tenti, G.; Parada, E.; León, R.; Egea, J.; Martínez-Revelles, S.; Briones, A. M.; Sridharan, V.; López, M. G.; Ramos, M. T.; Menéndez, J. C. *J. Med. Chem.* **2014**, 57, 4313–4323.
76. (a) Sridharan, V.; Maiti, S.; Menéndez, J. C. *Chem. Eur. J.* **2009**, 15, 4565–4572. (b) Tenti, G.; Ramos, M. T.; Menéndez, J. C. *Curr, Org, Synth,* **2013**, 10, 646–655.
77. Raja, A. V. P.; Tenti, G.; Perumal, S.; Menéndez, J. C. *Chem. Commun.* **2014**, 50, 12270–12272.
78. Bohlmann, F.; Rahtz, D. *Chem. Ber.* **1957**, 90, 2265–2272.
79. Bagley, M. C.; Glover, C.; Merritt, E. A. *Synlett* **2007**, 2459–2482.
80. Bagley, M. C.; Dale, J. W.; Bower, J. *Chem. Commun.* **2002**, 1682–1683.
81. Bagley, M. C.; Chapaneri, K.; Dale, J. W.; Xiong, X.; Bower, J. *J. Org. Chem.* **2005**, 70, 1389–1399.

82. Aulakh, V. S.; Ciufolini, M. A. *J. Org. Chem.* **2009**, 74, 5750–5753.
83. Bagley, M. C.; Xiong, X. *Org. Lett.* **2004**, 6, 3401–3404.
84. Bagley, M. C.; Hughes, D. D.; Sabo, H. M.; Taylor, P. H.; Xiong, X. *Synlett* **2003**, 1443–1446.
85. Gujar, J. B.; Chaudhari, M. A.; Kawade, D. S.; Shingare, M. S. *Tetrahedron Lett.* **2014**, 55, 6939–6942.
86. Khan, N.; Karamthulla, S.; Choudhury, L. H.; Faizib, S. H. *RSC Adv.* **2015**, 5, 22168–22172.
87. Maharani, S.; Ranjith Kumar, R. *Tetrahedron Lett.* **2015**, 56, 179–181.
88. Yang, J.; Li, J.; Hao, P.; Qiu, F.; Liu, M.; Zhang, Q.; Shi, D. *Dyes Pigments* **2015**, 116, 97–105.
89. Kumar Reddy, D. N.; Chandrasekhar, K. B.; Ganesh, Y. S. S.; Kumar, B. S.; Adepu, R.; Pal, M. *Tetrahedron Lett.* **2015**, 56, 4586–4589.
90. Sarkar, S.; Das, D. K.; Khan, A. T. *RSC Adv.* **2014**, 4, 53752–53760.
91. Davies, I. W.; Marcoux, J. F.; Corley, E. G.; Journet, M.; Cai, D. W.; Palucki, M.; Wu, J.; Larsen, R. D.; Rossen, K.; Pye, P. J.; Di Michele, L.; Dormer, P.; Reider, P. J. *J. Org. Chem.* **2000**, 65, 8415–8420.
92. Jiang, H.; Yang, J.; Tang, X.; Li, J.; Wu, W. *J. Org. Chem.* **2015**, 80, 8763–8771.
93 Vilkauskaitė, G.; Schaaf, P.; Šačkus, A.; Krystof, V.; Holzera, W. *Arkivoc* **2014**, *ii*, 135–149.
94. Domínguez, G.; Pérez-Castells, *J. Chem. Soc. Rev.* **2011**, 40, 3430–3444.
95. Bönnemann, H. *Angew. Chem. Int. Ed.* **1978**, 17, 505–515.
96. Wang, C.; Li, X.; Wu, F.; Wan, W. *Angew. Chem. Int. Ed.* **2011**, 50, 7162–7166.
97. Barluenga, J.; Jiménez-Aquino, A.; Fernández, M. A.; Aznar, F.; Valdés, C. *Tetrahedron* **2008**, 64, 778–786.

2 Isocyanide-Based Multicomponent Reactions Based upon Intramolecular Nitrilium Trapping

Susruta Majumdar, András Váradi, and Travis C. Palmer*

CONTENTS

2.1 INTRODUCTION

In multicomponent reactions (MCRs) more than two reactants form a single product. Typically, all of the atoms of the starting materials get incorporated into the adduct, making MCRs highly atom efficient. MCRs have recently gained popularity for this efficiency as more and more chemists focus on green chemistry and minimizing the waste

* Corresponding author: majumdas@mskcc.org.

their reactions generate. In addition to being atom efficient, these reactions can be carried out conveniently by simply mixing the starting materials in one pot. Furthermore, MCRs have long been known as methods to access an extreme diversity of complex molecules, including heterocycles with biological activity.[1] MCRs have been used in total synthesis, drug discovery, and bioconjugation. This advantage is made even greater because almost all MCRs make use of common functional groups that can be found on a wide variety of commercially available compounds. The integrative nature of MCRs is increasingly attractive when a rapid increase in molecular diversity is desired. Their convenient setup, one-pot nature, high atom efficiency, and product diversity make MCRs highly desirable to generate bioactive compound libraries. In the most preparatively useful MCRs, a sequence of reversible steps is concluded by an enthalpically driven and irreversible product.[2]

Additionally, MCRs tolerate a wide variety of functional groups. This allows for easy diversification beyond the MCR itself, without the need for protection and deprotection of reactive functional groups not involved in the MCR. Some MCRs generate uncommon, if not unique, scaffolds so the ability to further functionalize or modify them is key to exploring the utility of the scaffold in the biological realm. The unusual structure of many of these scaffolds makes them ideally suited for exploring biological targets that traditional scaffolds do not target. Novel scaffolds are becoming more sought after as pathogens mutate to become resistant to current medications, and as other diseases, such as Alzheimer's, cancer, and others, begin to be better understood by scientists. While MCRs may not be directly responsible for the drugs that treat these diseases in the future, they will certainly be involved as scientists search for tomorrow's remedies. Isocyanide-based reactions form the backbone of MCR chemistry; the most common examples are the Passerini reaction and the Ugi reaction.

2.2 PASSERINI REACTION

The Passerini reaction involves a carboxylic acid, aldehyde or ketone, and isocyanide, and yields α-acyloxy amides (**1**, Scheme 2.1). The reaction progresses quickly at room temperature and high reactant concentrations. The exact mechanism is a subject of some uncertainty; the mechanism postulated by Ugi suggests a non-ionic pathway, since the reaction is accelerated in aprotic solvents (Scheme 2.2).[3] The electrophilic activation of the carbonyl group is followed by a nucleophilic attack by the isocyanide. This creates a nitrilium intermediate which is then attacked by the carboxylate. The Passerini reaction has recently been used to synthesize a library of monomers, which were then polymerized. By varying the R groups in the Passerini reaction, the group was able to tune the properties of the resulting polymer.[4]

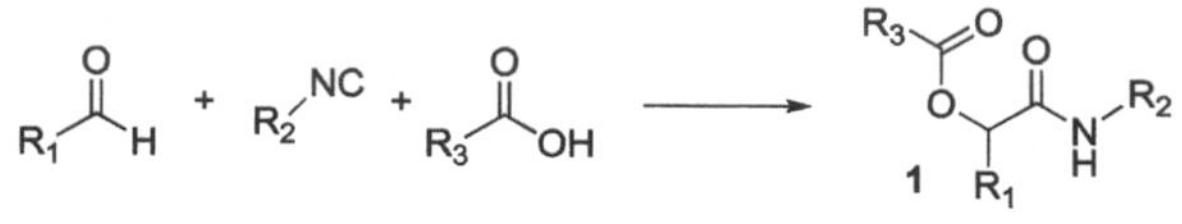

SCHEME 2.1 A general Passerini reaction yielding an α-acyloxy amide (**1**).

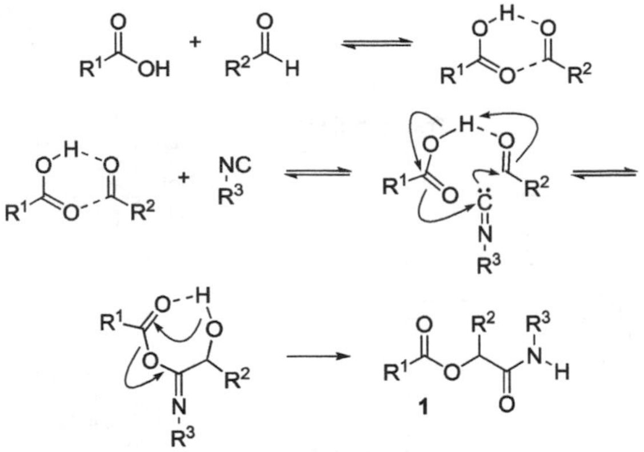

SCHEME 2.2 Possible mechanism of the Passerini reaction.

2.3 UGI REACTION

A traditional Ugi four-component reaction (U4CR, Scheme 2.3) employs a ketone or aldehyde, a carboxylic acid, an isocyanide, and an amine. The reaction is typically carried out in methanol or 2,2,2-trifluoroethanol in high reactant concentrations. The initial step is the formation of an imine from the amine and the carbonyl compound. This is followed by the nucleophilic attack of the isocyanide, resulting in the formation of the highly reactive nitrilium intermediate. The nitrilium is then attacked by the carboxylic acid, and as a result of intramolecular Mumm rearrangement, the reaction yields a central bis-amide (**2**, Scheme 2.4).

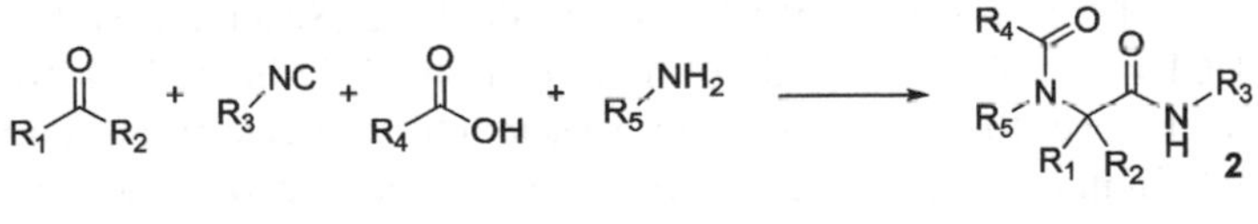

SCHEME 2.3 A standard U4CR. The reaction can tolerate a wide variety of R groups.

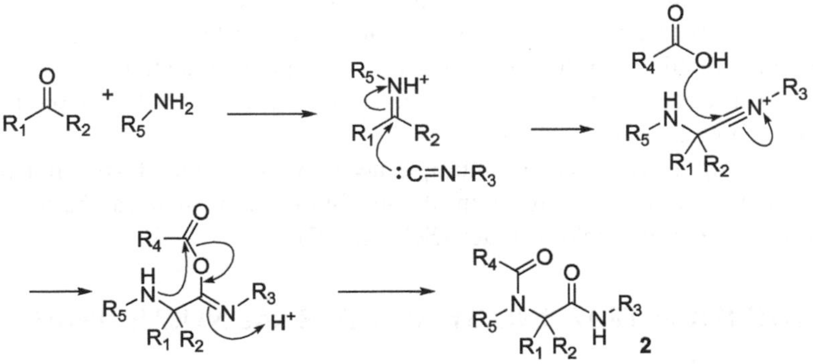

SCHEME 2.4 Mechanism of the Ugi four-component reaction.

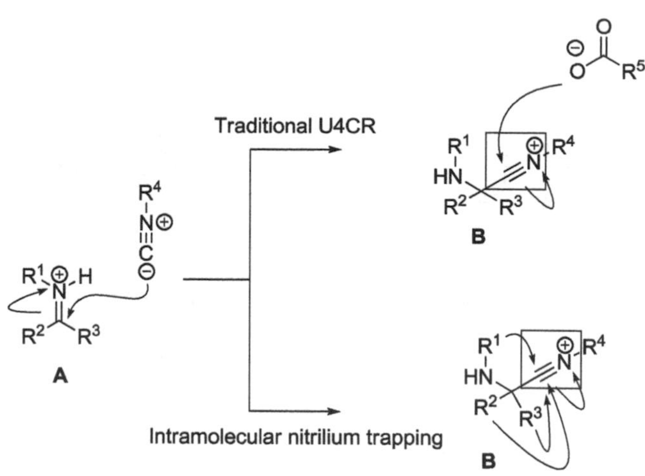

SCHEME 2.5 In nitrilium trapping reactions, an external or intramolecular nucleophile attacks the nitrilium species (**B**).

Depending upon the R groups, post-Ugi reactions have been reported. Most notable are the Ugi–Heck,[5] Ugi–Diels–Alder,[6] Ugi–click,[7] and Ugi–Buchwald–Hartwig[8] reactions. In these reactions, the Ugi bis-amide with reactive functional groups undergoes secondary reactions to form a ring. Linear bis-amides, on the other hand, are useful in the synthesis of peptides (linear and cyclic) and peptidomimetics.[9]

In the case of the Ugi and the Passerini reactions, an important intermediate in the mechanism is the reactive nitrilium. The mechanism of the Ugi reaction involves the nucleophilic attack of the isocyanide on the imine species (**A**) formed by the condensation of the amine and the ketone or aldehyde (Scheme 2.5). Lewis acids (such as $TiCl_4$) are effective in aiding the formation of the imine by activating the aldehyde or ketone, thereby facilitating a nucleophilic attack by an amine.[10] The resulting nitrilium intermediate (**B**) itself is highly reactive as well and readily reacts with nucleophiles.

In a traditional U4CR, the reaction partner of the nitrilium (**B**) is the carboxylic acid. Besides being a nucleophilic partner in the Ugi reaction, the carboxylic acid also plays a part in activating the nitrilium ion. Metal triflates have been shown to activate this nitrilium intermediate making them more susceptible to nucleophilic attack by carboxylic acids and thereby increasing rates of rates U4CR product formation by about two- to sevenfold.[11]

However, when reactants with multiple nucleophilic groups (bisfunctional components) are used in the Ugi reaction, the nitrilium intermediate can be trapped by an intramolecular nucleophilic attack (Scheme 2.5).

2.4 NITRILIUM TRAPPING BY AN EXTERNAL NUCLEOPHILE

The traditional Ugi reaction utilizes a carboxylic acid as the fourth component in trapping the nitrilium ion. In the absence of a carboxylic acid, water can act as the nucleophilic partner too. One well-known example of utilizing the Ugi reaction in bioactive molecule synthesis is the synthesis of the local anesthetic lidocaine (**3**, Scheme 2.6).[12] Trapping of nitrilium ions by water is usually carried out under acidic conditions.[13]

Besides carboxylic acids, water, hydrazoic acid, HOCN, HSCN, and H_2Se are some other external nucleophiles that participate as the fourth component in a U4CR (Scheme 2.7).[14]

Nitrilium trapping by primary amines leads to the synthesis of amidines[15] (**9**, Scheme 2.8A); electron deficient phenols lead to *N*-aryl amines[16] (Ugi–Smiles coupling, **10**, Scheme 2.8B).

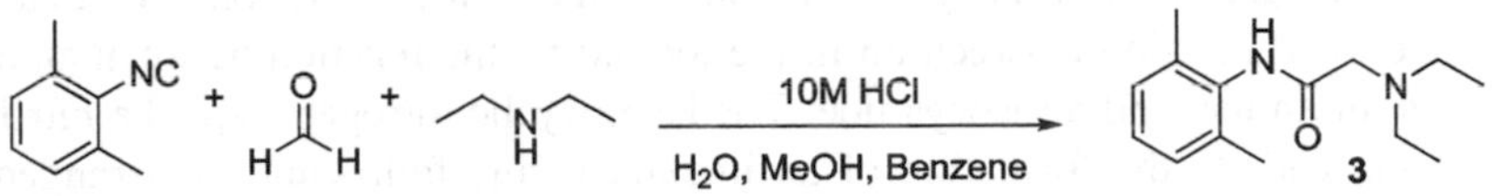

SCHEME 2.6 Lidocaine (**3**) is a local anesthetic on the WHO Model List of Essential Medicines that can be synthesized in one step via a U3CR.

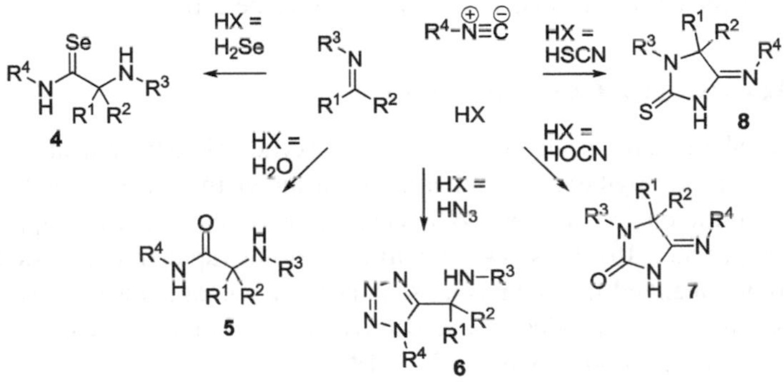

SCHEME 2.7 Replacement of the carboxylic acid in a U4CR.[14]

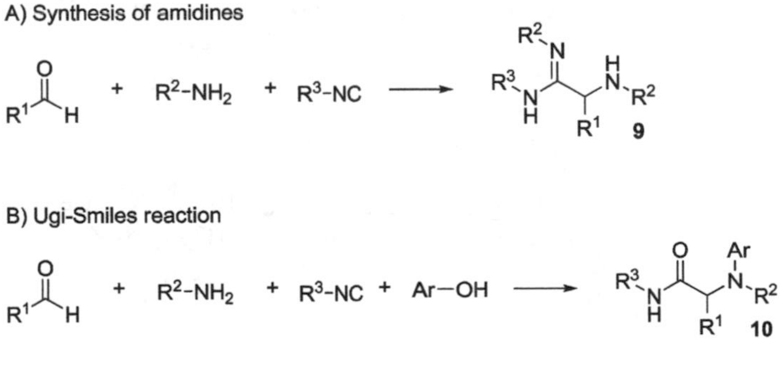

SCHEME 2.8 A: Reaction of amines with the nitrilium yields amidines (**9**). B: trapping with phenols leads to *N*-aryl amines (**10**, Ugi–Smiles reaction).

2.4.1 Intramolecular Trapping of Nitrilium Ion

This review will focus on the chemistry of nitrilium trapping and its possible application in the synthesis of biologically active heterocyclic compounds. For excellent reviews on the usefulness of MCRs in drug development, see the papers by Dömling and Hulme.[2,12,17] In the absence of an external nucleophile, the reactive nitrilium species can be trapped intramolecularly by a bisfunctional nucleophilic component. This component could be an amine, carboxylic acid, phenol, amide, imine, hydrazide, activated carbon, or oxime.

2.4.2 Trapping by an Imine

The Groebke–Blackburn–Bienaymé reaction[18] yields 3-aminoimidizoles (**11**) in a nonconcerted [4+1] reaction between an imine formed by the reaction between an aldehyde and an amine, and an isocyanide. The heterocyclic nitrogen traps the nitrilium ion, which leads to the formation of an imidazole ring following a rearrangement (Scheme 2.9). This reaction is widely used for the generation of highly diverse small molecule libraries[19]; the bioactive compounds synthesized this way include kinase inhibitors (**12**),[20] topoisomerase II inhibitors (**13**),[21] antibacterials (**14**) effective against methicillin-resistant *Staphylococcus aureus*,[22] fluorescent probes (**15**),[23] and HIV-1 (**16**) reverse transcriptase inhibitors (**17**, Scheme 2.10).[24]

2.4.3 Trapping by a Carboxylic Acid

In the case where both an aldehyde and a carboxylic acid are present on the same molecule, as in 2-formylbenzoic acid, it has been shown that the carboxylic acid will trap the nitrilium ion. Whereas a U4CR yields a linear scaffold, this trapping generates a cyclic product (**18**). This has been utilized for the rapid and high-yielding synthesis of functionalized isocoumarins, such as the anticoagulant warfarin (Scheme 2.11).[25] Treatment of the product with a catalytic amount of acid led to an isomerization to the expected product (Scheme 2.11, **19**).

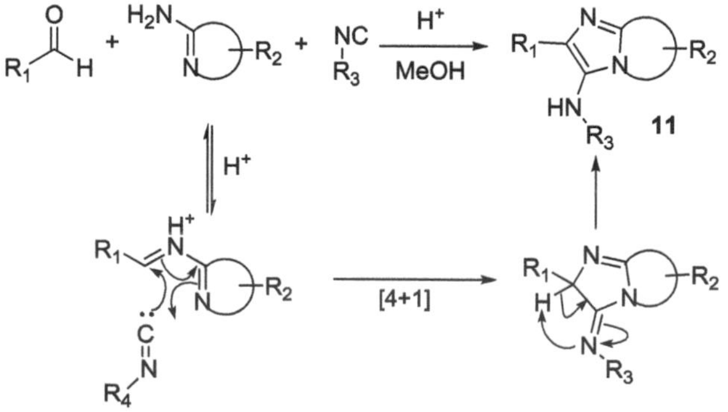

SCHEME 2.9 The isocyanide in the Groebke–Blackburn–Bienaymé reaction is trapped by the presence of an imine-like moiety.[18c]

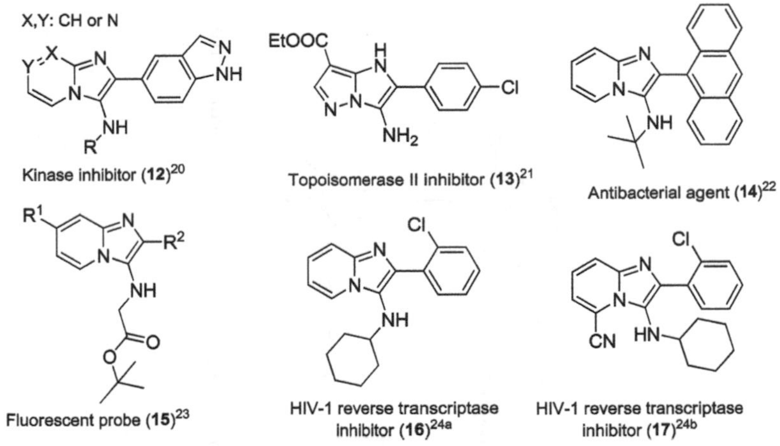

SCHEME 2.10 Bioactive compounds **12–17** synthesized via the Groebke–Blackburn–Bienaymé reaction.

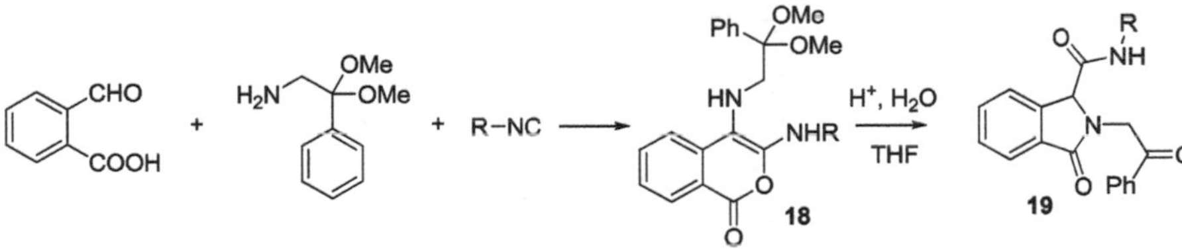

SCHEME 2.11 Trapping with an intramolecular carboxylic acid.[25] (Modified from Faggi, C. et al., *Org. Lett.* 2010, 12, 788.)

2.4.4 Trapping by a Secondary Amine

When an aldehyde and 4-morpholinepropionitrile are combined in the presence of trimethylsilyltriflate, the aldehyde is trapped by the trimethylsilyl (TMS) group and the morpholino nitrogen stabilizes the nitrilium ion (Scheme 2.12). The trimethylsilyltriflate coordinates to and activates the aldehyde to attack by an isocyanide, similarly to how the carboxylic acid normally activates the aldehyde through hydrogen bonding. If a ketone is present on the α-carbon of the isocyanide, the reaction can yield a cyclized product (**21**) as well.[26]

2.4.5 Trapping by Amide

Nitrilium ions can also be trapped by isocyanides containing a secondary amide. The nitrilium is trapped by the amide after attacking the imine, but contrary to what one would expect, the nitrogen does not directly attack the isocyanide carbon. Instead, the electrons flow from the nitrogen through the oxygen, and only then to the target atom (Scheme 2.13). The reaction proceeds even when R^{1-4} are bulky substituents, showing that the scope of the reaction is not limited by the trapping when compared with a traditional Ugi.[27]

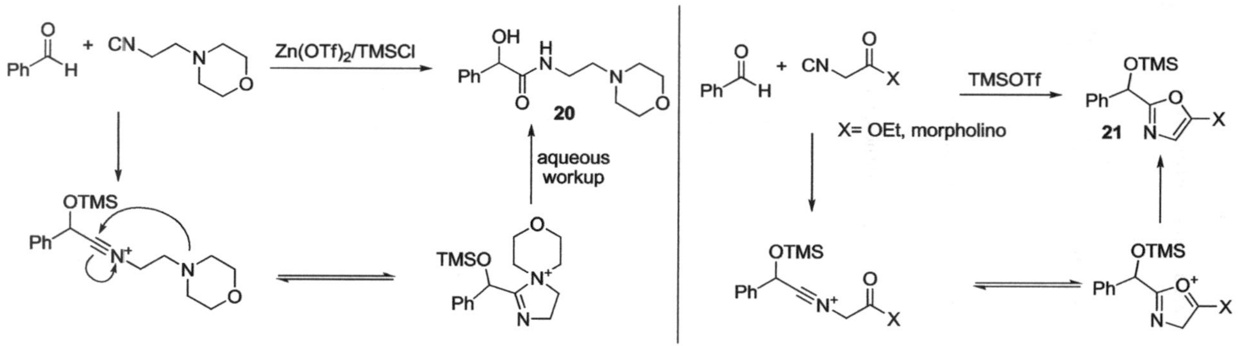

SCHEME 2.12 A morpholino isocyanide yields a linear product (**20**), while an α-keto morpholino isocyanide cyclizes to yield **21**.[26] (Modified from Ganem, B., *Acc. Chem. Res.* 2009, 42, 463.)

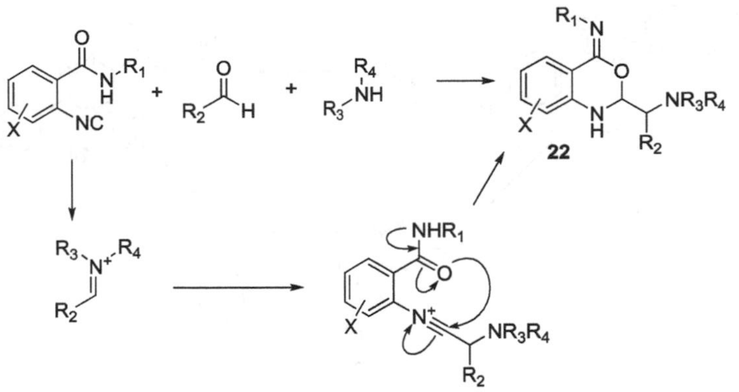

SCHEME 2.13 The isocyanide attacks the nitrilium ion, but is then trapped by the amide substituent before hydrolysis can occur to yield products such as **22**.[27] (Modified from Bonne et al., http://pubs.acs.org/doi/pdf/10.1021/ol052239w.)

2.4.6 Trapping by Amine

If the amine reagent is a diamine, one amino group will form an imine and the other will react with the nitrilium after it attacks that imine. This trapping, like many others shown here, forms a highly substituted heterocycle. By using 1,2-diaminobenzenes, the reaction yields quinoxalinones in a single step (**23**, Scheme 2.14). As noted by the authors, several quinoxalinone compounds have affinity for various central nervous system receptors, and also may be useful as antidiabetic agents and antiretrovirals.[28]

An intriguing way to ensure the generation of a heterocycle from a trapped MCR is to use a reactant that contains both the isocyanophile, which is an imine in a typical Ugi, and a nucleophile to trap the nitrilium after its attack (Scheme 2.15) to form a six-membered heterocycle (**24**). While similar to using a diamino reactant, this approach goes one step further and includes the "imine" with the nucleophile rather than forming it *in situ*.[29]

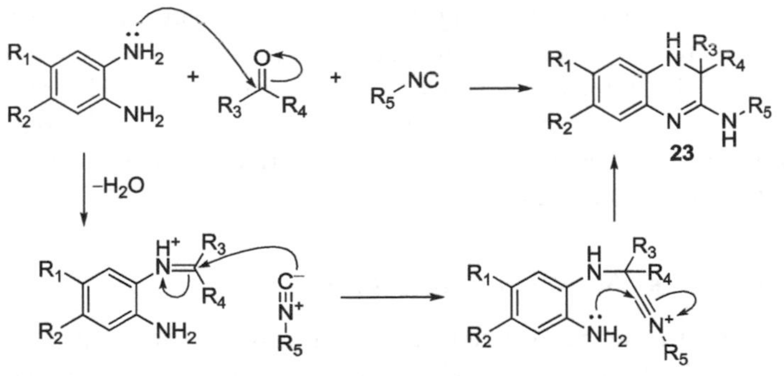

SCHEME 2.14 Synthesis of 3,4-dihydroquinoxalin-2-amine (**23**) and a library of substituted analogs using 1,2-diaminobenzenes.[28] (Modified from Shaabani, A. et al., *J. Comb. Chem. 2008*, 10, 2003.)

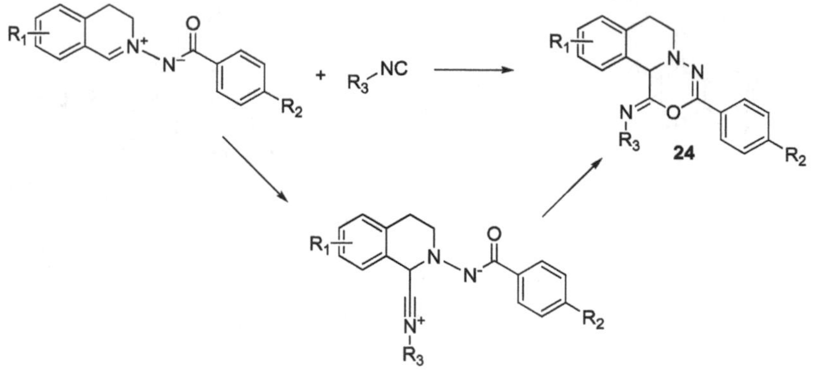

SCHEME 2.15 [5+1] Isocyanide cycloaddition yields six-membered heterocycles (**24**).

2.4.7 Trapping by Enamide

Lei and co-workers report a novel approach to synthesize pyridines.[30] The method features an unprecedented α-addition of aldehyde and enamide to an isocyanide. While not an Ugi reaction, this cycloaddition is an intriguing example of nitrilium trapping outside the realm of isocyanide-based MCRs. The cascade reaction involves a $Zn(OTf)_2$-promoted [1+5] cycloaddition of the isocyanide with a substituted enamide, followed by the aerobic oxidative aromatization and intramolecular acyl transfer and, finally, an acylation by an external acyl chloride to yield 2-substituted 4-acylamino-5-acyloxypyridines (**25**, Scheme 2.16). The method provides a straightforward route to diversely substituted pyridines that are analogs of acetylcholinesterase inhibitors.[31]

2.4.8 Trapping by Hydrazide

Achieving stereocontrol is among the current challenges of MCR chemistry. A recent example by Hashimoto et al. uses acyclic azomethine imines as prochiral electrophiles generated with catalytic amounts of axially chiral dicarboxylic acids.[32] The nitrilium intermediate following the attack of the isocyanide is trapped intramolecularly by the oxygen of the hydrazide (**26**). The imine is generated from benzaldehyde and *N′*-benzylbenzohydrazide in the presence of a chiral carboxylic acid. 2-Benzoyloxyphenyl isocyanides[33] are used for this reaction, and a one-pot basic hydrolysis is performed on the Ugi products to generate asymmetric benzoxazoles in high yield and enantiomeric excess (*ee*) (**27**, Scheme 2.17).

2.4.9 Trapping by Activated Aryl Carbon

Nitrilium trapping can be utilized in the synthesis of natural product analogs as demonstrated by the work of Kim et al.[34] In this work, the synthesis of 11-methoxymitragynine pseudoindoxyl (**28**), a derivative of the opioid natural product mitragynine, is reported. The electron-donating properties of the C-9 phenolic group are utilized to direct the annulation reaction through the intramolecular trapping

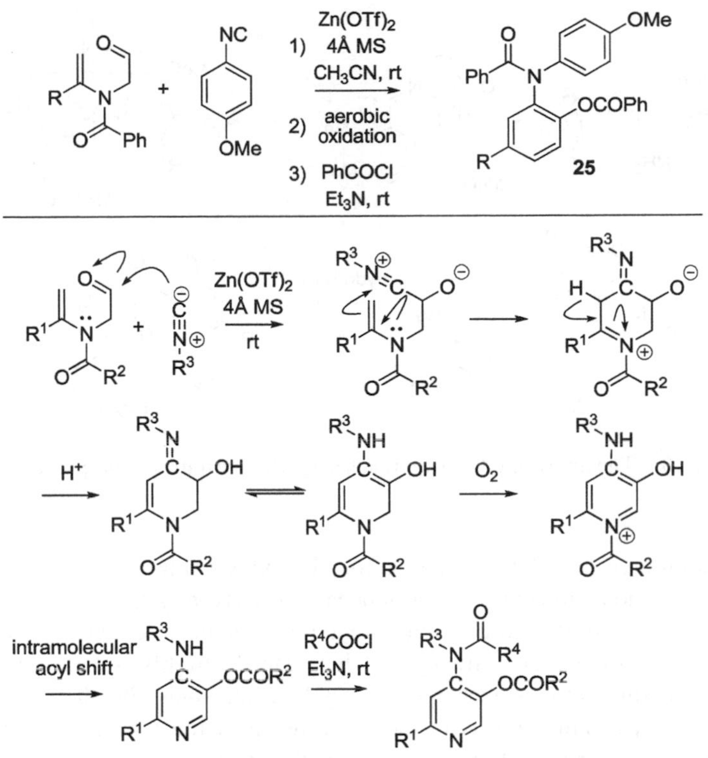

SCHEME 2.16 [1+5] Cycloaddition with isocyanides and enamides yields substituted pyridines.[30] (Modified from Lei, C.-H. et al., *J. Am. Chem.* Soc. 2013, 135, 4708.)

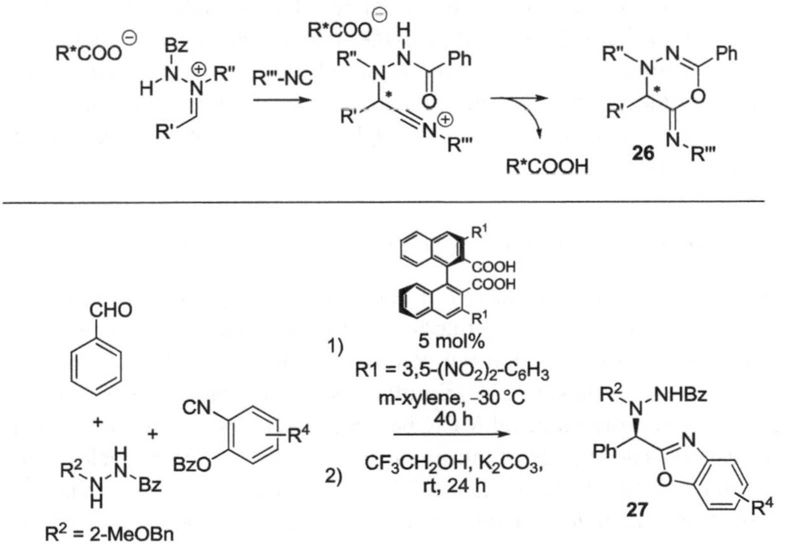

SCHEME 2.17 Chiral, acid catalyzed asymmetric Ugi reaction.[32]

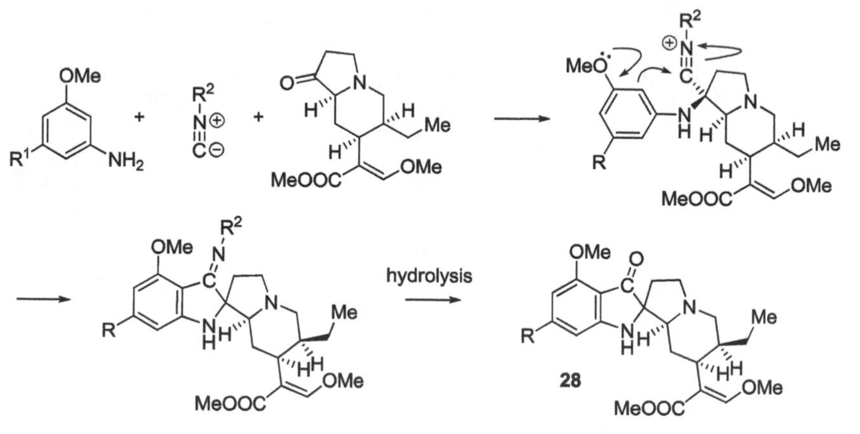

SCHEME 2.18 The interrupted Ugi approach to synthesize mitragynine pseudoindoxyl (**28**) analogs.[34]

of the nitrilium species, forming the spirocyclic indoxyl ring system. The Ugi three-component reaction is interrupted by a Houben–Hoesch type cyclization (Scheme 2.18).[35]

Kysil et al. report a multicomponent reaction between ethylenediamines, isocyanides, ketones, and aldehydes that affords highly substituted 3,4,5,6-tetrahydropyrazin-2-amines, including spirocyclic compounds (**29**, **30**, Scheme 2.19).[36] The tetrahydropyrazine ring is formed as a result of the intramolecular trapping of the nitrilium intermediate by the second ethylenediamine amino group and the subsequent tautomerization. The reaction is promoted by Lewis acids, particularly trimethylchlorosilane.

Xia and co-workers report an asymmetric Ugi-type reaction between α-isocyanoacetamides and chiral imines catalyzed by phenylphosphilic acid leading to 3-oxazolyl morpholin-2-one or piperazin-2-one derivatives (**31**).[37] The oxazol ring is formed following the intramolecular attack of the carbonyl oxygen of the isocyanoacetmide amide group. The subsequent reaction of amino-oxazoles with maleic anhydride led to fused heterocycles through a multiple domino reaction sequence (**32**, Scheme 2.20).

2.4.10 Intramolecular Nitrilium Trapping by Aminophenols

Heterocycles play an important role in the design and synthesis of bioactive small molecules: The vast majority of marketed drugs contain at least one heterocyclic ring. Numerous nitrogen-containing heterocycles, including dihydrothiazoles, benzoxazoles, benzothiazoles, and benzimidazoles, possess biological activity.[38]

We recently reported a novel MCR between α-ketones, 2-aminophenol, and isocyanide that takes advantage of intramolecular nitrilium trapping yielding benzoxazoles and other heterocycles (Scheme 2.21).[39] The reaction proceeds via a benzo[*b*][1,4]oxazine intermediate. Owing to the reactive nature of the trapped intermediate, it can be opened up by a second molecule of aminophenol or other bis-nucleophiles yielding benzoxazoles and other diverse heterocyclic scaffolds. Furthermore, the

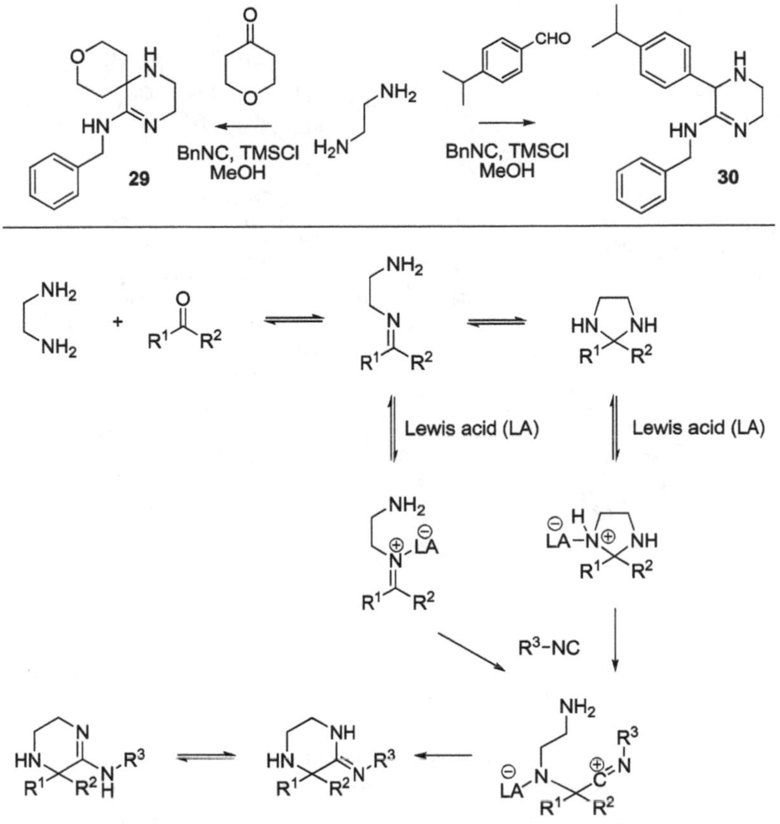

SCHEME 2.19 Synthesis of tetrahydropyrazines (**29**, **30**) using ethylenediamine and Lewis acid catalysis.[36]

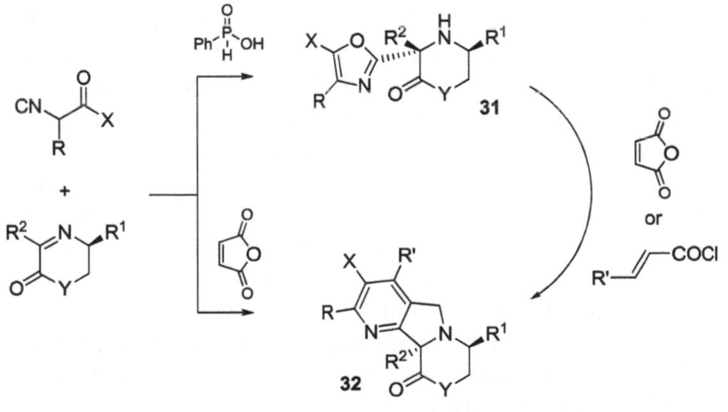

SCHEME 2.20 Synthesis of tetrahydropyrazines (**31**, **32**) using ethylenediamine and Lewis acid catalysis.[37] (Modified from Xia, L. et al., *J. Org. Chem.* 2013, 78, 3120.)

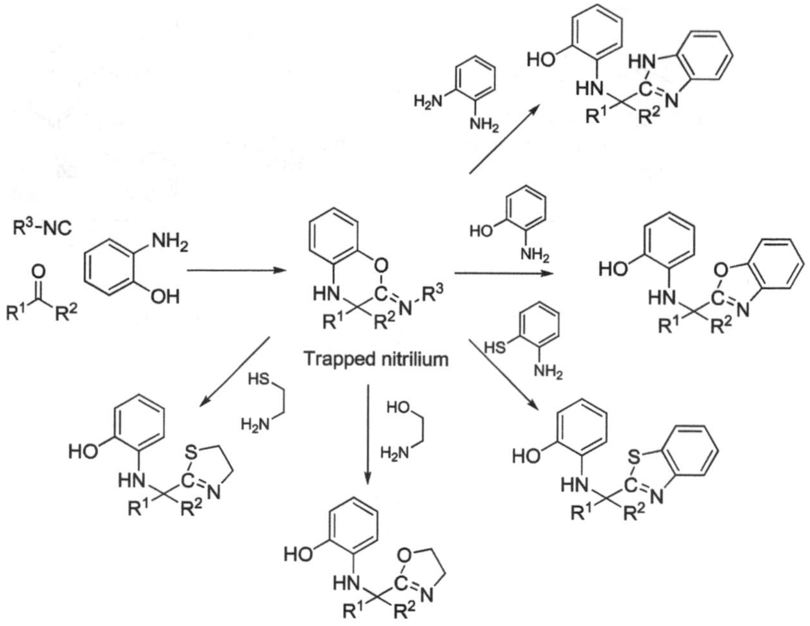

SCHEME 2.21 Synthesis of heterocyclic scaffolds using a three-component reaction.[39] (From Varadi, A. et al., *Org. Lett.* 2014, 16, 1668.)

reaction generates two additional diversity handles that were utilized to synthesize various substituted bis-heterocyclic derivatives. Synthesis of heterocycles is traditionally accomplished through multistep routes and/or harsh reaction conditions, particularly of benzoxazoles[40] and, while MCR approaches can also be taken,[14,41] there is a need for a more resource-efficient, preferably one-pot, method for their synthesis.[42] Metal catalyzed insertion of isocyanides leading to heterocycles is known and usually utilizes harsh conditions and has a somewhat limited scope in terms of diversity.[43]

We discovered this reaction while performing a routine U4CR in 2,2,2-trifluoroethanol (TFE) between 2-aminophenol, *N*-methyl-4-piperidone, 4-methoxyphenyl isocyanide, and acetic acid at 55 °C. We observed an unusual heterocyclic (benzo[*d*]oxazol-2-yl)-1-methylpiperidine product (benzoxazole, **34**) along with the expected U4CR product (**33**) seen only in trace amounts (Scheme 2.22).

We next used other carboxylic acids in the same reaction and again found that the carboxylic acid was not participating in the product formation. We thereby carried out the next reaction without any carboxylic acid. We again obtained the identical benzoxazole (**34**), suggesting that the product formation proceeds via a three-component MCR between 2-aminophenol, isocyanide, and ketone. Since two equivalents of the aminophenol reactant were used in product formation, we further optimized the formation of **34** by using three equivalents of 2-aminophenol, resulting in 84% yield for the isolated product.

The reaction failed in all solvents (including methanol, solvent of choice for traditional Ugi reactions) except TFE, suggesting that the slight acidity of the solvent (pKa = 12.4) played a crucial role in the reaction. Because MCRs have been shown

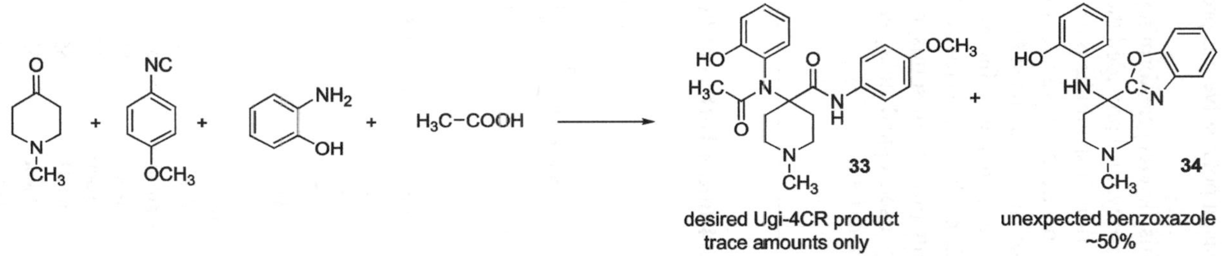

SCHEME 2.22 Our MCR yielded an unexpected benzoxazole (**34**) instead of the U4CR product (**33**).[39] (From Varadi, A. et al., *Org. Lett.* 2014, 16, 1668.)

to be catalyzed by acids,[11,44] we investigated the reaction using an array of Lewis and Brønsted acids. Transition metal triflates gave no advantage in yields. However, using 0.1 equivalent trifluoromethanesulphonic acid (HOTf) afforded quantitative yields of the product. The acid catalyzed reaction in toluene yielded 24% product, suggesting that besides acid catalysis, a polar solvent is preferred for this reaction to proceed to completion (Table 2.1).

We investigated the scope of this heterocycle-forming MCR using various isocyanides. While 4-methoxyphenyl isocyanide provided the best yields, other isocyanides were also effective in the reaction. It should be pointed out that the isocyanide contributes only one carbon atom toward benzoxazole; therefore, the reaction outcome is independent of the nature of isocyanide used (Table 2.2).

The scope of this reaction was evaluated by using different ketones and substituted 2-aminophenols (Scheme 2.23). The reaction was effective with a variety of ketones including the complex, multifunctional, semisynthetic natural product-like naloxone (product **40**), a clinically used opioid antagonist, highlighting the good functional group tolerance of this reaction.

TABLE 2.1
Yield of Benzoxazole 34 under Different Reaction Conditions[39]

HO, NH_2, NC, conditions, HN, 34

Catalyst (equiv.)	Solvent	Temp (°C)	Yield (%)
None	TFE	55	84
Acetic acid (0.1)	TFE	55	79
Zn(II)OTf (0.1)	TFE	55	62
Sc(III)OTf (0.1)	TFE	55	58
PTSA monohydrate (0.1)	TFE	55	92
None	MeOH	55	0
None	Toluene	55	0
None	Acetonitrile	55	0
None	IPA	55	0
None	Dioxane	55	0
HOTf (0.1)	MeOH	55	82
HOTf (0.1)	Toluene	55	65
HOTf (0.1)	TFE	RT	74
HOTf (0.1)	TFE	55	100

Source: Varadi, A. et al., *Org. Lett.* 2014, 16, 1668.

TABLE 2.2
Yield of Benzoxazole with Different Isocyanides[39]

0.1 equiv HOTf, TFE, 55°C, 24 h

Entry	R^1	Yield (%)
1	Cyclohexyl	70
2	4-OMe-Ph	100
3	4-F-Ph	28
4	Naphth-2-yl	59
5	*t*Bu	90
6	1-Adamantyl	61
7	2,6-diMe-Ph	0

Source: Varadi, A. et al., *Org. Lett.* 2014, 16, 1668.

Isocyanides play an unusual role in this MCR, contributing only one carbon atom to the benzoxazole. The aryl or alkyl amine moiety acts as the leaving group, leading to an insertion of only the isocyanide carbon into the benzoxazole moiety. While investigating the effect of various isocyanides, we found that 2,6-dimethylphenyl isocyanide, instead of the expected benzoxazole, yielded a spiro[benzo[*b*][1,4]oxazine]-imine (benzoxazine, **43**) scaffold when reacted with NMP and 2-aminophenol (Scheme 2.24). The isolation and structure elucidation of this benzoxazine derivative provided us the first clues about the mechanism of this reaction (Figure 2.1).

We next used another, albeit sterically somewhat less hindered isocyanide, 2-chloro-6-methyl isocyanide and isolated a similar benzoxazine. The reaction, however, also afforded **1**. Furthermore, the yield for the benzoxazine was significantly lower than that of 10, highlighting the importance of steric hindrance in stabilizing the benzoxazine structure. Benzoxazines (**12** and **13**) were isolated, even when sterically nonhindered isocyanides were involved, in lower yields. A mixture of benzoxazole **1** and the appropriate benzoxazine was observed in these cases (Table 2.3).

We hypothesized that the benzoxazine **43** was an intermediate in the reaction route leading to the final product benzoxazole **34**. We envisioned the following reaction mechanism for the formation of benzoxazoles (Scheme 2.25): The ketone first forms a Schiff base (imine, **A**) with 2-aminophenol, which in turn is activated by the acidity of the system (the acid catalyst HOTf and/or the solvent TFE), thereby facilitating an attack by the isocyanide yielding the highly reactive nitrilium **B**. The phenolic OH of 2-aminophenol then participates in an intramolecular nucleophilic attack on the reactive nitrilium intermediate (**B**) carbon, thus trapping the nitrilium to give the benzoxazine intermediate **C**, which is very susceptible to nucleophilic attack by a second molecule of the bis-nucleophile 2-aminophenol. It must be noted

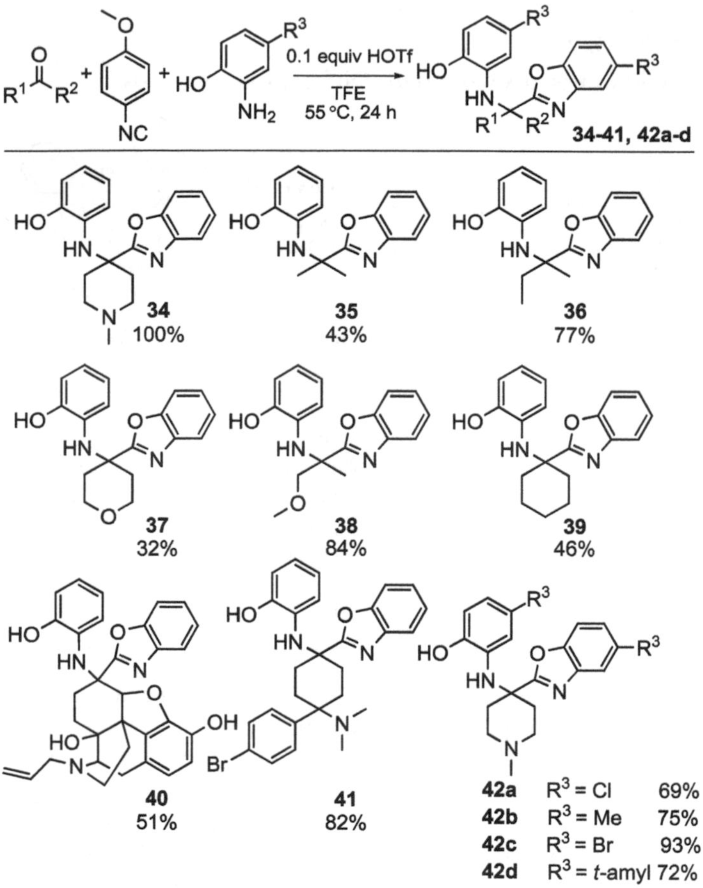

SCHEME 2.23 Scope of our MCR yielding benzoxazoles.[39] (From Varadi, A. et al., *Org. Lett.* 2014, 16, 1668.)

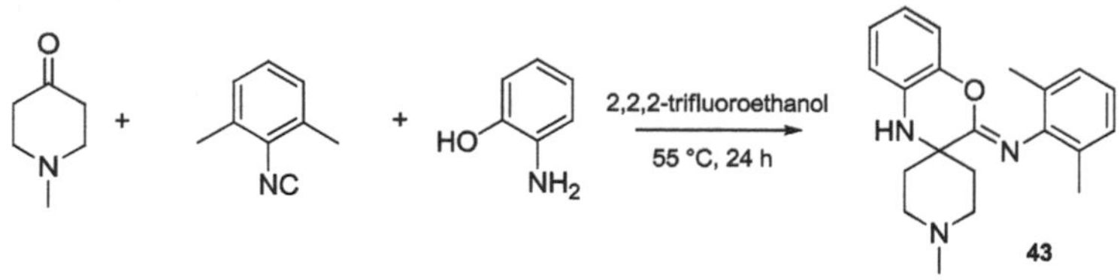

SCHEME 2.24 Sterically hindered isocyanide yields a [benzo[*b*][1,4]oxazine]-imine (**43**).

that products formed in most nitrilium trapping examples found in the literature are stable and easily isolable, while benzoxazines in the present case are extremely reactive to nucleophiles. This allowed us to utilize them as intermediates to synthesize substituted benzoxazoles and other heterocyclic systems as shown later. Acid catalyzed nucleophilic attack on the imine **A** with the phenol as a good leaving group is primarily responsible for the instability of the intermediate **C**.

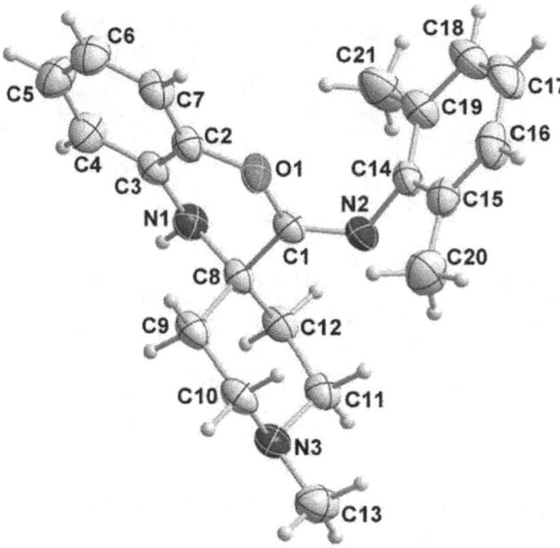

FIGURE 2.1 The molecular structure of benzoxazine intermediate **43**.

TABLE 2.3
Isolation of Trapped Nitrilium Intermediates, Benzoxazines 43–46[39]

O + Ar—NC + HO NH$_2$ conditions[a] HN O N Ar N **43-46**

Ar	Product	Yield (%)
2,6-diMe-Ph	**43**	79
2-Me-6-Cl-Ph	**44 + 34**	37 (4)[a]
4-OMe-Ph	**45 + 34**	39 (24)[a]
Naphth-2-yl	**46 + 34**	33 (12)[a]

Source: Varadi, A. et al., *Org. Lett.* 2014, 16, 1668.
[a] Yield of **1**.

To test our proposed mechanism that the benzoxazine **C** is an intermediate in the synthesis of benzoxazoles, we isolated the benzoxazine obtained using 4-methoxyphenylisocyanide (**45**). We reacted this intermediate with two equivalents of 2-aminophenol in the presence of triflic acid in TFE; mimicking our one-pot reaction, the appropriate benzoxazole (**34**) could be isolated in good yields. We enhanced the

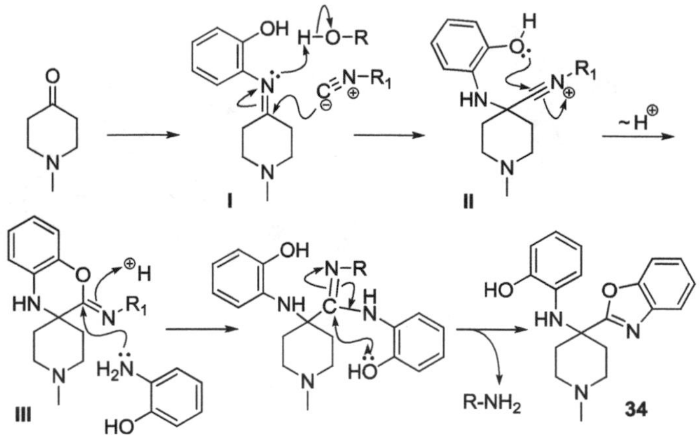

SCHEME 2.25 Proposed mechanism for the formation of benzoxazole **34**.[39] (From Varadi, A. et al., *Org. Lett.* 2014, 16, 1668.)

TABLE 2.4
Conversion of Benzoxazine Intermediate 45 to Diverse Heterocycles with bis-Nucleophiles[39]

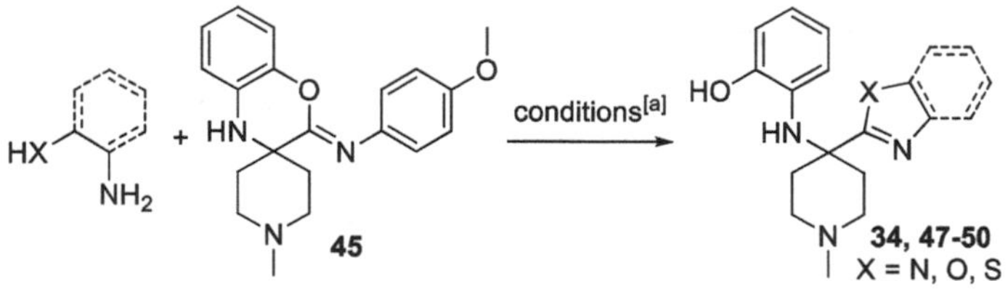

bis-Nucleophile	Product	Yield (%)
2-Aminophenol	**34**	87
2-Amino-4-chlorophenol	**47**	63
1,2-Diaminobenzene	**48**	98
2-Aminothiophenol	**49**	93
Cysteamine	**50**	44

Source: Varadi, A. et al., *Org. Lett.* 2104, 16, 1668.

scope of our MCR by reacting the isolated benzoxazine intermediate with a series of bis-nucleophiles including 1,2-diaminobenzene, 2-aminothiophenol, and cysteamine to synthesize benzimidazole, benzothiazole, and dihydrothiazole derivatives, respectively (Table 2.4). Owing to the hindered sterics of the system, benzoxazine **43** derived from sterically hindered 2,6-dimethylphenyl isocyanide was significantly more resistant to nucleophilic attack by bis-nucleophiles.

To enhance the diversity and thus the possible biological relevance of the benzoxazole scaffold we decided to exploit the free phenolic OH and secondary aromatic amine of the benzoxazole products to make second-generation derivatives (Figure 2.2). We cyclized the NH and the OH using carbonyldiimadazole to make the bis-heterocylic derivative with a benzoxazolone ring.[45] The phenolic OH was also alkylated, carbamoylated, acylated, and converted to triflate, which was in turn taken into Suzuki coupling[46] with 2-bromophenylboronic acid followed by Buchwald–Hartwig amination[47] to generate a carbazole scaffold. The versatile leaving group triflate could also be converted to a nitrile.[48]

In summary, we discovered a convenient, Brønsted acid catalyzed, isocyanide-based heterocycle-forming three-component reaction between 2-aminophenols, isocyanides, and ketones that leads to benzoxazoles and other heterocyclic products. The reaction progresses via a benzoxazine intermediate formed by intramolecular nucleophilic trapping of the reactive nitrilium intermediate by an adjacent phenolic group. The reactivity of benzoxazine to bis-nucleophiles was leveraged to synthesize an array of heterocyclic scaffolds including benzimidazoles, dihydrothiazoles and benzothiazoles. The reaction showed good functional group tolerance and is compatible with different ketones and aminophenols, ideal for the generation of molecule libraries under mild reaction conditions, without the use of transition metal catalysts. We have further exploited the reactivity of the phenolic hydroxyl and the aromatic secondary amine to diversify the scaffolds. Work is currently underway in our laboratory to further characterize opioid-benzoxazole conjugates as analgesics and

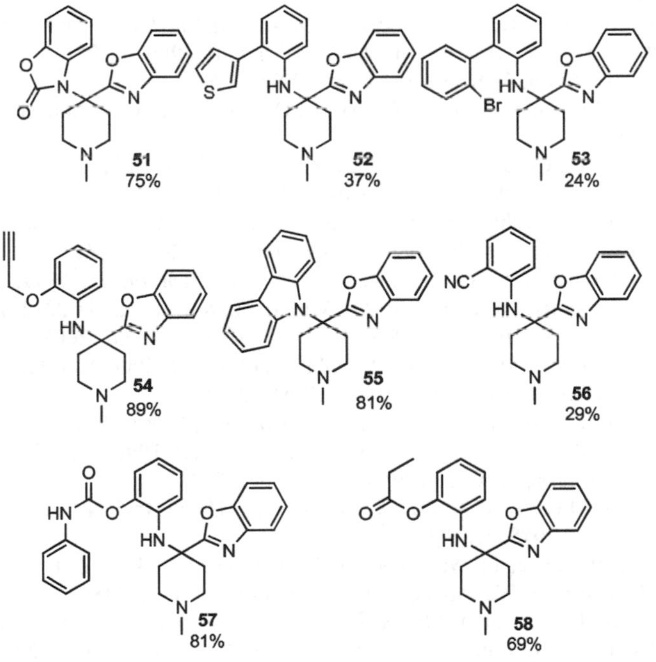

FIGURE 2.2 Diversification of the benzoxazole scaffold.[39] (From Varadi, A. et al., *Org. Lett.* 2104, 16, 1668.)

antiaddiction medications. Our preliminary examinations suggest the possibility of synthesizing dual MOR-DOR opioid agonists as analgesics using this MCR approach (Váradi and Majumdar, in preparation). Dual MOR-DOR agonists are known in the literature to be potent analgesics without the dangerous side effects of morphine.[49]

2.5 CONCLUSION

MCRs are an extremely useful and quickly evolving field of synthetic organic chemistry. In recent years, countless varieties of isocyanide-based MCRs have emerged, among which numerous intriguing reactions can be found that take advantage of nitrilium trapping, the nucleophilic attack on the extremely reactive nitrilium intermediate. These reactions have not only introduced us to novel chemical methodologies and reaction mechanisms, but also have made new areas of chemical space easily accessible. Based on the diversity of structures that can be accessed by MCRs that involve nitrilium trapping, these methodologies will have a bright future in drug design and synthesis.

REFERENCES

1. Ugi, I.; Werner, B.; Dömling, A., *Molecules* **2003**, 8, 53.
2. Dömling, A.; Ugi, I., *Angew. Chem. Int. Ed.* **2000**, 39, 3168–3210.
3. Ugi, I.; Meyr, R., *Chem. Ber.* **1961**, 94, 2229.
4. Sehlinger, A.; Kreye, O.; Meier, M. A. R., *Macromolecules* **2013**, 46, 6031.
5. Sunderhaus, J. D.; Martin, S. F., *Chem. Eur. J.* **2009**, 15, 1300.
6. Paulvannan, K., *Tetrahedron Lett.* **1999**, 40, 1851.
7. Koopmanschap, G.; Ruijter, E.; Orru, R. V. A., *Beilstein J. Org. Chem.* **2014**, 10, 544.
8. Bonnaterre, F.; Bois-Choussy, M.; Zhu, J., *Org. Lett.* **2006**, 8, 4351.
9. White, C. J.; Yudin, A. K., *Nature Chem.* **2011**, 3, 509.
10. Godet, T.; Bonvin, Y.; Vincent, G.; Merle, D.; Thozet, A.; Ciufolini, M. A., *Org. Lett.* **2004**, 6, 3281.
11. Okandeji, B. O.; Gordon, J. R.; Sello, J. K., *J. Org. Chem.* **2008**, 73, 5595.
12. Dömling, A.; Wang, W.; Wang, K., *Chem. Rev.* **2012**, 112, 3083.
13. (a) Shaabani, A.; Keshipour, S.; Shaabani, S.; Mahyari, M., *Tetrahedron Lett.* **2012**, 53, 1641; (b) Pan, S. C.; List, B., *Angew. Chem. Int. Ed.* **2008**, 47, 3622.
14. Ruijter, E.; Scheffelaar, R.; Orru, R. V. A., *Angew. Chem. Int. Ed.* **2011**, 50, 6234.
15. Khan, A. T.; Sidick Basha, R; Lal, M.; Mir, M. H., *RSC Advances* **2012**, 2, 5506.
16. (a) El Kaïm, L.; Grimaud, L.; Oble, J., *Angew. Chem. Int. Ed.* **2005**, 44, 7961; (b) El Kaim, L.; Gizolme, M.; Grimaud, L.; Oble, J., *Org. Lett.* **2006**, 8, 4019.
17. Hulme, C.; Gore, V., *Curr. Med. Chem.* **2003**, 10, 51.
18. (a) Groebke, K.; Weber, L.; Mehlin, F., *Synlett* **1998**, 6, 661; (b) Blackburn, C.; Guan, B.; Fleming, P.; Shiosaki, K.; Tsai, S., *Tetrahedron Lett.* **1998**, 39, 3635; (c) Bienaymé, H.; Bouzid, K., *Angew. Chem. Int. Ed.* **1998**, 37, 2234.
19. (a) Hulme, C.; Lee, Y.-S., *Mol. Divers* **2008**, 12, 1–15; (b) Devi, N.; Rawal, R. K.; Singh, V., *Tetrahedron* **2015**, 71, 183.
20. Akritopoulou-Zanze, I.; Wakefield, B. D.; Gasiecki, A.; Kalvin, D.; Johnson, E. F.; Kovar, P.; Djuric, S. W., *Bioorg. Med. Chem. Lett.* **2011**, 21, 1480.
21. Baviskar, A. T.; Madaan, C.; Preet, R.; Mohapatra, P.; Jain, V.; Agarwal, A.; Guchhait, S. K.; Kundu, C. N.; Banerjee, U. C.; Bharatam, P. V., *J. Med. Chem.* **2011**, 54, 5013.
22. Shukla, N. M.; Salunke, D. B.; Yoo, E.; Mutz, C. A.; Balakrishna, R.; David, S. A., *Bioorg. Med. Chem.* **2012**, 20, 5850.

23. Burchak, O. N.; Mugherli, L.; Ostuni, M.; Lacapère, J. J.; Balakirev, M. Y., *J. Am. Chem. Soc.* **2011**, 133, 10058.
24. (a) Elleder, D.; Baiga, T.; Russell, R.; Naughton, J.; Hughes, S.; Noel, J.; Young, J., *Virol. J.* **2012**, 9, 305; (b) Bode, M. L.; Gravestock, D.; Moleele, S. S.; van der Westhuyzen, C. W.; Pelly, S. C.; Steenkamp, P. A.; Hoppe, H. C.; Khan, T.; Nkabinde, L. A., *Bioorg. Med. Chem.* **2011**, 19, 4227.
25. Faggi, C.; García-Valverde, M. A.; Marcaccini, S.; Menchi, G., *Org. Lett.* **2010**, 12, 788.
26. Ganem, B., *Acc. Chem. Res.* **2009**, 42, 463.
27. Bonne, D.; Dekhane, M.; Zhu, J., *Org. Lett.* **2005**, 7, 5285.
28. Shaabani, A.; Maleki, A.; Mofakham, H.; Khavasi, H. R., *J. Comb. Chem.* **2008**, 10, 323.
29. Soeta, T.; Tamura, K.; Ukaji, Y., *Org. Lett.* **2012**, 14, 1226.
30. Lei, C.-H.; Wang, D.-X.; Zhao, L.; Zhu, J.; Wang, M.-X., *J. Am. Chem. Soc.* **2013**, 135, 4708.
31. Shutske, G. M.; Kapples, K. J.; Tomer, J. D.; Hrib, N. J.; Jurcak, J. J., Substituted-4-amino-3-pyridinols, a process for their preparation and their use as medicaments, **1991**, EP0477903A2.
32. Hashimoto, T.; Kimura, H.; Kawamata, Y.; Maruoka, K., *Angew. Chem. Int. Ed.* **2012**, 51, 7279.
33. Pirrung, M. C.; Ghorai, S., *J. Am. Chem. Soc.* **2006**, 128, 11772.
34. Kim, J.; Schneekloth, J. S.; Sorensen, E. J., *Chem. Sci.* **2012**, 3, 2849.
35. Schneekloth, J. S.; Jimin, K., Jr.; Sorensen, E. J., *Tetrahedron* **2009**, 65, 3096.
36. Kysil, V.; Tkachenko, S.; Khvat, A.; Williams, C.; Tsirulnikov, S.; Churakova, M.; Ivachtchenko, A., *Tetrahedron Lett.* **2007**, 48, 6239.
37. Xia, L.; Li, S.; Chen, R.; Liu, K.; Chen, X., *J. Org. Chem.* **2013**, 78, 3120.
38. (a) Chandrasekharappa, A. P.; Badiger, S. E.; Dubey, P. K.; Panigrahi, S. K.; Manukonda, S. R., *Bioorg. Med. Chem. Lett.* **2013**, 23, 2579; (b) Ulhaq, S.; Chinje, E. C.; Naylor, M. A.; Jaffar, M.; Stratford, I. J.; Threadgill, M. D., *Bioorg. Med. Chem.* **1999**, 7, 1787; (c) Horton, D. A.; Bourne, G. T.; Smythe, M. L., *Chem. Rev.* **2003**, 103, 893; (d) Hayashi, S.; Hirao, A.; Imai, A.; Nakamura, H.; Murata, Y.; Ohashi, K.; Nakata, E., *J. Med. Chem.* **2009**, 52, 610; (e) Siracusa, M. A.; Salerno, L.; Modica, M. N.; Pittalà, V.; Romeo, G.; Amato, M. E.; Nowak, M.; Bojarski, A. J.; Mereghetti, I.; Cagnotto, A.; Mennini, T., *J. Med. Chem.* **2008**, 51, 4529.
39. Varadi, A.; Palmer, T. C.; Notis, P. R.; Redel-Traub, G. N.; Afonin, D.; Subrath, J. J.; Pasternak, G. W.; Hu, C.; Sharma, I.; Majumdar, S., *Org. Lett.* **2014**, 16, 1668.
40. Kumar, R. V., *Asian J. Chem.* **2004**, 16, 1241.
41. Tempest, P.; Ma, V.; Thomas, S.; Hua, Z.; Kelly, M. G.; Hulme, C., *Tetrahedron Lett.* **2001**, 42, 4959.
42. (a) Boissarie, P. J.; Hamilton, Z. E.; Lang, S.; Murphy, J. A.; Suckling, C. J., *Org. Lett.* **2011**, 13, 6256; (b) Spatz, J. H.; Bach, T.; Umkehrer, M.; Bardin, J.; Ross, G.; Burdack, C.; Kolb, J., *Tetrahedron Lett.* **2007**, 48, 9030.
43. (a) El Kaim, L.; Grimaud, L., *Tetrahedron* **2009**, 65, 2153–2171; (b) Vlaar, T.; Ruijter, E.; Maes, B. U. W.; Orru, R. V. A., *Angew. Chem. Int. Ed.* **2013**, 52, 7084.
44. Dai, W.-M.; Li, H., *Tetrahedron* **2007**, 63, 12866.
45. Poupaert, J.; Carato, P.; Colacino, E.; Yous, S., *Curr. Med. Chem.* **2005**, 12, 877.
46. Miyaura, N.; Suzuki, A., *Chem. Rev.* **1995**, 95, 2457.
47. Hartwig, J. F., *Angew. Chem. Int. Ed.* **1998**, 37, 2046.
48. Kubota, H.; Rice, K. C., *Tetrahedron Lett.* **1998**, 39, 2907.
49. Varadi, A.; Hosztafi, S.; Le, R. V; Toth, G.; Urai, A.; Noszal, B.; Pasternak, G. W.; Grinnell, S. G.; Majumdar, S., *Eur. J. Med. Chem.* **2013**, 69C, 786.

3 The Role of Nanoporous Materials in Multicomponent Synthesis/Modification of Heterocyclic Compounds

Esmail Doustkhah and Sadegh Rostamnia**

CONTENTS

3.1 INTRODUCTION

In recent decades, a vast amount of research has been focused on the heterogeneous nanocatalysts due to their prevailing advantages compared to homogenous ones including their easy work-up and separation from the reaction mixture, their capability of sometimes being reused, their higher surface area, and new behaviors in nanoscale compared to bulk types. Among the nanocatalysts, nanoporous materials are becoming more interesting and preferable for the catalysis of many types of reactions, such as C–C and C–X coupling reactions. The tunable porosity and higher surface area have created a situation to control the chemoselectivity by pore type (e.g., hydrophobicity, array, and size of pore) and functions of surface.

On the other hand, multicomponent reactions have been promoted and known as an atom economically efficient, rapid, time-consuming and new reactions, which have led to chemoselective and valuable products within the favorable yields and products.

* Corresponding authors: rostamnia@maragheh.ac.ir and doustkhah.esmail@yahoo.com.

Nanocatalysis and multicomponent reactions (MCRs) together have generated a very efficient method for the facile and efficient synthesis of biologically and pharmaceutically active organic materials, which can be supposed to be extensively compatible and followed by green chemical protocols. Many types of nanoporous materials have been introduced as catalysts for a variety of MCRs. From zeolites, mostly natural and the first generation of nanoporous material, to periodic mesoporous silica materials (PMSMs), the synthetic types of mesopores, all take a place in the nanocatalysis of MCRs. In this chapter, there is a comprehensive collection of nanoporous materials that have been modified and applied as a nanocatalyst in the important classes of MCRs. Among these nanoporous catalysts, mesoporous silica materials (MSMs) are more diverse and apt to be modified and used in catalytic applications. Therefore, major attention will be directed to MSMs. Also, one reason that other types of nanoporous materials have not been studied in the catalysis of MCRs is that they are synthesized and characterized recently.

3.2 MESOPOROUS SILICA MATERIALS (MSMs) IN CATALYSIS

3.2.1 Metal Ion and Metal Oxide Supported MSMs

Tricomponent coupling of aldehyde, amine (preferably a secondary aliphatic amine), and acetylene is an alternative way to introduce a propargylic amine. This reaction can be performed by catalysis of various transition metals including Ag, Cu, and Au. This reaction is called A3-coupling, which undergoes primary formation of imine and then the acetylide anion (formed by oxidative addition of metal to a C–H bond) attack at the electrophilic carbon of imine to produce a final product.[1]

In 2010, mesoporous SBA-15 was applied for production of silver nanoparticles supported on the surface of the SBA-15 catalyst in the A3-coupling reaction of aldehydes, amines, and terminal alkynes using glycol as a green solvent (Scheme 3.1).[2] It was prepared by *in situ* reduction of $AgNO_3$ using hexamethylenetetramine as reducing agent to form Ag@SBA-15. In this reaction, it was found that Ag nanoparticles are crucial in the catalytic activity of Ag@SBA-15 and the higher reaction yield was found at 8 nm of Ag nanoparticles within SBA-15. Therefore, the reaction produced propargylamines in higher yield with 5 mol% loading of Ag@SBA-15 through A3-coupling reaction of various aliphatic aldehydes and amines with aromatic and aliphatic alkynes at 100 °C. The catalyst activity was reduced after five runs.

In 2013, the synthesis of a novel gold(0) stabilized in SBA-15 functionalized with disulfides was reported as an efficient and recyclable nanocatalyst in the A3-coupling reaction (Scheme 3.2).[3] For the preparation of the catalyst, pure mesoporous SBA-15

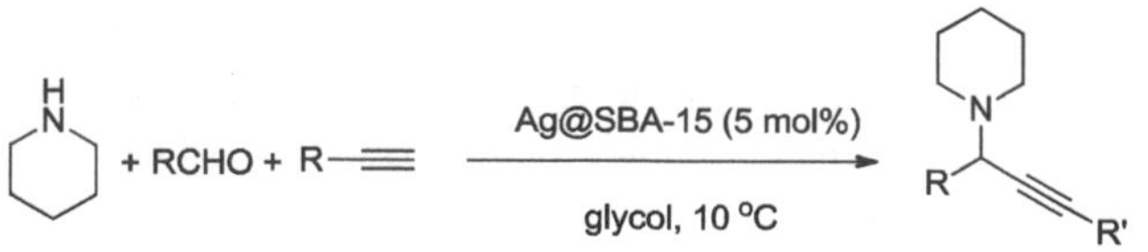

SCHEME 3.1 SBA-15 supported silver nanoparticles as catalyst for reaction of aldehydes, amines, and terminal alkynes.

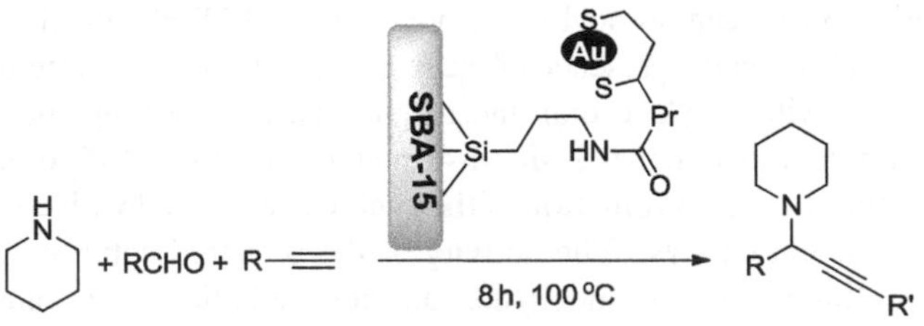

SCHEME 3.2 Au(0) NPs stabilized and supported in SBA-15 for A3-coupling reaction.

was initially functionalized by 3-(aminopropyl)triethoxysilane (APTES), and then it underwent an amidation with amine groups and a disulfide-based compound to introduce disulfide within pore channels. Subsequently, $HAuCl_4$ was adsorbed inside the SBA-15 for further reduction with $NaBH_4$ to produce the final form of catalyst. Under investigation, this designed catalyst exhibited a high performance for synthesis of propargylamines via the tricomponent coupling of aldehydes, piperidine, and alkynes in excellent yields under solvent-free conditions at 100 °C. Furthermore, the catalyst revealed five times recyclability without significant decrease in activity.

In addition, Cu species, specifically Cu(I), is an excellent transition metal for the catalysis of the A3-coupling reaction of amine, aldehyde, and acetylene. In this regard, a covalently bonded Cu(I)-Salen complex onto the surface of MCM-41 exhibited an effective and recoverable catalytic behavior for the A3-coupling reaction (Scheme 3.3).[4] For the production of the catalyst, the surface of MCM-41 was

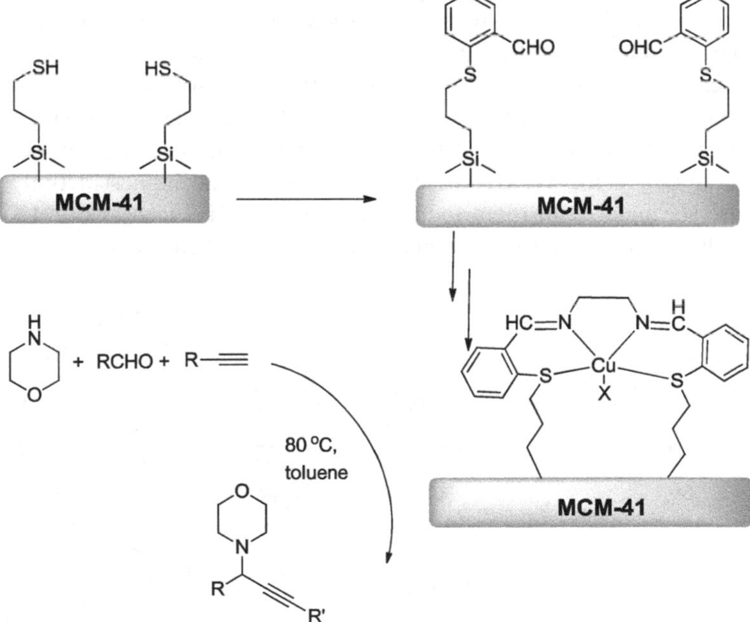

SCHEME 3.3 Cu(I)-salen complex covalently immobilized into MCM-41 as catalyst for the A3-coupling reaction.

functionalized by 3-(mercaptopropyl)trimethoxysilane (MPTMS) and then reacted with 2-*o*-nitrobenzaldehyde in the presence of magnesium methoxide in dimethylformamide (DMF), and then with ethylene diamine to produce bis[2-(phenylthio)benzylidene]-1,2-ethylenediamine functionalized MCM-41 material as a new N_2S_2 donor salen-type ligand. At the final stage, coordination of the synthesized ligand with copper(I) halide generated an efficient catalyst. The activity of the prepared catalyst was evaluated by the A3-coupling reaction of aldehydes, amines, and alkynes in toluene at 80 °C. Under these conditions, the corresponding propargylamines were achieved in good to excellent yields for a variety of aromatic and aliphatic substrates within 4–10 h of reaction time. Also, the catalyst was recycled at least for seven consecutive runs.

The Strecker reaction is another tricomponent reaction in which the product, α-aminonitrile, can easily be converted to its corresponding α-aminocarboxylic acid by a simple hydrolysis, which is a more valuable target in view of biochemistry. Recently, the synthesis of α-aminonitrile was achieved by a heterogeneous SBA-15 based heterogeneous catalyst containing Co(II)-salen complex, which was covalently anchored to the surface of SBA-15.[5] Anchoring of Co(II)-salen complex on the heterogeneous surface of mesoporous silica SBA-15 showed an environmentally friendly behavior owing to insignificant leaching in each run. Additionally, Co(II)-salen@SBA-15 was an easily recoverable catalyst for the synthesis of α-aminonitriles for a wide range of aldehydes/ketones and primary or secondary amines with good to excellent conversion yields at room temperature under solventless conditions. The catalyst recovery was performed by a simple filtration reusability test and showed at least 10 successful runs without remarkable loss in catalyst activity.

Additionally, the catalytic activity of Co(II)-salen complex in the reaction of Strecker was studied on the surface of various MSMs by anchoring it to three types of mesoporous materials (SBA-15, MCM-41, HMS) at both room temperature and low temperature microwave irradiation (LTMI) under solventless conditions. In comparison, all catalysts were also highly reusable under the investigated reaction conditions and could be reused at least 10 times without loss of catalytic activity. However, among them, Co(II)-salen@SBA exhibited the highest activity among other catalysts under both room temperature and LTMI. Furthermore, LTMI had better results compared to room temperature conditions (Scheme 3.4).[6]

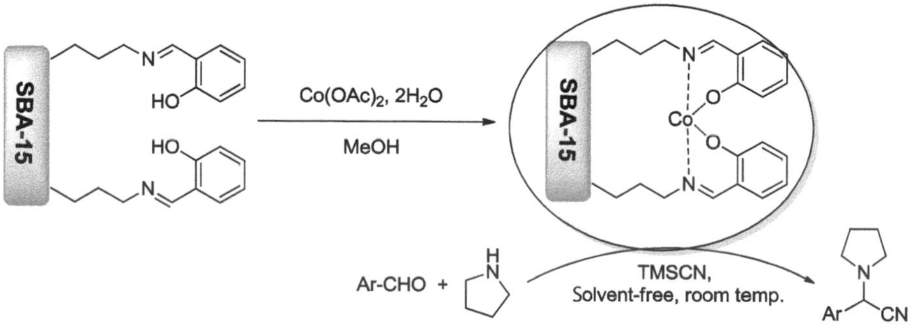

SCHEME 3.4 Supported Co(II) on SBA-15 functionalized by salen as a catalyst for the Strecker reaction.

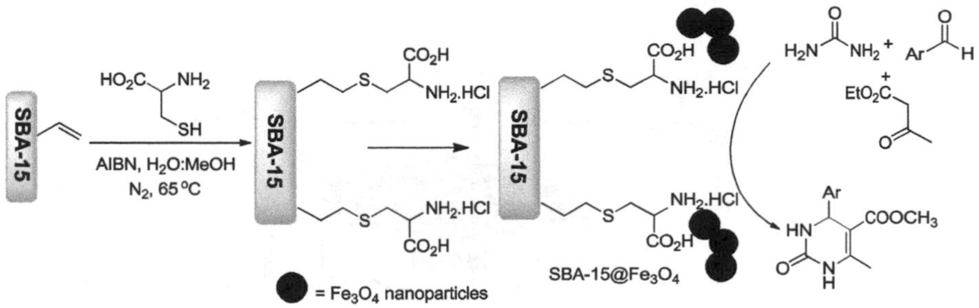

SCHEME 3.5 SBA-15@Fe_3O_4 synthesis and its catalytic use for the Biginelli reaction.

Conjugation of Fe_3O_4 nanoparticles to MSMs can raise the catalytic activity. For example, the catalyst can be recovered simply by a magnet, especially when it is supermagnetic material. On the other hand, Fe_3O_4 itself has catalytic activity in some reaction cases.[7] Conjugation ofmagnetic Fe_3O_4 to MSM like SBA-15 can be achieved by chemical conjugation of Fe_3O_4 onto surface of SBA-15 (Fe_3O_4@SBA) through active functions of carboxylic acid and amine (Scheme 3.5). SBA-15 is functionalization is performed by thiol-ene click reaction of vinyl groups and thiol group of cysteine. The catalytic activity of Fe_3O_4@SBA-15 was evaluated in the Biginelli reaction under mild conditions for the synthesis of 3,4-dihydropyrimidin-2(1*H*)-one derivatives. The separation and reuse of the Fe_3O_4@mesoporous SBA-15 nanocatalyst were simple, effective and economical.

3.2.2 Organocatalyst Supported MSMs

Organocatalysts, specifically those supported on heterogeneous surfaces, can bring an excellent opportunity for greening and sustaining the chemical procedure by removing transition metals.[8] We have recently reviewed supported organocatalysts on MSMs, which has recently attracted some attention.[9] One of our interests concerns the catalytic application of confined fluorous alcohols in mesoporous materials.[10] SBA-15 and 2,2,2-trifluoroethanol (SBA-15/TFE) exhibit a strange catalytic activity in the reaction of aldehydes, amines, and ammonia with benzil to produce substituted classes of imidazoles.[10a] Moreover, tri-component synthesis of indazolophthalazinetrione under the catalysis of SBA-15/TFE has been accomplished as a green protocol.[10b] In addition, the use of SBA-15/TFE accelerates the catalysis of aromatic and benzylic sulfides' oxidation in the presence of H_2O_2.[11] This reaction method is performed under neutral conditions without catalyst activation or modification steps. Moreover, the catalyst is recoverable and reusable and does not lose the catalytic activity (Figure 3.1).

In addition, this catalytic system was developed for the synthesis of rhodanine scaffolds, which are synthetically valuable in pharmaceuticals and industry (Figure 3.1). The catalyst (SBA-15/HFIP) was obtained by confining the hexafluproisopropanol (HFIP) inside SBA-15 at room temperature and adding no further solvent.[12] The other advantages of this method include its low catalyst loading, the simplicity of the procedure, waste-free and direct synthetic entry to excellent yields of rhodanines, high reusability of the catalyst, and short reaction time.

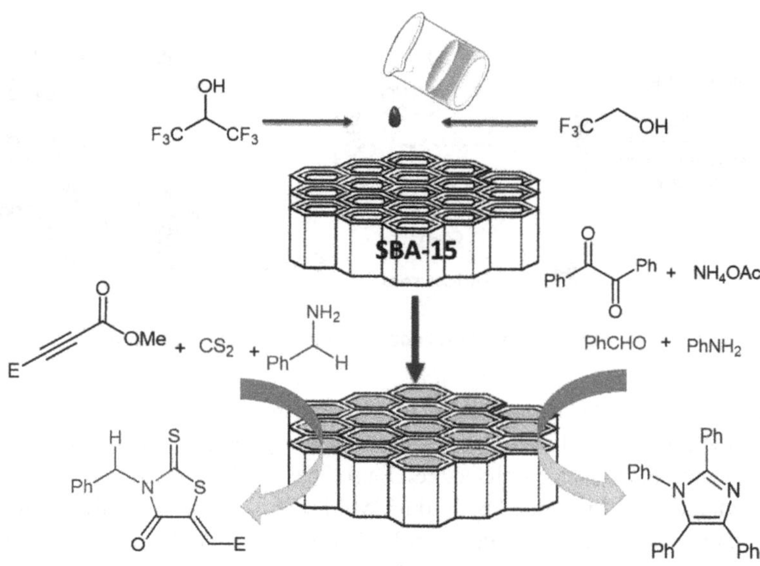

FIGURE 3.1 Confined fluoroalkyl chained alcohols in SBA-15 mesoporous materials.[9]

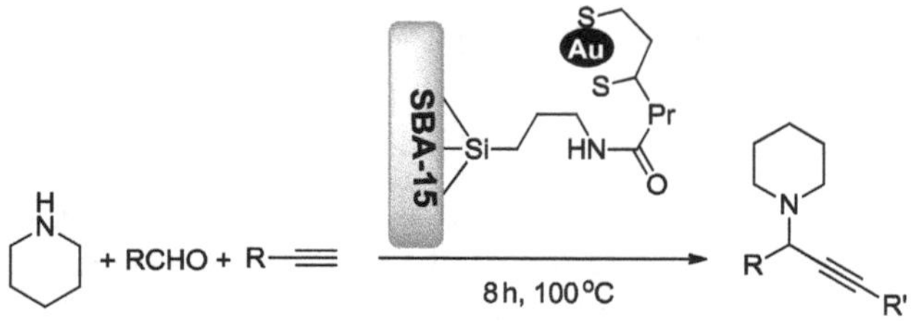

SBA-15 Si HN O S Au S Pr

H N + RCHO + R—≡ → 8h, 100°C N R R'

FIGURE 3.2 SBA-15 supported organocatalyst based on triazine in MCR.[13]

Functionalization of thiolated SBA-15 (SBA-15-SH) with triazine derivative can produce an organocatalytically supported nanocatalyst with efficient, recyclable nature. It is used for one-pot tricomponent synthesis of 2-amino-4*H*-chromenes through condensation reaction of aromatic aldehyde, malononitrile and activated phenols. An insight into details of catalyst structure reveals a post-modification step on SBA-15 with thiols followed by a thiol-ene click reaction using 2,4,6-triallyloxy-1,3,5-triazine. This novel mesoporous metal-free heterogeneous organocatalyst was obtained by clickable reaction of ene bonds of triazine and thiol functionalities on the surface of SBA-15 in the presence of azobisisobutyronitrile (AIBN) (Figure 3.2). The reaction was performed under solvent-free conditions with easy workup and efficient reusability of the catalyst.[13]

3.2.3 Organosulfonic Acid Supported MSMs

Organosulfonic acid supported mesoporous silica are a class of MSMs that have been extensively investigated and applied in catalysis during recent years. The first pioneering reports on sulfonated mesoporous silica materials were in 1998.[14] Postfunctionalization

MSMs and co-condensation of 3-mercaptopropylsilane (MPTS) with silica sources, such as tetraethylorthosilicate (TEOS) or tetramethylorthosilicate (TMOS), during the synthesis of MSM and further oxidation of thiol groups with an oxidizer like H_2O_2 are two general pathways to produce sulfonated MSMs.

Currently, in addition to MPTS, there are many precursors for the production of sulfonated MSMs. In this case, propylsulfonic acid-based precursors are the most common types of sulfonated MSMs. In 2006, Melero and co-workers[15] reviewed and discussed advances, developments, and catalytic applications of various types of sulfonated MSMs. They also overviewed and discussed new types of precursors. Two general pathways for the synthesis of MSM-SO_3Hs are depicted in Figure 3.3.

The silica frameworks in mesostructures have intrinsic hydrophilicity; therefore, adsorption of water may cause a catalyst surface deactivation for mass transfer of organic molecules as substrates during reaction times. To solve this problem, many investigations have been performed in this regard.[16a] One way is to confine a hydrophobic ionic liquid into SBA-15 to remove catalyst deactivation. This is a very efficient system. However, the ionic liquid preparation was costly and time consuming.[16c,17] Cofunctionalization of a hydrophobic function like phenyl or alkyl chain (methyl, ethyl, or even octyl) with an MPTS function is another approach to increase the hydrophobicity.[16e,18] Unfortunately, this method brings pore restriction by decreasing the real pore size and surface area. Intensive functionalization of a silica surface with MPTS produces a highly functionalized –SO_3H groups inside the MSM, which prevents catalyst deactivation by water.[16b] This method is efficient in the fabrication of highly acidic MSM-SO_3H. To prevent pore size decrease,[19] the use of phenylene- and ethylene-bridged periodic mesoporous organosilica (PMO) is suggested to increase hydrophobicity while meanwhile the keeping pore size intact.[16d] The other approach to preventing catalyst deactivation is to confine fluorous alcohols, which have the odd property of both hydrophobic and lipophilic behavior. Fluoroalkyl chains could cause an increase in the acidic strength and decrease catalyst poisoning

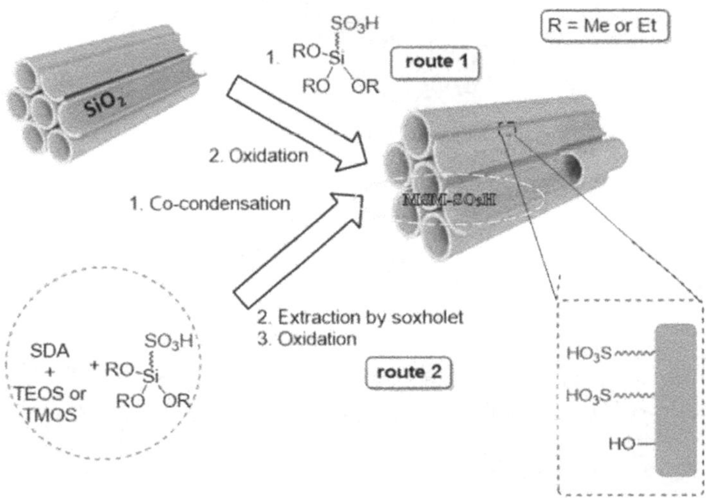

FIGURE 3.3 Sulfonation of MSM with two possible pathways.

by repelling the trapped water from the siliceous surface.[20] Therefore, studying the effect of TFE and HFIP on the surface of TFE/SBA-15-SO_3H led to the appearance of a powerful catalytic system. Compared to corresponding nonfluorous alcohols, fluorous alcohols (including TFE and HFIP) are more efficient in the catalysis of 1,2,4,5-tetrasubstituted imidazoles syntheses when added to SBA-SO_3H (Figure 3.2). This method needs no postmodification; covalent linking and weak interactions like hydrogen bonding are responsible. Therefore, by adding a few drops of R_FOH onto SBA-15, a convenient catalytic system can be promoted (Figure 3.4).[9]

It is found that simultaneous application of an ultrasound (US) system to SBA-15-SO_3H during the catalysis causes an increase in the efficiency and mass transfer inside the pores of SBA-15-SO_3H.[21] This strategy was investigated in the synthesis of three separate biologically interesting reactions including the syntheses of 2*H*-indazolo[2,1-*b*]phthalazine-triones, α-aminophosphonates, and polyhydroquinolines (Figure 3.5) under sonication. Under these conditions, higher selectivity and yields toward the products occur in very short reaction times (4–25 min).

Another method to increase the efficiency of MSM-SO_3H is to apply microwave irradiation during the catalysis of a bifunctional acid–base MSM that has both amine (as a base) and sulfonic acid as cooperative system. For the preparation of this catalyst, 3-aminopropyltrimethoxysilane as an amine precursor was used to functionalize the MCM-41. Then, the prepared MCM-41-NH_2 reacted with 1,4-butane-sultone to give MCM-41-$NHSO_3H$. Afterward, this catalytic system was used to catalyze a *tert*-butylation reaction of hydroquinone under microwave irradiation that had a synergistic effect. MCM-41-$NHSO_3H$ showed a high conversion (88.0%) and selectivity (93.1%) to 2-*tert*-butylhydroquinone (2-TBHQ) after 8 min[22] (Figure 3.6).

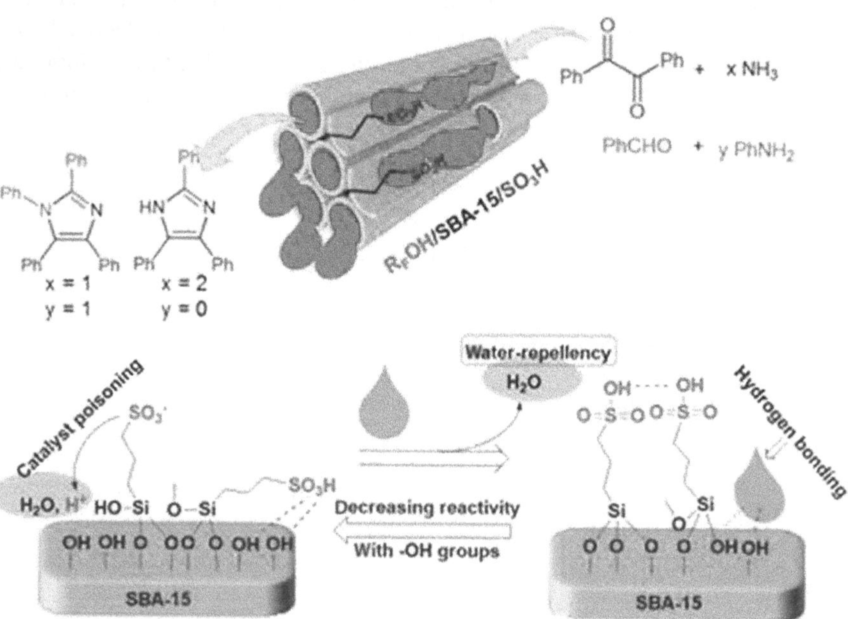

FIGURE 3.4 Silica/SO_3H catalyst poisoning with water.

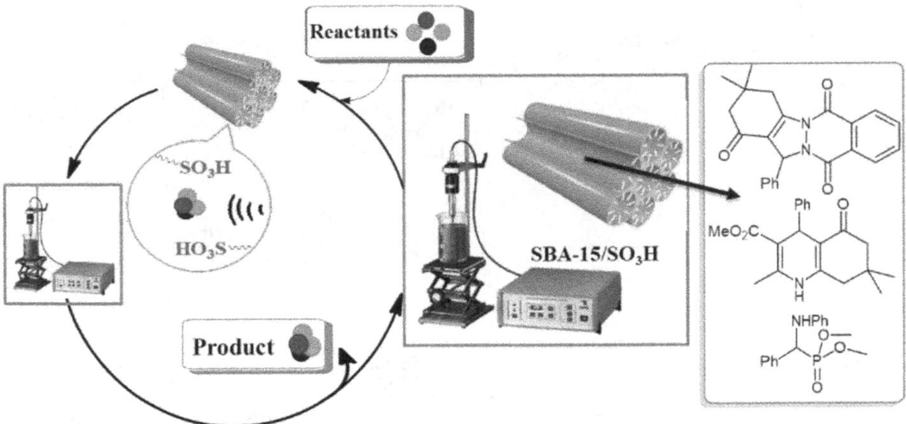

FIGURE 3.5 US/SBA-15-SO_3H system as simultaneous method.

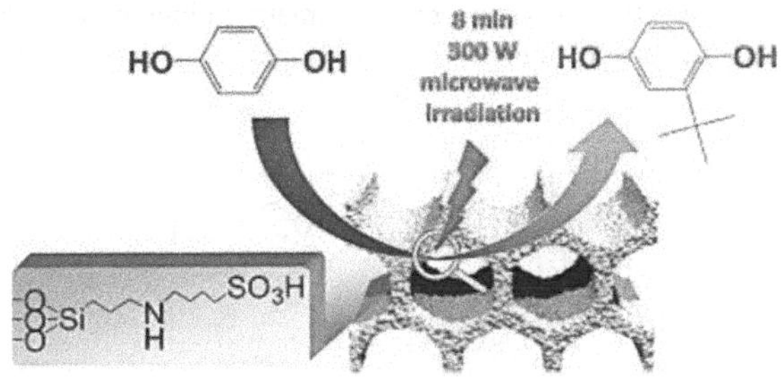

FIGURE 3.6 Alkylation of dihydroquinone by the catalysis of aminosulfonic acid modified mesoporous silica.

Bhaumik[23] designed a new carboxylic acid-modified SBA-15 using a postmodification approach. For the synthesis of the catalyst, the surface of SBA-15 was functionalized by aminopropylsilane (APS) followed by imination of amine moiety of SBA-NH_2 with *p*-formylbenzoic acid. The organocatalytic activity of SBA-CO_2H was subsequently studied in the synthesis of xanthenes under mild conditions in the absence of any other metal cocatalyst.

Phosphoric acid-modified MSMs are also interesting, but rarely developed as sulfonated MSMs are. Bhaumik and Pramanik[24] designed a new phosphoric acid-modified MSM (MSM-PO_3H). This hybrid material showed high surface area and ordered array of MSM with an average pore diameter of ca. 2.1 nm.

For the synthesis of MSM-PO_3H (Scheme 3.6), condensation of (triethoxysilyl)(propyliminomethyl)biphenylmethylphosphoester (PEFOS) as a phosphoester–organosilane precursor and TEOS happened in the presence of the cationic surfactant, cetyltrimethylammonium bromide (CTAB), as structure directing agent. Synthesis of phosphoester organosilane precursor (Scheme 3.7) was performed for the first time by this group through a simple SN_2 reaction of triethylphosphate with *p*-bromobenzylphosphate followed by Suzuki coupling with *p*-oxophenylboronic

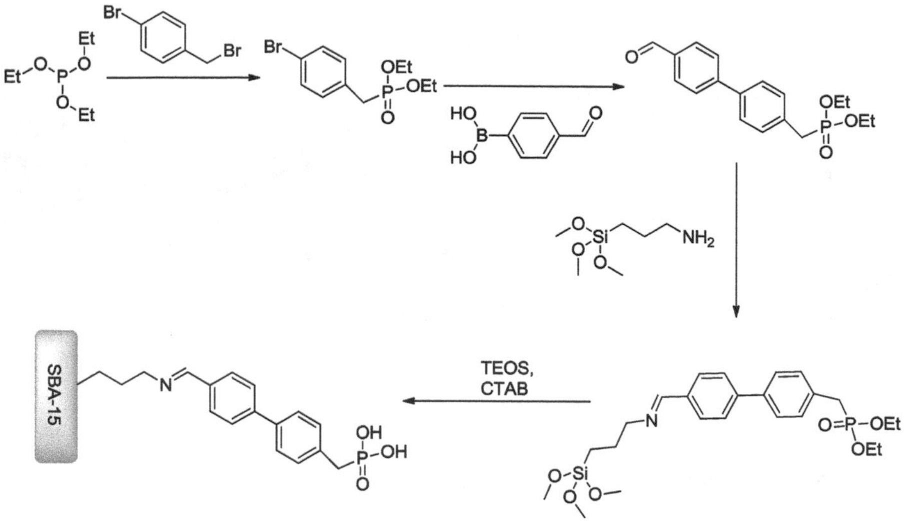

SCHEME 3.6 Use of (triethoxysilyl)(propyliminomethyl)biphenylmethylphosphoester as a precursor for MSM-SO_3H production.

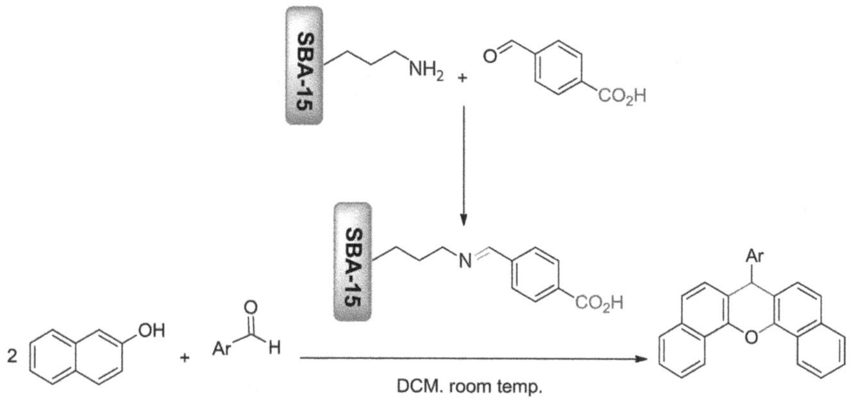

SCHEME 3.7 SBA-NH_2 grafted with *p*-formylbenzoic acid.

acid and imine formation with APS. To prove the presence of organic moieties (an aromatic biphenyl ring and an aliphatic side chain), phosphorous on the surface of the pore walls of MSM-PO_3H, solid-state magic-angle-spinning nuclear magnetic resonance (NMR), x-ray photoelectron, and Fourier transform infrared (FT-IR) spectroscopic resonance tools were applied for the analysis. Furthermore, FT-IR analysis of MSM-PO_3H over temperature-programmed pyridine adsorption study exhibited the surface acid strength. This new catalyst was used in the catalysis of biologically important 3,4-dihydropyridin-2-1*H*-(ones)/3,4-dihydropyridin-2-1*H*-(thiones) (DHPMs) through a tricomponent Biginelli reaction under solvent-free conditions at 60 °C. Compared to propylsulfonic acid-modified MCM-41 (MCM-SO_3H),

MSM-PO_3H exhibited much higher catalytic activity in this reaction. Moreover, it was easily recovered and reused for eight successful cycles.

3.3 PMO MATERIALS IN MULTICOMPONENT REACTIONS

Periodic mesoporous organosilica materials are new type of MSMs in which the organosilica motif (as a bridge) is formed inside the pore walls and therefore, it is part of structure of MSM. There have been developed many applications including catalysis, photocatalysis, adsorption, sensing, light harvesting, etc., for PMO materials.[25] Since the discovery in 1999, many types of organic bridges have been designed and developed. A general pathway for the synthesis of PMO materials is depicted in Scheme 3.8. It should be noticed that the addition of TEOS or TMOS as a silica source is necessary for some bridges. And in the step of surfactant removal, calcination is not applicable; instead, extraction with Soxhlet should be utilized.

An imidazolium bridge is an ionic liquid organosilica precursor that was first synthesized by Karimi for the fabrication of a new PMO.[26] In this PMO, imidazolium moieties with alkyl chains in each side are embedded in the frameworks of the PMO (Scheme 3.9). This PMO has been developed in many types of catalytic reactions by supporting various types of metals and metal oxides.[26,27] This catalyst has undergone postmodification with MPTS followed by thiol groups' oxidation to generate a PMO with ionic liquid framework and sulfonic acid postfunctionality (IL-PMO-SO_3H).[28] This efficient design was further applied in the tricomponent synthesis of 3,4-dihydropyrimidin-2(1*H*)-one derivatives through Biginelli reaction (Scheme 3.10). A TEM image of sulfonated IL-PMO indicates that the postmodification and oxidation with H_2O_2 did not have a destructive effect on the mesoscopic structure of IL-PMO.[29]

Moreover, the reactivity, reusability, and stability of IL-PMO-SO_3H were also investigated under optimized reaction conditions and showed that this catalyst can be recovered and reused at least for 10 successful cycles.[29]

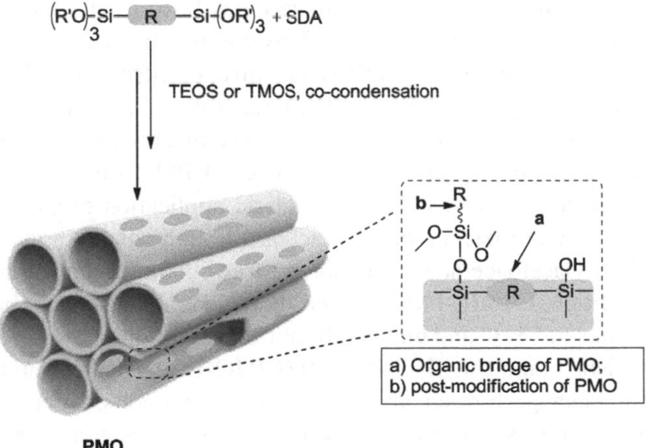

SCHEME 3.8 General pathway for the synthesis of PMO materials.

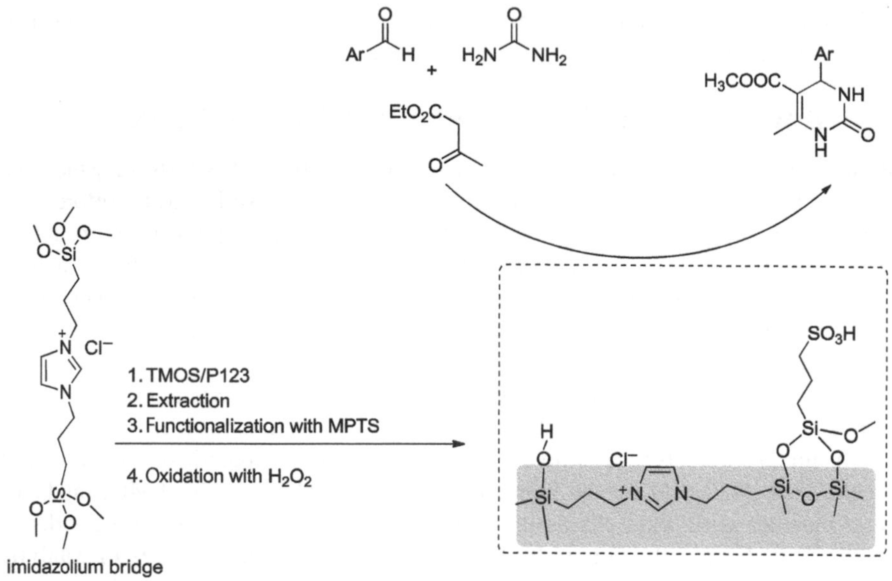

SCHEME 3.9 Imidazolium bridge is an ionic liquid organosilica precursor in PMO synthesis.

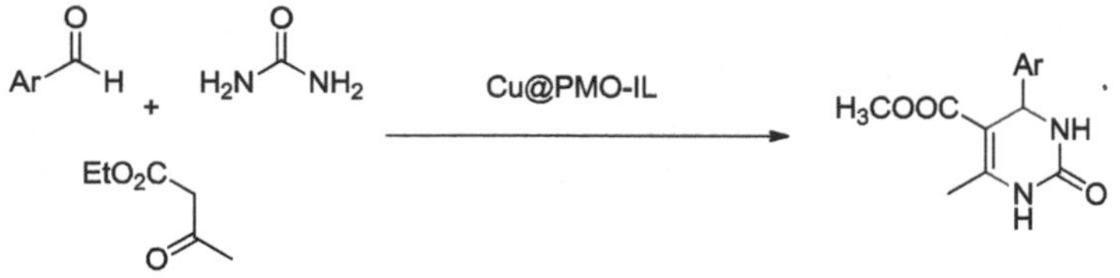

SCHEME 3.10 Tricomponent reaction of aldehyde, urea, and ethylacetoacetate catalyzed by Cu@PMO-IL.

This IL-PMO without sulfonic acid moiety has been used for Cu^{2+} supporting (Cu@PMO-IL)[30] through the ionic part. This catalyst has been characterized by transmission electron microscopy (TEM) and nitrogen adsorption–desorption analysis. The amount of organic bridge (dialkylimidazolium) in PMO and the thermal stability of the catalyst were determined by thermal gravimetric analysis (TGA). Oxidation state of copper inside the frameworks of PMO was confirmed by x-ray photoelectron spectroscopy (XPS). The catalytic application of Cu@IL-PMO was investigated in Biginelli tricomponent reaction through condensation of aldehydes with urea and alkylacetoacetates under solvent-free conditions and at moderate temperature. In addition, Cu@IL-PMO was easily recoverable and reusable at least for 16 successful runs at 75 °C and under solvent-free conditions.

Supporting Mn^{3+} onto this IL-PMO (Mn@PMO-IL) through ionic liquid framework has also been performed and used in two multicomponent reactions.[31] Its catalytic application was first investigated in a Biginelli reaction through condensation of urea, aromatic aldehyde, and alkylacetoacetate.[31] Furthermore, polyhydroquinoline derivatives were synthesized under the catalysis of Mn@PMO-IL over a four-component Hantzsch reaction of two unsymmetric diketones, ammonium acetate, and aromatic aldehyde under solvent-free conditions[31] (Scheme 3.11). The catalyst showed high reactivity and selectivity toward production of favored polyhydroquinolines under mild conditions and short reaction times. Moreover, the catalyst had the capability of being recovered and reused several times without important decreases in reactivity and yields in both multicomponent reactions. The oxidation state of the Mn ion was characterized by XPS analysis and indicated that it is Mn^{3+}.

Recently, an efficient and enantioselective method was developed for the direct asymmetric reaction of dibenzylmalonate with *N-tert*-butoxycarbonylaldimines in the presence of $Yb(OTf)_3$ and asymmetric *i*Pr-pybox complexes (pybox = pyridine bisoxazoline (**L**)) through a Mannich reaction.[32,33] Then, this asymmetric $Yb(OTf)_3$/*i*Pr-pybox catalytic system was immobilized on the aforementioned self-assembled IL-PMO (SAILP) and tested in the enantioselective Mannich reaction of malonate esters with *N*-Boc-aldimines (Scheme 3.12); this afforded the related products in good yields and enantioselectivity and recyclability of the catalyst were efficient. Of course, when the ratio of $Yb(OTf)_3$ and **L** in SALIP (10% of the imidazolium bridge) is 1:2, respectively, more enantioselectivities are seen in comparison with 1:1 $Yb(OTf)_3$/**L** on SAILP under the same reaction conditions. The catalyst with 1:2 $Yb(OTf)_3$/L was able to be recycled with only a slight decrease either in catalytic activity or in enantioselectivity.

A three-step postmodification for the preparation of Fe^{3+} supported PMO-IL bearing imidazolium within pore walls and Schiff base postfunctionality (Fe@PMO-IL) on the surface was obtained though primary functionalization with APTS and further imination with salicylaldehydein toluene at reflux conditions (Scheme 3.13).[34] In the next step, Fe@PMO-IL was applied as a catalyst in the one-pot, three-component synthesis of 3,4-dihydropyrimidinone/thione derivatives under solvent-free conditions. The catalyst exhibited high stability, reactivity, and reusability in the reaction conditions and was recovered and reused for several

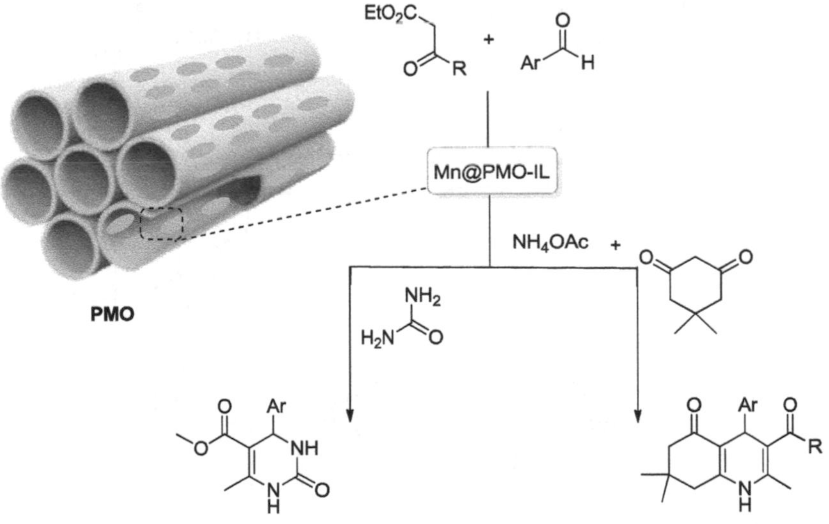

SCHEME 3.11 Mn@PMO-IL in the catalysis of four-component Hantzsch reaction.

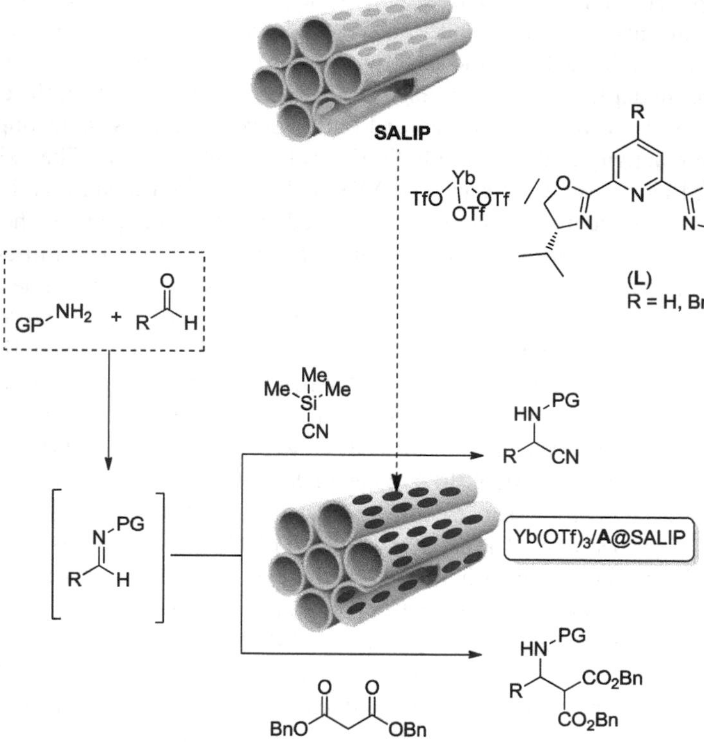

SCHEME 3.12 Asymmetric Yb(OTf)3/iPr-pybox immobilized on IL-PMO (SAILP) in enantioselective Mannich reaction.

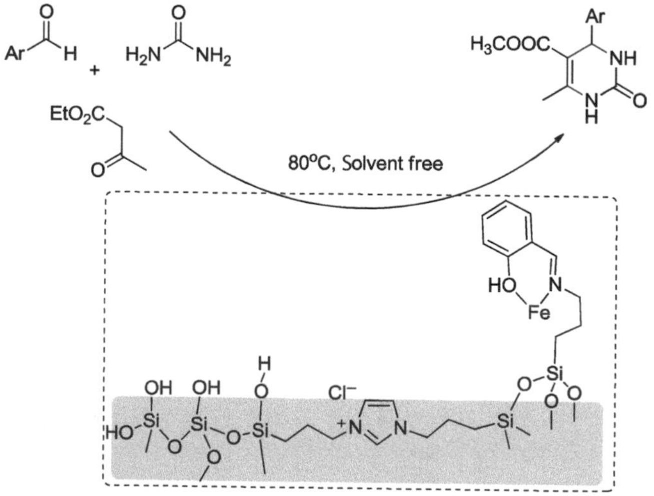

SCHEME 3.13 Fe@PMO-IL as a catalyst in synthesis of 3,4-dihydropyrimidinone/thione derivatives under solvent-free conditions.

successful runs. Bifunctional structure of the catalyst offers strong support for the immobilization of iron ions to minimize the iron leaching in each cycle of reaction, which could make a greener protocol while strong catalytic activity in each consecutive run occurs.

Karimi and co-workers[35] investigated three kinds of sulfonated MSMs including traditional SBA-15-SO_3H, phenyl-modified SBA-15-SO_3H, and ethylene-bridged PMO-SO_3H; in all three catalysts, sulfonation originated from MPTS functions. These three catalysts have shown different chemoselectivity in the one-pot, three-component reaction of urea, aromatic aldehyde, and methylacetoacetate toward selective synthesis of Biginelli products. This study exhibited that SBA-15-SO_3H prefers the production of a Hantzsch product against a Biginelli product unselectively. Catalysis of phenylated SBA-15-SO_3H leads to aldol condensation of methylacetoacetate and aldehyde. However, a Biginelli product with high yield and selectivity is generated when Et-PMO-SO_3H is the catalyst. This reusable catalyst could have higher chemoselectivity to produce Biginelli products of 3,4-dihydropyrimidin-2-one\thiones at 90 °C in 1.5 h under solvent-free conditions compared to other two sulfonated MSMs (Scheme 3.14).

An A3-coupling tricomponent reaction was also investigated and performed by a bifunctional PMO bearing both thiol groups and sulfonic acid groups on the surface and an ethylene bridge within the framework. Bifunctionalization of PMO is first undergone by MPTS and then subequivalent oxidation of thiol groups takes place. This bifunctional PMO is subsequently applied for Au nanoparticle generation and stabilization through Au^{3+} ion adsorption *in situ* reduction.[36] It was shown that Au nanoparticles inside the Et-PMO-SH/SO_3H are formed with uniform and narrow size distribution around 1–2 nm, which is very critical for essential catalytic activities (Scheme 3.15). Several types of reactions were investigated, such as three-component coupling for propargyl amine formation. The amphiphilic nature of Et-PMO-SO_3H/-SH/Au has enabled the organic reactions to proceed efficiently in a pure aqueous solution with no additional organic solvents added. The catalyst was recycled at least 10 times without significant loss in the catalytic activity of the model reaction.

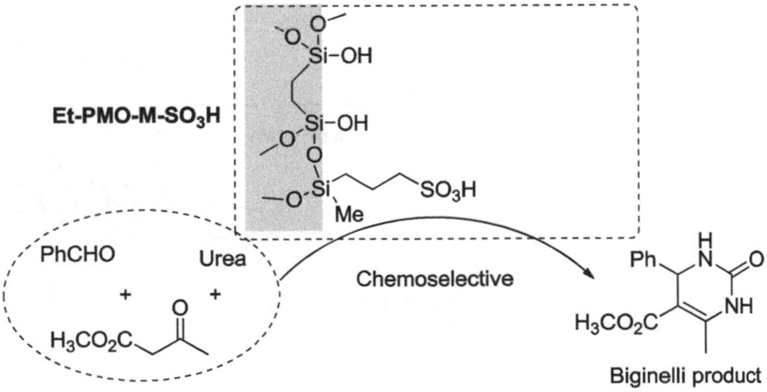

SCHEME 3.14 Catalysis of ethylene-based PMO-SO_3H.

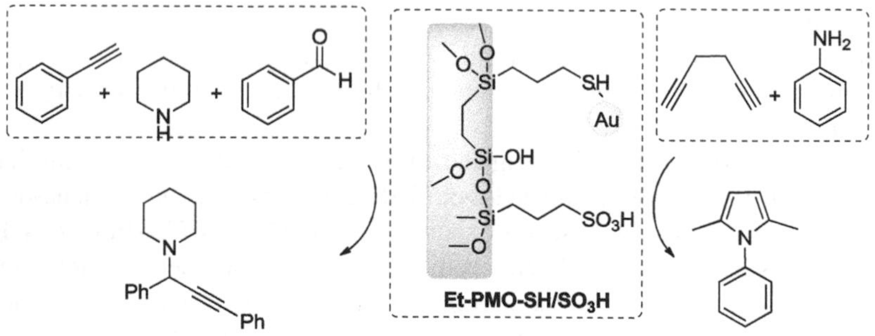

SCHEME 3.15 Et-PMO-SO_3H/-SH/Au catalysis in the organic reactions.

3.4 ORGANIC MESOPOROUS POLYMERS (OMPs)

Organic mesoporous polymers are recently developed materials with high surface area; they are obtainable synthetically via two general methods, including the use of a hard template to induce mesoporosity or of a soft-template. Owing to very recent development of OMPs, these materials are rarely applied in the catalysis, especially in multicomponent reactions among recent reports.[37] A new copper supported OMP has been investigated as a nanocatalyst in one-pot, tricomponent synthesis of propargylamines at room temperature (Scheme 3.16). Synthesis of the catalyst has been achieved by *in situ* radical polymerization of triallylamine in the presence of organic–organic self-assembly of anionic surfactant SDS, followed by grafting of Cu(II) at room temperature under inert atmosphere. Furthermore, the catalyst is easily recoverable and reusable for six successful runs with insignificant loss in the activity of the catalyst. It is claimed that highly dispersed Cu(II) sites in the Cu-grafted OMP could be responsible for the higher activity of the Cu@OMP catalyst in the coupling reactions, since no copper ions were leached from the support during and after the reaction completion.

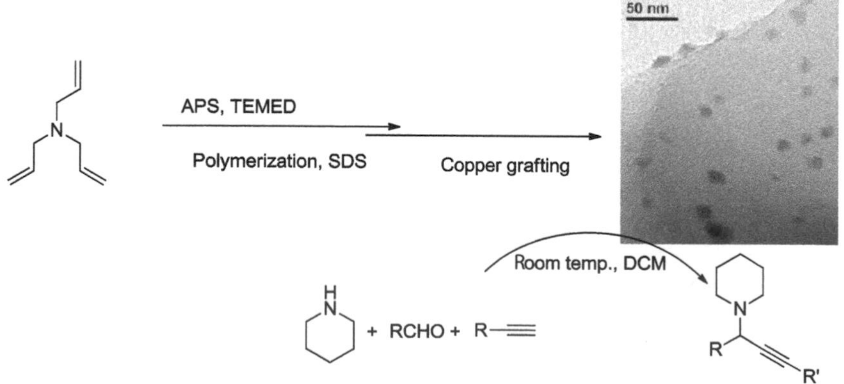

SCHEME 3.16 Nanocatalysis in one-pot tri-component synthesis of propargylamines at room temperature.

3.5 METAL–ORGANIC FRAMEWORKS (MOFs) AS NANOPOROUS CATALYSTS IN MCRs

Metal–organic frameworks (MOFs) have received a great deal of attention in various fields of chemistry such as gas adsorption and separation, sensing, nanoelectronics, and catalysis due to their exceptional surface area and their tunable frameworks according to type of MOF role. In general, MOFs are crystalline materials that have two-dimensional (2D) and three-dimensional (3D) structures containing metal ions that coordinate to one or two organic ligands with at least two dentates that could form a coordinative polymer—namely, MOF (Scheme 3.17). Moreover, MOFs are capable of being postmodified through a metal center or linker to produce an advanced nanomaterial. The other advantage of MOFs is that they can be used as a catalyst without postmodification and also as support for additional metal nanoparticle and ion grafting. In addition, MOFs can be fabricated in the form of a composite to have additional features and properties.[38] MOFs can be prepared by two main solvo- and hydrothermal routes.[39] In each case, some advantages and disadvantages are dominant in which the final target for catalysis has some features. For example, MOFs prepared by a hydrothermal route have more stability to water and aqueous media.[40]

In addition, MOFs have played a significant role in the catalysis of multicomponent reactions through C–C and C–X (X = O, N, S) bond formation between reactants. Therefore, in this section we will discuss the application of MOFs in multicomponent reactions. Recently, efficient MOF-based catalysts in the field of multi-component reaction catalysis have opened a huge research domain for catalysis issues. In the case of MCRs, MOFs have pore limitation because the products of MCRs are relatively bulky. In this case, when the substrate is unable to diffuse into the pores of MOFs, the reactants can continue to react on the external surface of MOFs, and therefore, the external surface will play a major role in the catalysis. On the other hand, crystalline size of MOFs can have a drastic effect on the catalysis of such reactions.[41]

Zn- and Cd-based MOFs were prepared as catalysts by a hydrothermal method with a 2D structure—highly efficient for the green synthesis of dihydropyrimidinone derivatives through the Biginelli reaction (Scheme 3.18) ofaldehydes, urea, and acetoacetate.[42] The same MOFs also exhibited efficient catalytic activity in tricomponent reaction between aldehydes, malononitrile, and thiophenols, which

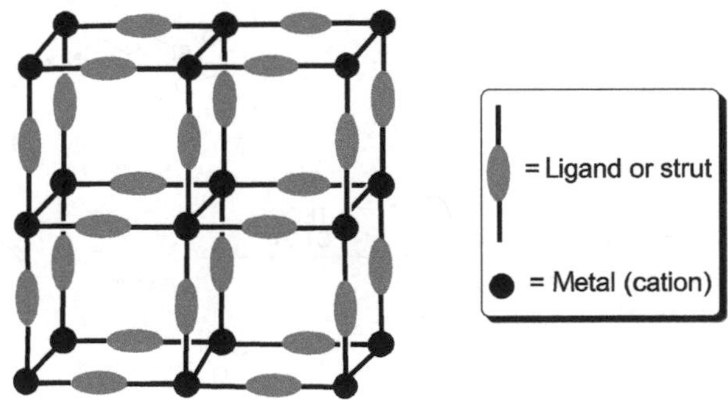

SCHEME 3.17 Typical MOF structure.

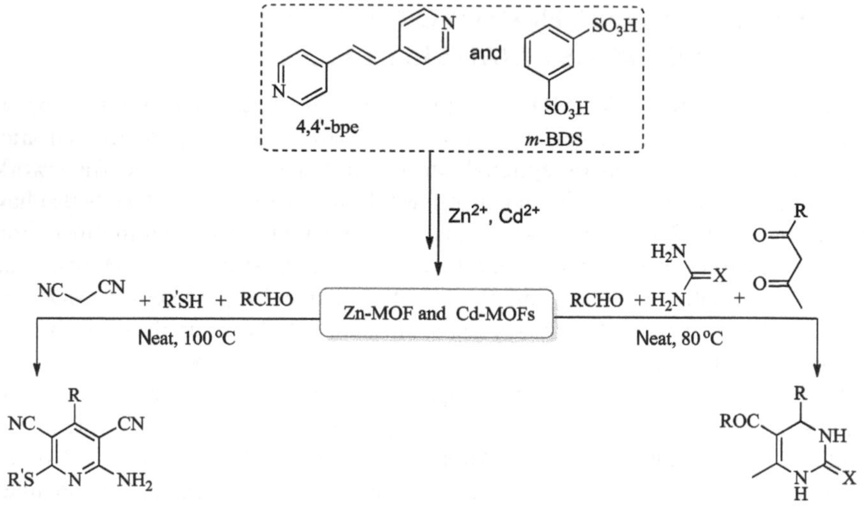

SCHEME 3.18 Cd-Mof and Zn-Mof catalyzed MCRs.

provided the biologically active molecules of 2-amino-6-(arylthio)pyridine-3,5-dicarbonitriles in high yields (Scheme 3.18). Both Zn and Cd MOFs demonstrated similar activity and they yielded best results in solvent-free conditions. This protocol limits the volatile organic compounds (VOCs) and toxic catalysts, which are very important from the view point of green chemistry.[43]

The MIL-101(Cr)/phosphotungstic acid (PTA) composite as a catalyst reveals an excellent catalytic activity in the three-component condensation of benzaldehyde, β-naphthol, and acetamide toward production of amidonaphthols (Figure 3.7). MIL-101/PTA has catalyzed the reaction by two different strategies including autoclave and

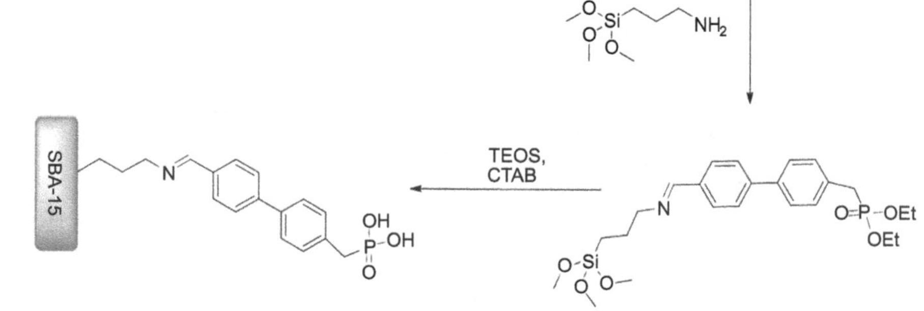

FIGURE 3.7 MIL-101 as catalyst for MCR.[44]

microwave methods. For this microwave-assisted MCR, almost complete conversion of the reactants was observed after 5 min. However, leaching of PTA was observed during the course of the reaction and catalyst recovery by filtration was problematic.[44]

Likewise, IRMOF-3, NH_2-MIL-53 is a MOF with amine functionality that has a basic nature and can be used for other applications such as postmodification and metal coordination. NH_2-MIL-53(Al) material can be obtained by reacting 2-aminoterephthalic acid (H_2ATA) with aluminum chloride in DMF. The nanoporous NH_2-MIL-53 has a crystalline structure with diamond-shaped and one-dimensional (1D) pores (Figure 3.8).

We investigated chemo- and size selectivity of the NH_2-MIL-53 in a three-component reaction of cyclohexylisocyanide and 2-aminopyridine with mixture of two aldehydes, *p*-nitrobenzaldehyde and benzaldehyde, at 55 °C. It was observed that the other reactants under these conditions favor the reaction of benzaldehyde (44%) compared to *p*-nitrobenzaldehyde (11%) (Scheme 3.19). This can be attributed to the larger molecular size of *p*-nitrobenzaldehyde, which therefore results in less diffusion. However, a small amount of *p*-nitrobenzaldehyde was incorporated to yield the relative product.[45]

Recently, we found that amine groups of IRMOF-3 have basic natures within the structure and therefore we used it as a base catalyst, because of free NH_2 groups, for one-pot, tricomponent Kabachnik-fields reaction of amine, aldehyde, and phosponate, which yields α-aminophosphonates. This reaction resulted in excellent yields using IRMOF-3 as catalyst. The catalyst revealed high selectivity toward the main product.[46] Moreover, some other MCRs have been developed in our laboratory. We used IRMOF-3 as a suitable green catalyst for the one-pot synthesis of dihydropyrimidinone and dihydropyridine derivatives through the Biginelli and Hantzsch reactions, which are both important MCRs. The corresponding products were produced in high yields and short reaction times under solvent-free and mild conditions. The IRMOF-3 in all three reactions revealed recoverable and reusable features.[47] IRMOF-3 was also studied in the catalysis of tetrahydrochromene synthesis, a bioactive heterocycle, using a multicomponent method (Scheme 3.20). In this work, the basic activity of IRMOF-3 was sufficient to catalyze the reaction in efficient, selective, and green conditions. Tricomponents in this reaction involved aromatic aldehydes, malononitrile, and dimedone. The catalyst was reusable at least five times.[48]

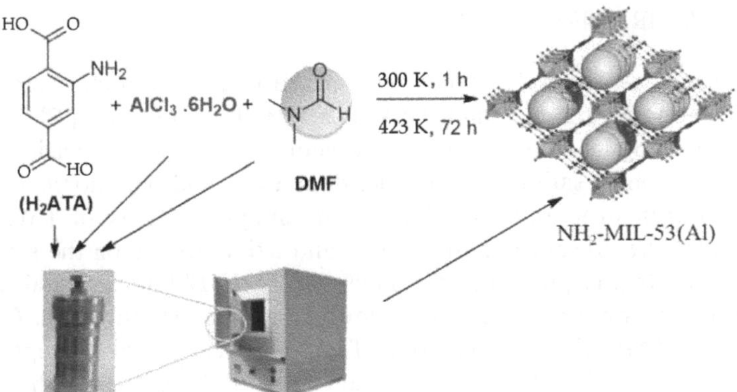

FIGURE 3.8 Preparation process of NH_2-MIL-53(Al).

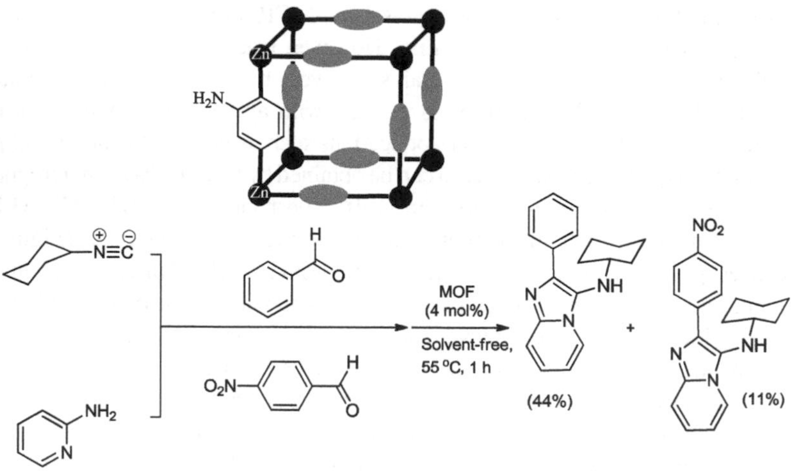

SCHEME 3.19 Chemo-and size selectivity of the MOF in an Ugi-tpe reaction.

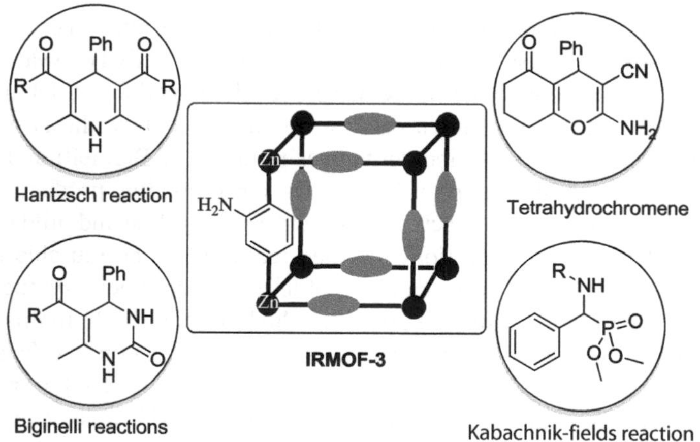

SCHEME 3.20 IRMOF-3 for MCRs.

Morsali and co-workers[48] synthesized a mixed-ligand, twofold, interpenetrated microporous metal-organic framework, [Zn(NH_2-BDC)(4-bpdb)] 2DMF(TMU-17-NH_2), for using as an efficient heterogeneous catalyst with a basic nature for tricomponent condensation of aldehyde, dimedone, and malononitrile for production of tetrahydrochromenes. The major catalytic activity of TMU-17-NH_2 is affected by the presence of free amine and azine groups on the structure of TMU-17-NH_2. In the presence of 6 mol% of TMU-17-NH_2 as catalyst, excellent conversions were obtained under mild conditions (Figure 3.9). Compared to amine-free TMU-17 in this reaction, TMU-17-NH_2 exhibited a superior catalytic activity, which highlights the critical role of –NH_2 groups in the catalysis. Avoiding the utilization of harmful solvents in this environmentally friendly process is particularly appealing.

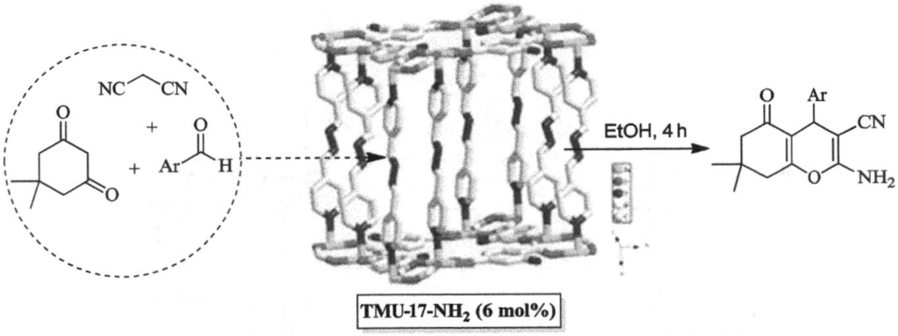

FIGURE 3.9 MCR among benzaldehyde, malononitrile, and dimedone in the presence of TMU-17-NH_2.

As mentioned before, MOFs can undergo postmodification through metal centers or organic linkers. For example, IRMOF-3 can undergo postmodification through amine groups with glyoxal to form a bidentate ligand on the surface of IRMOF-3. Treating with CuI can support and stabilize it through a bidentate N–O ligand on IRMOF-3. This nanoporous material exhibits an excellent catalytic study toward A3-coupling tricomponent coupling reactions of aldehyde, alkyne, and amine (Scheme 3.21).[49]

IRMOF-3 was also postmodified by salicylaldehyde and coordinated with Au^{3+} ions. Then, Au ions were reduced to convert Au(o) nanoparticles and stabilized inside the nanopores of IRMOF-3 (IRMOF-3-SI-Au) (Scheme 3.22). It was incorporated as a green and heterogeneous catalyst for A3-coupling synthesis of propargyl amines within a tricomponent reaction (Scheme 3.22).[50] Corma studied this catalyst in multicomponent domino coupling-cyclization of *N*-protected ethynyllaniline, aldehyde, and amine (Scheme 3.18), which was superior to its homogeneous counterpart and Cu(I) salts.[51] The catalyst exhibited much higher activity when compared to its heterogeneous and homogeneous counterparts (e.g., Au/ZrO_2, $AuCl_3$, and Au(III) Schiff base complex).[52] The reaction in the presence of ethynyllaniline bearing electron withdrawing groups yielded highest conversion.[50]

Copper-based MOFs (e.g., Cu(BDC) and $Cu(2\text{-pymdo})_2$) are found to be active, stable, and reusable solid catalysts for (a) three-component couplings of amines,

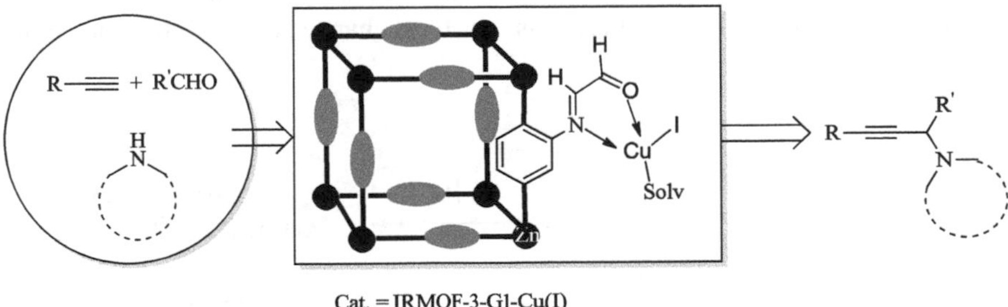

Cat. = IRMOF-3-G1-Cu(I)

SCHEME 3.21 IRMOF-3-G1-C4 catalyzed A^3-coupling reaction.

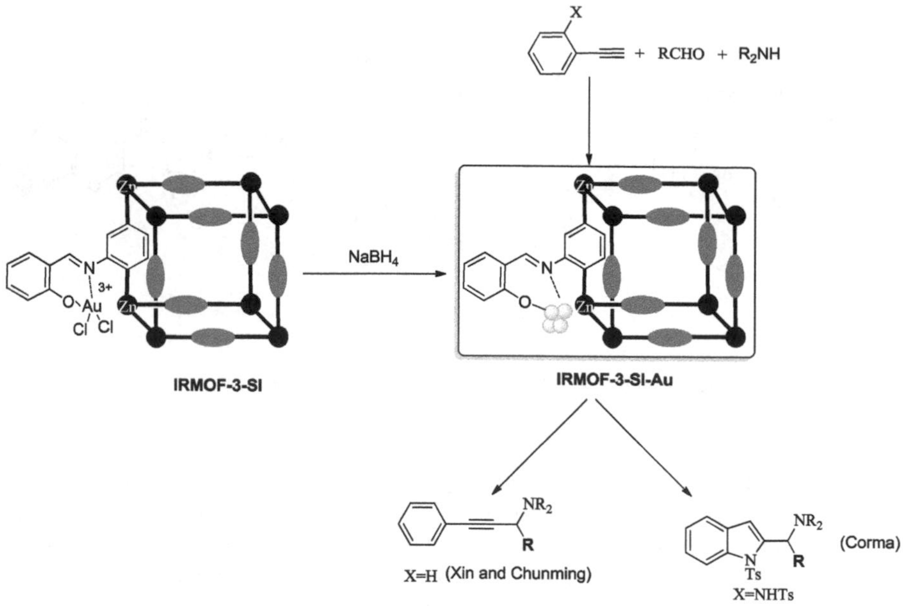

SCHEME 3.22 Postmodified IRMOF-3 for Au nanoparticle synthesis and stabilization in IRMOF-3 for two types of simple and domino-type A3 coupling.

aldehydes, and alkynes to form the corresponding propargylamines; (b) three-component coupling 2-aminopyridine, aldehydes, and acetylenes tandem A3-coupling-cyclization reaction to form imidazopyridine compounds; and (c) tri-component tandem coupling cyclization reaction to form indole derivatives (Scheme 3.23). Between these two catalysts, Cu(BDC) was highly efficient for the preparation of imidazopyridines. However, a progressive structural change of the catalyst to a catalytically inactive species was observed (Scheme 3.24). In the case of [Cu(2-pymo)$_2$], high yield of the corresponding propargylamine was obtained. In this reaction, the released H_2O from imination of aldehydes and amines did not have any negative effects on the reaction but decreasing the reaction rate was observed for aromatic alkynes larger than phenylacetylene, which obviously indicates that the components of the reaction react together inside the pores as a nanoreactor. After two consecutive runs, the catalyst underwent catalyst deactivation by deforming in the native structure of the catalyst. Fortunately, this structural change and catalytic inactivation

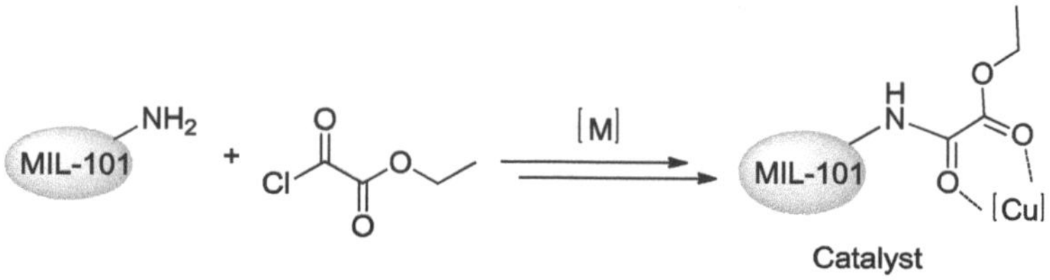

SCHEME 3.23 A3-coupling reaction investigation by MIL-101-oxamate@Cu.

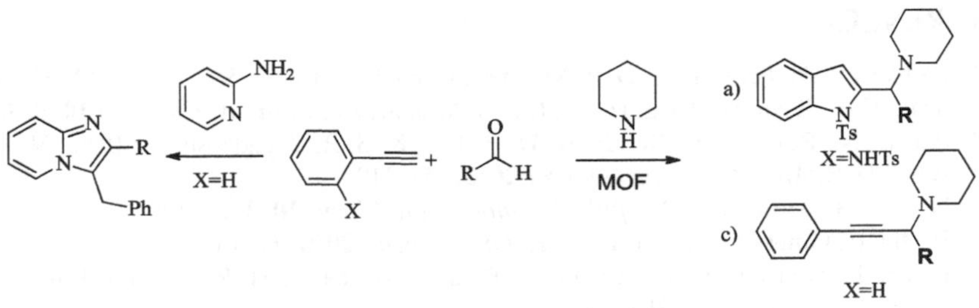

SCHEME 3.24 MOFs for MCR couplings.

could be retrieved by treating and recrystallization with DMF. Additionally, Glaser's oxidative coupling of phenylacetylene (homocoupling of acetylene) as the oxidative coupling by-products even in the presence of air was not observed.[53]

The A3-coupling reaction has also been investigated by MIL-101-oxamate@Cu in which the oxamate acts as a ligand to coordinate MIL-101 which is postmodified to MIL-101, and its catalytic activity in A3-coupling reaction is comparable with that of CuBDC or CuBTC, which has Cu species in the framework (Scheme 3.23).[54]

Cu(I)-MOF has been proved to be an active catalyst in an A3-coupling reaction under solvent-free conditions from multicomponent coupling of pyrolidine or pipyridine with aldehydes and phenyl acetylene.[55] This catalyst exhibits recyclability for five consecutive runs and insignificant loss in the activity.

CuBTC (the organic linker is benzenetricarboxylate) is also applied as a copper-based nanoporous catalyst in a four-component synthesis of hydrazonosulfonamide through *in situ* generated ketenimine. The generation of the ketenimine intermediate is first achieved by using a copper-based catalyst. Aldehyde, alkyne, tosylazide, and phenylhydrazine were four components of the reaction in this work, which was achieved in a one-pot process at room temperature during 5 min in the presence of 1 mol% of Cu. CuBTC successfully acted as a heterogeneous nanoporous catalyst with recoverability and reusability for six runs (Scheme 3.25).[56]

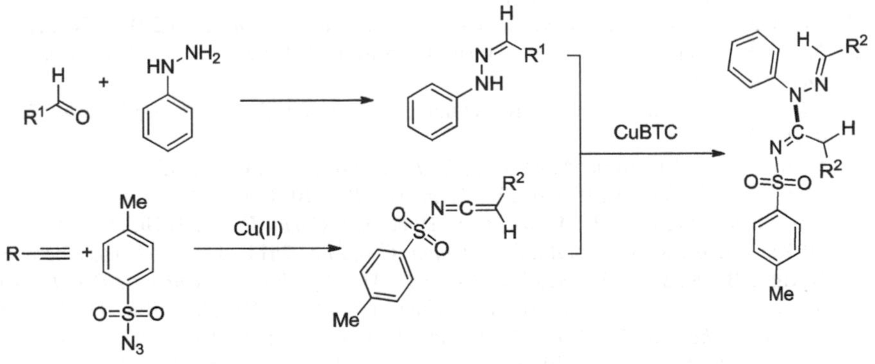

SCHEME 3.25 Four-component synthesis of hydrazano-sulfonamide catalyzed by CuBTC.

REFERENCES

1. Peshkov, V. A.; Pereshivko, O. P.; Van der Eycken, E. V. *Chem. Soc. Rev.* **2012**, *41*, 3790.
2. Yong, G.-P.; Tian, D.; Tong, H.-W.; Liu, S.-M. *J. Molec. Catal. A: Chem.* **2010**, *323*, 40.
3. Anand, N.; Ramudu, P.; Reddy, K. H. P.; Rao, K. S. R.; Jagadeesh, B.; Babu, V. S. P.; Burri, D. R. *Appl. Catal. A: General* **2013**, *454*, 119.
4. Naeimi, H.; Moradian, M. *Appl. Organometallic Chem.* **2013**, *27*, 300.
5. Rajabi, F.; Ghiassian, S.; Saidi, M. R. *Green Chem.* **2010**, *12*, 1349.
6. Rajabi, F.; Nourian, S.; Ghiassian, S.; Balu, A. M.; Saidi, M. R.; Serrano-Ruiz, J. C.; Luque, R. *Green Chem.* **2011**, *13*, 3282.
7. Mondal, J.; Sen, T.; Bhaumik, A. *Dalton Trans.* **2012**, *41*, 6173.
8. Wan, J.; ding, L.; Wu, T.; Ma, X.; Tang, Q. *RSC Adv.* **2014**, *4*, 38323.
9. Rostamnia, S.; Doustkhah, E. *RSC Adv.* **2014**, *4*, 28238.
10. (a) Rostamnia, S.; Zabardasti, A. *J. Fluorine Chem.* **2012**, *144*, 69; (b) Rostamnia, S.; Doustkhah, E. *Tetrahedron Lett.* **2014**, *55*, 2508.
11. Rostamnia, S.; Doustkhah, E.; Bahrami, K.; Amini, S. *J. Mol. Liquids* **2015**, *207*, 334.
12. Rostamnia, S.; Doustkhah, E.; Nuri, A. *J. Fluorine Chem.* **2013**, *153*, 1.
13. Mondal, J.; Modak, A.; Nandi, M.; Uyama, H.; Bhaumik, A. *RSC Adv.* **2012**, *2*, 11306.
14. (a) Lim, M. H.; Blanford, C. F.; Stein, A. *Chem. Mater.* **1998**, *10*, 467; (b) Rhijn, W. V.; De Vos, D. D.; Sels, B.; Bossaert, W. *Chem. Commun.* **1998**, 317; (c) Rhijn, W. V.; Vos, D. D.; Bossaert, W.; Bullen, J.; Wouters, B.; Grobet, P.; Jacobs, P. In *Studies in Surface Science and Catalysis*; Bonneviot, F. B. C. D. S. G. L.; Kaliaguine, S., Eds.; Elsevier. **1998**; vol. 117, p. 183 .
15. Melero, J. A.; van Grieken, R.; Morales, G. *Chem. Rev.* **2006**, *106*, 3790.
16. (a) Melero, J. A.; Bautista, L. F.; Iglesias, J.; Morales, G.; Sánchez-Vázquez, R.; Wilson, K.; Lee, A. F. *Appl. Catal. A: General,* **2014**, *488*, 111; (b) Dacquin, J.-P.; Cross, H. E.; Brown, D. R.; Duren, T.; Williams, J. J.; Lee, A. F.; Wilson, K. *Green Chem.* **2010**, *12*, 1383; (c) Karimi, B.; Vafaeezadeh, M. *Chem. Commun.* **2012**, *48*, 3327; (d) Pirez, C.; Lee, A. F.; Jones, C.; Wilson, K. *Catal. Today* **2014**, *234*, 167; (e) Karimi, B.; Zareyee, D. *Org. Lett.* **2008**, *10*, 3989.
17. Karimi, B.; Vafaeezadeh, M. *RSC Adv.* **2013**, *3*, 23207.
18. Karimi, B.; Mirzaei, H. M. *RSC Adv.* **2013**, *3*, 20655.
19. (a) Li, C.; Yang, J.; Shi, X.; Liu, J.; Yang, Q. *Microporous Mesoporous Mater.* **2007**, *98*, 220; (b) Tucker, M. H.; Crisci, A. J.; Wigington, B. N.; Phadke, N.; Alamillo, R.; Zhang, J.; Scott, S. L.; Dumesic, J. A. *ACS Catal.* **2012**, *2*, 1865.
20. (a) Jeong, H.-J.; Kim, D.-K.; Lee, S.-B.; Kwon, S.-H.; Kadono, K. *J. Coll. Interface Sci.* **2001**, *235*, 130; (b) Qiu, Z. M.; Google patents, 2003; (c) Ghaffarzadeh, M.; Ahmadi, M. *J. Fluorine Chem.* **2014**, *160*, 77.
21. Rostamnia, S.; Xin, H.; Liu, X.; Lamei, K. *J. Mol. Catal. A: Chemical* **2013**, *374–375*, 85.
22. Ng, E.-P.; Mohd Subari, S. N.; Marie, O.; Mukti, R. R.; Juan, J.-C. *Appl. Catal. A: General* **2013**, *450*, 34.
23. Nandi, M.; Mondal, J.; Sarkar, K.; Yamauchi, Y.; Bhaumik, A. *Chem. Commun.* **2011**, *47*, 6677.
24. Pramanik, M.; Bhaumik, A. *ACS Appl. Mater. Interfaces* **2014**, *6*, 933.
25. Mizoshita, N.; Tani, T.; Inagaki, S. *Chem. Soc. Rev.* **2011**, *40*, 789.
26. Karimi, B.; Elhamifar, D.; Clark, J. H.; Hunt, A. J. *Chem. Eur. J.* **2010**, *16*, 8047.
27. (a) Karimi, B.; Kabiri Esfahani, F. *Chem. Commun.* **2011**, *47*, 10452; (b) Abedi, S.; Karimi, B.; Kazemi, F.; Bostina, M.; Vali, H. *Org. Biomol. Chem.* **2013**, *11*, 416; (c) Karimi, B.; Khorasani, M.; Bakhshandeh Rostami, F.; Elhamifar, D.; Vali, H. *ChemPlusChem* **2015**, *80*, 990; (d) Karimi, B.; Esfahani, F. K. *Adv. Synth. Catal.* **2012**, *354*, 1319; (e) Karimi, B.; Elhamifar, D.; Yari, O.; Khorasani, M.; Vali, H.; Clark, J. H.; Hunt, A. J. *Chem. Eur. J.* **2012**, *18*, 13520.

28. Elhamifar, D.; Karimi, B.; Moradi, A.; Rastegar, J. *ChemPlusChem* **2014**, *79*, 1147.
29. Elhamifar, D.; Nasr-Esfahani, M.; Karimi, B.; Moshkelgosha, R.; Shábani, A. *ChemCatChem* **2014**, *6*, 2593.
30. Elhamifar, D.; Hosseinpoor, F.; Karimi, B.; Hajati, S. *Microporous and Mesoporous Mater.* **2015**, *204*, 269.
31. (a) Nasr-Esfahani, M.; Elhamifar, D.; Amadeh, T.; Karimi, B. *RSC Adv.* **2015**, *5*, 13087; (b) Elhamifar, D.; Shábani, A. *Chem. Eur. J.* **2014**, *20*, 3212.
32. Karimi, B.; Jafari, E.; Enders, D. *Chem. Eur. J.* **2013**, *19*, 10142.
33. (a) Karimi, B.; Jafari, E.; Enders, D. *Eur. J. Org. Chem.* **2014**, *2014*, 7253; (b) Karimi, B.; Maleki, A.; Elhamifar, D.; Clark, J. H.; Hunt, A. J. *Chem. Commun.* **2010**, *46*, 6947.
34. Elhamifar, D.; Nazari, E. *ChemPlusChem* **2015**, *80*, 820.
35. Karimi, B.; Mobaraki, A.; Mirzaei, H. M.; Zareyee, D.; Vali, H. *ChemCatChem* **2014**, *6*, 212.
36. Zhu, F.-X.; Wang, W.; Li, H.-X. *J. Am. Chem. Soc.* **2011**, *133*, 11632.
37. Salam, N.; Kundu, S. K.; Roy, A. S.; Mondal, P.; Roy, S.; Bhaumik, A.; Islam, S. M. *Catal. Sci. Technol.* **2013**, *3*, 3303.
38. (a) Bradshaw, D.; Garai, A.; Huo, J. *Chem. Soc. Rev.* **2012**, *41*, 2344; (b) Evans, J. D.; Sumby, C. J.; Doonan, C. J. *Chem. Soc. Rev.* **2014**, *43*, 5933; (c) Chughtai, A. H.; Ahmad, N.; Younus, H. A.; Laypkov, A.; Verpoort, F. *Chem. Soc. Rev.* **2015**, *44*, 6804; (d) McGuire, C. V.; Forgan, R. S. *Chem. Commun.* **2015**, *51*, 5199.
39. Mahata, P.; Prabu, M.; Natarajan, S. *Inorg.Chem.* **2008**, *47*, 8451.
40. Forster, P. M.; Stock, N.; Cheetham, A. K. *Angew. Chem. Int. Ed.* **2005**, *44*, 7608.
41. Nguyen, L. T.; Le, K. K.; Phan, N. T. *Chin. J. Catal.* **2012**, *33*, 688.
42. Li, P.; Regati, S.; Butcher, R. J.; Arman, H. D.; Chen, Z.; Xiang, S.; Chen, B.; Zhao, C.-G. *Tetrahedron Lett.* **2011**, *52*, 6220.
43. Thimmaiah, M.; Li, P.; Regati, S.; Chen, B.; Zhao, J. C.-G. *Tetrahedron Lett.* **2012**, *53*, 4870.
44. Bromberg, L.; Diao, Y.; Wu, H.; Speakman, S. A.; Hatton, T. A. *Chem. Mater.* **2012**, *24*, 1664.
45. Rostamnia, S.; Karimi, Z. submitted.
46. Rostamnia, S.; Xin, H.; Nouruzi, N. *Microporous Mesoporous Mater.* **2013**, *179*, 99.
47. Rostamnia, S.; Morsali, A. *RSC Adv.* **2014**, *4*, 10514.
48. (a) Rostamnia, S.; Morsali, A. *Inorg. Chim. Acta.* **2014**, *411*, 113; (b) Safarifard, V.; Beheshti, S.; Morsali, A. *Cryst. Eng. Comm.* **2015**, *17*, 1680.
49. Yang, J.; Li, P.; Wang, L. *Catal. Commun.* **2012**, *27*, 58.
50. Zhang, X.; Llabres i Xamena, F.; Corma, A. *J. Catal.* **2009**, *265*, 155.
51. Ohno, H.; Ohta, Y.; Oishi, S.; Fujii, N. *Angew. Chem. Int. Ed.* **2007**, *46*, 2295.
52. Gonzalez-Arellano, C.; Corma, A.; Iglesias, M.; Sanchez, F. *Chem. Commun.* **2005**, 3451.
53. Luz, I.; Llabrés i Xamena, F.; Corma, A. *J. Catal.* **2012**, *285*, 285.
54. Juan-Alcañiz, J.; Ferrando-Soria, J.; Luz, I.; Serra-Crespo, P.; Skupien, E.; Santos, V. P.; Pardo, E.; Llabrés i Xamena, F. X.; Kapteijn, F.; Gascon, J. *J. Catal.* **2013**, *307*, 295.
55. Li, P.; Regati, S.; Huang, H.-C.; Arman, H. D.; Chen, B.-L.; Zhao, J. C. G. *Chin. Chem. Lett.* **2015**, *26*, 6.
56. Mahendran, V.; Shanmugam, S. *RSC Adv.* **2015**, *5*, 20003.

4 Multicomponent Reactions of Amino Acids and Their Derivatives in Heterocycle Chemistry

*Nahid S. Alavijeh, Somayeh Ahadi, and Saeed Balalaie**

CONTENTS

4.1 INTRODUCTION

Heterocyclic compounds constitute the largest and most varied family of organic compounds and are of immense importance biologically and industrially. Many natural products and synthetic drugs are partly heterocycles.[1] Consequently, the preparation of heterocycles in a convenient one-pot procedure has emerged as a topic of great interest.[2] Nowadays one-pot, multicomponent reactions (MCRs), involving three or more reactants in a one-pot reaction, are a topic of great interest both from organic synthesis and combinatorial chemistry points of view.[3] Easily automated one-pot reactions, such as Ugi, Petasis, Passerini, Hantzsch, and Gewald reactions represent an excellent methodology for the synthesis of diverse arrays of heterocycles offering the advantage of simplicity and synthetic efficiency over conventional chemical reactions.[4]

Amino acids—composed of amine (–NH_2) and carboxylic acid (–COOH) functional groups along with a side chain specific to each amino acid—and their derivatives are important building blocks of proteins.[5] Despite the pivotal role of amino acids and their derivatives, which can be played in one-pot MCRs due to their bifunctionality and optical purity, the study of MCRs based on amino acids and their derivatives

* Corresponding author: balalaie@kntu.ac.ir.

is still in its infancy. The aim of this chapter is to summarize recently multicomponent reactions based on amino acids and their derivatives in the heterocycles chemistry.

We classified multicomponent reactions of amino acids and derivatives according to the ring size of heterocycle compounds in five categories:

- Four-membered heterocycles
- Five-membered heterocycles
- Six-membered heterocycles
- Seven-membered heterocycles
- Fused heterocycles

4.2 FOUR-MEMBERED HETEROCYCLES

Beta-amino acids, which have their amino group bonded to the β-carbon, are key components of a wide variety of pharmaceutically important compounds (natural products, peptides, and peptidomimetics).[6,7] They have been extensively applied in the synthesis of biologically active β-lactams, which have been proved to be powerful antibacterials and enzyme inhibitors (serine and cysteine proteases).[8,9] Therefore, exploring the new synthetic methods for the generation of β-lactams has attracted a great deal of attention of synthetic chemists and pharmacologists in recent years.[10–15]

Chiral β-amino acids constitute interesting building blocks for the synthesis of chiral β-lactams. One of the recent reports on this concept is the synthesis of conformationally constrained tricyclic β-lactam enantiomers through Ugi four-center, three-component reactions of a monoterpene-based β-amino acid (Scheme 4.1).[16]

The Ugi reaction was attempted in MeOH, in water, and under solvent-free conditions to devise a green methodology. The yields in water or under solvent-free conditions were similar to those in methanol with some restrictions, indicating that the water solubility of the Schiff base formed in the first step of the Ugi reaction limits the application of water as solvent.

Another example in this series is employing chiral *N*-protected amino alkyl isonitriles as C-terminally modified amino acid derivatives in Ugi multicomponent reactions (Ugi 4C-3CR) to obtain functionalized β-lactam peptidomimetics.[17] All the reactions were complete within 24–28 h under room temperature with diastereomeric ratio varying from 100:0 to 85:15 (Scheme 4.2).

In a more recent study, the combination of solid-phase synthesis and the Petasis–Borono–Mannich multicomponent reaction has been utilized for the rapid generation of libraries of biologically promising β-lactams with nonproteinogenic α-amino acids (Scheme 4.3), offering advantages such as easy automatization, parallelization, and purification.[18]

4.3 FIVE-MEMBERED HETEROCYCLIC RING

Peptidomimetics typically are small protein-like molecules designed to mimic natural peptides or proteins with more pharmacological activity than their peptide role models. The replacement of amide bonds by 1,5-disubstituted tetrazoles as surrogate for the *cis* amide bonds displays an important and general tool in design of peptidomimetics.[19–26]

Entry	Product	Solvent	Reaction time	R^1	R^2	d.r. (a:b)	Isolated yield (%)
1	4	MeOH	24	^{t}Bu	^{t}Bu	100:0	75:—
2	4	H_2O	12	^{t}Bu	^{t}Bu	100:0	51:—
3	4	—	6	^{t}Bu	^{t}Bu	100:0	53:—
4	5	MeOH	24	^{t}Bu	CH_2Ph	88:12	52:7
5	5	H_2O	12	^{t}Bu	CH_2Ph	87:13	57:9
6	5	—	6	^{t}Bu	CH_2Ph	87:13	44:4
7	6	MeOH	48	Ph	^{t}Bu	87:13	53:9
8	6	H_2O	12	Ph	^{t}Bu	74:26	7:2
9	6	—	6	Ph	^{t}Bu	—	—
10	7	MeOH	48	Ph	CH_2Ph	90:10	54:6
11	7	H_2O	12	Ph	CH_2Ph	—	—
12	7	—	6	Ph	CH_2Ph	—	—
13	8	MeOH	48	Et	^{t}Bu	74:26	74[a]
14	8	H_2O	12	Et	^{t}Bu	67:33	82[a]
15	8	—	6	Et	^{t}Bu	75:25	53[a]

[a] Combined yield of the minor and major products (separation of the isomers was not successful).

SCHEME 4.1 Synthesis of conformationally constrained tricyclic β-lactam enantiomers through U4CRs of a monoterpene-based β-amino acid.

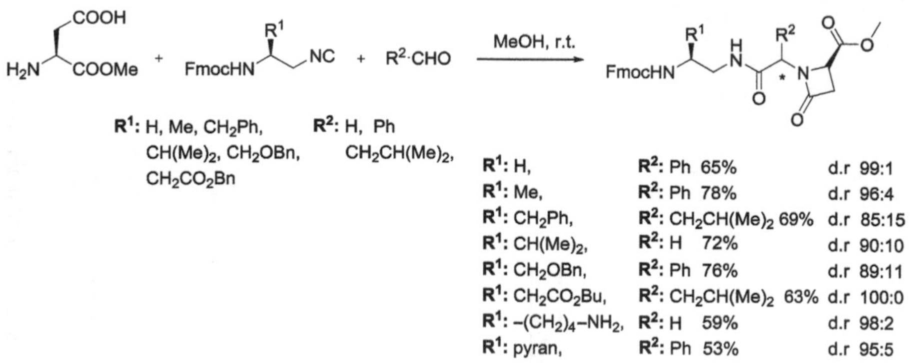

MeOH, r.t.

R^1: H, Me, CH_2Ph, $CH(Me)_2$, CH_2OBn, CH_2CO_2Bn

R^2: H, Ph, $CH_2CH(Me)_2$,

R^1	R^2	Yield	d.r
R^1: H,	R^2: Ph	65%	d.r 99:1
R^1: Me,	R^2: Ph	78%	d.r 96:4
R^1: CH_2Ph,	R^2: $CH_2CH(Me)_2$	69%	d.r 85:15
R^1: $CH(Me)_2$,	R^2: H	72%	d.r 90:10
R^1: CH_2OBn,	R^2: Ph	76%	d.r 89:11
R^1: CH_2CO_2Bu,	R^2: $CH_2CH(Me)_2$	63%	d.r 100:0
R^1: $-(CH_2)_4-NH_2$,	R^2: H	59%	d.r 98:2
R^1: pyran,	R^2: Ph	53%	d.r 95:5

SCHEME 4.2 Synthesis of β-lactam peptidomimetics through Ugi MCR.

In this concept, and by using *N*-Boc-α-amino aldehydes as the aldehyde partners in the $TMSN_3$-modified Passerini three-component reaction followed by deprotection and *N*-capping with TFP (tetrafluoro phenyl) esters,[27–29] *cis*-constrained norstatine mimetic libraries with three diversity points have been synthesized (Scheme 4.4).[30] The TMS ether-1 side product was observed in up to 40% of the crude mixture by high-performance liquid chromatography (HPLC) (UV 215 nm).

In continuation of this approach, a PADAM (Passerini reaction–amine deprotection–acyl migration) reaction sequence followed by a TFA-mediated microwave-assisted cyclization utilizing *ortho*-*N*-Boc-phenylisonitrile, *N*-Boc-α-aminoaldehydes, and supporting carboxylic acids has been introduced to generate the benzimidazole isosteres of the norstatine scaffold in moderate to good yields (Scheme 4.5).[31] The increased $pK_a \sim 5.2$ of the benzimidazole isosteres dramatically alters the physicochemical properties of the molecules.[32–35]

The Boc-protected chiral α-amino aldehydes can also be considered as a variant of the three-component Gewald reaction in the presence of cyanoacetamides and sulfur for the synthesis of arrays of rendering of 2-aminothiophene-3-carboxamides.[36] This chemistry is potentially useful for the synthesis of peptidomimetics with potential biological activity owing to their thiophene moieties, which are used in the pharmaceutical industry to produce bioactive compounds (Scheme 4.6).[37–43]

Further progress in the development of MCRs involving chiral α-amino aldehydes resulted in the synthesis of influenza neuramidase inhibitor A-315675 featuring a pyrrolidine scaffold[44–46] in one pot via an *exo*-selective asymmetric [C+NC+CC] coupling reaction (Scheme 4.7).[47]

In addition, isonitriles of the type α-isocyano ester obtained by converting the amino group of amino acid ester into an isocyano group also play an important role in stereoselective heterocyclic synthesis (Scheme 4.8). One of the most prominent examples in this case is the synthesis of chiral 2-imidazolines suited for combinatorial application[48–63] by the three-component condensation between amines, aldehydes, and α-acidic isocyanides in a one-pot reaction under mild conditions (Scheme 4.9).[64]

Notably, solvent has no effect on the antidiastereomer **a** favorability.

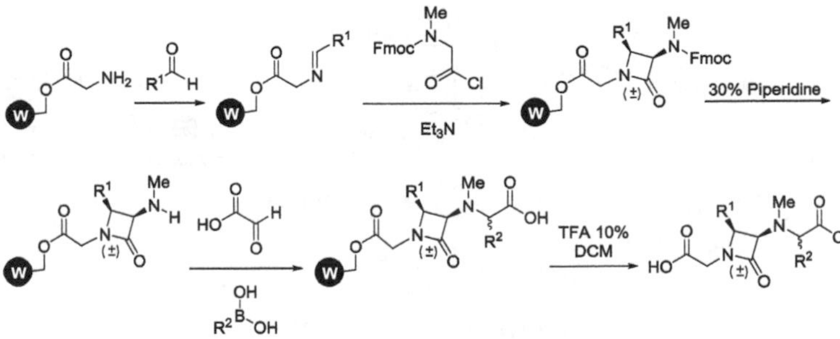

Entry	R^1	R^2	Purity (%)[a]	Yield (%)[b]	d.r[a]
1	Ph	Ph	95	>95	3.4:1
2	4-MePh	Ph	100	>95	2.6:1
3	4-MeOPh	Ph	100	53	5.5:1
4	(*E*)-PhCH=CH	Ph	85	>95	–[c]
5	3,4-$(MeO)_2$Ph	Ph	100	70	6:1
6	4-Br-Ph	Ph	61	74	2:1
7	Ph	(*E*)-PhCH=CH	74	86	10:1
8	4-Me-Ph	(*E*)-PhCH=CH	100	>95	7.5:1
9	4-MeO-Ph	(*E*)-PhCH=CH	75	45	8:1
10	(*E*)-PhCH=CH	(*E*)-PhCH=CH	100	>95	6.2:1
11	3,4-$(MeO)_2$Ph	(*E*)-PhCH=CH	100	65	3:1
12	4-Br-Ph	(*E*)-PhCH=CH	84	67	14:1
13	4-Me-Ph	(*E*)-Me$(CH_2)_5$CH=CH	100	>95	5.3:1
14	4-MeO-Ph	(*E*)-Me$(CH_2)_5$CH=CH	100	55	6:1
15	(*E*)-PhCH=CH	(*E*)-Me$(CH_2)_5$CH=CH	100	>95	3.6:1
16	3,4-$(MeO)_2$Ph	(*E*)-Me$(CH_2)_5$CH=CH	99	57	4:1
17	4-Br-Ph	(*E*)-Me$(CH_2)_5$CH=CH	84	92	2:1
18	4-MeO-Ph	4-MeO-Ph	49	47	9:1
19	3,4-$(MeO)_2$Ph	4-MeO-Ph	60	30	2.6:1
20	4-Br-Ph	4-MeO-Ph	32	54	5:1
21	Ph	6-Methoxynaphthalen-2-yl	52	46	3.3:1
22	4-MeO-Ph	6-Methoxynaphthalen-2-yl	47	34	4.5:1
23	3,4-$(MeO)_2$Ph	6-Methoxynaphthalen-2-yl	25	61	3:1
24	4-Br-Ph	6-Methoxynaphthalen-2-yl	67	45	3.4:1

[a] Determined by HPLC.

[b] Yields are based on the weight of the crude product and are relative to the initial loading level of the resin.

[c] Only one diastereoisomer was obtained, absolute configuration not determined.

SCHEME 4.3 Solid-phase Petasis multicomponent reaction for the generation of ß-lactams 3-substituted with nonproteinogenic α-amino acids.

In a more recent study, ethyl isocyanoacetate has been used in the one-pot, four-component regioselective synthesis of pyrrolo[1,2-α][1,10]phenanthrolines by 1,3-dipolar cycloaddition in the presence of aldehydes and malononitrile with 1,10-phenanthroline under solvent and catalyst-free conditions investigated as chromogenic and fluorescent sensors for Cu^{2+} ions.[65] These compounds have been of

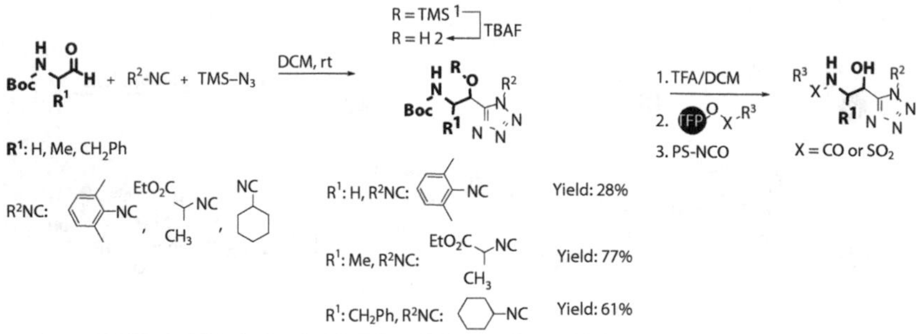

R = TMS 1; R = H 2 (TBAF)

Boc-NH-CH(R^1)-CHO + R^2-NC + TMS–N_3 →(DCM, rt) 1 → 1. TFA/DCM; 2. TFP-O-X-R^3; 3. PS-NCO → X = CO or SO_2

R^1: H, Me, CH_2Ph

R^2NC:

R^1: H, R^2NC: Yield: 28%

R^1: Me, R^2NC: EtO_2C, NC, CH_3 Yield: 77%

R^1: CH_2Ph, R^2NC: NC Yield: 61%

Entry	R^1	R^2	R^3-X	A%[a]	A%[b]	MH^+
1	CH_2Ph	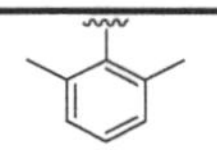	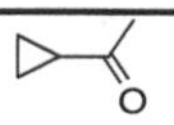	73	83	406
2	CH_2Ph	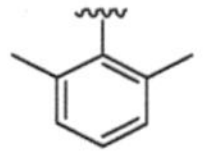	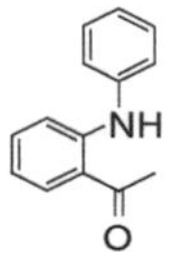 NH, O	71	76	519
3	CH_2Ph	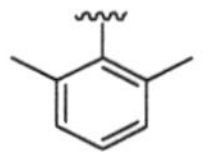	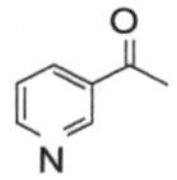 O, N	68	77	429
4	CH_2Ph	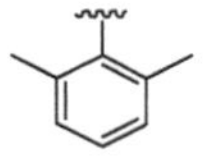	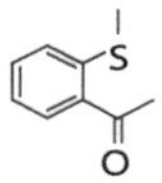 S, O	69	80	474
5	CH_2Ph	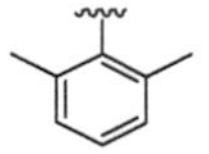	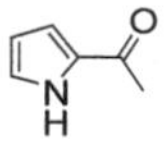 O, NH	66	80	431
6	CH_2Ph	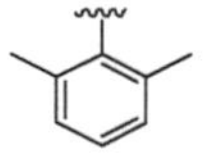	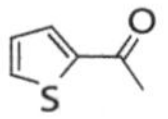 O, S	30	69	422
7	CH_2Ph	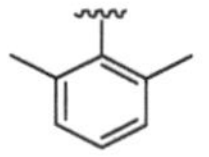	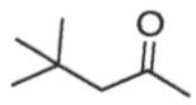 O	36	77	434
8	CH_2Ph	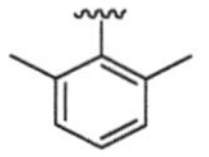	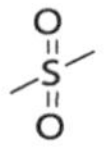 O=S=O	56	84	402
9	CH_2Ph	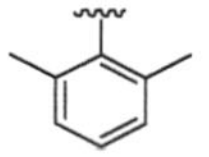	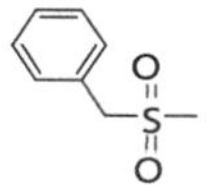 O=S=O	67	79	508
10	CH_2Ph	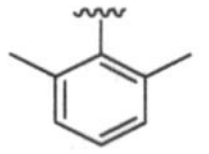	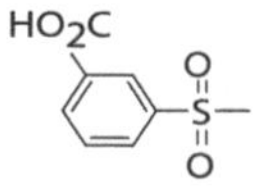 HO_2C, O=S=O	74	83	478

[a] A% by LC/MS.

[b] A% following PS-NCO scavenging.

SCHEME 4.4 Synthesis of *cis*-constrained norstatine analogs using a $TMSN_3$-modified Passerini MCC/N-capping strategy.

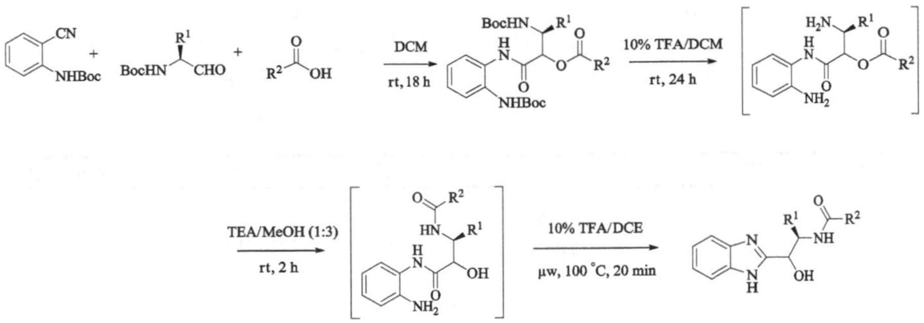

R^1	R^2	Yield (%)
Ph	$CH(Me)_2$	84%, 35%
$CH_3CH_2CH_2$	$CH(Me)_2$	67%, 38%
p-OMe-Ph	$CH(Me)_2$	63%, 41%
p-Br-Ph	$CH(Me)_2$	70%, 27%
α-naphtole	Me	76%, 23%
2-pyridine	Me	56%, 23%
5-indole	Me	66%, 26%
3-Me-benzyle	Me	63%, 25%
α-naphtole	CH_2Ph	70%, 38%
2-pyridine	CH_2Ph	41%, 38%
5-indole	CH_2Ph	47%, 21%
3-Me-benzyle	CH_2Ph	61%, 29%

x% = Passerini yield, x% = yield from products.

SCHEME 4.5 Synthesis of novel norstatine analogs via PADAM cyclization methodology.

great interest because of their importance in biological and environmental processes (Scheme 4.10).[66–77]

The first step is believed to be the Knoevenagel condensation between the aldehyde and malononitrile to generate 2-arylidenemalononitrile, which reacts with ethyl isocyanoacetate to give a highly reactive zwitterion intermediate, followed by its trapping with 1,10-phenanthroline to afford the intermediate **II**. Next **III** and **IV** were obtained by *N*-cyclization and aromatization with loss of hydrogen cyanide (HCN). Finally, the desired products were received (Scheme 4.11).

In another approach, cascade 1,3-dipolar cycloaddition/Michael addition reaction of aryl diazoesters derived from α-amino acid esters with β-nitrostyrenes and cinnamaldehydes led to the synthesis of fully substituted tetrahydrofurans in moderate to high yields with high diastereoselectivity (Scheme 4.12).[78]

4.4 SIX-MEMBERED HETEROCYCLES

Nonproteinogenic heterocyclic amino acids possessing a side carbon chain with a chiral glycinyl moiety are of prime interest in medicinal chemistry due to having

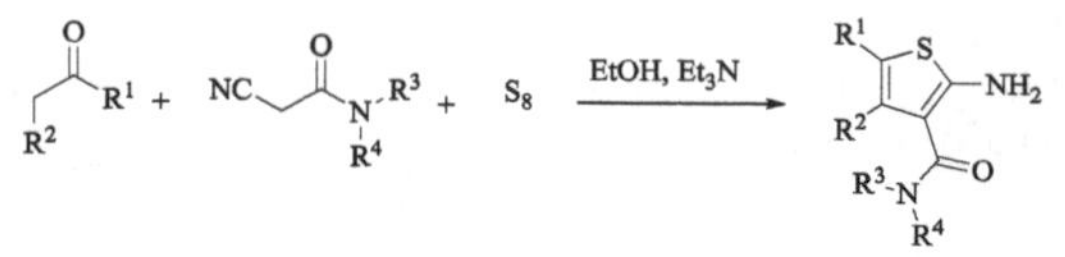

Entry	R^1	R^2	R^3	R^4	Yield (%)
1	H	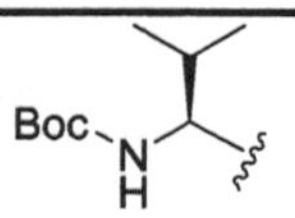 Boc-NH-	H	$CH(Me)_2CO_2Me$	84
2	H	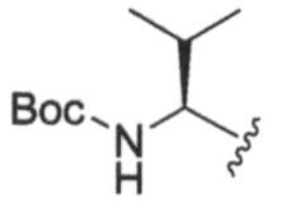 Boc-NH-	H	CH_2CH_2OH	86
3	H	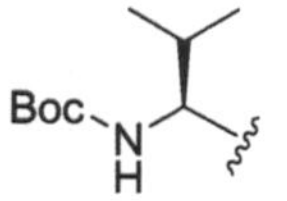 Boc-NH-	H	$CH_2CH_2NMe_2$	71
4	H	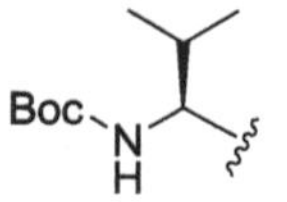 Boc-NH-	H	CH(Me)(Ph)	90
5	H	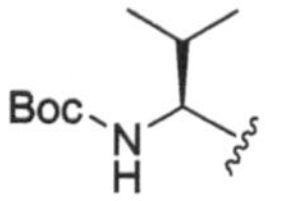 Boc-NH-	H	Morpholine	75
6	H	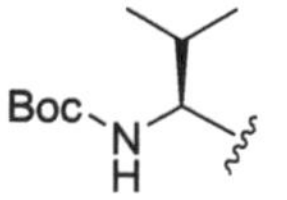 Boc-NH-	H	3-CH_2CH_2-Indole	84
7	H	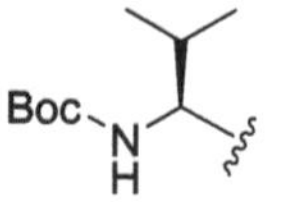 Boc-NH-	H	Cyclopropan	90
8	H	Ph-	H	$CH(Me)_2CO_2Me$	78

SCHEME 4.6 Synthesis of arrays of 2-aminothiophene-3-carboxamides via a Gewald-3CR variation.

a wide range of biological activities.[79] In this respect, intensive research has been focused on the design and synthesis of nonproteinogenic heterocyclic amino acids, especially pyridyl-α-alanines and ring substituted derivatives; these have been proved to function as antagonists of phenylalanines and inhibitors of histidine decarboxylase and possess anti-inflammatory, antitumor and antibiotic activities.[80–84]

One of the most recent research methods on this concept is the synthesis of enantiopure highly functionalized β-(2-pyridyl)- and β-(4-pyridyl)alanines and

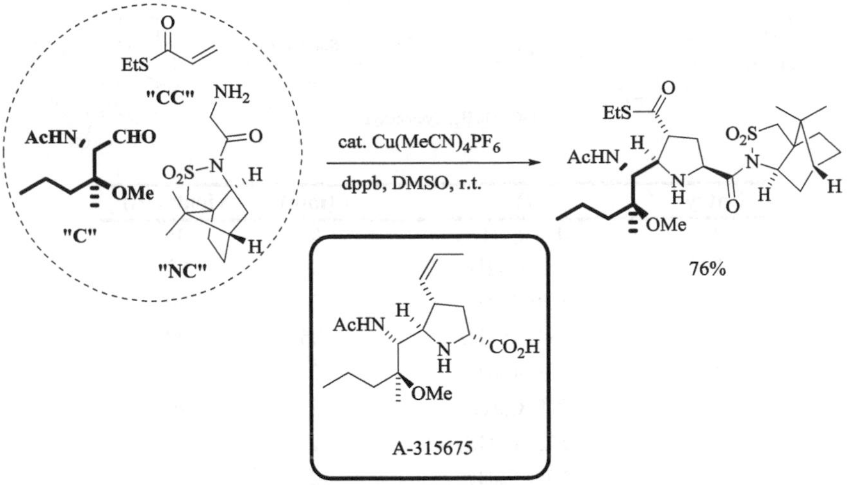

SCHEME 4.7 Synthesis of influenza neuramidase inhibitor A-315675 via an exoselective asymmetric [C+NC+CC] coupling reaction.

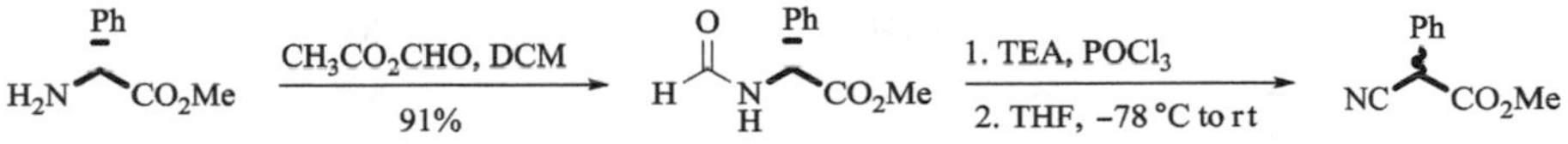

SCHEME 4.8 Synthesis of 2-phenylisocyanoacetate.

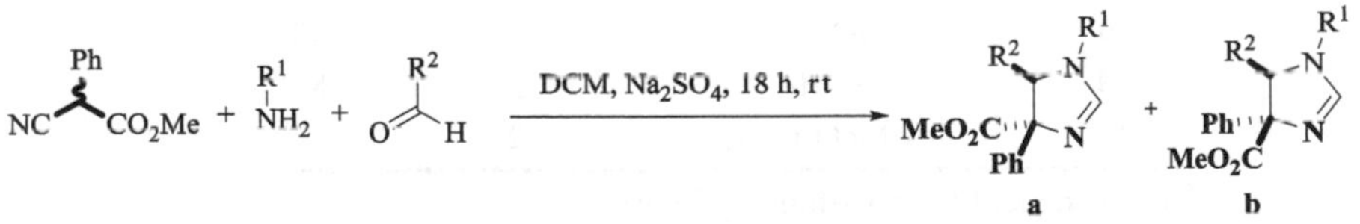

R^1	R^2	Yield	d.r
$CHPh_2$	$CH(CH_3)_2$	13	66:34
PMB	$CH(CH_3)_2$	73	76:24
Ph	$CH(CH_3)_2$	53	70:30
PMB	*p*-OMe-Ph	77	70:30
$CH(CH_3)_2$	*p*-OMe-Ph	90	55:45
$CH(CH_3)_2$	2-Me-furan	81	76:24
PMB	2-Me-furan	91	70:30
$CH(CH_3)_2$	2-pyridin	71	84:16
CH_2-furan	$-CO_2Et$	47	67:33

SCHEME 4.9 Synthesis of 2-imidazolines via a novel multicomponent reaction between an amine, an aldehyde, and an α-acidic isocyanide.

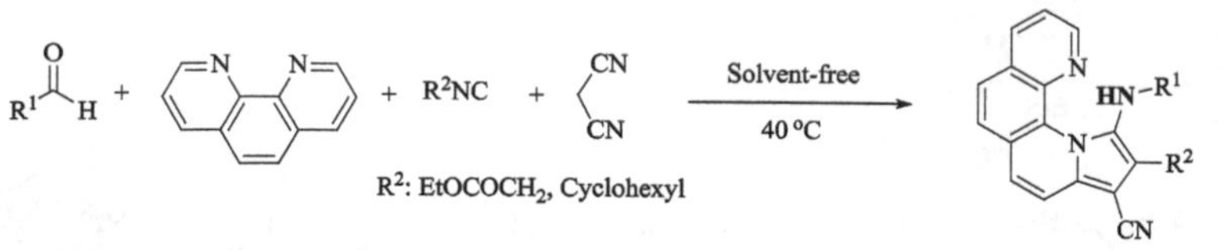

R^2: $EtOCOCH_2$, Cyclohexyl

Entry	R^1	t (min)	Yield[a] (%)
1	4-F-C_6H_4	2	95
2	4-Cl-C_6H_4	2	93
3	4-Br-C_6H_4	2	90
4	4-NO_2-C_6H_4	2.5	90
5	3-Br-C_6H_4	2	88
6	2-F-C_6H_4	2	90
7	2-Cl-C_6H_4	2	90
8	2-Br-C_6H_4	2	86
9	2-NO_2-C_6H_4	2	88
10	C_6H_5	2.5	85
11	4-Me-C_6H_4	3	84
12	4-MeO-C_6H_4	3	82
13	2-Me-C_6H_4	3	83
14	4-OH-C_6H_4	3	80[b]
15	2-furan	3	90
16	2-Cl-C_6H_4	2	93
17	4-Cl-C_6H_4	2	94
18	4-Br-C_6H_4	2	92
19	4-NO_2-C_6H_4	2	90
20	C_6H_5	2	89
21	4-MeO-C_6H_4	2	87

[a] Isolated yield after washing by EtOH.

[b] Yield after purification via silica gel column chromatography.

SCHEME 4.10 Direct solvent-free regioselective construction of pyrrolo[1,2-α] [1,10] phenanthrolines.

the corresponding 1,4-dihydro and *N*-oxide derivatives through one-pot thermal Hantzsch-type cyclocondensation of aldehyde–ketoester–enamine systems in which one of the reagents (aldehyde or ketoester) is carrying the chiral glycinyl moiety in a suitably protected form (Scheme 4.13).[85]

The use of polymer supported reagents and sequestrants leads to high-yield isolation of the products in high purity without any chromatography (Scheme 4.14).

1,8-Naphthyridines and quinolines attracted considerable attention from both synthetic and medicinal chemists due to their wide range of biological properties, such as anticancer, anti-inflammatory, antibacterial, antimycobacterial, and anticonvulsant

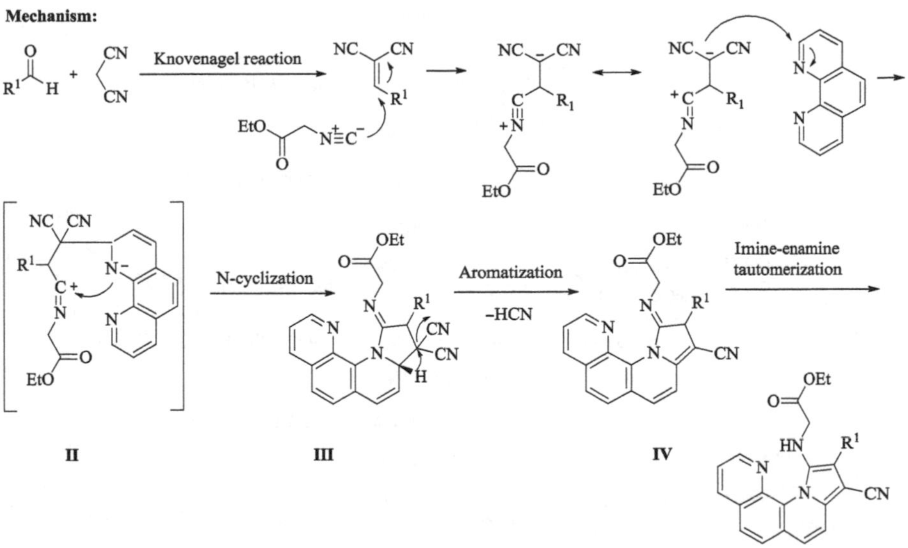

Mechanism:

SCHEME 4.11 Plausible mechanism for the synthesis of pyrrolo[1,2-α][1,10]phenanthrolines.

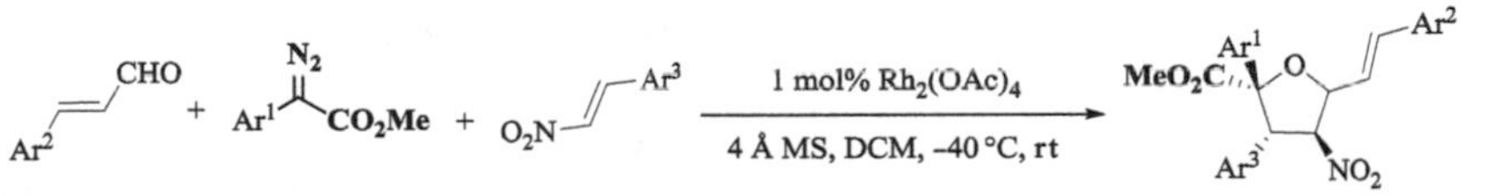

Entry	Ar^1	Ar^2	Ar^3	d.r.[b]	Yield (%)[c]
1	Ph	Ph	Ph	90:10	75
2	4-$MeOC_6H_4$	Ph	Ph	>95:5	90
3	4-BrC_6H_4	Ph	Ph	93:7	71
4	4-FC_6H_4	Ph	Ph	>95:5	65
5	4-ClC_6H_4	Ph	Ph	>95:5	76
6	Ph	Ph	2-$MeOC_6H_4$	>95:5	48
7	Ph	Ph	3-ClC_6H_4	>95:5	70
8	Ph	Ph	2-BrC_6H_4	83:17	43
9	Ph	Ph	4-BrC_6H_4	88:12	58
10	Ph	Ph	4-$NO_2C_6H_4$	>95:5	45
11	Ph	Ph	2-furyl	92:8	71
12	Ph	2-furyl	Ph	93:7	40
13	Ph	4-$MeOC_6H_4$	Ph	>95:5	45

[a] Unless otherwise noted, the reaction was carried out on a 0.2 mmol scale in DCM (2 mL) in the presence of $Rh_2(OAc)_4$ (1 mol%), and the ratio of 1/2/3 was 2.0/1.5/1.0.
[b] Determined by ^1H NMR spectroscopy of the reaction mixtures.
[c] Isolated yield of product.

SCHEME 4.12 Stereoselective synthesis of fully substituted tetrahydrofurans through 1,3-dipolar cycloaddition with cinnamaldehydes.

properties.[86,87] In parallel with the increasing interest in the synthesis of biologically potent 1,8-naphthyridine and quinoline derivatives, numerous approaches to their synthesis have been described.[88–100]

An elegant approach to the synthesis of 2-aryl-4-amino quinolines and 2-aryl-4-amino[1,8]-naphthyridines involves palladium-mediated multicomponent domino

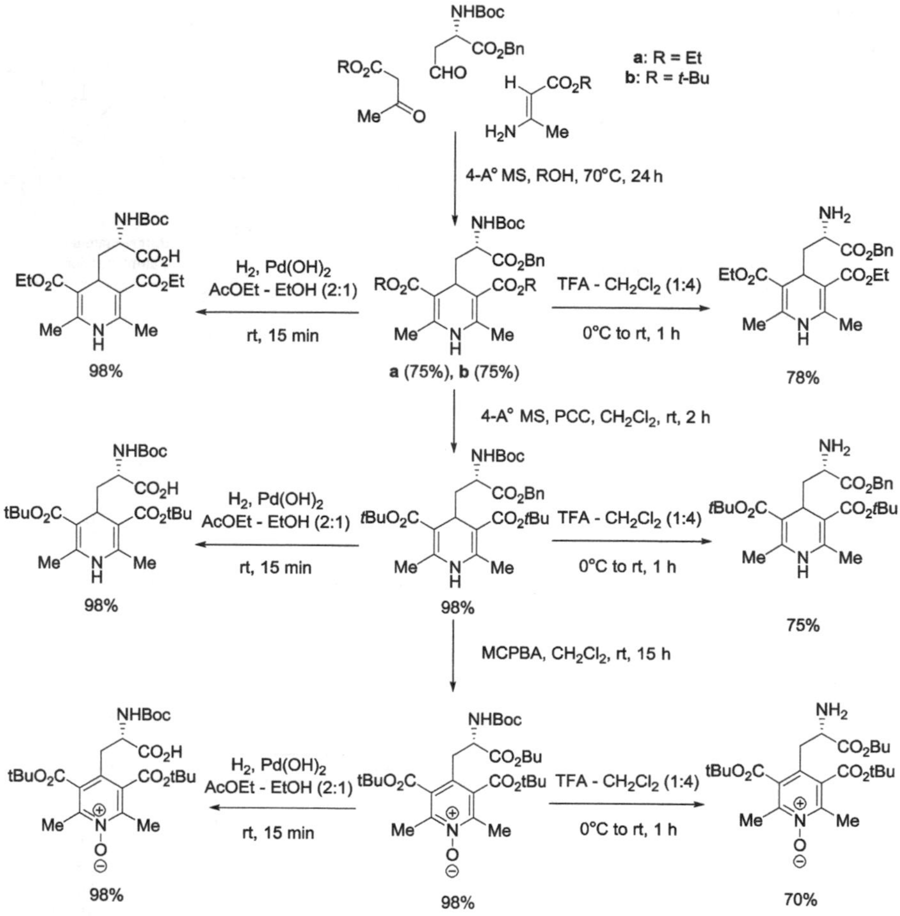

SCHEME 4.13 Synthesis of highly functionalized 2- and 4-dihydropyridylalanines, 2- and 4-pyridylalanines, and their *N*-oxides by one-pot thermal Hantzsch-type cyclocondensation.

reaction of 2-ethynylarylamines, aryl iodides, α-amino esters, and carbon monoxide (Scheme 4.15).[101]

The proposed reaction mechanism involves two catalytic cycles. Both catalytic cycles entail the oxidative addition of Pd (0) to the aryl iodide followed by coordination of carbon monoxide to give the arylcarbonyl palladium species. 1,2-Aryl migration from palladium to coordinated CO generates the aroyl palladium complex ArCOPdI. Subsequent addition of the incipient acetylide anion gives rise to a δ-alkynyl-δ-acylorganopalladium complex ArCCPdCOAr, whereas reductive elimination is the final step. The domino process of inter-/intramolecular nucleophilic attack between **I** and **II**, followed by elimination of water, leads to the formation of the desired products (Scheme 4.16).

Owing to their prolific bioactivity, diketopiperazine derivatives (DKPs)—the smallest cyclic peptides known—that are featured in natural products, bioactive compounds,

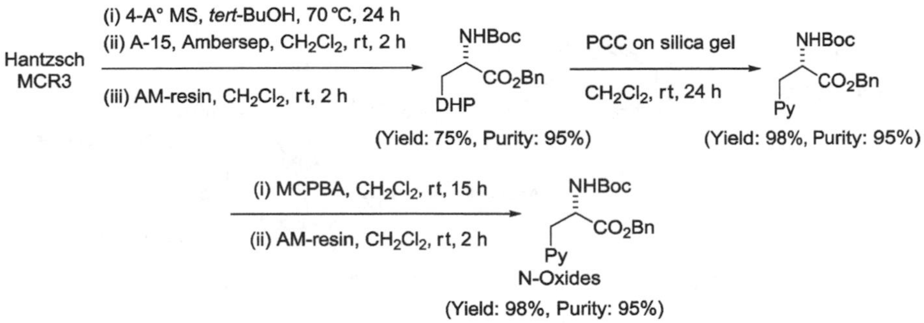

SCHEME 4.14 Synthesis of highly functionalized 2- and 4-dihydropyridylalanines, 2- and 4-pyridylalanines, and their *N*-oxides via a polymer-assisted solution-phase approach.

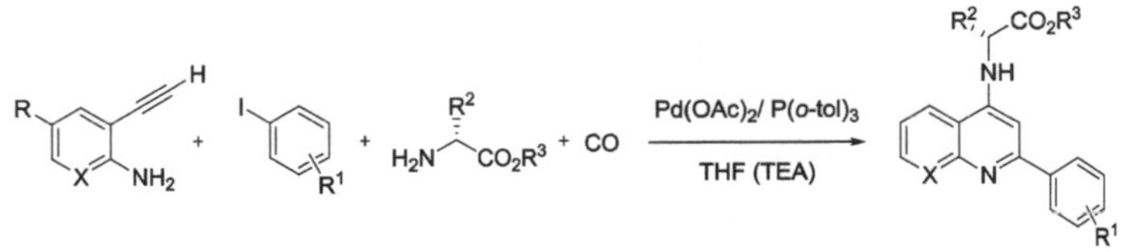

Entry	R	R¹	R²	R³	Yields (%)
1	H	Cl	H	Et	42
2	H	Cl	Ph	Et	59
3	H	COMe	$CH(Me)_2$	Et	57
4	H	Cl	$BuO_2CNH(CH_2CH_2CH_2CH_2)$	Et	48
5	Me	H	H	Et	64

Reaction performed in THF (4 mL) and in the presence of 15 equiv of triethylamine using the following molar ratios: **13/14/15**/Pd catalyst/phosphine ligand = 1:1:3:0.05:0.07.

SCHEME 4.15 Palladium-assisted multicomponent synthesis of 2-aryl-4-aminoquinolines and 2-aryl-4-amino[1,8]naphthyridines.

and in several therapeutic agents have emerged as promising biologically active scaffolds.[102,103] An early synthesis of these systems involved a SPOT-synthesis approach based on an Ugi deprotection cyclization (UDC) strategy introducing six DKPs capable of inhibiting luminescence by at least 80% at 500 μM.[104] *N*-Boc-protected amino acid components lead to cyclative cleavage (Scheme 4.17). The strength of this process lies in the reliable and easy experimental procedure, inexpensive equipment needs, and a rapid and highly flexible array and library formatting in high purities.

Though the preparation of 2,5-diketopiperazines based on the UDC strategy is an established method, very few reports have described MCR-based methods to access pyrazin-2(1*H*)-ones considered as biologically important active scaffolds in drug discovery.[105–108] One of the most prominent pathways in the synthesis of diverse pyrazin-2(1*H*)-one chemotypes follows the UDC strategy employing glyoxals and *N*-Boc-protected amino acids as key bifunctional precursors (Scheme 4.18).[109]

The use of enantiopure *N*-Boc amino acids successfully afforded enantiomerically pure 3,4-dihydropyrazin-2(1*H*)-ones without racemization at the stereocenter

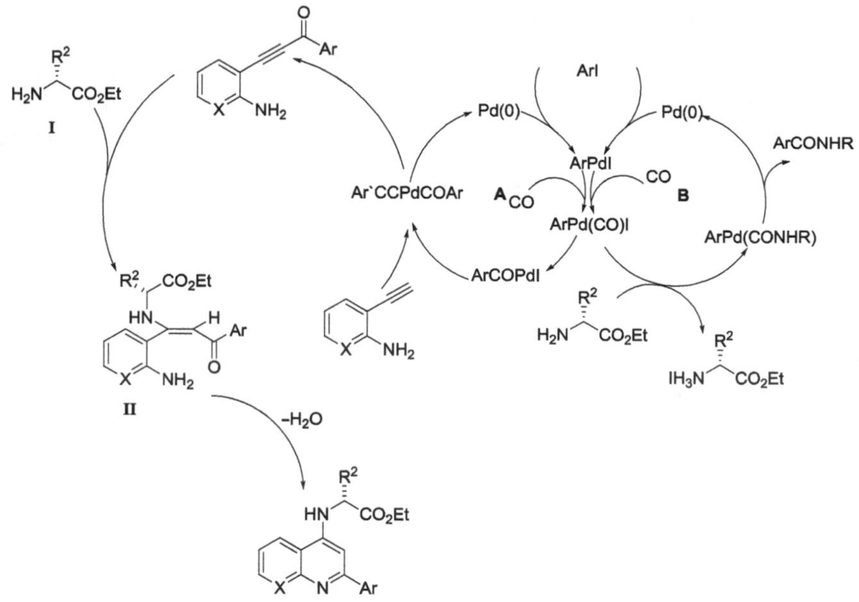

SCHEME 4.16 Plausible mechanism for the synthesis of 2-aryl-4-aminoquinolines.

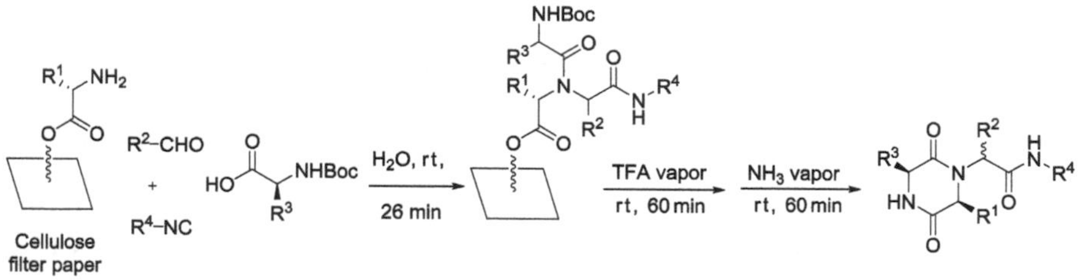

R^1	R^2	R^3	R^4
H	CH_2CH_2Ph	$CH_2CH(Me)_2$	nC_5H_{11}
$CH_2CH(Me)_2$	CH_2CH_2Ph	H	nC_5H_{11}
CH_2Ph	CH_2OCH_2Ph	H	CH_2Ph
H	CH_2OCH_2Ph	CH_2Ph	CH_2Ph
$CH_2CH(Me)_2$	CH_2OCH_2Ph	H	CH_2Ph
H	CH_2OCH_2Ph	$CH_2CH(Me)_2$	CH_2Ph

SCHEME 4.17 Efficient construction of diketopiperazine macroarrays via general UDC strategy on ester-bound amine-functionalized cellulose support.

(*ee* > 95%), as determined by chiral HPLC analysis. Two remarkable features that appear here are the outstanding bond-forming efficiency and step economy with operational simplicity in a time- and cost-effective manner. Furthermore, for assembly of 3,3-disubstituted-3,4-dihydropyrazin-2(1*H*)-ones and 4-substituted-pyrazin-2,3-diones, glycine Boc derivatives were submitted to U-4CR conditions and subsequently deprotected (Scheme 4.19).

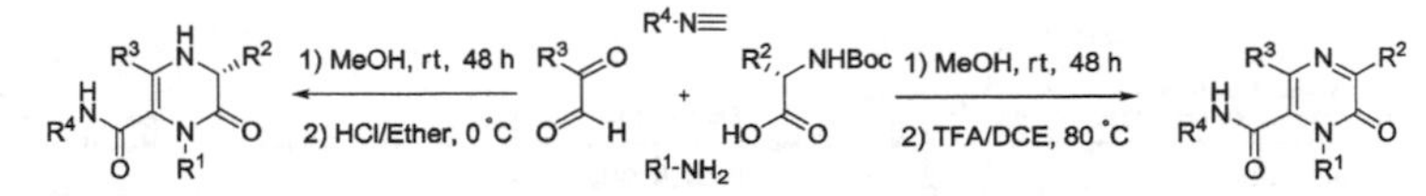

R^1	R^2	R^3	R^4	Yields (%)
$CH_2CH_2CH_2$-morpholine	H	Ph	^{t}Bu	41
$-CH_2CH_2CH_2NMe_2$	H	Ph	^{t}Bu	42
CH_2Ph	4-MeO-$PhCH_2$	4-MeO-Ph	Cyclohexyle	43
4-MeO-Ph	Ph	Ph	$PhCH_2$	62
$CH_3CH_2CH_2$	Ph	Ph	Cyclohexylamine	57
Me	Ph	Ph	$PhCH_2$	56
$PhCH_2$	$CH_2CH_2CO_2CH_2Ph$	Ph	^{t}Bu	48
$PhCH_2$	Me	Ph	^{t}Bu	58
$PhCH_2$	$CH_2CH(Me)_2$	Ph	^{t}Bu	66
$PhCH_2$	CH_2CH_2SMe	Ph	^{t}Bu	52
$CH_3CH_2CH_2$	$PhCH_2$	Ph	$PhCH_2$	74
$PhCH_2$	Me	tBu	$PhCH_2$	57
$PhCH_2$	CH_2imidazole	Ph	^{t}Bu	32
$PhCH_2$	CH_2indole	Ph	^{t}Bu	38
$CH_3CH_2CH_2$	$CH(Me)_2$	2-furan	$PhCH_2$	62
$PhCH_2$	$CH(Me)_2$	Ph	^{t}Bu	72

SCHEME 4.18 Ugi-based assembly of pyrazin-2(1*H*)-ones and their 3,4-dihydro-analogs.

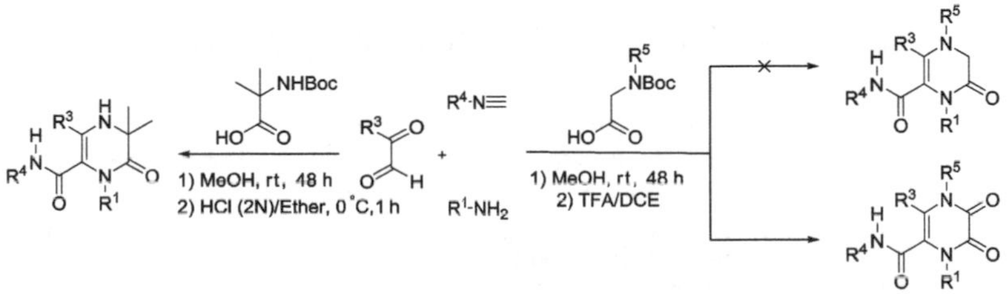

Entry	R^1	R^3	R^4	R^5	Yields (%)
1	Me	4-MeO-Ph	*t*Bu	H	56
2	CH_2Ph	Cyclopropane	CH_2Ph	H	58
3	$CH_2CH_2CH_3$	Ph	CH_2Ph	Me	74
4	$CH_2CH_2CH_3$	Ph	*t*Bu	Me	67
5	$CH_2CH_2CH_3$	Ph	CH_2Ph	Et	64
6	$CH_2CH_2CH_3$	Ph	*t*Bu	Et	70

SCHEME 4.19 Ugi-based assembly of 3,3-disubstituted-3,4-dihydropyrazin-2(1*H*)-ones and 4-substituted-pyrazin-2,3-diones.

The occurrence of an oxidative pathway mediated by highly reactive 2(1*H*)pyrazinoium salts leads to the first direct preparative method for the straightforward assembly of diversely decorated pyrazin-2,3-(1*H*,4*H*)-diones.

On the same line of reasoning, the employment of α-amino aldehydes and α-isocyanoacetates in a sequential Ugi reaction/cyclization two-step strategy leads to the synthesis of three structurally distinct piperazine-based scaffolds bearing appropriate functionalities to be easily applied in peptide chemistry (Scheme 4.20).[110]

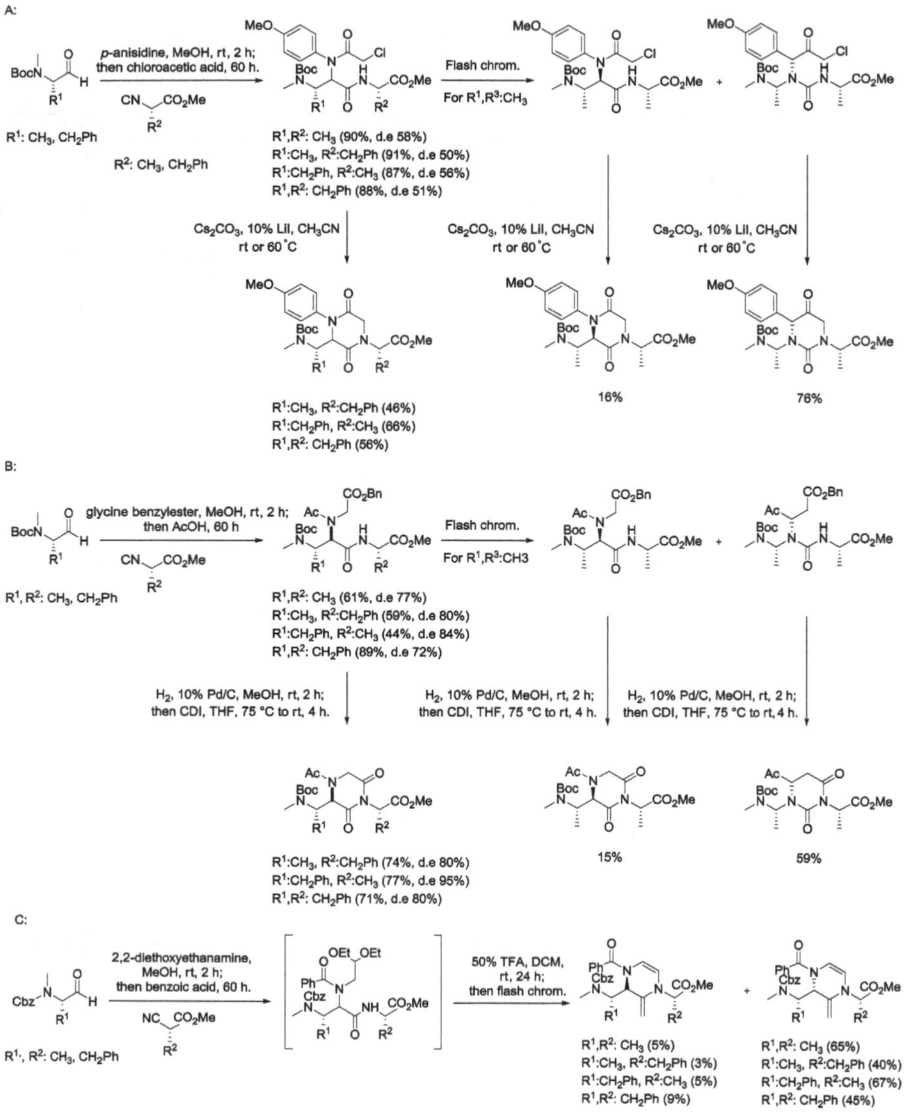

SCHEME 4.20 Synthesis of three structurally distinct piperazine-based minimalist peptidomimetics via sequential Ugi reaction/cyclization two-step strategy.

At the first approach, the use of chloroacetic acid and *p*-anisidine provides a small family of 2,5-diketopiperazine-based peptidomimetics in good yields but as inseparable diastereoisomeric mixtures—with one exception, which could be separated by flash chromatography. In the second approach, the use of glycine benzylester and acetic acid results in access to 2,6-diketopiperazine-based peptidomimetics in good yields (71%–77%) and high d.e. (80%–95%, from ^{1}H NMR). Finally, employing bifunctional amine was applied to the synthesis of 3,4-dihydropyrazin-2(1*H*)-one-based peptidomimetics with high diastereoisomeric ratios, up to >93:7.

4.5 SEVEN-MEMBERED HETEROCYCLIC RING

Alpha-amino acid derivatives have been demonstrated to be suitable bifunctional starting materials for multicomponent reactions producing novel biologically active medium ring-sized heterocycles. Among all the medium ring-sized heterocyclic compounds, 1,4-benzodiazepines (BDZs) are one of the most important heterocycles exhibiting remarkable pharmacological activities.[111–121] Some drugs and bioactive compounds derived from benzodiazepin scaffolds are demonstrated in Figure 4.1.

Due to the influence of conformation of the 1,4-diazepine ring and its substituents, the propensity of hydrogen bond donor and acceptor, and the electrostatic profile on its biological activity, synthetic methodologies targeting this structural platform bearing new substituents have attracted considerable attention.

In this respect, a two-step, one-pot synthesis employing UDC strategy between 2-aminophenylketones as amine inputs, *N*-Boc-α-amino acids, isocyanides, and a carbonyl partner has been utilized to obtain functional, skeletal, and stereochemically diverse 1,4-benzodiazepin-2-ones (Scheme 4.21).[122]

In another approach based on UDC methodology employing α-amino esters, perfluorooctanesulfonyl protected 4-hydroxy benzaldehydes used as the limiting agent

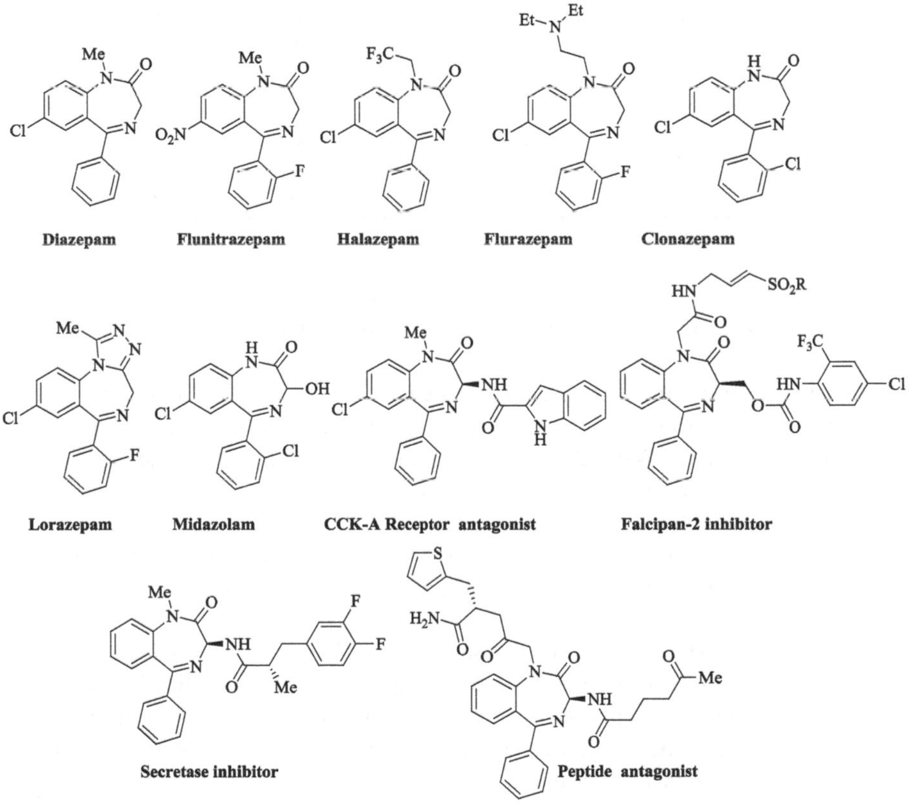

Diazepam Flunitrazepam Halazepam Flurazepam Clonazepam

Lorazepam Midazolam CCK-A Receptor antagonist Falcipan-2 inhibitor

Secretase inhibitor Peptide antagonist

FIGURE 4.1 Representative drugs and bioactive compounds featuring benzodiazepin scaffolds.

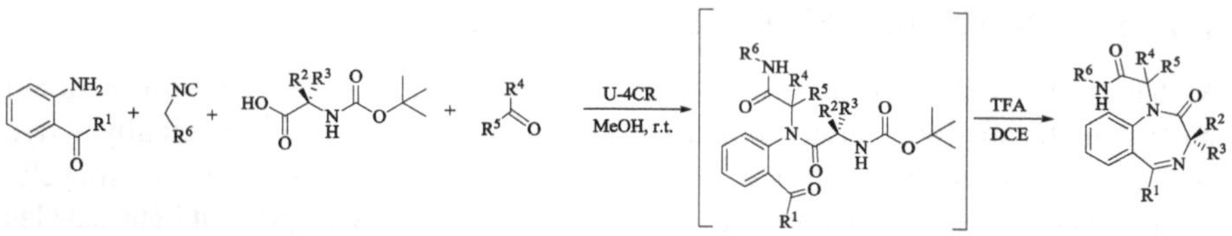

Entry	R^1	R^4	R^5	R^6	Yields (%)
1	Me	H	H	$PhCH_2$	60
2	Me	CH_2CH_2Me	H	$PhCH_2$	52
3	Me	Ph	H	$PhCH_2$	67
4	Me	Me	Me	$PhCH_2$	44
5	Me	Cyc	Cyc	$PhCH_2$	59
6	Ph	H	H	$PhCH_2$	70
7	Ph	CH_2CH_2Me	H	$PhCH_2$	63
8	Ph	Ph	H	$PhCH_2$	62
9	Ph	Me	Me	$PhCH_2$	45
10	Ph	Cyc	Cyc	$PhCH_2$	47
11	Me	H	H	Cyc	62
12	Me	CH_2CH_2Me	H	Cyc	54
13	Me	Ph	H	Cyc	65
14	Me	Me	Me	^{t}Bu	49
15	Me	Ph	H	^{t}Bu	60
16	Ph	H	H	Cyc	52
17	Ph	CH_2CH_2Me	H	Cyc	67
18	Ph	Ph	H	Cyc	56
19	Ph	Me	Me	^{t}Bu	46

Cyc = Cyclohexyl.

SCHEME 4.21 1,4-Benzodiazepin-2-one chemotypes employing UDC strategy between 2-aminophenylketones, *N*-Boc-α-amino acids, isocyanides, and a carbonyl partner.

for the Ugi four-component reaction (U-4CR). Suzuki coupling reactions further derivatized the benzodiazepine ring by removing the fluorous tag and introducing the biaryl functionality.[123] When 2-nitrobenzoic acid replaced Boc-protected anthranilic acids, fluorous-tagged Ugi products were converted to the BDZs by zinc-promoted nitro reductions/cyclizations (Scheme 4.22).

Resin-bound α-amino acids have also served as important building blocks in the UDC methodology, allowing the preparation of highly pure and diverse arrays of 1,4-benzodiazepines (Scheme 4.23).[124]

The effective UDC strategy based on α-amino aldehydes employing anthranilic acid derivatives has also been elaborated and applied for the modification of diverse classes of 1,4-benzodiazepine-6-ones. Serving aminophenylketones as amine components for the Ugi-4CR *under microwave irradiation*, a small, focused library of 1,4-benzodiazepines with four points of diversity was obtained (Scheme 4.24).[125]

These groups conceived novel application of *N*-Boc-α-amino aldehydes as bifunctional starting material for the synthesis of a benzodiazepine scaffold. In addition, they introduced *N*-Boc-α-amino acids as anchor fragments, which can

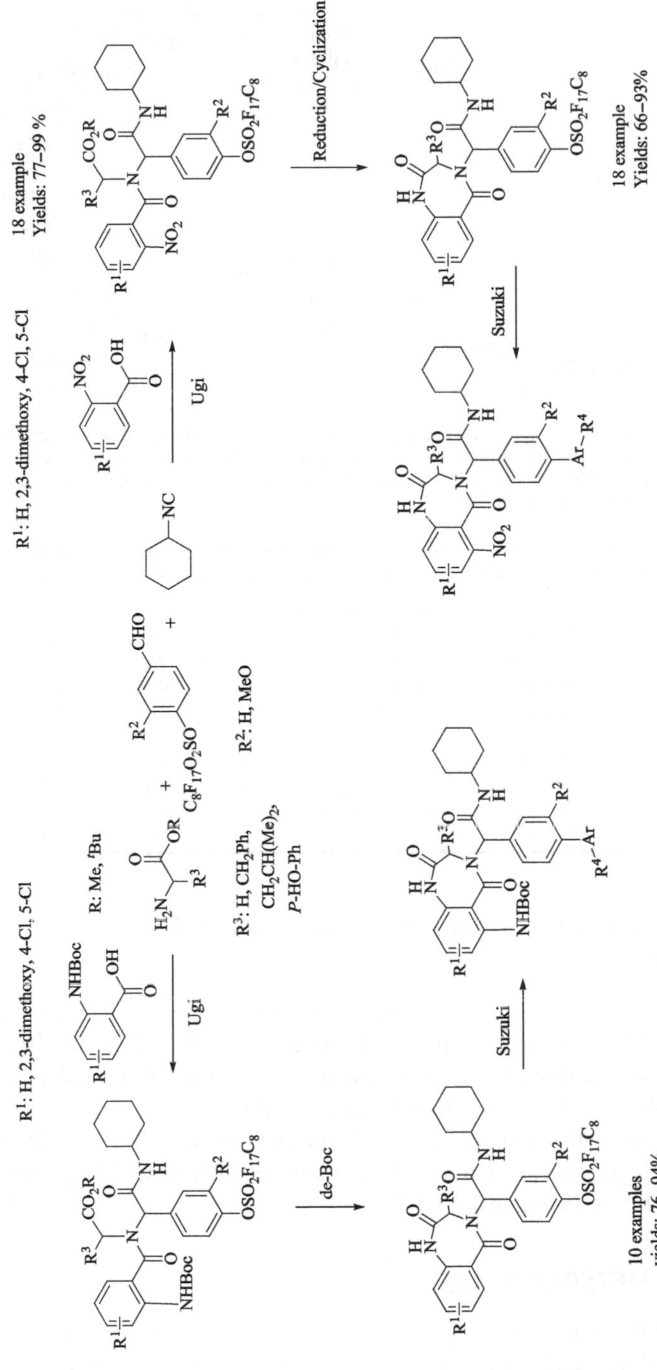

SCHEME 4.22 Microwave-assisted fluorous synthesis of a 1,4-benzodiazepine-2,5-dione library.

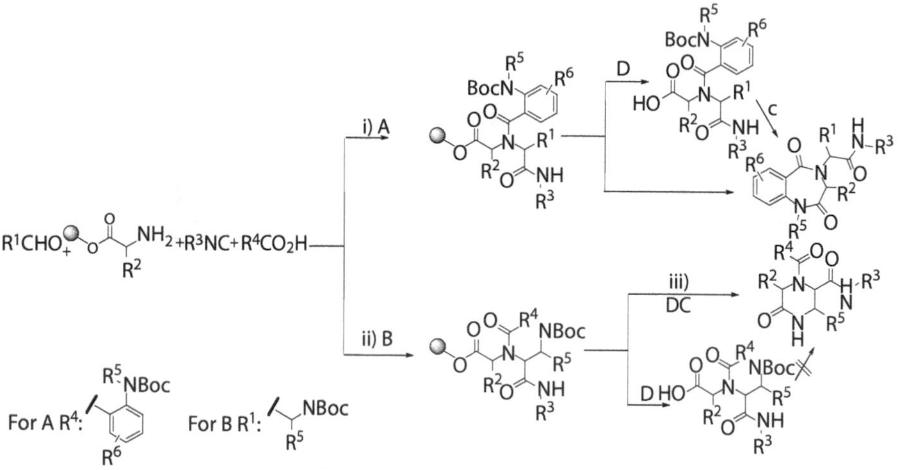

Reagent and conditions: RCHO (3 equiv.), A (3 equiv.), RNC (3 equiv), all 0.5 M solutions (MeOH:CH_2Cl_2, 1:1), **2** Wang Resin, room temperature, 24 h. Wash resin CH_2Cl_2 (3), MeOH (3); (ii) **B** (3 equiv.) R^3NC (3 equiv.), R^4CO_2H (3 equiv.), all 0.5 M solution (MeOH:CH_2Cl_2, 1:1), 2 hydroxymethyl resin. Wash resin CH_2Cl_2 (3), MeOH (3); (iii) 10% TFA in CH_2Cl_2, wash resin CH_2Cl_2(2)

R^1	R^2	R^3	R^5	R^6	ELS (A%)	UV (%)
CH_2CH_2Ph	CH_2Ph	CH_2Ph	H	H	100	100
Ph	4-HO-$C_6H_4CH_2$	CH_2-tetrahydrofuran	H	H	91	78
Et	CH_2Ph	$CH(Me)_2$	Me	H	92	81
2-pyridine	$CH_2CH_2CONH_2$	Cyclohexyl	H	7,8-MeO	80	47
Et	CH_2CO_2H	-$(CH_2)_3CH_3$	H	7-OH	90	80
2-pyridine	$CH_2CH_2CONH_2$	Cyclohexyl	H	H	83	48
2-pipyridine	Ph	CH_2CO_2Me	H	H	100	92
2-tetrahydropyran	CH_2CH_2SMe	-$C(Me)_2CH_2Me_3$	H	H	100	78
3-MeO-Ph	Ph	$CH(Me)_2$	H	Benzen	100	65
CH_2CH_2Ph	$CH_2CH_2CO_2H$	4-morpholin- C_6H_4	H	8-Cl	100	90
$CH_2CH(Me)_2$	CH_2Ph	Cyc	H	H	100	100

Cyc = Cyclohexyl.

SCHEME 4.23 Applications of resin bound-amino acids for the synthesis of benzodiazepines (via Wang resin).

be incorporated into drug-like compounds via multicomponent reactions. Notably, Ugi–azide reaction between α-amino aldehydes, aminophenylketones, isocyanides, and trimethyl azide resulted in the generation of 2-tetrazole substituted 1,4-benzodiazepines with three points of diversity (Scheme 4.25).

A further approach toward access to libraries of benzodiazepines is based on combination of Ugi condensation of *N*-Boc-α-amino aldehydes with a postcondensation S_NAr cyclization (Scheme 4.26).[126]

4.6 FUSED HETEROCYCLES

One major challenge in current organic chemistry is achieving the synthesis of polycyclic structured molecules since the molecules of fused, bridged, or spiro polycyclic structures are found in numerous biologically active natural products.[127] However,

[b]Method A: (i) MeOH, rt, 2 days;
(ii) DCE (10% TFA), rt, 2 days
(iii) THF, Et_3N, triazabicyclodecene (TBD), 40 °C, overnight

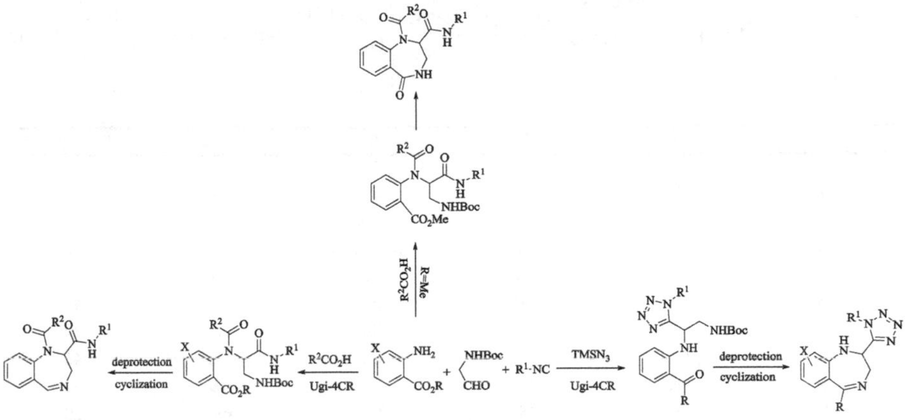

[b]Method B: (i) MeOH, rt, 2 days;
(ii) DCE (10% TFA), 40 °C, overnight.
[c]Method C: (i) MeOH, microwave irradiation (100 °C, 30 min);
(ii) DCE (10 % TFA), 40 °C, overnight

[b]Method D: (i) MeOH, rt, 2 days;
(ii) DCE (10% TFA), 40 °C, overnight.
[c]Method E: (i) MeOH, microwave irradiation (100 °C, 30 min);
(ii) DCE (10 % TFA), 40 °C, overnight

Entry	X	Y	R^1	R^2	Yields (%)
1	–	–	^{t}Bu	Me	41[b]
2	–	–	^{t}Bu	Cyclohexyl	28[b]
3	–	–	mesityl	Me	16[b]
4	–	–	^{t}Bu	^{n}Pr	20[b]
5	–	–	^{t}Bu	cyclopropenyl	38[b]
6	–	–	^{t}Bu	P-F-C_6H_4	22[b]
7	H	Ph	^{t}Bu	Me	47[c]
8	H	Ph	cyclohexyl	Me	36[c]
9	H	Ph	cyclohexyl	Me	43[c]
10	H	Ph	benzyl	Me	40[d]
11	4-Cl	Ph	^{t}Bu	CH_2OH	22[d]
12	4-Cl	Ph	^{t}Bu	^{n}Pr	24[d]
13	4-Cl	Ph	^{t}Bu	cyclobutyl	25[d]
14	4-Cl	Ph	benzyl	Me	13[d]
15	H	Ph	^{t}Bu	–	29[c]
16	H	Ph	cyclohexyl	–	35[c]
17	H	Me	cyclohexyl	–	49[c]
18	4-Cl	Ph	^{t}Bu	–	32[d]
19	4-Cl	o-F-C_6H_4	benzyl	–	12[d]
20	H	Ph	benzyl	–	26[d]

[a] Isolated yields (over two steps).
[b] Method A.
[c] Method B.
[d] Method C.

SCHEME 4.24 Ugi–4CR route to synthesis of 1,4-benzodiazepine-6-ones, 1,4-benzodiazepines, and 2-tetrazole substituted 1,4-benzodiazepines.

Entry	X	R	R1	anchor	Yields (%)[a,b]
1	H	Ph	tBu	Ph	41
2	H	Ph	tBu	Leu	41
3	H	Ph	cyclohexyl	Ph	50
4	H	Ph	cyclohexyl	Leu	69
5	H	Me	tBu	Ph	44
6	H	Me	tBu	Leu	46
7	4-Cl	Ph	tBu	Ph	32
8	4-Cl	Ph	tBu	Leu	43
9	H	Ph	tBu	Trp	26
10	H	Ph	tBu	Tyr	29
11	H	Ph	cyclohexyl	Trp	65
12	H	Ph	cyclohexyl	Tyr	60
13	H	Me	benzyl	Trp	22

[a] Isolated yields (over two steps).
[b] Method F: (i) MeOH, rt, 2 days; (ii) DCE (10% TFA), 40 °C, overnight

SCHEME 4.25 Synthesis of anchor-directed 1,4-benzodiazepines.

synthesis of these target molecules has been found to be rather tough work, mainly due to their inherent structural complexity. Multicomponent reactions based on amino acids and their derivatives have been found to be amazingly efficient for generating libraries of polycyclics.

In this respect, Ugi-azide four-component reaction between *N*-Boc-α-amino aldehydes, secondary amines, methyl isocyanoacetates, and trimethylsilylazide has been disclosed for the preparation of fused azepine-tetrazoles (Scheme 4.27).[128]

In a recent study, fused lactams—popular starting materials for many chemical transformations—have been prepared using a multicomponent reaction between chiral amino alcohols, gaseous reagents (CO and H_2) that are scarcely exemplified, and an unsaturated carboxylic acid in moderate to good yields using an autoclave or under MW irradiations with good control of the diastereoselectivity (Scheme 4.28).[129]

In addition to chiral amino alcohols, chiral α-amino acids have also been studied as possible binucleophilic partners in the proposed MCR providing only the corresponding trans adducts as single diastereomers, indicating a thermodynamic control. Both autoclave and microwave irradiation heating produced similar results (Scheme 4.29).

In a more recent study, glycine amino alcohol has been used for the highly diastereoselective construction of tetrahydropyridine fused bicyclic structures via a three-component domino reaction (Scheme 4.30).[130]

First, the addition of secondary amine DEA to the alkynes led to generation of the reactive enamine intermediate (**I**), which was proposed to undergo transamination with amino alcohols to provide (**II**) *in situ* and release DEA. In the presence of

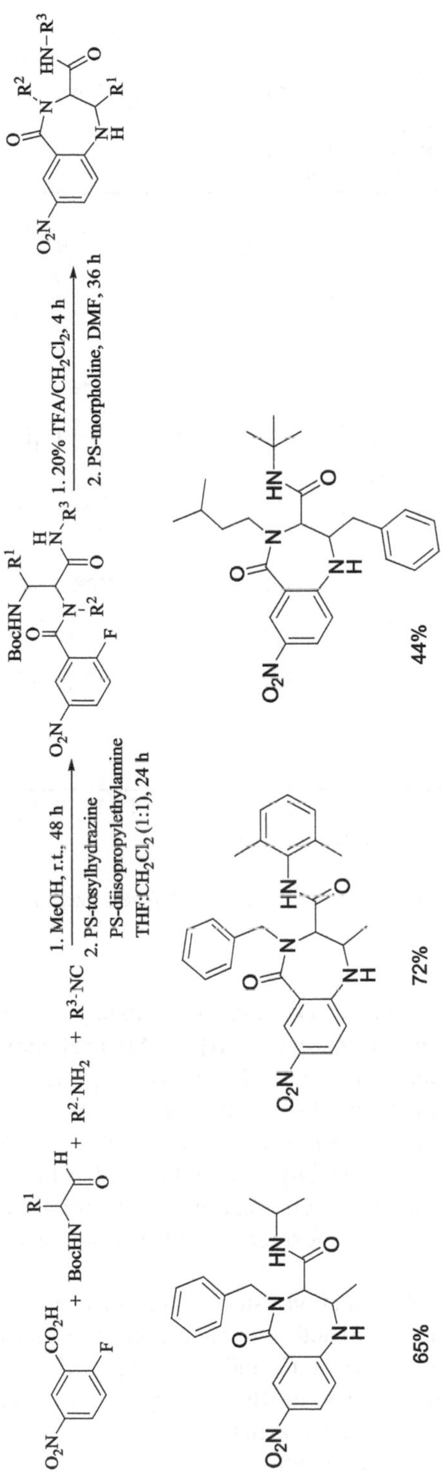

SCHEME 4.26 Access to libraries of benzodiazepines through the Ugi condensation reaction, followed by a secondary S_NAr cyclization.

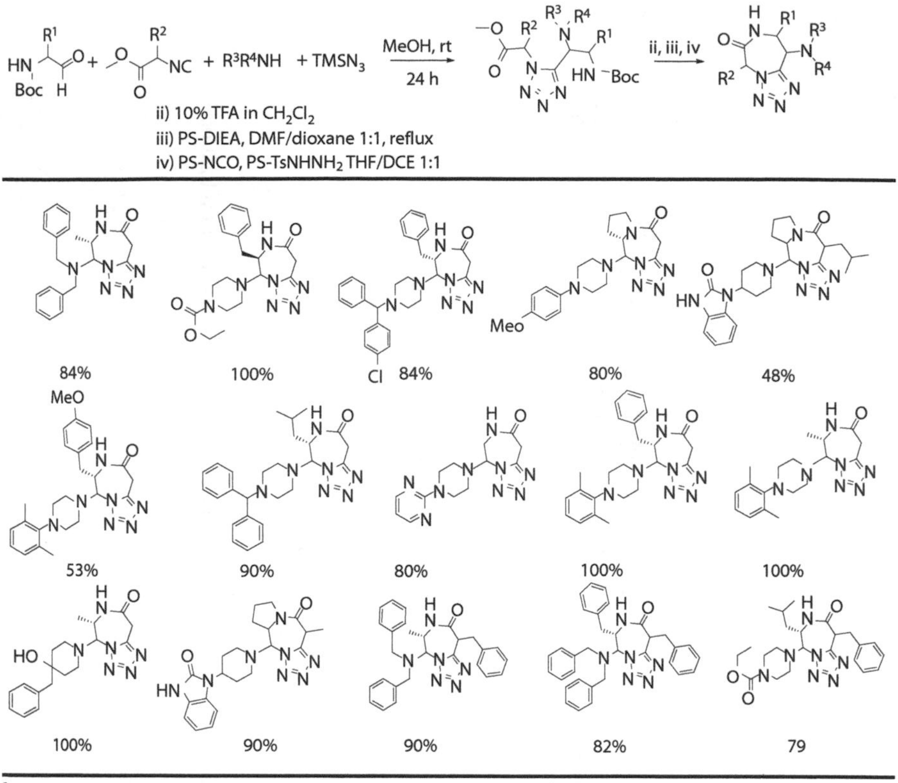

[a] Yield% purities as judged by l c/ms UV215 after step (iv) PS-NCO and PS-TsNHNH$_2$.

[b] Isolated yield after column chromatography (no scavenging).

SCHEME 4.27 Short solution phase preparation of fused azepine-tetrazoles via a UDC strategy.

the enal component, intermediates (**II**) were immediately captured by the iminium ion-activated enal to form intermediates (**III**) via Michael addition. Subsequently, the intramolecular aza-adolization on (**III**) led to the production of hydroxylated tetrahydropyridine intermediates (**IV**). In the presence of a proton acid, the dehydration on (**IV**) proceeded to iminium ion transition state TS-I, which quickly transformed to target products through the addition of hydroxyl in the structure. The attack of hydroxyl from the reverse side of R^1 in favored TS-I might be the key factor in determining the selective formation of the trans-C_2 and C_4 diastereoisomer (Scheme 4.31).

In another study, L-proline methyl esterhydrochloride was subjected to Petasis three-component condensation reaction in the presence of glyoxylic acid monohydrate and a variety of boronic acids. Coupling the obtained products with a variety of *N*-1-Boc-*N*-2-(alkyl)-hydrazines, de-Boc, and cyclization conditions resulted in the synthesis of 4,5-bridged 1,2,5-triazepine-3,7-diones suitable for elaboration into larger peptides at amino termini (Scheme 4.32).[131]

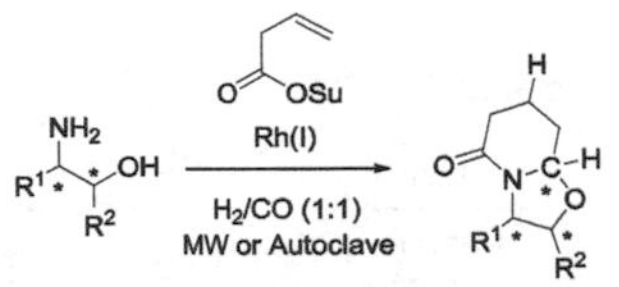

Entry	Substrate	Condition[a]	Product[b]	Yield[b] (*trans/cis*)[c]
1	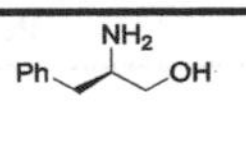	A	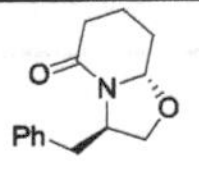	81% (83/17)
2	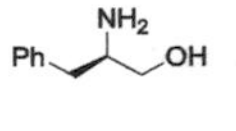	B	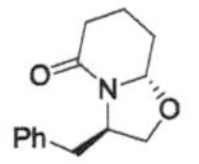	75% (83/17)
3	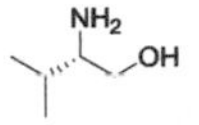	A	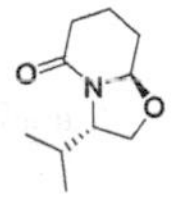	63% (75/25)
4	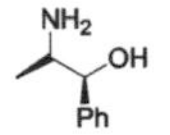	A	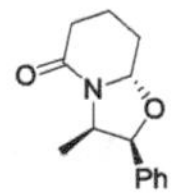	81% (83/17)
5	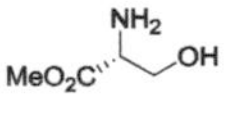	B	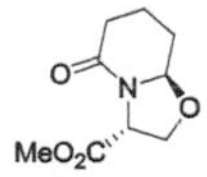	54% (84/16)
6	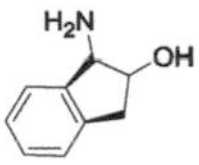	B	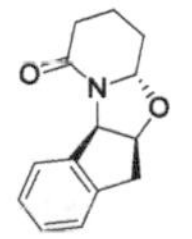	98% (83/17)

[a] A: reactions carried out in autoclave (ratio unsaturated carboxylic acid: amino alcohol = 1.2:1, $Rh(CO)_2$acac 1 mol %, biphephos 2 mol %, [amino-alcohol]) 0.04 M in THF, PPTS 5 mol %, 5 bar H_2/CO (1:1), 70 °C, 12 h. B: reactions carried out in MW (ratio unsaturated carboxylic acid: amino alcohol = 1.2:1; 7 bar H_2/CO (1:1), $Rh(CO)_2$ acac 1 mol %, biphephos 2 mol %, 150 W, max internal temperature and pressure 70 °C and 10 bar, *p*TSA 10 mol %, [amino-alcohol]) 0.04 M in THF, 1 h.

[b] Isolated yields.

[c] Diastereoselectivity determined by 1H NMR of the crude reaction mixtures.

SCHEME 4.28 Synthesis of fused lactams using four-component reaction.

The second and arguably the most important proline-based MCR in this direction is the metal- and catalyst-free decarboxylative α,β-difunctionalization of secondary α-amino acids via pseudo-four-component coupling of proline, aldehyde, and 1,3-diketone under microwave irradiation to generate multifunctionalized pyrano[2,3-*b*]pyrroles and pyrrolizinones (Scheme 4.33).[132]

Surprisingly, an excess of thiophene or furan carbaldehyde (4 equiv.) led to formation of pyrrolizinone derivatives, which could only be observed when thiophene or furan carbaldehyde was used as the substrate (Scheme 4.34).

Spiro compounds are another prominent class of natural products due to their notable biological activities.[133,134] Among all the spiro compounds, spiro pyrrolidines

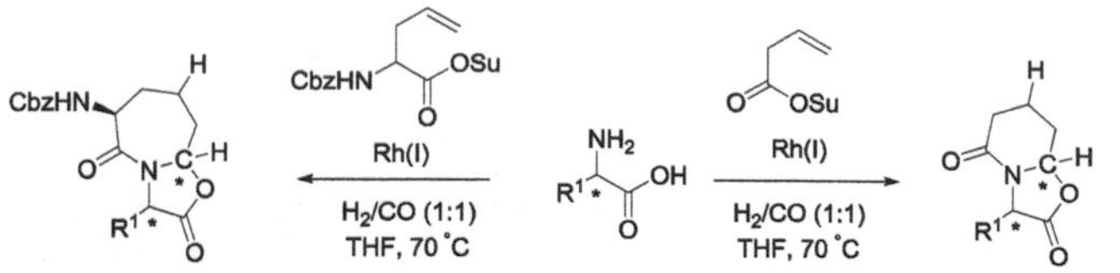

Entry	Amino acid	Condition[a]	Product[b]	Yield[c]
1	D-Phenylalanine	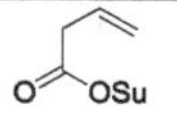 Autoclave, PPTS[c] 5% mol	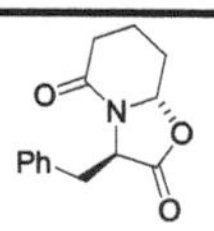	69%
2	(R)-Phenylglycine	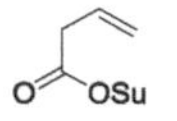 Autoclave, PPTS 10% mol	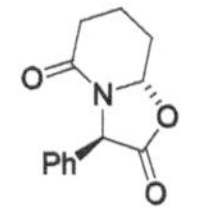	54%[b]
3	L-Valine	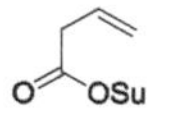 Autoclave, p-TSA 10% mol	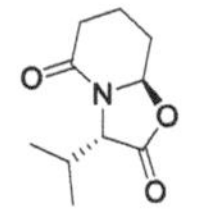	70%
4	L-Methionine	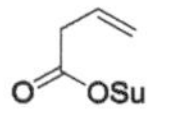 Autoclave, PPTS 5% mol	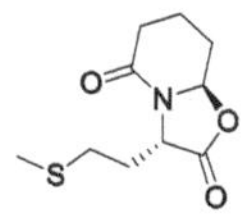	71%
5	L-Methionine	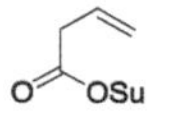 MW, p-TSA 10% mol	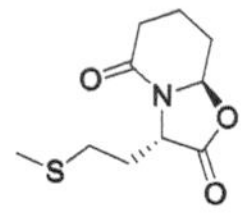	70%
6	L-Phenylalanine	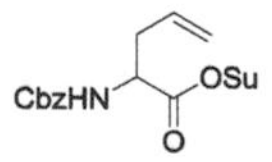 MW, p-TSA 10% mol	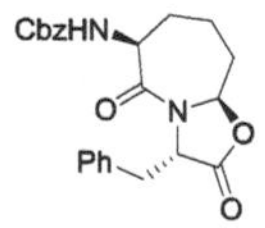	65%

[a] Conditions as in the previous table.
[b] Reaction carried out in toluene. *trans/cis* > 95/5 in all cases, determined by ^{1}H NMR of the crude reaction mixture.
[c] Diastereoselectivity
[d] Pyridinium p-toluenesulfonate (PPTS).

SCHEME 4.29 Synthesis of fused lactams through four-component reaction of H_2, CO, an unsaturated carboxylic acid, and chiral amino acids.

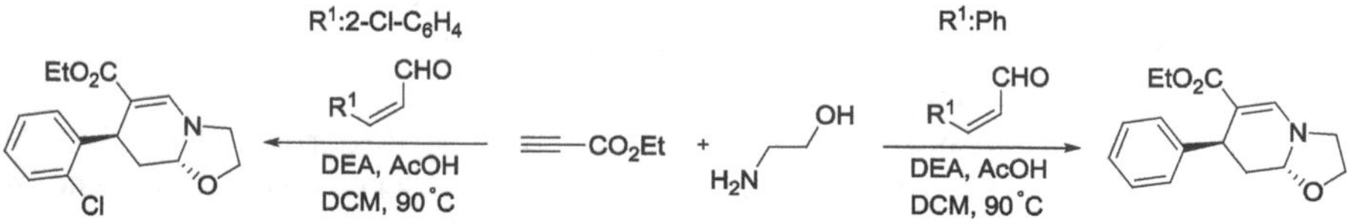

SCHEME 4.30 Diastereoselective construction of tetrahydropyridine fused bicyclic structures via three-component domino reaction.

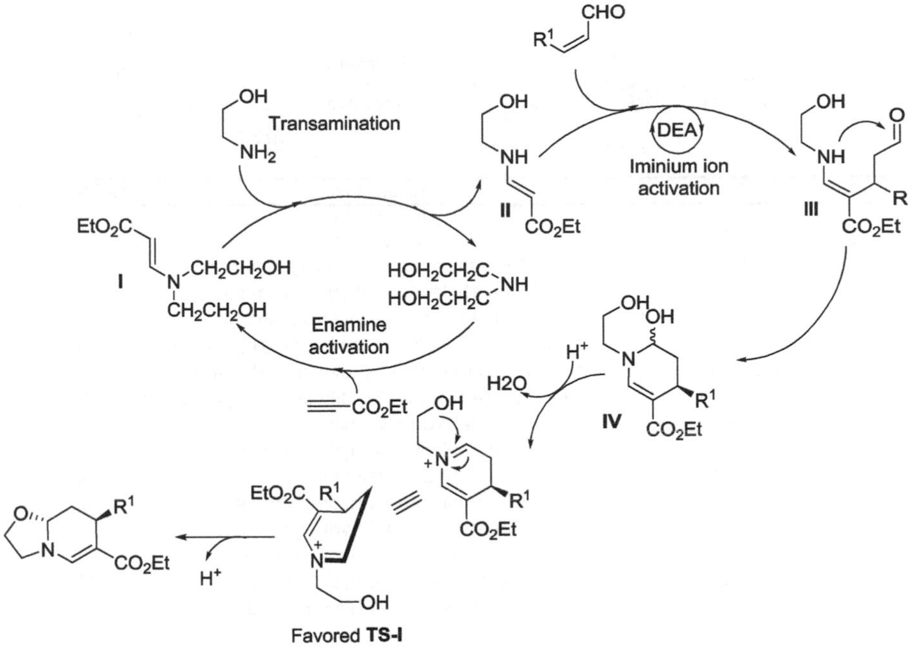

SCHEME 4.31 Plausible mechanism for the synthesis of tetrahydropyridine fused bicyclic structures.

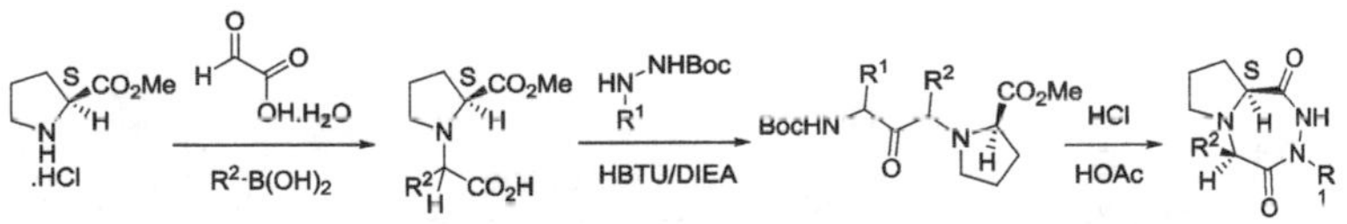

Entry	R^1	R^2	Yields (%)[a]
1	CH_2Ph	α-naphtol	48
2	CH_2Ph	4-MeO-C_6H_4	34
3	CH_2Cyclohexyl	Ph	32
4	CH_2Cyclohexyl	4-MeO-C_6H_4	32

[a] All yields refer to pure, isolated products; the reactions proceed through the Petasis, coupling, de-Boc, and cyclization in one pot without purification until the final product is isolated.

SCHEME 4.32 One-pot synthesis of new fused 4,5-bridged 1,2,5-triazepine-3,7-diones heterocycles by Petasis reaction.

are the most important ones, exhibiting remarkable pharmacological activities such as antiviral[135] and local anesthetic[136] activities and as potential antileukemic[137] and anticonvulsant agents.[138–140] A number of methods have been reported for the preparation of spiro oxindole fused heterocycles. Nowadays, [3+2] cyclization has usually been used for the generation of these valuable compounds; the more recent ones have been summarized in Scheme 4.35.[141–149]

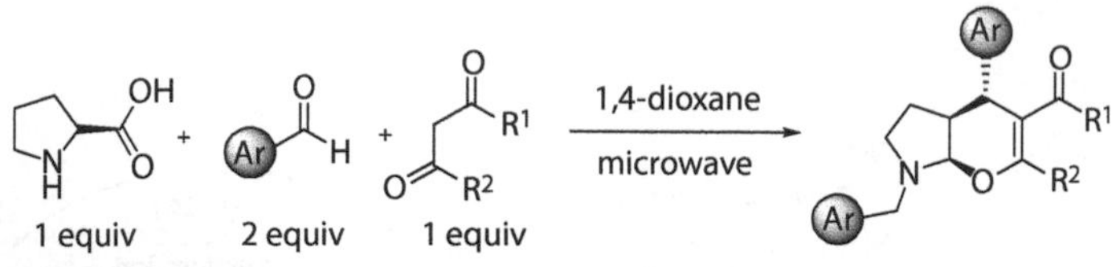

Entry	Ar	1,3-dicarbonyl	Yields (%)
1	Ph	4-Hydroxycumarin	53
2	4-Br-C_6H_4	4-Hydroxycumarin	50
3	Ph	7-Dimethylamino-4-hydroxycumarin	56
4	4-MeO-C_6H_4	7-Dimethylamino-4-hydroxycumarin	64
5	4-NO_2-C_6H_4	7-Dimethylamino-4-hydroxycumarin	60
6	Ph	4-hydroxy-6,6-dimethyl-5,6-dihydro-2H-pyran-2-one	43
7	4-Br-C_6H_4	4-hydroxy-6,6-dimethyl-5,6-dihydro-2H-pyran-2-one	45
8	4-NO_2-C_6H_4	4-hydroxy-6,6-dimethyl-5,6-dihydro-2H-pyran-2-one	46
9	Thienyl	4-Hydroxycumarin	67
10	Thienyl	7-Dimethylamino-4-hydroxycumarin	76
11	Furyl	4-Hydroxycumarin	62
12	Furyl	7-Dimethylamino-4-hydroxycumarin	67
13	4-Ph-C_6H_4	7-Dimethylamino-4-hydroxycumarin	52
14	4-F-C_6H_4	7-Dimethylamino-4-hydroxycumarin	55
15	3-HO-C_6H_4	7-Dimethylamino-4-hydroxycumarin	45
16	3-F_3C-C_6H_4	7-Dimethylamino-4-hydroxycumarin	72
17	5-Me-thienyl	7-Dimethylamino-4-hydroxycumarin	66
18	5-Br-Thienyl	7-Dimethylamino-4-hydroxycumarin	62

SCHEME 4.33 Synthesis of multifunctionalized pyrano[2,3-*b*]pyrroles.

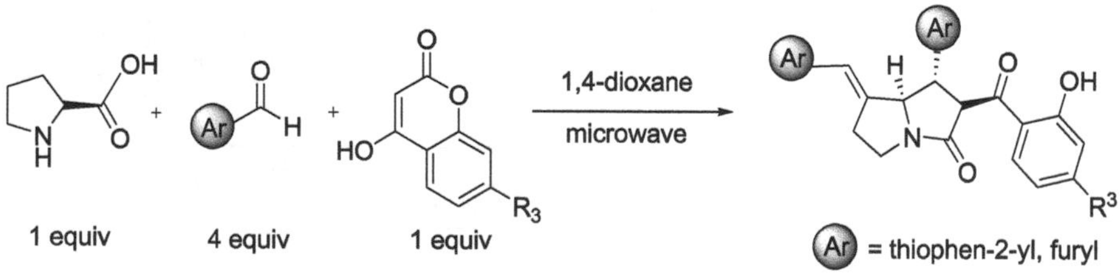

Entry	Ar	R^3	Yields (%)
1	Thienyl	H	65
2	Furyl	H	62
3	5-Me-Thienyl	NMe_2	53
4	Thienyl	NMe_2	52
5	Furyl	NMe_2	58

SCHEME 4.34 Synthesis of multifunctionalized pyrano[2,3-*b*] pyrrolizinones.

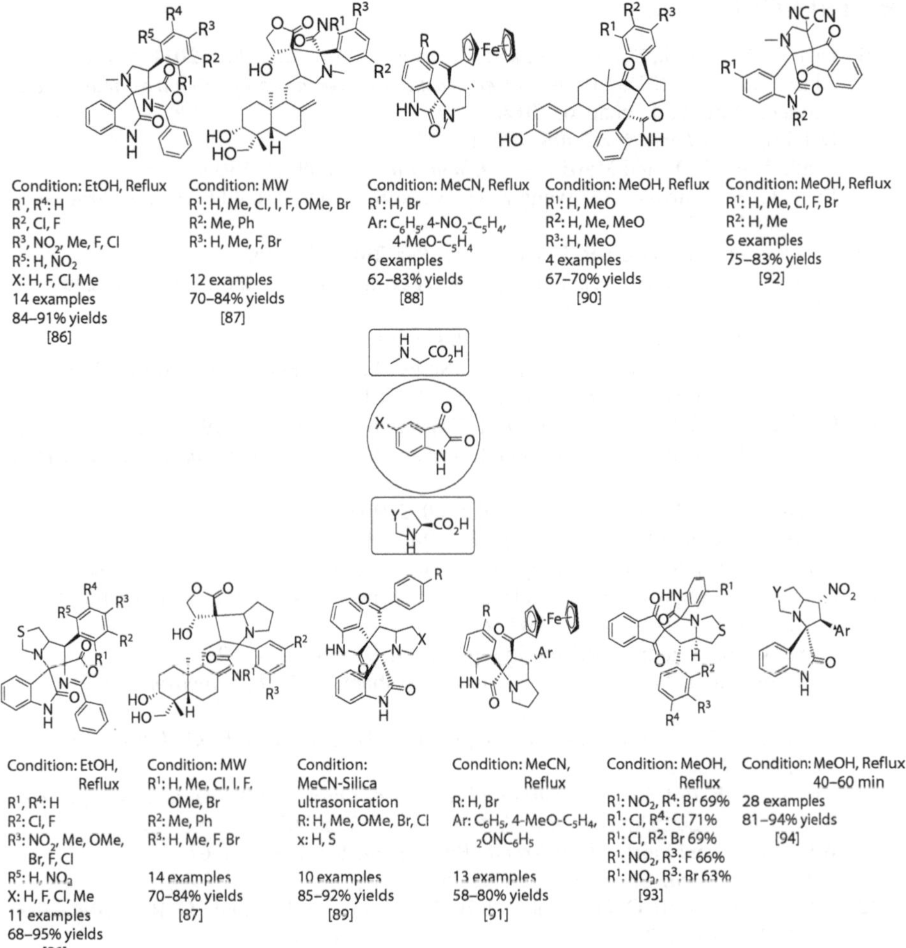

SCHEME 4.35 Preparation of various spiro oxindole fused heterocycles via [3+2] cyclization.

4.7 SUMMARY

The selected examples presented in this chapter demonstrate amino acids as flexible platforms in the short synthesis of versatile heterocycles via one-pot MCRs due to their bifunctionality and optical purity. Although amino acid side chains induce specific effects on the activities of designed compounds to make them suitable for applied application, the use of amino acids in MCRs has just scratched the surface. It is obvious that efficient and novel multicomponent routes for the synthesis of heterocycles based on amino acids has a bright future.

REFERENCES

1. Dua, R., Suman S., Sonwane S., and Srivastava S. *Adv. Biol. Res.* **2011**, *5*, 120.
2. Ameta, K. L., and Dandia A. *Green chemistry: Synthesis of bioactive heterocycles.* Springer, New York, chap. 6, **2014**.
3. Dömling, A. *Chem. Rev.* **2006**, *106*, 17.
4. Sunderhaus, J. D., and Martin, S. F. *Chem. Eur. J.* **2009**, *15*, 1300.
5. Voigt, C. A., Martinez, C., Wang Z.-G., Mayo S. L., and Arnold F. H. *Nat. Struct. Mol. Biol.* **2002**, *9*, 553.
6. Seebach, D., and Gardiner, J. *Acc. Chem. Res.* **2008**, *41*, 1366.
7. Weiner, B., Baeza, A., Jerphagnon, T., and Feringa, B. L. *J. Am. Chem. Soc.* **2009**, *131*, 9473.
8. Singh, G. S. *Mini-Rev. Med. Chem.* **2004**, *4*, 69.
9. Moreira, R., Santana, A. B., Iley, J., Neres, J., Douglas, K. T., Horton, P. N., and Hursthouse, M. B. *J. Med. Chem.* **2005**, *48*, 4861.
10. Dömling, A., Kehagia, K., and Ugi, I. *Tetrahedron* **1995**, *51*, 9519.
11. Ruhland, B., Bhandari, A., Gordon, E. M., and Gallop, M. A. *J. Am. Chem. Soc.* **1996**, *118*, 253.
12. Schunk, S., and Enders, D. *Org. Lett.* **2000**, *2*, 907.
13. Linder, M. R., and Podlech, J. *Org. Lett.* **2001**, *3*, 1849.
14. France, S., Weatherwax, A., Taggi, A. E., and Lectka, T. *Acc. Chem. Res.* **2004**, *37*, 592.
15. Palomo, C., Aizpurua, J. M., Balentová, E., Jimenez, A., Oyarbide, J., Fratila, R. M., and Miranda, J. I. *Org. Lett.* **2007**, *9*, 101.
16. Szakonyi, Z., Sillanpää, R., and Fülöp, F. *Mol. Div.* **2010**, *14*, 59.
17. Vishwanatha, T. M., Narendra, N., and Sureshbabu, V. V. *Tetrahedron Lett.* **2011**, *52*, 5620.
18. Cornier, P. G., Delpiccolo, C. M., Boggián, D. B., and Mata, E. G. *Tetrahedron Lett.* **2013**, *54*, 4742.
19. Spatola, A. F. In *Chemistry and biochemistry of amino acids, peptides, and proteins*; B. Weinstein, Ed.; Dekker: New York, **1983**, vol. 7, p. 267.
20. West, M. L., and Fairlie, D. P. *Trends Pharmacol. Sci.* **1995**, *16*, 67.
21. Huff, J. R. *J. Med. Chem.* **1991**, *34*, 2305.
22. Zabrocki, J., Smith, G. D., Dunbar, J. B., Iijima, H., and Marshall, G. R., *J. Am. Chem. Soc.* **1988**, *110*, 5875.
23. Yu, K.-L., and Johnson, R. L. *J. Org. Chem.* **1987**, *52*, 2051.
24. Marshall, G. R., Humblet, C., Van Opdenbosch, N., and Zabrocki, J. *Proc. Seventh Am. Peptide Sym.*, Rich, D. H., Gross, E., Eds.; Pierce Chemical: Rockford, IL, **1981**, 669.
25. May, B. C. H., and Abell, A. D. *Tetrahedron Lett.* **2001**, *42*, 5641.
26. Abell, A. D., and Foulds, G. J. *J. Chem. Soc., Perkin Trans.* **1997**, *17*, 2475.
27. Salvino, J. N., Kumar, N. V., Orton, E., Airey, J., Kiesow, T., Crawford, K., Rose, M., Krolikowski, P., Drew, M., Engers, D., Krolinkowski, D., Herpin, T., Gardyan, M., McGeehan, G., and Labaudiniere, R. *J. Comb. Chem.* **2000**, *2*, 691.
28. Drew, M., Orton, E., Krolikowski, P., Salvino, J., and Kumar, N. V. *J. Comb. Chem.* **2000**, *2*, 8.
29. Jones, W., Overland, D., Poppe, L., Cardenas, J., Pate, M., and Hulme, C. *Lab Automation* **2002**, T002.
30. Thomas, N., and Hulme, C. *Tetrahedron Lett.* **2002**, *43*, 6833.
31. Shaw, A. Y., Medda, F., and Hulme, C. *Tetrahedron Lett.* **2012**, 53, 1313.
32. Wells, J. I. In *Pharmaceutical preformulation*; Ellis Horwood Ltd: London, **1998**; p. 25.
33. Clark, D. E. *Drug Discov. Today*, **2003**, *8*, 927.
34. Upthagrove, A. L., and Nelson, W. L. *Drug Metab. Dispos.* **2001**, *29*, 1377.
35. Wan, H., and Ulander, J. *Opin. Drug Metab. Toxicol.* **2006**, *2*, 139.

36. Wang, K., Kim, D., and Dömling, A. *J. Comb. Chem.* **2010**, *12*, 111.
37. Barnes, D. M., Haight, A. R., Hameury, T., McLaughlin, M. A., Mei, J. Z., Tedrow, J. S., and Toma, J. D. R. *Tetrahedron* **2006**, *62*, 11311.
38. Mkrtchyan, A. P., Noravyan, A. S., and Petrosyan, V. M. *Chem. Heterocycl. Compd.* **2002**, *38*, 238.
39. Zavarzin, I. V., Smirnova, N. G., Chernoburova, E. I., Yarovenko, V. N., and Krayushkin, M. M. *Russ. Chem. Bull.* **2004**, *53*, 1257.
40. Doss, S. H., Mohareb, R. M., Elmegeed, G. A., and Luoca, N. A. *Pharmazie* **2003**, *58*, 607.
41. Roman, G., and Andrei, M. *Glasnikna Hemicaritei Tehnolozitena Makedonija* **2001**, *20*, 131.
42. Froehlich, J., Chowdhury, A. Z. M. S., and Hametner, C. *Arkivoc* **2001**, *ii*, 163.
43. Elkholy, Y. M. Phosphorus, *Sulfur Silicon Relat. Elem.* **2002**, *177*, 115.
44. Garner, P., Kaniskan, H. U., Hu, J., Youngs, W. J., and Panzner, M. *Org. Lett.* **2006**, *8*, 3647.
45. Garner, P., Hu, J., Parker, C. G., Youngs, W. J., and Medvetz, D. *Tetrahedron Lett.* **2007**, *48*, 3867.
46. Garner, P., and Kaniskan, H. U. *Curr. Org. Synth.* **2010**, *7*, 348.
47. Garner, P., Weerasinghe, L., Youngs, Y. J., Wrigh, B., Wilson, D., and Jacobs, D. *Org. Lett.* **2012**, *14*, 1326.
48. Hassner, A., and Fischer, B. *Heterocycles* **1993**, *35*, 1441.
49. Betschart, C., and Hegedus, L. S. *J. Am. Chem. Soc.* **1992**, *114*, 5010.
50. Hsiao, Y., and Hegedus, L. S. *J. Org. Chem.* **1997**, *62*, 3586.
51. Schäfer, U., Burgdorf, C., Engelhardt, A., Kurz, T., and Richardt, G. *J. Pharmacol. Exp. Ther.* **2002**, *303*, 1163.
52. Rondu, F., le Bihan, G., Wang, X., Lamouri, A., Touboul, E., Dive, G., Bellahsene, T., Pfeiffer, B., Renard, P. B., Manechez, D., Penicaud, L., Ktorza, A., and Godfroid, J.-J. *J. Med. Chem.* **1997**, *40*, 3793.
53. Gust, R., Keilitz, R., and Schmidt, K. *J. Med. Chem.* **2001**, *44*, 1963.
54. Gust, R., Keilitz, R., Schmidt, K., and Von Rauch, M. *J. Med Chem.* **2002**, *45*, 3356.
55. Dunn, P. J., Haner, R., and Rapoport, H. *J. Org. Chem.* **1990**, *55*, 5017–5025 and references therein.
56. Han, H., Yoon, J., and Janda, K. D. *J. Org. Chem.* **1998**, *63*, 2045.
57. Jones, R. C. F., Howard, K. J., and Snaith, J. S. *Tetrahedron Lett.* **1996**, *37*, 1707.
58. Jones, R. C. F., Howard, K. J., and Snaith, J. S. *Tetrahedron Lett.* **1996**, *37*, 1711.
59. Dalko, P. I., and Langlois, Y. *Chem. Commun.* **1998**, *3*, 331–332.
60. Sisko, J., Kassick, A. J., Mellinger, M., Filan, J. J., Allen, A., and Olsen, M. A. *J. Org. Chem.* **2000**, *65*, 1516.
61. Hayashi, T., Kishi, E., Soloshonok, V. A., and Uozumi, Y. *Tetrahedron Lett.* **1996**, *37*, 4969.
62. Lin, Y.-R., Zhou, X.-T., Dai, L.-X., and Sun, J. *J. Org. Chem.* **1997**, *62*, 1799.
63. Zhou, X.-T., Lin, Y.-R., Dai, L.-X., Sun, J., Xia, L.-J., and Tang, M.-H. *J. Org. Chem.* **1999**, *64*, 1331.
64. Bon, R. S., Hong, C., Bouma, M. J., Schmitz, R. F., de KanterLutz, F. J. J., Spek, M. A. L., and Orru, R. V. A. *Org. Lett.* **2003**, *5*, 3759.
65. Li, M., Lv, X-L., Wen, L-R., and Hu, Z-Q. *Org. Lett.* **2013**, *15*, 1262.
66. Huang, H., Jiang, H., Chen, K., and Liu, H. *J. Org. Chem.* **2009**, *74*, 5476.
67. Bolink, H. J., Cappelli, L., Coronado, E., and Gratzel, M. *J. Am. Chem. Soc.* **2006**, *128*, 46.
68. Zong, R., and Thummel, R. P. *J. Am. Chem. Soc.* **2004**, *126*, 10800.
69. White, T. A., Higgins, S. L. H., Arachchige, S. M., and Brewer, K. J. *Angew. Chem. Int. Ed.* **2011**, *50*, 12209.
70. Leontie, L., Druta, I., Danac, R., and Rusu, G. I. *Synth. Met.* **2005**, *155*, 138.

71. Prelipceanu, M., Prelipceanu, O. S., Leontie, L., and Danac, R. *Phys. Lett. A* **2007**, *368*, 331.
72. Prelipceanu, M., Leontie, L., Danac, R., and Prelipceanu, O. S. *Proc. Romanian Conf. Adv. Mater.*, September 4–11. Bucharest-Magurele, Romania, **2006**, p. 68.
73. Dumitrascu, F., and Mitan, C. I. *Tetrahedron Lett.* **2001**, *42*, 8379.
74. Danac, R., Rotaru, A., Drochioiu, G., and Druta, I. *J. Heterocycl. Chem.* **2003**, *40*, 283.
75. Dumitrascu, F., Caira, M. R., Draghici, C., Caproiu, M. T., and Badoiu, A. *Molecules* **2005**, *10*, 321.
76. Dumitraşcu, F., Caira, M. R., Drâghici, C., Câproiu, M. T., Barbu, L., and Bâdoiu, A. *J. Chem. Crystallography* **2005**, *35*, 361.
77. Dumitrascu, F., Caira, M. R., Draghici, C., Caproiu, M. T., and Brrbu, L. *Revue Roumaine Chimie* **2008**, *53*, 183.
78. Qiu, L., Guo, X., Zhou, J., Liu, S., Yang, L., Wu, X., and Hu. W. *RSC Adv.* **2013**, *3*, 20065.
79. Rosenthal, G. *Plant nonprotein amino and imino acids: Biological, biochemical, and toxicological properties.* Academic Press, Inc. **1982**.
80. Chang, G. W., and Snell, E. E. *Biochemistry* **1968**, *7*, 2005.
81. Shimeno, H., Soeda, S., and Nagamatsu. S. H. *Chem. Pharm. Bull.* **1977**, *25*, 2983.
82. Tanase, S., Guirard, B. M., and Snell, E. E. *J. Biol. Chem.* **1985**, *260*, 6738.
83. Izawa, M., Takayama, S., Shindo-Okada N., Doi, S., Kimura, M., Katsuki, M., and Nishimura, S. *Cancer Res.* **1992**, *52*, 1628.
84. Adamczyk, M., Akireddy, S. R., and Reddy, R. E. *Org. Lett.* **2001**, *3*, 3157.
85. Dondoni, A., Massi, A., Minghini, E., and Bertolasi, V. *Tetrahedron* **2004**, *60*, 2311.
86. Litvinov, V. P. *Russ. Chem. Rev.* **2004**, *73*, 637.
87. Kumar, S., Bawa, S., and Gupta, H. *Mini Rev. Med. Chem.* **2009**, *9*, 1648.
88. Bouzard, D., Cesare, P. D., Essiz, M., Jacquet, J. P., Remuzon, P., Weber, A., Oki, T., and Masuyoshi, M. *J. Med. Chem.* **1989**, *32*, 537.
89. Ferrarini, P. L., Mori, C., Primofiore, G., and Calzolari, L. *J. Heterocycl. Chem.* **1990**, *27*, 881.
90. Di Cesare, P., Bouzard, D., Essiz, M., Jacquet, J. P., Ledoussal, B., Kiechel, J. R., Remuzon, P., Kessler, R. E., Fung-Tomc, J., and Desiderio, J. *J. Med. Chem.* **1992**, *35*, 4205.
91. Movassaghi, M., Hill, M. D., and Ahmad, O. K. *J. Am. Chem. Soc.* **2007**, *129*, 10096.
92. Huang, W., Wang, L., Tanaka, H., and Ogawa, T. *Eur. J. Inorg. Chem.* **2009**, 1321.
93. Mohammadpoor-Baltork, I., Tangestaninejad, S., Moghadam, M., Mirkhani, V., Anvar, S., and Mirjafari, A. *Synlett* **2010**, *20*, 3104.
94. Huo, Z., Gridnev, I. D., and Yamamoto, Y. *J. Org. Chem.* **2010**, *75*, 1266.
95. Shan, G., Sun, X., Xia, Q., and Rao, Y. *Org. Lett.* **2011**, *13*, 5770.
96. Sakai, N., Tamura, K., Shimamura, K., Ikeda, R., and Konakahara, T. *Org. Lett.* **2012**, *14*, 836.
97. Wang, Z., Li, S., Yu, B., Wu, H., Wang, Y., and Sun, X. *J. Org. Chem.* **2012**, *77*, 8615.
98. Liu, B., Gao, H., Yu, Y., Wu, W., and Jiang, H. *J. Org. Chem.* **2013**, *78*, 10319.
99. Yan, R., Liu, X., Pan, C., Zhou, X., Li, X., Kang, X., and Huang, G. *Org. Lett.* **2013**, *15*, 4876.
100. Fu, L., Lin, W., Hu, M. H., Liu, X. C., Huang, Z. B., and Shi, D. Q. *ACS Comb. Sci.* **2014**, *16*, 238.
101. Abbiati, G., Arcadi, A., Canevari, V., Capezzuto, L., and Rossi, E. *J. Org. Chem.* **2005**, *70*, 6454.
102. P de Carvalho, M., and Abraham. W. R. *Curr. Med. Chem.* **2012**, *19*, 3564.
103. Borthwick, A. D. *Chem. Rev.* **2012**, *112*, 3641.
104. Campbell, J., and Blackwell, H. E. *J. Comb. Chem.* **2009**, *11*, 1094.
105. Kleemann, A., Engel, J., Kutscher, B., and Reichert, D. *Pharmaceutical substances: Syntheses, patents, applications of the most relevant APIs.* Stuttgart. **2008**, Thieme.

106. Rhoden, C. R. B., Rivera, D. G., Kreye, O., Bauer, A. K., Westermann, B., and Wessjohann, L. A. *J. Comb. Chem.* **2009**, *11*, 1078.
107. El Kaïm, L., Grimaud, L., and Purumandla, S. R. *J. Org. Chem.* **2011**, *76*, 4728.
108. Basso, A., Banfi, L., Guanti, G., Riva, R., and Tosatti, P. *Synlett* **2011**, *14*, 2009–2012. DOI: 10.1055/s-0030-1260807.
109. Azuaje, J., Maatougui A. E., Pérez-Rubio, J. M., Coelho, A., Fernández, F., and Sotelo, E. *J. Org. Chem.* **2013**, *78*, 4402.
110. Stucchi, M., Cairati, S., Cetin-Atalay, R., Christodoulou, M. S., Grazioso, G., Pescitelli, G., Silvani, A., Yildirim, D. C., and Lesma, G. *Org. Biomol. Chem.* **2015**, *13*, 4993.
111. Duarte, C. D., Fraga, E. J., and Barreiro, C. A. M. *Mini-Rev. Med. Chem.* **2007**, *7*, 1108.
112. Welsch, M. E., Snyder, S. A., and Stockwell, B. R. *Curr. Opin. Chem. Biol.* **2010**, *14*, 347.
113. Loew, G. H., Nienow, J. R., and Poulsen, M. *Mol. Pharmacol.* **1984**, *26*, 19.
114. Bateson, A. N. *Sleep Med.* **2004**, *5*, 9.
115. Bolli, M. H., Marfurt, J., and Grisostomi, C. *J. Med. Chem.* **2004**, *47*, 2776.
116. De Clerq, E. *Antiviral Res.* **1998**, *38*, 153.
117. Dourlat, J., Liu, W., Gresh, N., and Garbay, C. *Biorg. Med. Chem. Lett.* **2007**, *17*, 2527.
118. McDowell, R. S., Gadek, T. R., Barker, P. L., Burdick, D. J., Chan, K. S., Quan, C. L., Skelton, N., Struble, M., Thorsett, E. D., Tischler, M., Tom, J. Y. K., Webb, T. R., and Burnier, J. P. *J. Am. Chem. Soc.* **1994**, *116*, 5069.
119. Marugan, J. J., Leonard, K., Raboisson, P., Gushue, J. M., Calvo, R., Koblish, H. K., Lattanze, J., Zhao, S. Y., Cummings, M. D., Player, M. R., Schubert, C., Maroncy, A. C., and Lu, T. B. *Bioorg. Med. Chem. Lett.* **2006**, *16*, 3115.
120. Leonard, K., Marugan, J. J., Raboisson, P., Calvo, R., Gushue, J. M., Koblish, H. K., Lattanze, J., Zhao, S. Y., Cummings, M. D., Player, M. R., Lu, A. C., and Maroney, T. B. *Bioorg. Med. Chem. Lett.* **2006**, *16*, 3463.
121. Leonard, K., Marugan, J. J., Koblish, H. K., Calvo, R., Raboisson, P., Gushue, J. M., Lattanze, J., Zhao, S. Y., Cummings, M. D., Lu, T. B., Player, M. R., and Maroney, A. *Clin. Cancer Res.* **2005**, *11*, 9152S.
122. Azuaje, J., Perez-Rubio, J. M., Yaziji, V., El Maatougui, A., González-Gómez, J. C., Sánchez-Pedregal, V., Navarro-Vázquez, A., Masaguer, C. F., Teijeira, M., and Sotelo, E. *Org. Lett.* **2015**, *80*, 1533.
123. Liu, A., Zhou, H., Su, G., Zhang, W., and Yan, B. *J. Comb. Chem.* **2009**, *11*, 1083.
124. Hulme, C., Ma, L., Kumar, N. V., Krolikowski, P. H., Allen, A. C., and Labaudiniere, R. *Tetrahedron Lett.* **2000**, *41*, 1509.
125. Huang, Y., Khoury, K., Chanas, T., and Domling, A. *Org. Lett.* **2012**, *14*, 5916.
126. Tempest, P., Pettus, L., Gorea, V., and Hulmea, C. *Tetrahedron Lett.* **2003**, *44*, 1947.
127. Ulaczyk-Lesanko, A., and Hall, D. G. *Curr. Opin. Chem. Biol.* **2005**, *9*, 266.
128. Nixey, T., Kelly, M., Semin, D., and Hulme, C. *Tetrahedron Lett.* **2002**, *43*, 3681.
129. Airiau, E., Girard, N., Mann, A., Salvadori, J., and Taddei, M. *Org. Lett.* **2009**, *11*, 5314.
130. Wan, J. P., Lin, Y., Huang, Q., and Liu, Y. *J. Org. Chem.* **2014**, *79*, 7232.
131. Neogi, S., Roy, A., and Naskar, D. *J. Comb. Chem.* **2009**, *12*, 75.
132. Manjappa, K. B., Jhang, W. F., Huang, S. Y., and Yang, D. Y. *Org. Lett.* **2014**, *16*, 5690.
133. Kobayashi, J., Tsuda, M., Agemi, K., Shigemori, H., Ishibashi, M., Sasaki, T., and Mikami, Y., Purealidins B and C. *Tetrahedron* **1991**, *47*, 6617.
134. James, D. M., Kunze, H. B., and Faulkner, D. J. *J. Nat. Prod.* **1991**, *54*, 1137.
135. Kornett, M. J., and Thio, A. P. *J. Med. Chem.* **1976**, *19*, 892.
136. Lundahl, K., Schut, J., Schlatmann, J. L. M. A., Paerels, G. B., and Peters, A. *J. Med. Chem.* **1972**, *15*, 129.
137. Abou-Gharbia, M. A., and Doukas, P. H. *Heterocycles* **1979**, *12*, 637.
138. Cravotto, G., Giovenzana, G. B., Pilati, T., Sisti, M., and Palmisano, G. *J. Org. Chem.* **2001**, *66*, 8447.
139. Onishi, T., Sebahar, P. R., and Williams, R. M. *Org. Lett.* **2003**, *5*, 3135.

140. Grigg, R., Millington, E. L., and Thornton-Pett, M. *Tetrahedron Lett.* **2002**, *43*, 2605.
141. Yang, J.-M., Hu, Y., Li, Q., Yu, F., Cao, J., Fang, D., Huang, Z.-B., and Shi, D.-Q. *ACS Comb. Sci.* **2014**, *16*, 139.
142. Hazra, A., Bharitkar, Y. P., Chakraborty, D., Mondal, S. K., Singal, N., Mondal, S., Maity, A., Paira, R., Banerjee, S., and Mondal, N. B. *ACS Comb. Sci.* **2013**, *15*, 41.
143. Suresh Babu, A. R., and Raghunathan, R. *Tetrahedron Lett.* **2008**, *49*, 4487.
144. Suresh Babu, A. R., Raghunathan, R. *Tetrahedron Lett.* **2007**, *48*, 6809.
145. Suresh Babu, A. R., Raghunathan, R. *Tetrahedron* **2009**, *65*, 2239.
146. Suresh Babu, A. R., and Raghunathan, R. *Tetrahedron Lett.* **2008**, *49*, 4618.
147. Lakshmi, N. V., Thirumurugan, P., and Perumal, P. T. *Tetrahedron Lett.* **2010**, *51*, 1064.
148. Maheswari, S. U., Balamurugan, K., Perumal, S., Yogeeswari, P., and Sriram, D. *Bioorg. Med. Chem. Lett.* **2010**, *20*, 7278.
149. Rajesh. S. M., Perumal, S., Menéndez, J. C., Yogeeswari, P., and Sriram, D. *Med. Chem. Commun.* **2011**, *2*, 626.

5 Multicomponent Reactions for Generation of Molecular Libraries in Anticancer Drug Discovery

Mohammad Saquib, Mohammad Faheem Khan, Mohammad Imran Ansari, Irfan Khan, Mohammad Kamil Hussain, and Jagdamba Singh**

CONTENTS

5.1 INTRODUCTION

Multicomponent reactions (MCRs) have emerged as a powerful new strategy in synthetic organic chemistry and drug discovery. They provide a quicker and more efficient way to synthesize chemical compounds than have methods used in traditional chemistry. MCRs can be defined as flexible chemical reactions in which three or more substrates react in one synthetic operation to form highly selective products with high atom efficiency.[1] MCRs offer great possibilities for obtaining molecular diversity and complexity in fewer steps and less time. Most multicomponent

* Corresponding authors: mkhcdri@gmail.com and dr.jdsau@gmail.com.

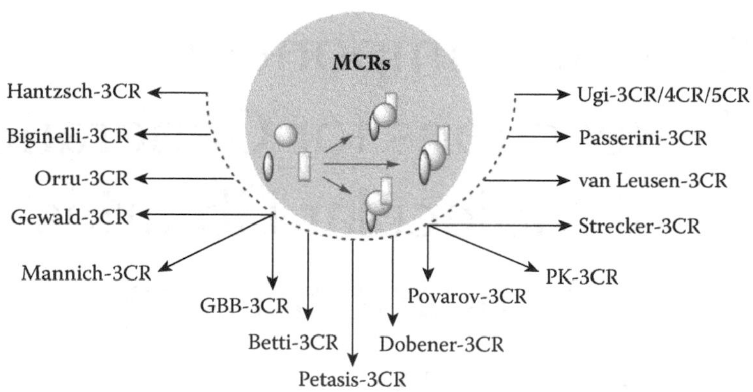

FIGURE 5.1 Some named reactions based on MCR strategy.

reactions are named reactions, for example, Ugi, Passerini, van Leusen, Strecker, Orru, Gewald, Povarov, Dobener, Pauson–Khand, Groebke–Blackburn–Bienaymé, Mannich, Staudinger, Petasis, Betti, Biginelli, Hantzsch dihydropyridine synthesis, Hantzsch pyrrole synthesis, Radziszewski imidazole synthesis etc. (Figure 5.1).

Multicomponent reactions can be subcategorized into two general classes, isocyanide-based multicomponent reactions (IMCRs) and nonisocyanide-based multicomponent reactions (NIMCRs).

5.2 ISOCYANIDE-BASED MULTICOMPONENT REACTIONS (IMCRs)

IMCRs are the more versatile and multifarious of the MCRs. These reactions often show high levels of regio-, chemo-, and stereoselectivity. Nowadays, most multicomponent reactions being performed with isocyanides are based on the classical Passerini and Ugi reactions. The Passerini (P-3CR) reaction is a well-known isocyanide-based MCR involving reaction between carboxylic acids and aldehydes or ketones and isocyanides, to afford α-acyl carboxamides in one step.[2,3] The second most important IMCR is the Ugi reaction. This elegant four-component reaction is a reaction between isocyanides, carboxylic acids, ketones or aldehydes, and primary amines to afford dipeptide-like structures.[4] The mechanism of Ugi involves the reaction of a Schiff base or an enamine with a nucleophile and an isocyanide, followed by Mumm rearrangement. Fascinatingly, the Ugi reaction as compared to the Passerini reaction is much more versatile, not only in terms of library size but also in terms of scaffold diversity. Both the Passerini and Ugi reactions lead to interesting peptidomimetic compounds, which are potentially bioactive. The products of these reactions can constitute interesting lead compounds for further development into more active compounds (Figure 5.2).

Both reactions offer cost-effective and rapid access to generate molecular libraries. Since a wide variety of isocyanides are commercially available, an equivalently diverse spectrum of products may be obtained. Variations in the starting compounds may also lead to totally new scaffolds. Other important isocyanide-based multicomponent reactions are Groebke–Blackburn–Bienayme (GBB-3CR), Van Leusen

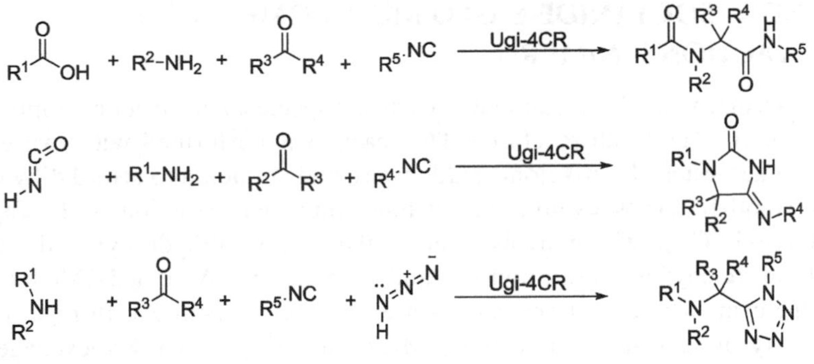

FIGURE 5.2 Isocyanide-based multicomponent reactions involving four components.

(VL-3CR), and Orru-3CR. All these reactions are further modifications of Ugi reactions. The Groebke–Blackburn–Bienayme reaction (GBBR) involves *in situ* formation of iminium species followed by a nonconcerted cycloaddition with the isocyanide to give the corresponding fused imidazoles. GBBR-3CR is used for the one-pot synthesis of fused imidazole, bridgehead nitrogen heterocyclic compounds from aldehyde, isocyanide, and amidine building blocks.[5] Van Leusen-MCR (VL-3CR) is very useful for the synthesis of polysubstituted imidazoles in one pot from aryl substituted tosylmethylisocyanide (TosMIC) reagents and imines generated *in situ*.[6] Orru-3CR is the three-component condensation between an amine, an aldehyde, and an α-acidic isocyanide, which efficiently provides highly substituted 2-imidazolines in a one-pot reaction[7] (Figure 5.3).

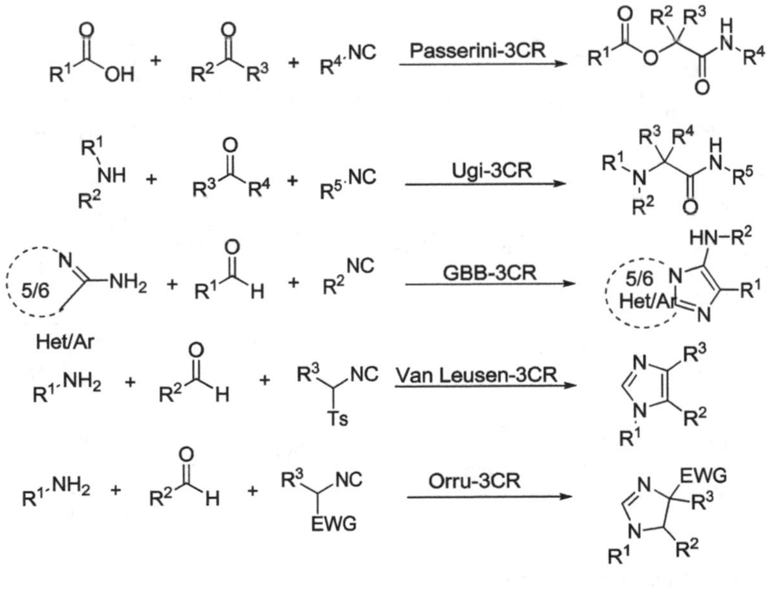

FIGURE 5.3 Isocyanide-based multicomponent reactions involving three components.

5.3 NON-ISOCYANIDE-BASED MULTICOMPONENT REACTIONS (NIMCRs)

NIMCRs usually involve an activated carbonyl species. An earlier example of this reaction is the Hantzsch reaction.[8] The reaction is performed with two equivalents of β-ketoesters/1,3 diketones, aldehydes, and ammonia to afford dihydropyridines. Another non-isocyanide-based multicomponent reaction is the Biginelli reaction (BG-3CR),[9] which involves the synthesis of 3,4-dihydropyrimidin-2(1*H*)-ones from aldehydes, urea/thiourea, and β-ketoesters. Among NIMCRs, those involving cyanoacetic acid derivatives are extremely versatile with regard to the multiplicity of scaffolds. Often these MCRs involve primary Knoevenagel-type condensations of the cyanoacetic acid derivative with an aldehyde or ketone, followed by Michael attack of a nucleophile and subsequent ring closure via a second nucleophile attack involving a nitrile. A well-known MCR of this class is the Gewald-3CR (G-3CR), which has recently gained ground by the usage of cyanoacetamides. The Gewald 3-CR (G-3CR) of cyanoacetic acid derivatives, active methylene carbonyls, and elemental sulfur is a popular MCR often used in drug discovery yielding 2-amino-3-carbonyl thiophenes.[10–12] Additionally, Gewald products can be easily transformed into more complex scaffolds by secondary transformations[13] (Figure 5.4).

The Pauson–Khand reaction (PK-3CR) is used to synthesize α-,β-cyclopentenone through [2+2+1] cycloaddition of an alkyne, an alkene, and carbon monoxide.[14] For generation of scaffold diversity, other non-isocyanide-based important multicomponent reactions are Pavarov-3CR, Dobener-3CR, Betti-3CR, and Petasis-3CR (Figure 5.5).

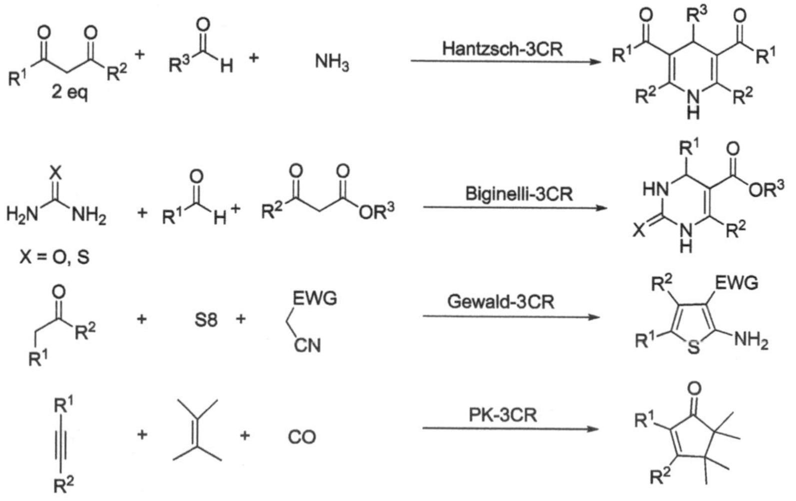

FIGURE 5.4 Non-isocyanide-based multicomponent reactions.

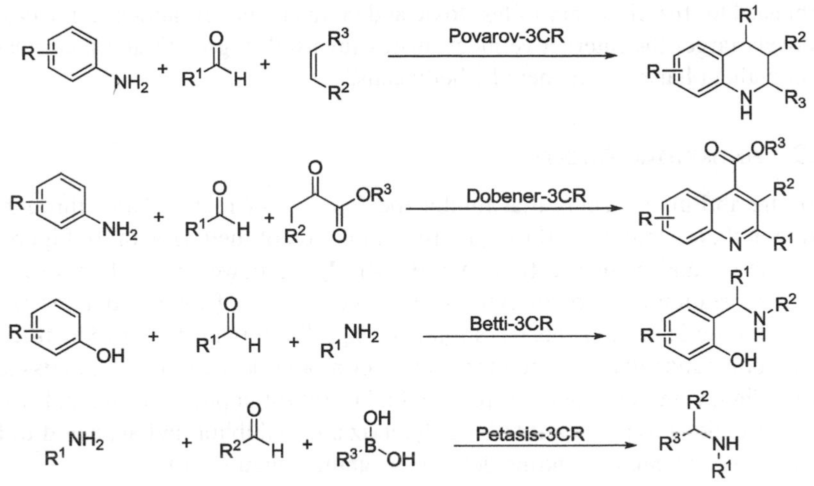

FIGURE 5.5 Some more types of non-isocyanide-based multicomponent reactions.

5.4 BIOLOGY OF MULTICOMPONENT REACTIONS

Drugs are molecules designed to combat pathological processes and repair the ensuing damage. The key to the success of these therapeutics lies in their pharmacophores, components that are biologically active and can elicit a desired effect in the body. MCR is a powerful new strategy, having enormous applications in drug discovery and development. In it, medicinal chemists now have a tool that opens the door to a more efficient and rapid synthesis of biologically important molecules that could potentially be developed into drugs.

In recent years, MCR-based synthetic strategies are finding extensive application in the field of drug discovery and development. Several drugs synthesized through MCR are now available in the market and thousands more are being developed. Drugs prepared through the MCR pathway and presently in the market include calcium blockers such as Amilodipine.

5.4.1 CANCER

"Cancer is a word that drives fear into people's hearts, producing a profound sense of helplessness."[15a] It is a major public health problem worldwide, having a profound impact on society. Cancer is a group of diseases characterized by the uncontrolled growth and spread of abnormal cells. If the spread is not controlled, it can result in death. The most common cancers in 2016 were projected to be breast cancer, lung and bronchus cancer, prostate cancer, colon and rectum cancer, bladder cancer, melanoma of the skin, non-Hodgkins lymphoma, thyroid cancer, kidney and renal pelvis cancer, leukemia, endometrial cancer, and pancreatic cancer.[15b] With emerging drug resistance as a major new impediment in cancer treatment, combined with the problems of low tumor selectivity, diversity of cancer types, and drug toxicity, there is an urgent need for the discovery of less toxic and more potent new anticancer drugs that selectively target the interactive mechanisms involved in growth and metastasis of cancer without harming the healthy body cells.[16]

5.4.2 Antimitotic Agents

There are antitumor agents that inhibit the function of microtubules through the binding of their subunits or through direct inhibition of their growth. In Figure 5.5, colchicine (**1**) and combretastatin A-4 (CA4) (**2**) are powerful inhibitors of tubulin polymerization and potent cytotoxins.[17] The success of tubulin polymerization inhibitors as anticancer agents has stimulated significant interest in the identification of new compounds that may be more potent or more selective in targeted tissues or tumors. Pinney and co-workers[18] reported a benzo[*b*]-thiophene compound (**3**) that exhibited some activity as a tubulin polymerization inhibitor and appeared to bind weakly to the colchicine-binding domain of tubulin (Figure 5.6).

Flynn and co-workers reported multicomponent coupling of *o*-iodophenols (or *o*-iodoacetanilides) with terminal alkynes and aryl iodides providing rapid access to potent benzo[*b*]furan and indole-based tubulin polymerization inhibitors. These systems represent valuable new leads in the pursuit of anticancer chemotherapies (Scheme 5.1).[19]

Podophyllotoxin (PPT) has a well established lead in the development of antineoplastic agents for the treatment of cancer. Modifications were made at several positions in its skeleton with the aim of either improving potency or overcoming drug resistance. In recent years, the structurally modified podophyllotoxins have been investigated for their apoptosis-inducing ability. Podophyllotoxin-mimetic libraries rival the parent natural product by exhibiting nanomolar antiproliferative activities against human cancer cell lines and manifesting potent apoptosis inducing.

At present such libraries include compounds based on dihydropyridopyrazole, dihydropyridonaphthalene, dihydropyridoindole, and dihydropyridopyrimidine scaffolds.[20] Magedov et al. described a convergent approach employing three MCRs to synthesize the podophyllotoxin-mimetic, dihydropyridopyrazole library that

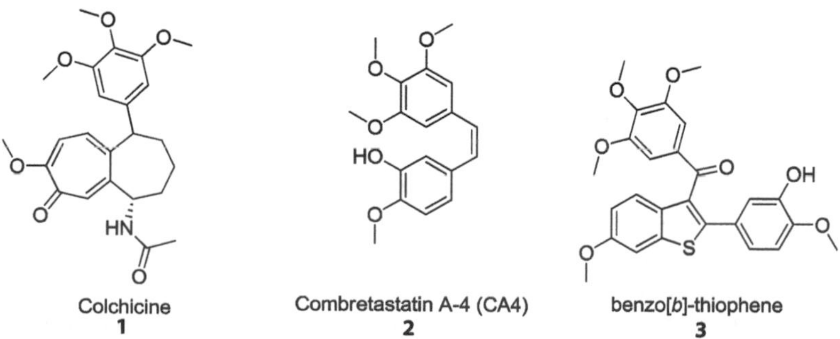

FIGURE 5.6 Structures of antitumor agents as tubulin polymerization inhibitors.

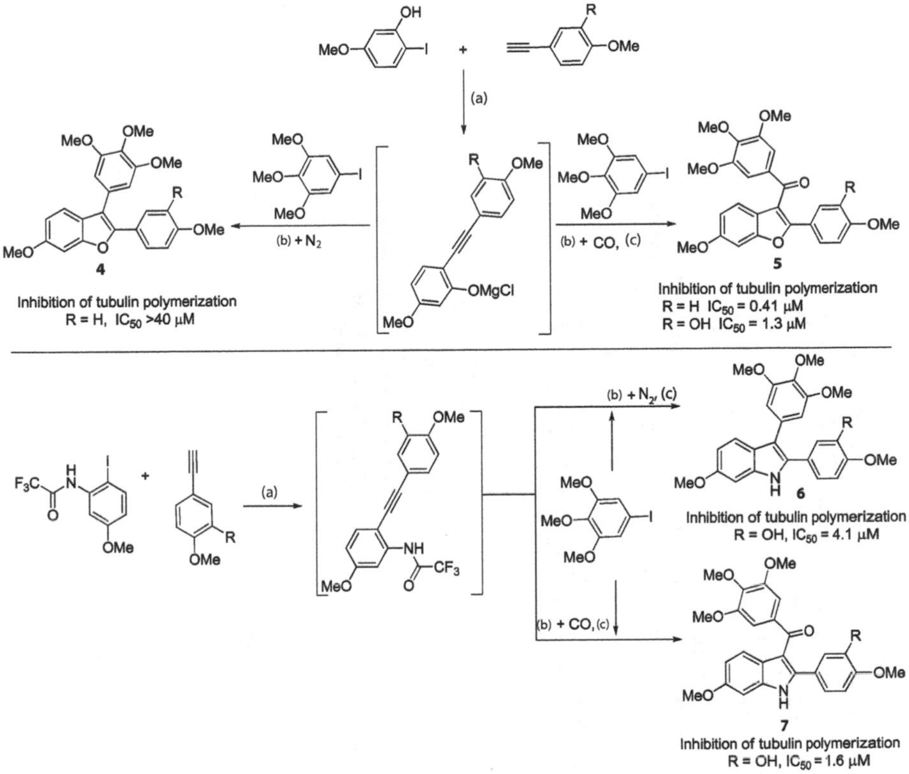

(a)

(b) + N_2

(b) + CO, (c)

4

Inhibition of tubulin polymerization
R = H, IC_{50} >40 μM

5

Inhibition of tubulin polymerization
R = H IC_{50} = 0.41 μM
R = OH IC_{50} = 1.3 μM

(b) + N_2, (c)

6

Inhibition of tubulin polymerization
R = OH, IC_{50} = 4.1 μM

(b) + CO, (c)

7

Inhibition of tubulin polymerization
R = OH, IC_{50} = 1.6 μM

SCHEME 5.1 Reagents and conditions: (a) MeMgCl, $Pd(PPh_3)_2Cl_2$ 3 mol%, THF, 65 °C, N_2; (b) DMSO, then heat to 80 °C, 16–18 h; (c) $AlCl_3$ 3 equiv., DCM.

retain the antitubulin mode of action of podophyllotoxin. These dihydropyridopyrazole compounds were found to inhibit *in vitro* tubulin polymerization and to disrupt the formation of mitotic spindles in dividing cells at low nanomolar concentrations (GI_{50} = 10 nM, MCF-7), in a manner similar to podophyllotoxin itself. Dihydropyridonaphthalene derivative **4** showed antimitotic activity against MCF-7 and HeLa human cancer cell lines in nanomolar concentarations.[21]

4-Aza-2,3-didehydropodophyllotoxin **3** is more than twice as cytotoxic as natural podophyllotoxin against the P-388 leukemia cancer cell line.[22] For the synthesis of the library of structurally modified podophyllotoxin (Scheme 5.2), a Knoevenagel-initiated, MCR-based, three-component reaction was designed to introduce diversity in scaffold. This process involves the reaction of tetronic acid, aromatic benzaldehydes, and activated anilines.[23,24]

Human kinesin Eg5, which plays an essential role in mitosis by establishing the bipolar spindle, has proven to be an interesting drug target for the development of cancer chemotherapeutics.[25a] Monastrol inhibits the basal and the microtubule-stimulated ATPase activity of the kinesin Eg5 motor domain.[25b,c] Monastrol, a BG-3CR product, was synthesized from ethyl-3-oxobutanoate, thiourea, and

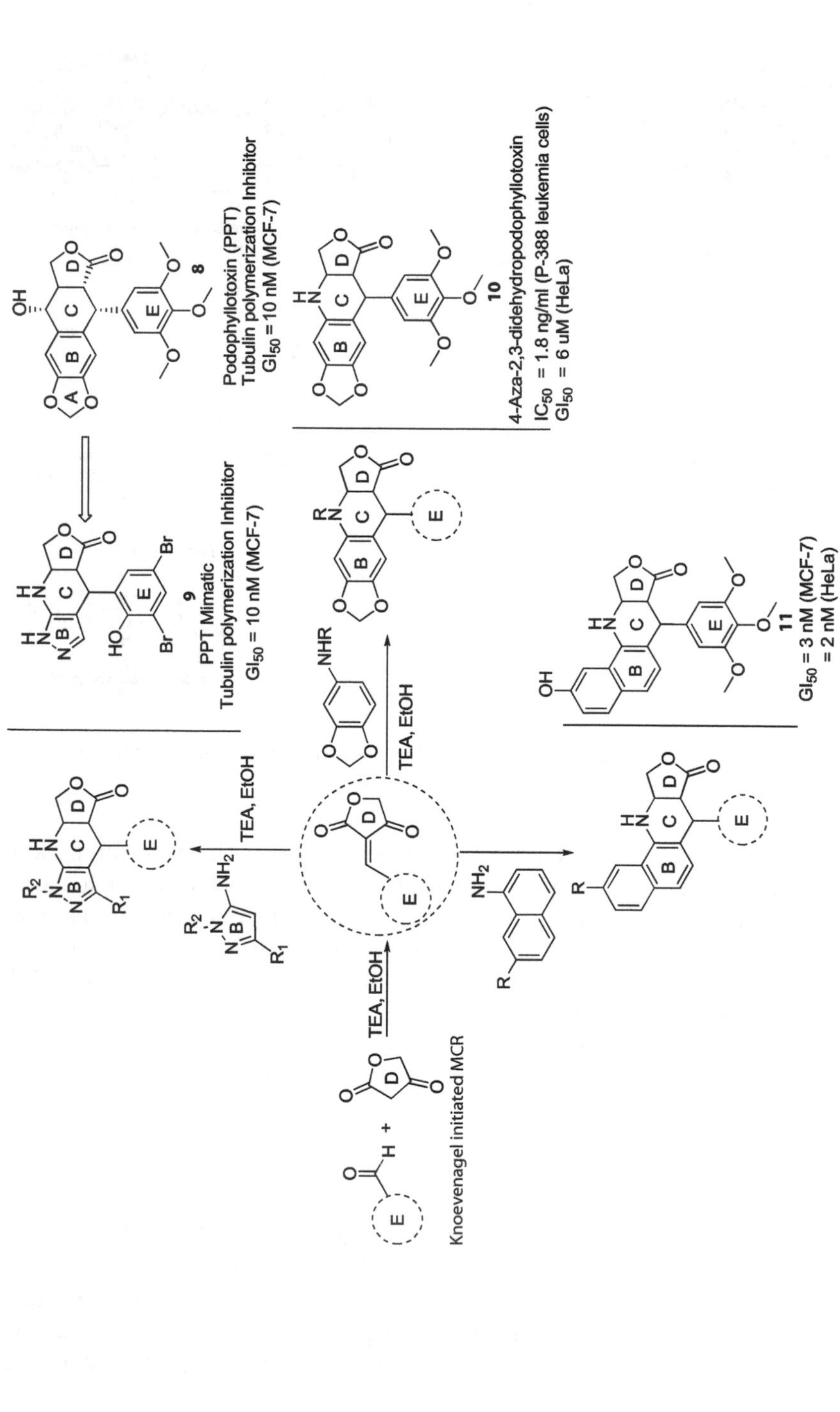

Knoevenagel initiated MCR

TEA, EtOH

9
PPT Mimatic
Tubulin polymerization Inhibitor
GI_{50} = 10 nM (MCF-7)

8
Podophyllotoxin (PPT)
Tubulin polymerization Inhibitor
GI_{50} = 10 nM (MCF-7)

10
4-Aza-2,3-didehydropodophyllotoxin
IC_{50} = 1.8 ng/ml (P-388 leukemia cells)
GI_{50} = 6 uM (HeLa)

11
GI_{50} = 3 nM (MCF-7)
= 2 nM (HeLa)

SCHEME 5.2 Synthesis of structurally modified podophyllotoxins.

3-hydroxybenzaldehydes.[26] Other kinesin inhibitors are enestron, dimethyl enestron, mon-97, and fluorastrol[27a,b] (Scheme 5.3).

A series of thiophene-based kinesin spindle protein inhibitor was synthesized from Gewald-3CR of cyclic ketone, ethyl cyanoacetate, and sulfur as the key step. (Scheme 5.4). These molecules have been found to exhibit submicromolar activity in secondary cellular assays and a cell phenotype consistent with KSP inhibition.[28]

Pyridone and quinolones are versatile building blocks for the construction of many bioactive natural products, synthetic pharmaceuticals, and therapeutic leads. The unique structural features and a wide range of biological activities of these heterocycles make them valuable privileged scaffolds in medicinal chemistry. Pyridones and quinolones are extremely diverse in structure due to the presence of different types of substitutions in their ring structure, which can influence their biological, pharmacological, and therapeutic actions. MCR is the most important tool for generation of diversity in these motifs. A Knoevenagel-initiated, MCR-based, three-component reaction was designed to introduce diversity in a pyridone scaffold. This reaction proceeds via formation of dicyano adduct, which undergoes Michael type addition followed by ring closure as a result of nucleophilic attack of the alcohol moiety to afford pyrano-pyridones hybrid molecules of the general formula **18**.[29] The synthesized library was evaluated for its antiproliferative activity against HeLa cells

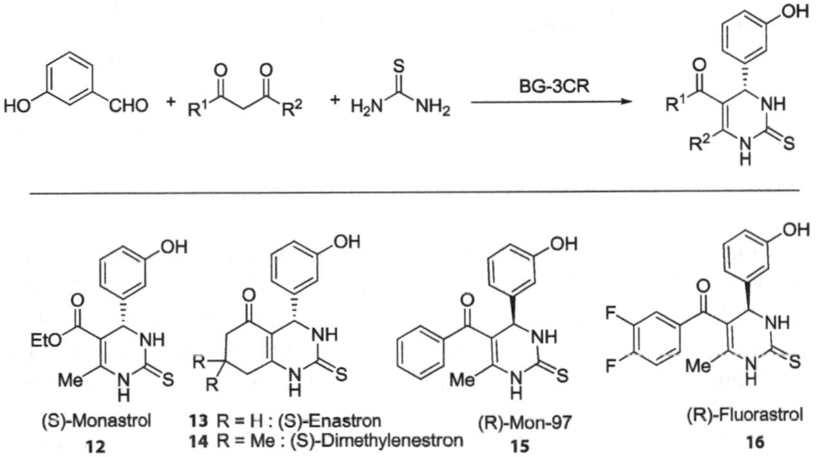

HO CHO + R1 R2 + H2N NH2 (S) BG-3CR → OH, O, R1, NH, R2, N H, S

OH, O, EtO, NH, Me, N H, S — (S)-Monastrol **12**

OH, O, NH, R, R, N H, S — **13** R = H : (S)-Enastron; **14** R = Me : (S)-Dimethylenestron

OH, O, NH, Me, N H, S — (R)-Mon-97 **15**

OH, F, F, O, NH, Me, N H, S — (R)-Fluorastrol **16**

SCHEME 5.3 Biginelli reaction involving three-component reaction of aldehyde, substituted 1,3-diketone, and thiophenol.

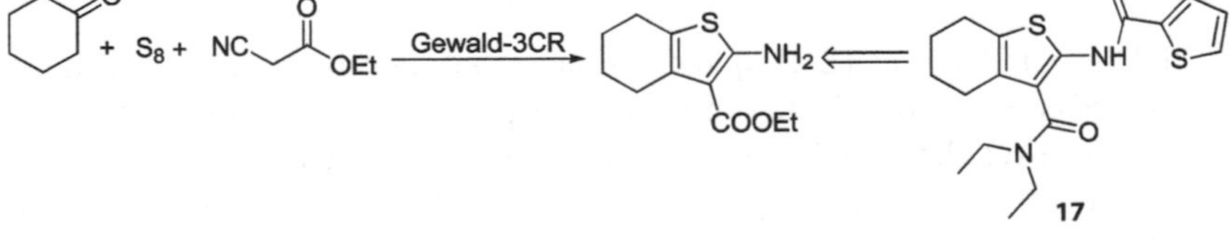

O + S_8 + NC O OEt — Gewald-3CR → S, NH_2, COOEt ⟸ S, NH, O, S, N, O — **17**

SCHEME 5.4 Gewald 3CR synthesis of thiophene-based KSP inhibitors.

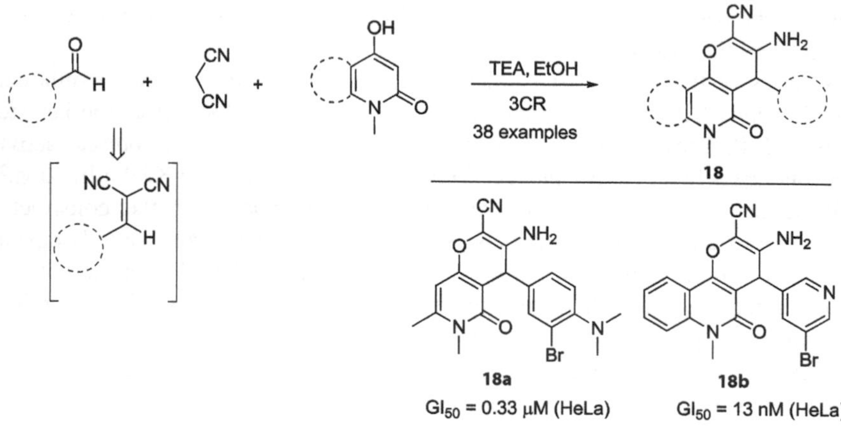

SCHEME 5.5 3CR toward antimitotic pyrano[2,3-*c*]pyridones and pyrano[2,3-*c*]quinolone.

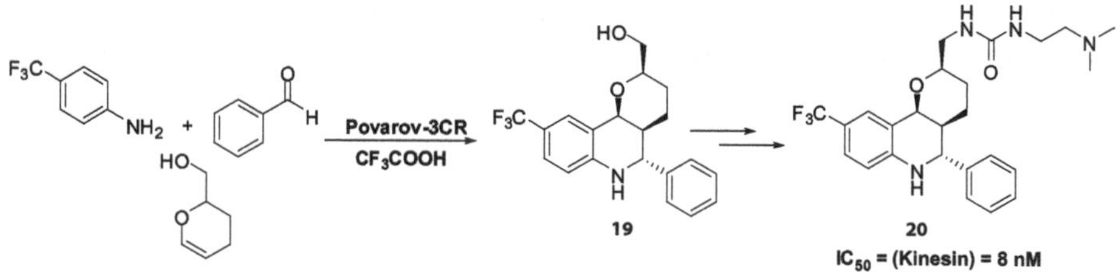

SCHEME 5.6 Pyrano[3,2-*c*]quinoline derivative formation through Povarov-3CR.

by assessing the viability of the cells. The most potent compounds of the pyrano[3,2-*c*]pyridone and pyrano[3,2-*c*]quinolone series were **18a** (GI_{50} = 0.33 μM) and **18b** (GI_{50} = 13 nM) (Scheme 5.5). However, compound 3-bromopyridine derivative **18b** showed much higher potency. Preliminary assays indicate that these compounds acted as tubulin polymerization inhibitors, which is a clinically validated anticancer drug target.[30–32]

A pyrano[3,2-*c*]quinoline derivative **20** obtained through a Povarov-3CR between benzaldehyde, aniline, and an electron-rich olefin was found to be a kinesin-5 inhibitor exhibiting promising potency in an *in vivo* xenograft model of COLO 205 cells and is currently undergoing early investigation in clinical cancer trials[33] (Scheme 5.6).

5.4.3 BET Inhibitors

The BET (bromodomains [BRDs] and extraterminal) family proteins regulate transcription and cell growth. The inhibition of BET protein by small molecules has been shown to down-regulate c-MYC expression in a number of hematopoietic cancer cell lines, resulting in a potent antiproliferative effect. Recently, two fused diazepine BET inhibitors have entered phase I clinical trials. These inhibitors selectively target BRDs of the BET. The GSK compound I-BET762 for treatment of NUT midline

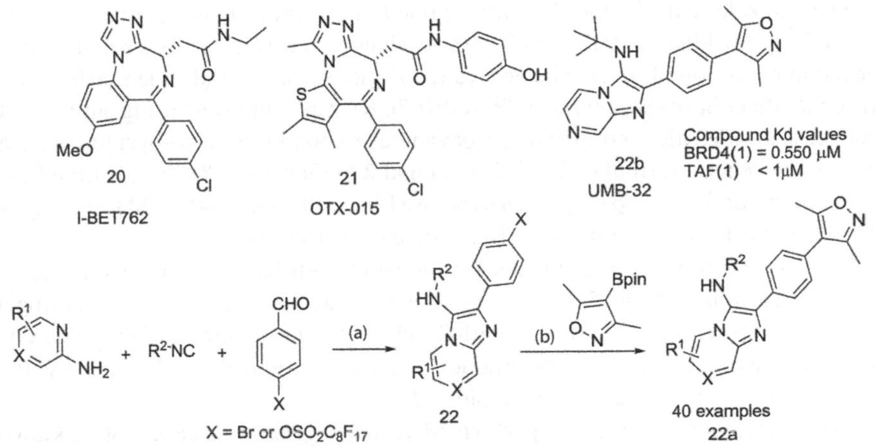

SCHEME 5.7 Synthesis of imidazo[1,2-*a*]pyrazine scaffold as BET protein inhibitors. Reagents and conditions: (a) $Sc(OTf)_3$, MW, DCM/MeOH, 150 °C; (b) $Pd(dppf)_2Cl_2$, K_2CO_3 MW, 150 °C.

carcinoma (NMC) while the Oncoethix compound OTX-015 is used for treatment of acute leukemia and other hematological malignancies.[34,35] Beyond BETs, there are 38 additional bromodomain-containing proteins for which high-quality, small-molecule probes are urgently needed. Transcription initiation factor TFIID subunits 1 (TAF1) and 1L (TAF1L) are two such proteins. As a component of the STAGA complex, which contains TRRAP, GCN5, TFIID, CBP/P300, mediator,[36] and Sp1,[37] TAF1 is susceptible to oncogenic activation by MYC. Moreover, TAF1 has been shown to block p53 activity,[38] and inactivation of TAF1 triggers a DNA damage response.[39]

Using fluorous-tagged multicomponent reactions, Bradner et al. developed a chemical library based on imidazo[1,2-*a*]pyrazine scaffold of bromodomain inhibitors around a 3,5-dimethylisoxazole biasing element with micromolar biochemical IC50. Lead compound UMB-32 was cocrystallized with BRD4, yielding a 1.56 Å resolution crystal structure. The lead compound (UMB-32) binds BRD4 with a K_d of 550 nM and 724 nM cellular potency in BRD4-dependent lines. Additionally, compound **22b** shows potency against TAF1[40] (Scheme 5.7).

5.4.4 Kinase Inhibitors

Synthesis of kinase inhibitors is another area of the successful application of MCR reactions for large-scale production of clinical candidates. Kinases are enzymes that transfer a phosphate group from a high-energy donor molecule, such as ATP, to a target substrate (most often a protein).[41] Protein phosphorylation is fundamental to many cellular events and heavily involved in cell division, making it an attractive target for the treatment of cancer. Several kinase inhibitors have been developed as successful anticancer drugs.[42]

The indazole scaffold was the starting point in this project, as it is known to possess affinity for kinases. Indazole-5-carboxaldehyde was selected as a suitable input for the introduction of diverse heterocyclic substituent on the indazole scaffold. The Groebke–Blackburn–Bienayme-3CR (GBB-3CR) was employed using amino heterocycles, aryldehyde, and different isocyanides to obtain imidazo-pyridine derivatives of the general formula **23**.[43–45] Compound **23a** (Scheme 5.8) was identified as a potent inhibitor for Gsk3β (glycogen synthase kinase 3β), (Ki = 0.01 μM), an enzyme that has been shown to be involved in prostate cancer growth.[46]

Pirali and co-workers prepared a series of compounds based on the imidazo[1,2-*a*] pyrazine scaffold (Scheme 5.9), which is known to interact with kinases via hydrogen bonding. Employing the GBB-3CR with 2-amino-3-chloropyrazine, they prepared a library consisting of two types of products—namely, imidazo[1,2-*a*]pyrazine with a primary, **25**, or a secondary amine moiety, **26**.[47]

The biological screening was performed against a general cascade system known to be involved in tumor formation. It involved a screen toward the FLT3 receptor, a member of the tyrosine kinase family, which after stimulation dimerizes, autophosphorylates, and activates a signaling cascade leading to cancer.[47] It is known that this system is involved in 25% of all myeloid leukemias through a mutation of the FLT3 receptor, causing it to be constitutively active. The compound collection was screened against this target leading to the identification of four hits (**25a**, **26a–c**) that were able to inhibit the signaling cascade significantly at 100 nM concentrations. For these compounds, complete dose–response curves were measured, which resulted in full inhibition of STAT5-dependent signaling at 10 μM concentration.

P38α MAP kinases are heavily involved in cytokine-mediated progression of rheumatoid arthritis, autoimmune diseases, and, most importantly, cancer.[48] Abadi and co-workers generated a small library by using a multicomponent reaction between an acetophenone, an aldehyde, ammonium acetate, and malononitrile or ethyl cyanoacetate to obtain 2-imino-1,2-dihydropyridine-carbonitriles **27** or the corresponding pyridone derivatives **28** (Scheme 5.10). The synthesized molecules were subsequently evaluated *in vitro* for their inhibitory activity against the p38a

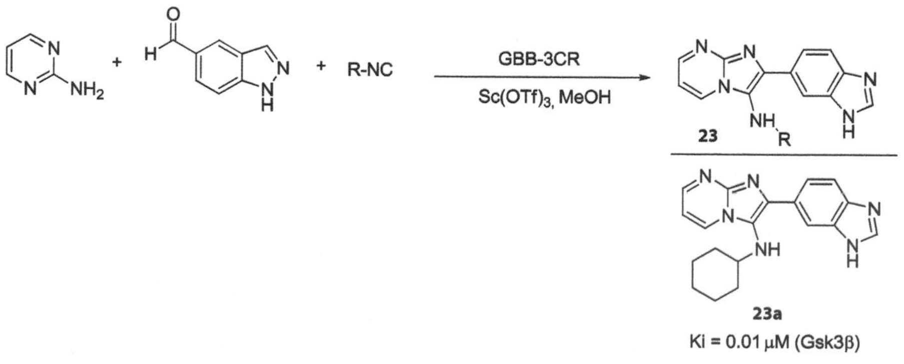

+ R-NC GBB-3CR $Sc(OTf)_3$, MeOH

23

23a

Ki = 0.01 μM (Gsk3β)

SCHEME 5.8 Synthesis of imidazo-pyridine derivatives by using amino heterocycles, aryldehyde, and different isocyanides.

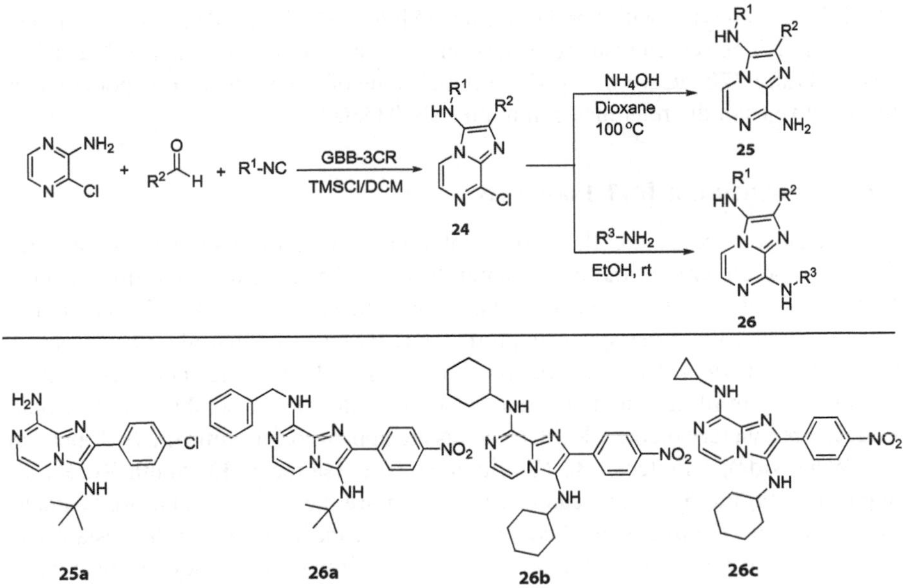

SCHEME 5.9 GBB-3CR synthesis of imidazole[1,2-*a*]pyrazines and postcondensation products showing inhibition of STAT5 signaling cascade.

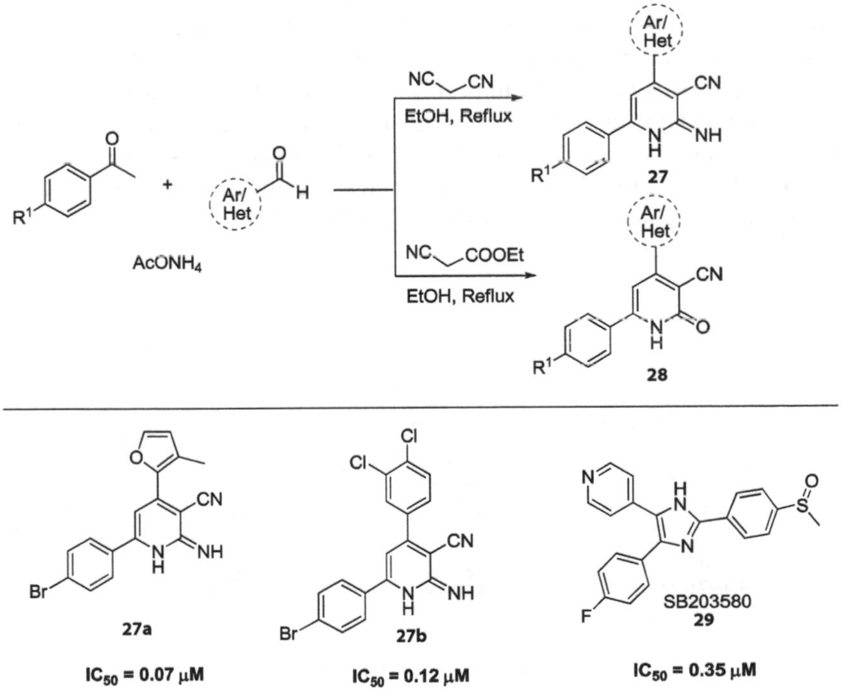

SCHEME 5.10 Synthesis of p38a MAP kinase enzyme inhibitors via 3-CR.

MAP kinase enzyme with known inhibitor SB203580 (IC_{50} = 0.035 μM) as a positive control. The screen resulted in several promising inhibitors, such as **27a** (IC_{50} = 0.07 μM) and **27b** (IC_{50} = 0.12 μM), although none of the produced compounds were more active than the reference compound SB203580.[49]

5.4.5 Anticancer Top1 Inhibitors

DNA topoisomerases are the targets of important anticancer and antibacterial drugs. These enzymes wind and unwind the DNA for replication and transcription. Interference of this process usually leads to apoptosis, which makes them an attractive target for cancer therapy. Camptothecin (CPT) 30 was first identified from the Chinese tree *Camptotheca acuminate.* It poisons TOP1 cleavage complex (Toplcc) by reversibly inhibiting its relegation. Due to limited water solubility and toxicity issues, three water-soluble CPT derivatives are approved for clinical use: Topotecan **31** (Worldwide), Irinotecan **32** (Worldwide), and Belotecan **33** (South Korea). As depicted in Figure 5.7, Topotecan and Irinotecan are the two FDA-approved camptothecins used for the treatment of colon and ovarian cancer. In spite of their established anticancer activity, camptothecins have a major limitation as they get inactivated within minutes at physiological pH by lactone E-ring opening.[50]

Magedov and co-workers synthesized a series of a small library of simplified tetracyclic camptothecin analogs (topoisomerases II inhibitors) **34**, **35**, and **36** by multicomponent reaction between aldehydes, 1,3-indanedione, and amino heterocycle invariable yields (Scheme 5.11). The reaction proceeds via a mechanism involving the formation of a Knoevenagel adduct followed by Michael addition/ring closure.

All the synthesized compounds were evaluated for their antiproliferative activities and apoptosis induction, leading to the discovery of the compound **35a**[51] exhibiting cytotoxic and apoptosis inducing potencies, which compared favorably with the clinical anticancer agent etoposide (56% inhibition of cell growth at 25 mM after 48 h). The **36a** and **36b** displayed high activities against several cancer cell lines, even higher than camptothecin,[51] as shown in Scheme 5.11.

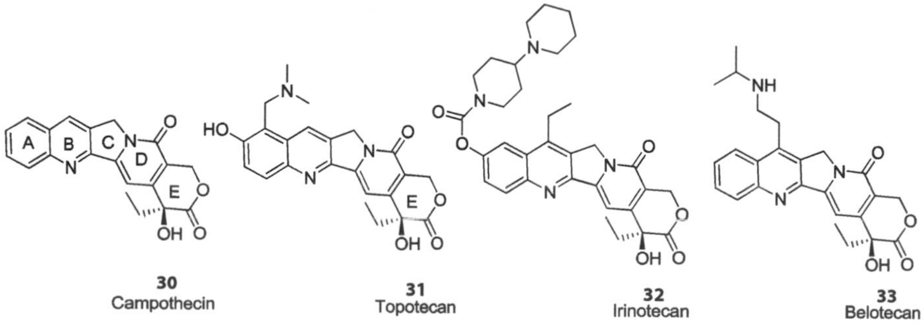

30 Campothecin

31 Topotecan

32 Irinotecan

33 Belotecan

FIGURE 5.7 Camptothecins as anticancerous TOP1 inhibitors.

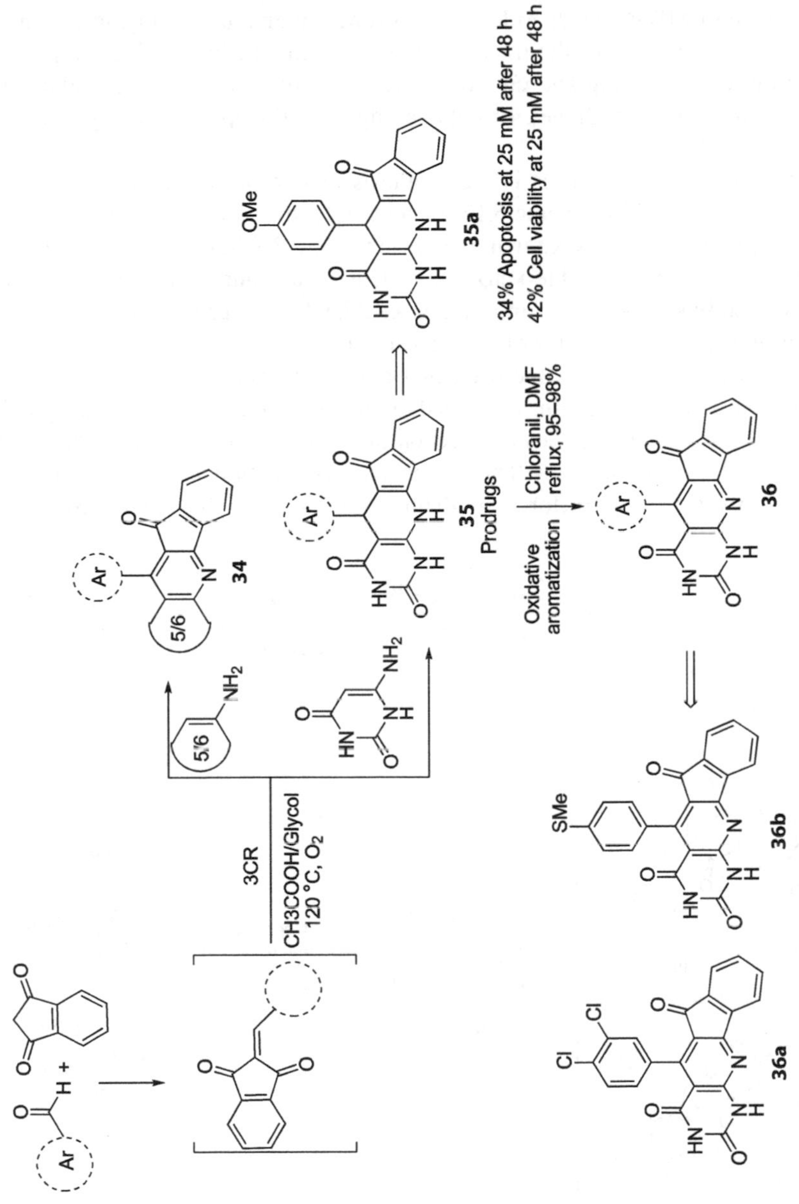

SCHEME 5.11 3-CR-based synthesis of uracil-based camptothecin analogs.

5.4.6 Inhibitors of p53–MDM2 Interaction

The p53 protein is a tumor suppressor protein that is activated upon cellular stress or damage. When activated it stimulates DNA repair, cell cycle arrest, or apoptosis. MDM2 is regulator of p53. It binds to p53 and targets it for proteasomal degradation. Overexpression of MDM2 leads to loss of p53 function and, ultimately, to carcinogenesis. Thus, suppression of the interaction between these two proteins is a promising drug target.[52] Domling and co-workers reported the synthesis and evaluation of 1,4-thienodiazepine-2,5-diones **40** obtained by an Ugi–deprotection–cyclization (UDC) approach.[53]

The key step in this synthesis is the synthesis of tetrasubstituted thiophene **37** by a U-4CR as depicted in Scheme 5.12. Interestingly, the carboxylic acid unit (**38**) is generated by another MCR known as the Gewald-3CR (G-3CR), which involves a reaction between an enolizable ketone or aldehyde, a cyanoacetate, and elemental sulfur to obtain substituted aminothiophenes **37**, which after Boc-protection and saponification are used as input for the Ugi reaction.[53,54]

The Ugi 4CR between the Boc-protected aminothiophenecarboxylic acids **38**, aldehydes, isocyanides, and amines afforded the linear products **39** in average yield. Initially the authors attempted the deprotection step without purification of **39**, which was not successful in all the cases. The subsequent deprotection and cyclization proceeded smoothly in one pot to afford a small library of 1,4-thienodiazepine-2,5-diones

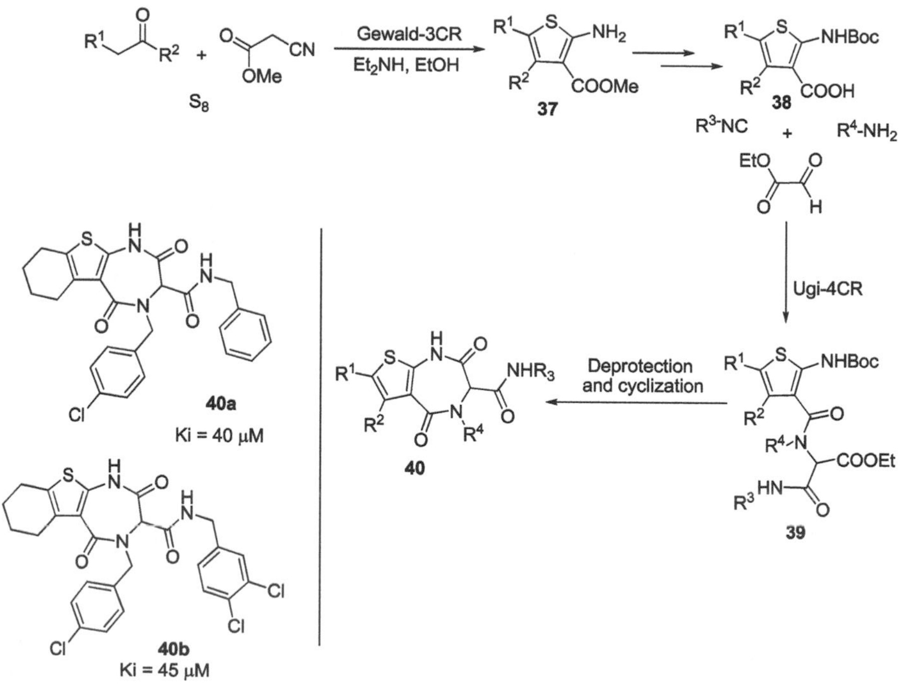

SCHEME 5.12 Synthesis of 1,4-thienodiazepine-2,5-diones (**40**) through the Ugi–deprotection–cyclization reaction.

40 in good to marginal yields (12%–70%), with variations possible on the amine, isocyanide, and thiophene carboxylic acid components. The authors screened the resulting compounds for inhibition of p53–MDM2 interaction and obtained two competitive inhibitors (**40a** and **40b**, which were similar in structure and activity, with an inhibition constant (Ki) in the range of 40 μM). The *p*-chlorobenzyl substituent appears to be playing an important role in mediating the activity as compounds that do not possess this moiety were inactive.[53]

A series of novel tricyclic 5*H*-thiochromenoyridines **41** were synthesized through 3-CR strategy involving the one-pot, trimethylamine catalyzed cyclization of 4-methoxy-salicylaldehyde, 2-amino-1,1,3-propenetricarbonitrile and different aromatic thiols in ethanol under reflux (Scheme 5.13).[55]

All newly prepared 5*H*-thiochromenopyridines showed good to moderate cytotoxicity against three melanoma and two glioma cell lines (3–15 μM) with compounds **41b** and **41c** showing the best activity (IC_{50} ~ 3.6 μM) against melanoma, while compounds **1e** and **1f** were the most potent for glioma cell lines (IC_{50} ~ 3 μM). J-M Yang et al. reported a one-pot, regioselective 3CR [3+2] cycloaddition reaction of azomethineylides, 1,2 diketo compounds, and amino acid derivatives to obtain novel dispiropyrrolizidines **42** (Scheme 5.14). The synthesized derivatives **42a–42y**

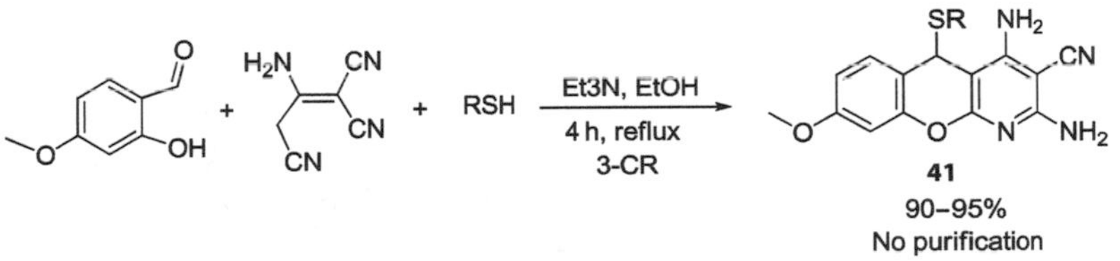

Et3N, EtOH
4 h, reflux
3-CR
41
90–95%
No purification

SCHEME 5.13 One-pot synthesis of 5*H*-thiochromenoyridines through 3-CR strategy.

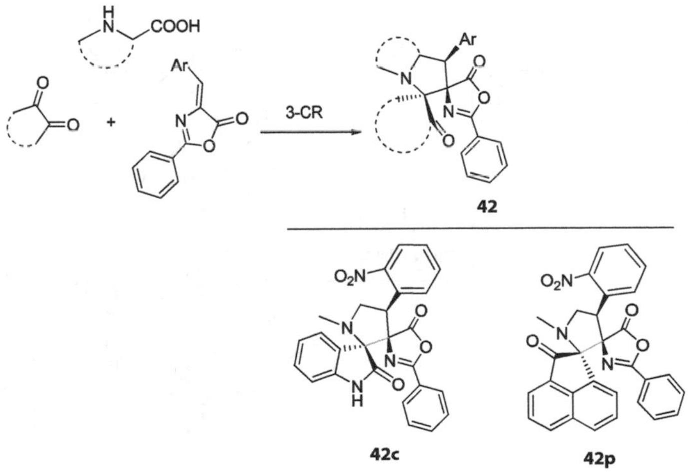

3-CR
42
42c
42p

SCHEME 5.14 Formation of novel dispiropyrrolizidines and their derivatives via 3CR [3+2] cycloaddition reaction.

were screened *in vitro* against hepatic carcinoma (HepG2) cells for anticancer activity. Most of the compounds were found to exhibit cytotoxicity. Compounds **42c** and **42p** were the most promising of the lot, having IC_{50} values of 10.5 ± 0.3 and 11.7 ± 1.4 µM, respectively.[56]

5.5 OTHER MISCELLANEOUS ANTICANCER MOLECULES

Novel 4-hydroxyquinolone grafted spiro pyrrolidines **43**, **45** and spiro thiapyrrolizidines **44** (Scheme 5.15) were prepared through a one-pot 3-CR of 3-cinnamoyl-4-hydroxyquinolin-2(1*H*)-one dipolarophiles, isatin and sarcosine/thiaproline/phenylglycine. Selected molecules were screened *in vitro* for cytotoxicity against cervical and colon cancer cell lines and were found to show good anticancer activity. Compounds **43c**, **43d**, and **45c** showed very promising antiproliferative activity (IC_{50} > 10 µM) against human colon cancer cell line (HCT116).[57]

A one-pot convenient synthesis of a novel series of spiro oxindole pyrrolidines **46** and **47** was accomplished through a 3-CR of isatin and sarcosine or thioproline with the dipolarophile, 3-(1*H*-imidazol-2-yl)-2-(1*H*-indole-3-carbonyl) acrylonitrile. The disclosed synthetic strategy involves an initial 1,3-dipolar cycloaddition of isatin and sarcosine or thioproline to generate azomethineylides. This then reacts with

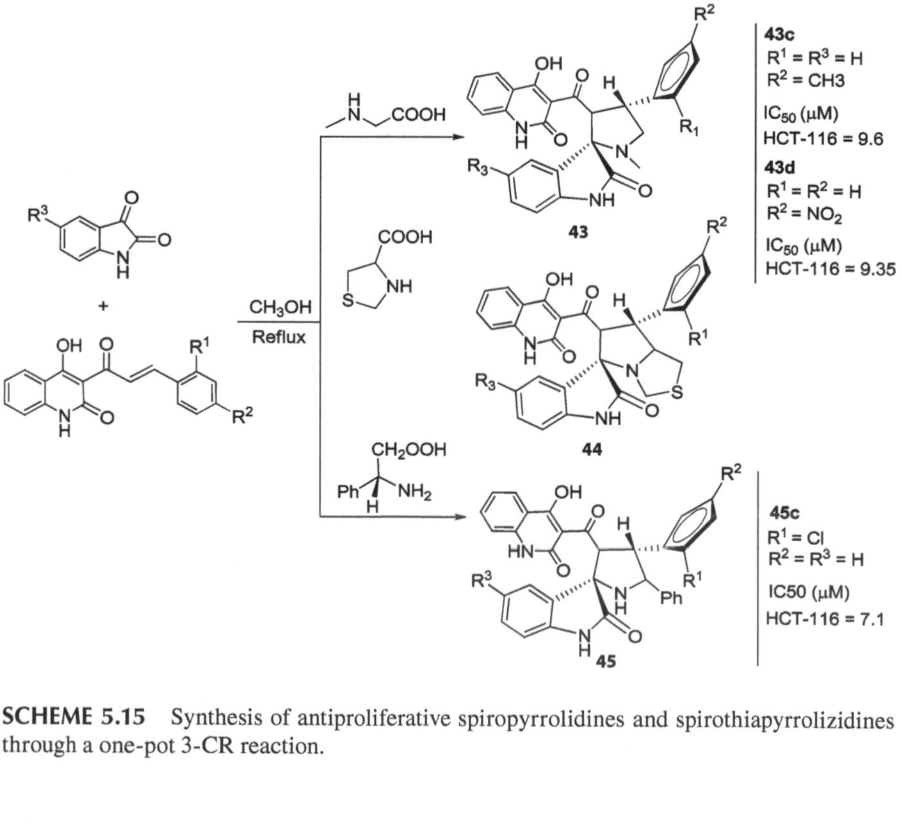

SCHEME 5.15 Synthesis of antiproliferative spiropyrrolidines and spirothiapyrrolizidines through a one-pot 3-CR reaction.

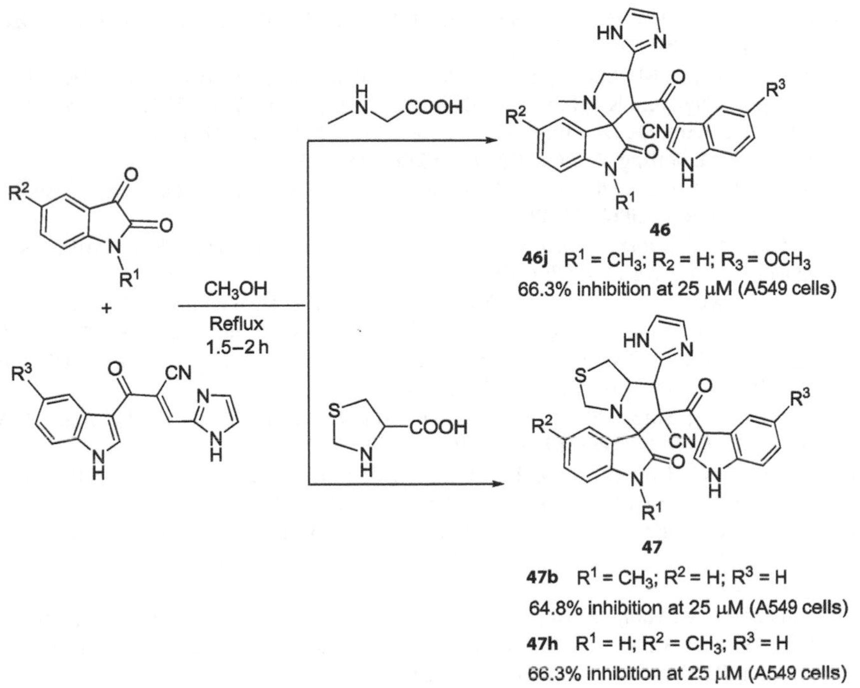

46

46j $R^1 = CH_3$; $R_2 = H$; $R_3 = OCH_3$
66.3% inhibition at 25 μM (A549 cells)

47

47b $R^1 = CH_3$; $R^2 = H$; $R^3 = H$
64.8% inhibition at 25 μM (A549 cells)

47h $R^1 = H$; $R^2 = CH_3$; $R^3 = H$
66.3% inhibition at 25 μM (A549 cells)

SCHEME 5.16 A one-pot 3-CR synthesis of spirooxindole-pyrrolidines showing cytotoxicity against A549 human lung adenocarcinoma cancer cell line.

the dipolarophile to furnish the target compounds (Scheme 5.16).[58] The synthesized spiro oxindole pyrrolidines were evaluated for *in vitro* cytotoxicity against the A549 human lung adenocarcinoma cancer cell line. Though all the 29 molecules were found to be active, three molecules—**46j**, **47b**, and **47h**—showed very promising cytotoxicity: 66.3%, 64.8%, and 66.3% at 25 μg/mL concentrations, respectively. Compounds **47b** and **47h** were also found to cause apoptotic DNA fragmentation with compound **47h** also inducing apoptosis in the A549 cancers.

REFERENCES

1. Ugi, I.; Dömling, A.; Horl, W. *Endeavour.* **1994**, *18*, 115–122.
2. Passerini, M. *Gazz. Chim. Ital.* **1921**, *51*, 126–129.
3. Passerini, M. *Gazz. Chim. Ital.* **1921**, *51*, 181–188.
4. Dömling, A. *Chem. Rev.* **2006**, *106*, 17–89.
5. Parenty, A. D.; Song, Y. F.; Richmond, C. J.; Cronin, L. *Org Lett.* **2007**, *9*, 2253–2256.
6. Joseph, S.; Andrew, J. K.; Mark, M.; John, J. F.; Andrew, A.; Mark, A. O. *J. Org. Chem.* **2000**, *65*, 1516–1524.
7. Robin, S. B.; Chongen, H.; Marinus, J. B.; Rob, F. S.; Frans, J. J.; de K.; Martin, L.; Anthony, L. S.; Romano, V. A. O. *Org. Lett.* **2003**, *5*, 3759–3762.
8. Hantzsch, A. *Ber. Dtsch. Chem. Ges.* **1881**, *14*, 1637–1638.

9. (a) Biginelli, P. *Ber. Dtsch. Chem. Ges.* **1891**, *24*, 1317–1319; (b) Biginelli, P. *Ber. Dtsch. Chem. Ges.* **1891**, *24*, 2962–2967.
10. Shestopalov, A. M.; Shestopalov, A. A.; Rodinovskaya, L. A. *Synthesis* **2008**, *1*, 1–25.
11. Wang, K.; Nguyen, K.; Huang, Y.; Dömling, A. *J. Comb. Chem.* **2009**, *11*, 920–927.
12. Wang, K.; Kim, D.; Dömling, A. *J. Comb. Chem.* **2010**, *12*, 111–118.
13. Huang, Y.; Dömling, A. *Mol. Diversity* **2011**, *15*, 3–33.
14. Blanco-Urgoiti, J.; Añorbe, L.; Pérez-Serrano, L.; Domínguez, G.; Pérez-Castells, J. *Chem. Soc. Rev.* **2004**, *33*, 32–42.
15. (a) Goel, N. S. *South Asian J. Cancer* **2013**, *2*, 285–287; (b) Siegel, R. L.; Miller, K. D.; Jemal, A. *CA. Cancer J. Clin.* **2016**, *66*, 7–30.
16. Hussain, M. K.; Ansari, M. I.; Yadav, N.; Gupta, P. K.; Gupta, A. K.; Saxena, R.; Fatima, I.; Manohar, M.; Kushwaha, P.; Khedgikar, V.; Gautam, J.; Kant, R.; Maulik, P. R.; Trivedi, R.; Dwivedi, A.; Kumar, K. R.; Saxena, A. K.; Hajela, K. *RSC Adv.* **2014**, *4*, 8828–8845 and references cited therein.
17. Pettit, G. R.; Singh, S. B.; Boyd, M. R.; Hamel, E.; Pettit, R. K.; Schmidt, J. M.; Hogan, F. *J. Med. Chem.* **1995**, *38*, 1666–1672.
18. Pinney, K. G.; Bounds, A. D.; Dingeman, K. M.; Mocharla, V. P.; Pettit, G. R.; Bai, R.; Hamel, E. *Bioorg. Med. Chem. Lett.* **1999**, *9*, 1081–1086.
19. Bernard, L. F.; Ernest, H.; Katherine, J. M. *J. Med. Chem.* **2002**, *45*, 2670–2673.
20. Ahmed, K.; Hussaini, S. M. A.; Malik, S. M. *Anticancer Agents Med. Chem.* **2015**, *15*, 565–574.
21. Magedov, I. V.; Manpadi, M.; Ogasawara, M. A.; Dhawan, S. A.; Rogelj, S.; Slambrouck, S. V.; Wim, F. A.; Nikolai, S.; Evdokimov, M.; Uglinskii, P. Y.; Elias, E. M.; Knee, E. J.; Tongwa, P.; Yu, M. A.; Kornienko, A. *J. Med. Chem.* **2011**, *54*, 4234–4246.
22. Hitotsuyanagi, Y.; Fukuyo, M.; Tsuda, K.; Kobayashi, M.; Ozeki, A.; Itokawa, H.; Takeya, K. *Bioorg. Med. Chem. Lett.* **2000**, *10*, 315–317.
23. Christophe, T.; Giorgi-Renault, S.; Henri-Philippe, H. *Org. Lett.* **2002**, *4*, 3187–3189.
24. Isambert, N.; Lavilla, R. *Chem. Eur. J.* **2008**, *14*, 8444–8454.
25. (a) Sawin, K. E.; Mitchison, T. *J. Proc. Natl. Acad. Sci. USA.* **1995**, *92*, 4289–4293; (b) Mayer, T. U.; Kapoor, T. M.; Haggarty, S. J.; King, R. W.; Schreiber, S. L.; Mitchison, T. J., *Science* **1999**, *286*, 971–974; (c) Kapoor, M. Z.; Mitchison, T. J. *Chem Biol.* **2002**, *9*, 989–996.
26. (a) Bose, D. S.; Sudharshan, M.; Chavhan, S. W. *Arkivoc*, **2005**, *iii*, 228–236; (b) Doris, D. D.; Kappe, C. O. *Nat. Protoc.* **2007**, *2*, 317–321.
27. (a) Garcia-Saez, I.; De-Bonis, S.; Lopez, R.; Trucco, F.; Rousseau, B.; Kozielski, F.; Thuéry, P. *J. Biol. Chem.* **2007**, *282*, 9740–9747; (b) Sarli, V.; Giannis, A. *Chem. Med. Chem.* **2006**, *1*, 293–298.
28. Pinkerton, A. B.; Lee, T. T.; Hoffman, T. Z.; Wang, Y.; Kahraman, M.; Cook, T. G.; Severance, D.; Gahman, T. C.; Noble, S. A.; Shiau, A. K.; Davis, R. L. *Bioorg. Med. Chem. Lett.* **2007**, *17*, 3562–3569.
29. Magedov, I. V.; Manpadi, M.; Ogasawara, M. A.; Dhawan, S. A.; Rogelj, S.; Slambrouck, S. V.; Wim, F. A.; Nikolai, S.; Evdokimov, M.; Uglinskii, P. Y.; Elias, E. M.; Knee, E. J.; Tongwa, P.; Yu, M. A.; Kornienko, A. *J. Med. Chem.* **2008**, *51*, 2561–2570.
30. Kamperdick, C.; Van, N. H.; Sung, T. V.; Adam, G. *Phytochemistry* **1999**, *50*, 177–181.
31. Chen, I.-S.; Wu, S.-J.; Tsai, I.-L.; Wu, T.-S.; Pezzuto, J. M.; Lu, M. C.; Chai, H.; Suh, N.; Teng, C.-M. *J. Nat. Prod.* **1994**, *57*, 1206–1211.
32. McBrien, K. D.; Gao, Q.; Huang, S.; Klohr, S. E.; Wang, R. R.; Pirnik, D. M.; Neddermann, K. M.; Bursuker, I.; Kadow, K. F.; Leet, J. E. *J. Nat. Prod.* **1996**, *59*, 1151–1153.
33. Schiemann, K.; Finsinger, D.; Zenke, F.; Amendt, C.; Knöchel, T.; Bruge, D.; Buchstaller, H.-P.; Emde, U.; Stähle, W.; Anzali, S. *Bioorg. Med. Chem. Lett.* **2010**, *20*, 1491–1495.

34. Jean-Marc, G.; Phillip, P. S.; Christopher, J. *Burns Expert Opin. Ther. Patents* **2014**, *24*, 185–199.
35. Muller, S.; Filippakopoulos, P.; Knapp, S. *Expert Rev. Mol. Med.* **2011**, *13*, e29.
36. Liu, X.; Vorontchikhina, M.; Wang, Y. L.; Faiola, F.; Martinez, E. *Mol. Cell. Biol.* **2008**, *28*, 108–121.
37. Parisi, F.; Wirapati, P.; Naef, F. *Nuc. Acids Res.* **2007**, *35*, 1098–1107.
38. Li, H. H.; Li, A. G.; Sheppard, H. M.; Liu, X. *Mol. Cell* **2004**, *13*, 867–878.
39. Buchmann, A. M.; Skaar, J. R.; De-Caprio, J. A. *Mol. Cell. Biol.* **2004**, *24*, 5332–5339.
40. McKeown, M. R.; Shaw, D. L.; Fu, H.; Liu, S.; Xu, X.; Marineau, J. J.; Huang, Y.; Zhang, X.; Buckley, D. L.; Kadam, A.; Zhang, Z.; Blacklow, S. C.; Qi, J.; Zhang, W.; Bradner, J. E. *J. Med. Chem.* **2014**, *57*, 9019–9027.
41. Manning, G.; Whyte, D. B.; Martinez, R.; Hunter, T.; Sudarsanam, S. *Science* **2002**, *298*, 1912–1934.
42. Zhang, J.; Yang, P. L.; Gray, N. S. *Nat. Rev. Cancer* **2009**, *9*, 28–39.
43. Groebke, K.; Weber, L. *Synlett* **1998**, *51*, 661–663.
44. Guan, B.; Fleming, B. P.; Shiosaki, K.; Tsai, S. *Tetrahedron Lett.* **1998**, *39*, 3635–3638.
45. Bienaym, H.; Bouzid, K. *Angew. Chem., Int. Ed.* **1998**, *37*, 2234–2237.
46. Wang, L.; Lin, H.-K.; Hu, Y.-C.; Xie, S.; Yang, L.; Chang, C. *J. Biol. Chem.* **2004**, *279*, 32444–32452.
47. Guasconi, M.; Lu, X. Y.; Massarotti, A.; Caldarelli, A.; Ciraolo, E.; Tron, G. C.; Hirsch, E.; Sorba, G.; Pirali, T. *Org. Biomol. Chem.* **2011**, *9*, 4144–4149.
48. Miwatashi, S.; Arikawa, Y.; Kotani, E.; Miyamoto, M.; Naruo, K.-I.; Kimura, H.; Tanaka, T.; Asahi, S.; Ohkawa, S. *J. Med. Chem.* **2005**, *48*, 5966–5979.
49. Serry, A. M.; Luik, S.; Laufer, S.; Abadi, A. H. *J. Comb. Chem.* **2010**, *12*, 559–565.
50. Pommier, Y.; Leo, E.; Zhang, H. L.; Marchand, C. *Chem. Biol.* **2010**, *17*, 421–433.
51. Evdokimov, N. M.; Slambrouck, S. V.; Heffeter, P. L.; Le Calve, T. B.; Lamoral-Theys, D.; Hooten, C. J.; Uglinskii, P. Y.; Rogelj, S.; Kiss, R.; Steelant, W. F. A.; Berger, W. J.; Yang, J.; Bologa, C. G.; Kornienko, A.; Magedov, I. V. *J. Med. Chem.* **2011**, *54*, 2012–2021.
52. Vassilev, L. T.; Vu, T. B.; Graves, D.; Carvajal, D.; Podlaski, F.; Filipovic, Z.; Kong, N.; Kammlot, C.; Lukacs, C.; Klein, C.; Fothouhi, E.; Liu, A. *Science* **2004**, *303*, 844–848.
53. Huang, Y.; Wolf, S.; Blsta, M.; Meireles, L.; Camacho, C.; Holak, T. A., Domling, A. *Chem. Biol. Drug Des.* **2010**, *76*, 116–129.
54. Sabnis, R. W.; Rangnekar, D. W.; Sonawane, N. D. *J. Heterocycl. Chem.* **1999**, *36*, 333–345.
55. Banerjee, S.; Wang, Jin.; Pfeffer, S.; Dejian, M.; Pfeffer, L. M., Patil, S. A., Wei, L.; Miller, D. D. *Molecules* **2015**, *20*, 17152–17165.
56. Yang, J.-M.; Shi, D.-Q.; Hu, Y.; Li, Q.; Yu, F.; Cao, J.; Fang, D.; Huang, Z.-B. *ACS Comb. Sci.* **2014**, *16*, 139–145.
57. Sankaran, M.; Uvarani, C.; Chandraprakash, K.; Lekshmi, S. U.; Suparna, S.; Platts, J.; Mohan, P. S. *Mol. Divers.* **2014**, *18*, 269–283.
58. Arun, Y.; Saranraj, K.; Balachandran, C.; Perumal, P. T. *Eur. J. Med. Chem.* **2014**, *74*, 50–64.

6 Transition Metal Catalyzed Synthesis of Heterocycles via Multicomponent Reactions

*Satavisha Sarkar, Arghya Banerjee, and Bhisma K. Patel**

CONTENTS

* Corresponding author: patel@iitg.ernet.in.

6.1 INTRODUCTION

Heterocycles have grabbed the attention of the scientific fraternity for centuries as they are not only the essential constituents of biologically active molecules but also important building blocks of pharmaceutical compounds and organic and various other functional materials. In fact, two-thirds of organic compounds are made of heterocyclic compounds. Even the most significant molecules of our mundane life contain heterocyclic moieties such as DNA, RNA, chlorophyll, heme, and vitamins. The biological activities of heterocycles have found applications in antitumor, anti-HIV, antibiotic, antiviral, antimicrobial, antibacterial, antifungal, antidiabetic, antidepressant, anti-inflammatory, antimalarial, herbicidal, fungicidal, and insecticidal activities. Apart from their role as drug molecules, the heterocyclic skeletons acts as synthons for further organic transformations or serve as organocatalysts, chiral auxiliaries, and metal ligands in asymmetric catalysis. Therefore, new, versatile, and productive strategies for the synthesis of heterocyclic scaffolds are always in high demand.

While designing an efficient strategy to synthesize targeted complex heterocyclic skeletons in a simple yet selective pathway, multicomponent reactions or MCRs evolved as a revolution in the synthetic organic community. Blending the concepts of domino reactions, sequential reactions as well as consecutive reactions in conventional

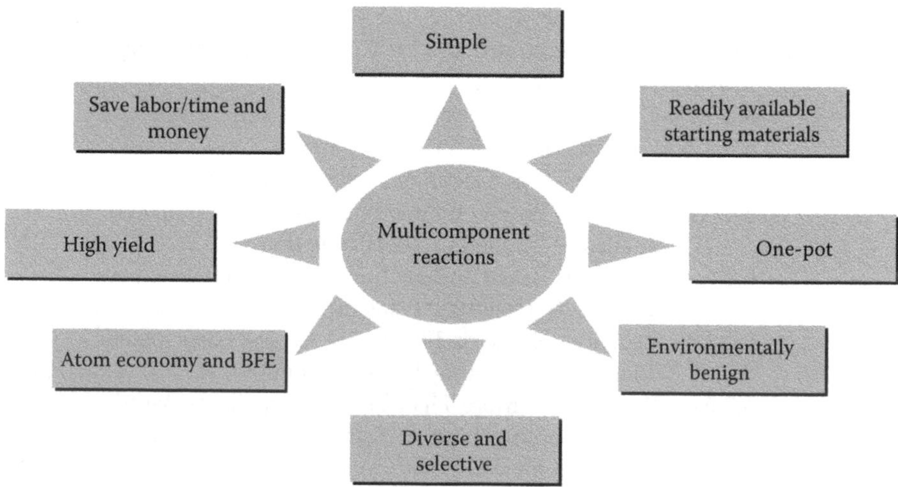

FIGURE 6.1 An overview of multicomponent reactions.

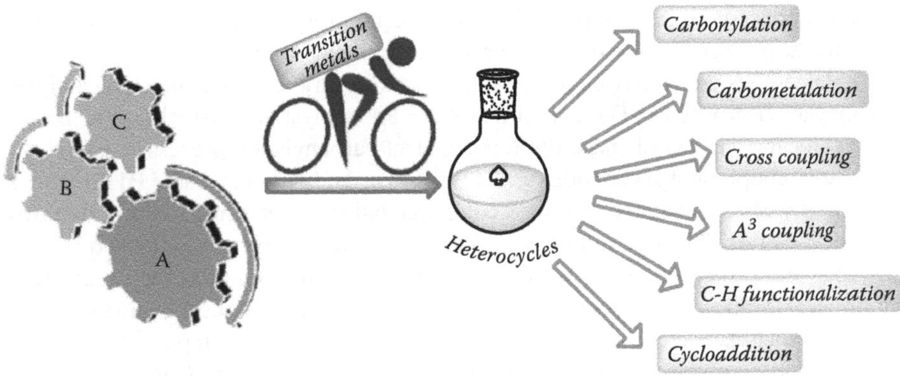

FIGURE 6.2 Versatile role of transition metals during the synthesis of heterocycles by MCRs.

concept, MCRs have blanketed a vast area of organic transformation.[1] The reasons behind the increase in the popularity of MCRs are summarized in Figure 6.1.

The development of efficient strategies for the synthesis of complex building blocks from simple and readily available starting materials in a single-step is one of the challenges in synthetic organic chemistry. The growing relevance of multi-component reactions is thus the subject of interest from the past several decades as it requires multiple starting materials to react selectively, generating a wide array of relevant products via diversity-oriented synthesis. In this context, transition metal catalyzed multicomponent reactions came up as the central theme due to their versatile role for designing new synthetic routes for the synthesis of bioactive heterocycles.[2] The content of this chapter has been divided into many sections and subsections based on the nature of the reaction, the type of intermediate formed, and the mode of cyclization during the formation of heterocyclic scaffolds. Some of the strategies delineated are depicted in Figure 6.2.

6.2 CYCLIZATION VIA COUPLING REACTIONS

Various coupling reactions (viz. Heck, Sonogashira, Suzuki and many related carbon–carbon (C–C) and carbon–heteroatom (C–X) bond-forming reactions are the foremost examples for the synthesis of a variety of bioactive heterocycles using transition metal catalyzed multicomponent processes. These two component coupling reactions are extended to multicomponent reactions via two different pathways, either employing the *in situ* generated cross-coupled products in subsequent cyclization reactions or merging transition metal-containing oxidative addition intermediates (organometallic intermediates) with any other transformations.

6.2.1 Cyclization via Carbonylation Process

Carbonylative coupling reactions were started during 1974 by Heck et al.[3]; however, their implications in heterocyclic synthesis were realized much later. This transformation inserts carbon monoxide during the multicomponent process to produce carbonyl containing heterocycles. To date, the majority of carbonylation reactions reported have been carried out in the presence of a palladium catalyst.

o-Haloarenes are employed as useful synthons in various palladium (Pd) catalyzed carbonylation processes. The use of a suitable Pd catalyst and also choice of base is crucial for the success of these three component carbonylation processes. Initial carbonylative coupling of aryl halides with alkynes via the *in situ* generated Pd (0) catalyst followed by a base mediated intramolecular cyclization, generating a range of bioactive heterocycles (viz. flavones, quinolones and many more as shown in Scheme 6.1).[4]

Apart from these three-component carbonylative cyclization processes, many four-component cyclocarbonylations are also reported in the literature. 1,2-Dihalo substituted benzenes such as *o*-bromoiodobenzene and *o*-diiodobenzene are reported to undergo Pd catalyzed, four-component reactions with alkynes, amines, and CO to form indolin-1-ones by the Sonogashira coupling–aminocarbonylation–hydroamination sequence in the presence of phosphonium salt-based ionic liquids (IL) (Scheme 6.2).[5] Both Sonogashira and aminocarbonylation are the initial two steps in this process, generating alkyne-tethered intermediate (**I**), which undergoes further intramolecular cyclization to generate the desired product. A similar four-component MCR reaction

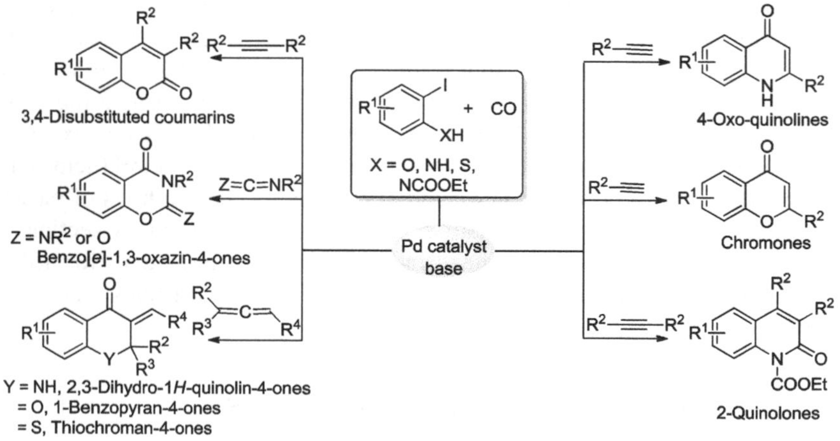

SCHEME 6.1 Cyclization via a multicomponent carbonylation process.

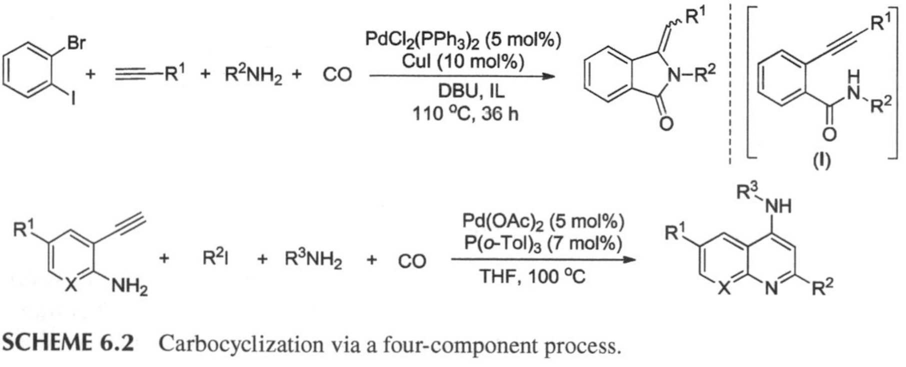

SCHEME 6.2 Carbocyclization via a four-component process.

is reported by Rossi et al. using *o*-alkynyl substituted anilines, CO, amines, and iodoarenes. A carbonylative Sonogashira reaction of *o*-alkynyl amine and iodoarene generates alkynone, which further undergoes subsequent cyclization (Scheme 6.2).[6]

6.2.2 Use of *ortho*-Substituted Arene for the Synthesis of Fused Ring Heterocycles

6.2.2.1 Monometallic/Bimetallic Catalysis

ortho-Substituted arenes and heteroarenes are the common building blocks for the synthesis of a variety of fused ring heterocycles via multicomponent reactions using transition metal catalysts. A palladium catalyst typically activates the aryl halide bonds of *o*-substituted haloarenes toward new bond formation and subsequent cyclization to generate fused ring products.[7] These *ortho*-substituted aromatic or heteroaromatic compounds can be coupled with terminal alkynes via Sonogashira coupling to generate an *o*-alkynylated intermediate for subsequent cyclization. By changing the nucleophile and even by using a non-nucleophilic unsaturated unit, the final cyclization step can be modulated suitably to generate fused ring heterocycles. After the early examples by Cacchi,[8] the Flynn group reported three-component coupling reactions of *ortho*-haloanilines or *ortho*-halophenols, terminal alkynes, and aryl iodides to synthesize indoles or benzofurans (Scheme 6.3).[9,10] This reaction occurs in two steps, where the initial Sonogashira type coupling of *o*-haloarene or heteroarene generates alkynyl intermediate (**II**) and subsequent palladium catalyzed intramolecular cyclization and reductive elimination, with organic halides generating the desired benzo-fused ring.

In addition to this, indoles can be synthesized using *ortho*-dihaloarenes, terminal alkynes, and amines via a one-pot, two-step Sonogashira coupling/amination protocol (Scheme 6.4).[11] Sonogashira coupling of the C–I bond, followed by C–N bond formation on the C–Cl bond provides the desired product after an intramolecular cyclization. Reaction of *ortho*-dihaloarenes with amines and ketones can also generate indoles via the Buchwald–Hartwig amination and subsequent Heck coupling (Scheme 6.4).[12]

Using tribrominated styrene derivatives as the *o*-halo substituted arene source, substituted indoles can be synthesized by reacting the arene sources with terminal alkynes and amines in the presence of a Pd catalyst (Scheme 6.5).[13] This process consist of assembling two C–N and a C–C bond forming reactions in a cascade manner.

o-Halo substituted arenes have been utilized to synthesize a range of other fused ring heterocycles utilizing the alkynylation and subsequent nucleophilic cyclization as demonstrated in Scheme 6.6.[14]

SCHEME 6.3 *o*-Halophenol/*o*-haloaniline for benzofuran/indole synthesis.

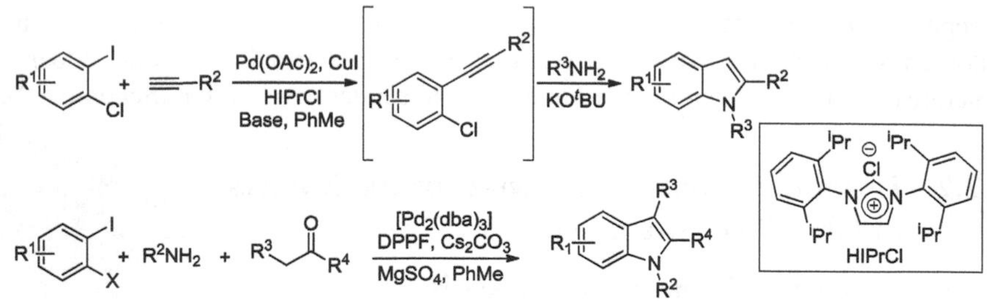

SCHEME 6.4 Synthesis of indoles from *o*-dihaloarenes.

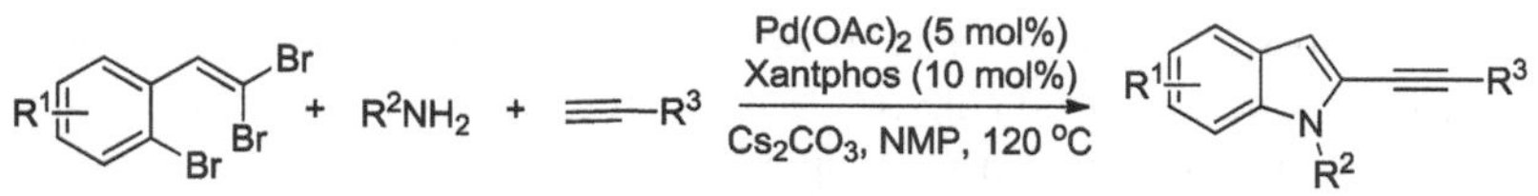

SCHEME 6.5 Synthesis of indoles from tribromostyrene.

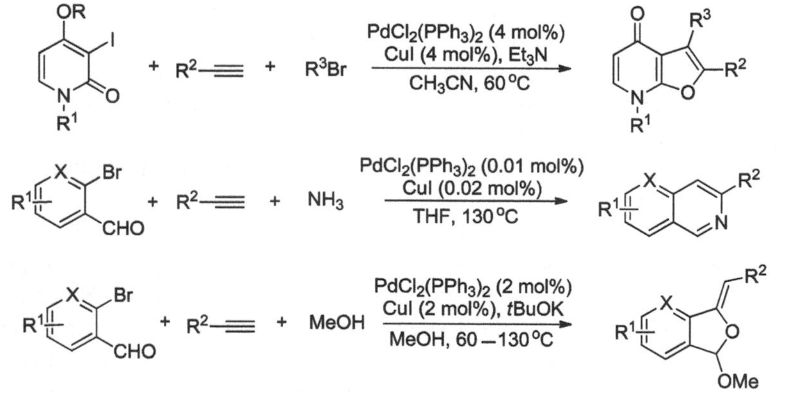

SCHEME 6.6 Fused ring heterocycles from various alkynylarenes.

Rather than alkyne containing substrates, similar fused ring heterocycles were synthesized using *o*-halo arenes and dihaloarenes in the presence of enolizable ketones, allenes, and even amines. The Donohoe group reported the synthesis of isoquinolines from protected *ortho*-substituted aryl bromides, methyl ketones, electrophiles, and ammonium chloride. This reaction proceeds via the Pd catalyzed α-arylation reaction of an enolizable ketone with a haloarene, followed by *in situ* trapping with an electrophile, and subsequent aromatization with ammonium chloride (Scheme 6.7).[15] The Grigg group described a range of metal catalyzed multicomponent coupling reactions via allene insertion for the synthesis of fused ring heterocycles.[16] A one-pot, four-component synthesis of substituted indolines and dihydrobenzofurans hase been reported via allene insertion. In this reaction the *in situ* generated π–allyl palladium species from aryl/heteroaryl iodide and allene further couples with an nucleophile-tethered second aryl halide, and after the

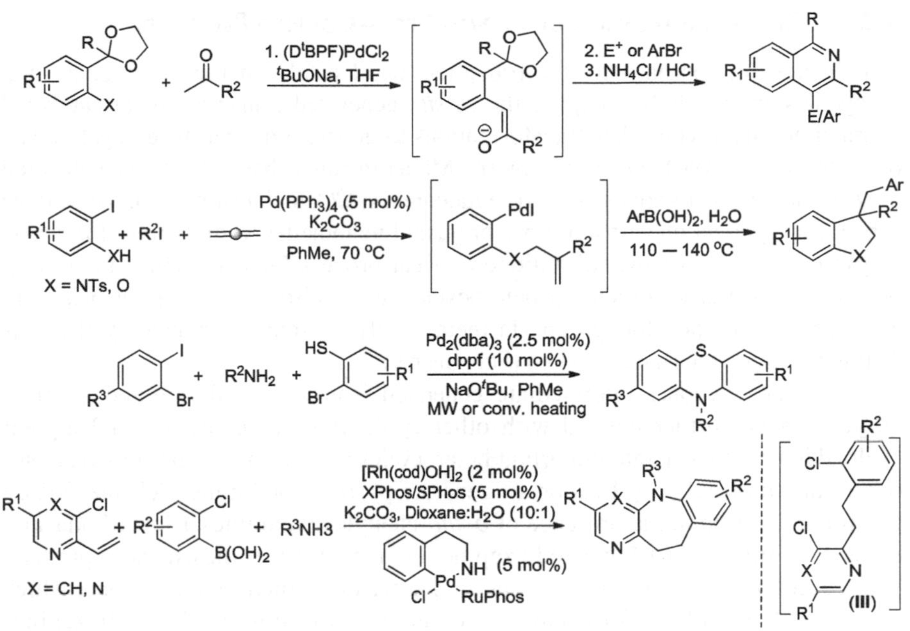

SCHEME 6.7 Pd catalyzed synthesis of fused ring heterocycles from *o*-halo and dihaloarenes.

cyclization step and subsequent Suzuki coupling afforded the desired indolines or dihydrobenzofurans (Scheme 6.7).[17] Jorgensen et al. also reported a Pd catalyzed synthesis of phenothiazines (Scheme 6.7) using 1-bromo-2-iodobenzenes, primary amines, and 2-bromothiophenols following two sequential C–N and one C–S coupling steps.[18a] The Lautens group demonstrated the synthesis of aza-benzazepines using *o*-halostyrene, *o*-haloboronic acid, and amine derivatives in the presence of Rh(I)/Pd(0) catalytic combinations.[18b] This diversity-oriented synthesis proceeds via initial rhodium catalyzed coupling of boronic acid with alkene forming intermediate (**III**) (Scheme 6.7), which is followed by a palladium catalyzed double amidation to provide the desired aza-benzazepines.

6.2.2.2 Multimetallic Catalysis

More than two metal catalysts can be used in similar one-pot, multicomponent reactions for the synthesis of fused ring heterocycles. In a pioneering study, the Lautens group demonstrated a strategy for the synthesis of dihydroquinolinones by suitably combining three transition metals [Rh(I)/Pd(0)/Cu(I)] (Scheme 6.8).[19]

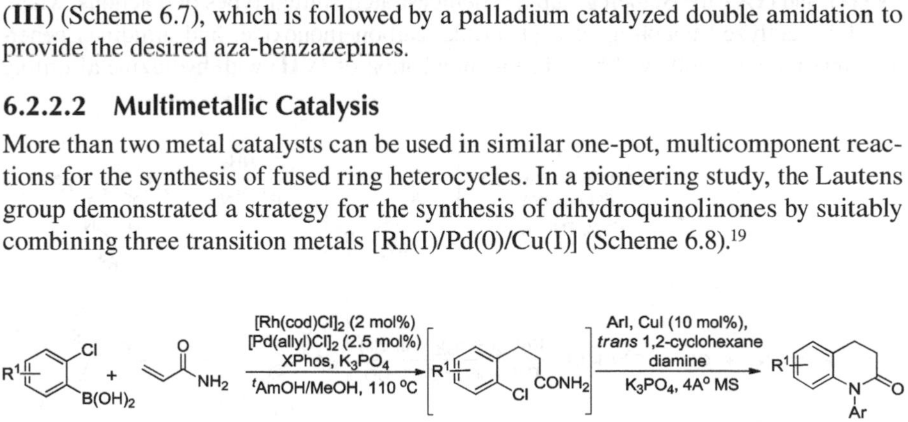

SCHEME 6.8 Fused ring heterocycles from haloarenes via multimetallic catalysis.

6.2.3 Cyclocondensation of *In Situ* Cross-Coupled Products

One interesting approach to this transition metal catalyzed multicomponent heterocyclic synthesis is by merging the *in situ* generated transition metal catalyzed coupled products with other transformations to generate desired heterocycles. The most general approach, developed by the Müller group, is based on the Pd catalyzed *in situ* generation of ynones (**IV**) from haloarenes, CO, and terminal alkynes and its subsequent cyclization with amidines or guanidine moieties leading to the formation of pyrimidines via an overall multicomponent process (Scheme 6.9).[20] Interesting protocols on synthesis of similar heterocycles such as isoxazoles or pyrazoles were also reported by the Mori group via metal catalyzed four-component cyclizations following ynone intermediate (**IV**) (Scheme 6.9).[21]

The reactive ynones intermediate generated from aroyl chloride and terminal alkyne can be further merged with other cyclization processes (viz. 1,3-dipolar cycloadditions) to generate important heterocycles such as indolizines or isoxazoles or via intramolecular cyclization to generate pyrroles or furans (Scheme 6.10).[22] A similar reaction in the presence of *ortho*-phenylene diamines afforded pharmacologically important 1,5-benzodiazepines (Scheme 6.10) as the exclusive product. This reaction is likely to proceed via the formation of intermediate ynone, which reacts further with *ortho*-phenylene diamines affording 1,5-benzodiazepines (Scheme 6.10).[23]

This *in situ* generated ynone intermediate from acid chloride and alkynes can also be coupled with tryptamine derivatives (**V**) and α,β-unsaturated acid chlorides (**VI**) in a four-component reaction during the one-pot synthesis of tetrahydro-β-carbolines. This reaction clearly demonstrates the potential of this methodology for the rapid construction of highly substituted, complex heterocycles where five new σ-bonds and four new stereocenters are installed in a one-pot process (Scheme 6.11). The final key step of this sequence is an aza-annulation reaction that presumably generates an acyliminium ion, which concludes the sequence by a Pictet–Spengler cyclization.[24]

Not only the ynones intermediate but utilizing similar intermediates such as (**VII**), (**VIII**), and (**IX**), the Staben group has demonstrated similar types of reactions, where the Pd- catalyzed coupling of aryl iodide, carbon monoxide, and amidines generates intermediate, followed by subsequent addition of (**VII**) with hydrazine affording

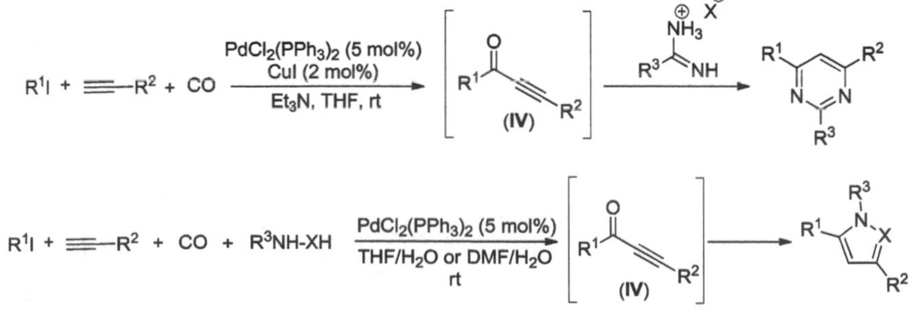

SCHEME 6.9 Pd catalyzed ynone cyclization for the synthesis of heterocycles.

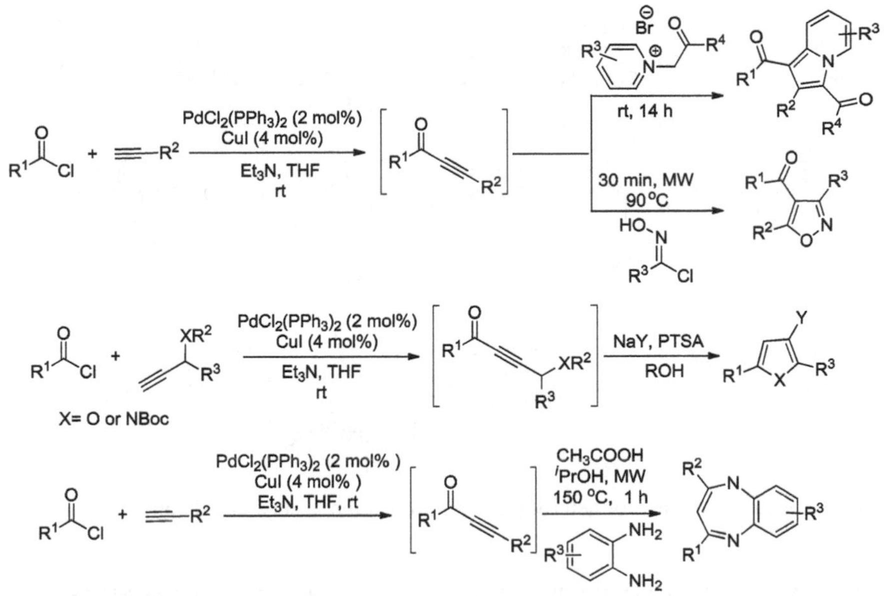

SCHEME 6.10 Fused ring heterocycles via ynone cyclization.

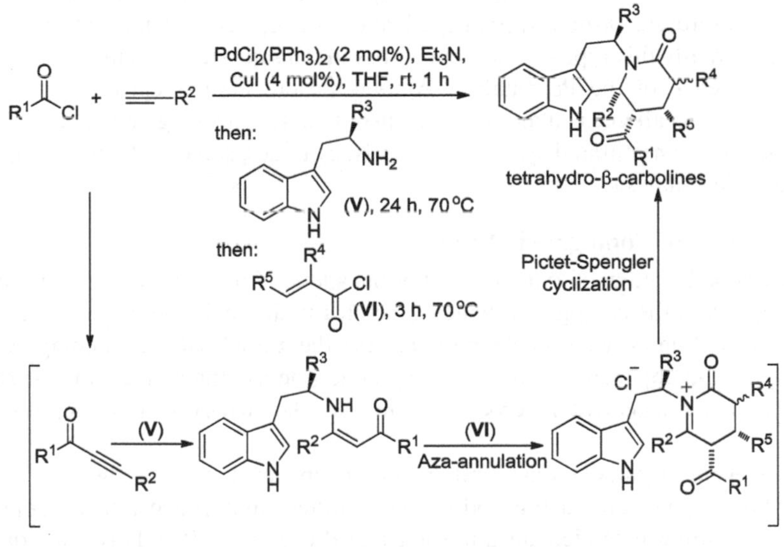

SCHEME 6.11 One-pot, four-component synthesis of tetrahydro-β-carbolines via ynone intermediate.

triazoles (Scheme 6.12).[25a] A number of related multicomponent heterocycle syntheses have been reported by the Malinakova group using Pd and Cu catalysts. The coupling of imines, unsaurated acid chlorides with dienyl stannanes, or terminal alkynes with spontaneous Diels–Alder cycloaddition afforded the desired polycyclic isoindolones via intermediates (**VIII**) and (**IX**) (Scheme 6.12).[25b]

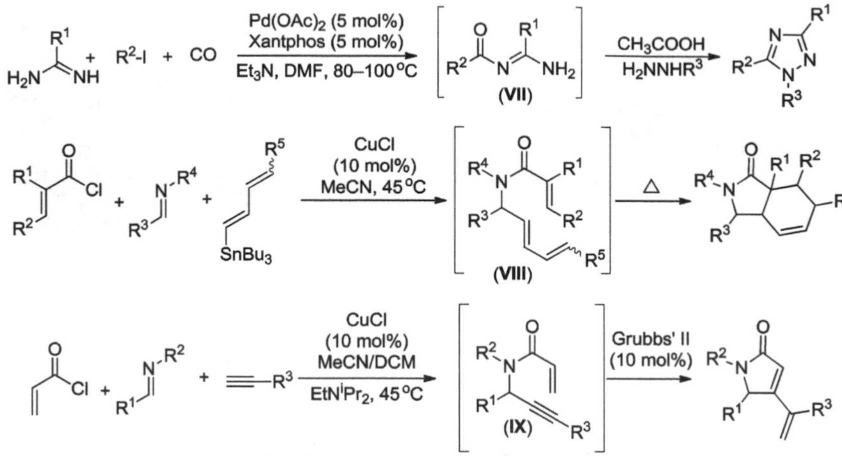

SCHEME 6.12 Synthesis of heterocycles via cyclization of substituted amides.

6.2.4 Cyclization via Insertion across Unsaturated π–Components

Metal catalyzed, three-component coupling/cyclization involves the oxidative addition step of a low valent metal across an organic halide and its subsequent insertion with the unsaturated π-bonds; final cyclization is triggered by a nuclophilic component. Commonly this type of insertion across the unsaturated π–components occurs in the presence of the Pd catalyst. Among various carbon–carbon unsaturation sources, such as alkynes, alkenes, conjugated dienes, nonconjugated dienes, allenes, and even other unsaturated systems such as isocyanides, some of them participate in heterocyclic synthesis.

6.2.4.1 Use of Conjugated Dienes

The carbopalladation of oxidative addition adduct (from organic halide and Pd catalyst) across the conjugated dienes generates π-allyl palladium species. Further, carbonucleophile or heteronucleophile attacks the π-allyl palladium complex, generating three-component linear or cyclic products depending on reaction conditions. As this chapter deals with heterocyclic synthesis, the final cyclization step is crucial here.

An elegant intramolecular strategy has been developed for the synthesis of α-alkylidene-γ-lactams using iodoarene, amines, and predesigned conjugated diene containing extended unsaturation (**X**) (Scheme 6.13).[26] This reaction proceeds via the regioselective addition of arylpalladium halide across the extended triple bond of acyclic diene substrate (**X**) generating species (**XI**). Further intramocular carbopalladation of (**XI**) forms π-allyl palladium complex (**XII**), which further cyclizes to afford the desired product after the nucleophilic attack of the amine component. The by-product can be obtained if aryl amines are involved in this process.

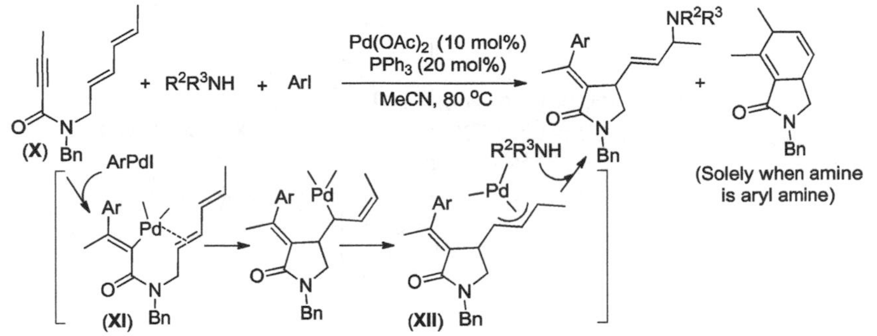

SCHEME 6.13 Synthesis of lactams via diene carbopalladation and cyclization.

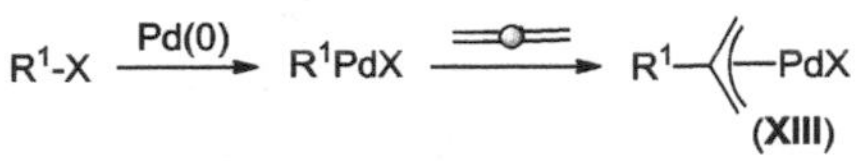

SCHEME 6.14 Formation of π-allyl palladium complex.

6.2.4.2 Use of Allenes

In this process, the π-allyl palladium complex is generated via the carbopalladation of allene substrates with an organopalladium species. In contrast to other carbopalladation, here the addition of organopalladium species occurs at the central carbon atom, generating a π-allyl palladium complex (**XIII**) (Scheme 6.14).

6.2.4.2.1 Cyclization via π-Allyl Palladium Complex from External Nucleophile Attack

Based on this concept of allene carbopalladation, Grigg and co-workers demonstrated an interesting protocol for the synthesis of fused isoxazolidine using allene, iodoarene, and phenolic moiety containing a nitrone moiety. Here the π-allyl palladium complex generated via the insertion of an organopalladium species to an allene unit is attacked by the phenolic nucleophile (**XIV**), forming intermediate (**XV**). In the final step, 1,3-dipolar cycloaddition of nitrone moiety (**XV**) with the newly formed olefin gave the fused isoxazolidine product (Scheme 6.15).[27a]

An analogous Pd catalyzed multicomponent process was reported by Grigg and co-workers for the formation of benzo-fused heterocycles. This process involves

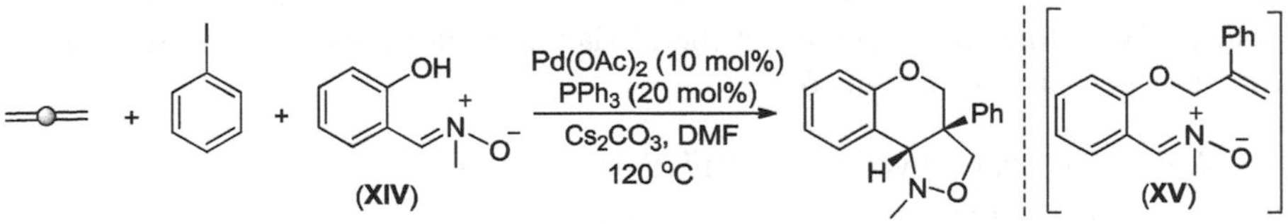

SCHEME 6.15 Fused isoxazolidine from cyclization of π-allyl Pd complex.

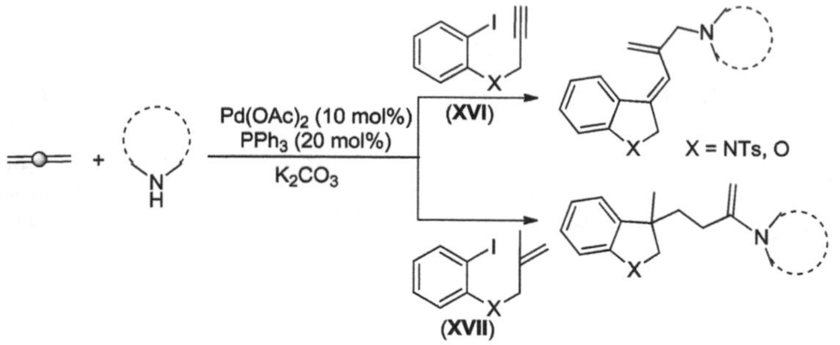

SCHEME 6.16 Synthesis of benzo-fused rings via cyclization–anion capture using allene as relay.

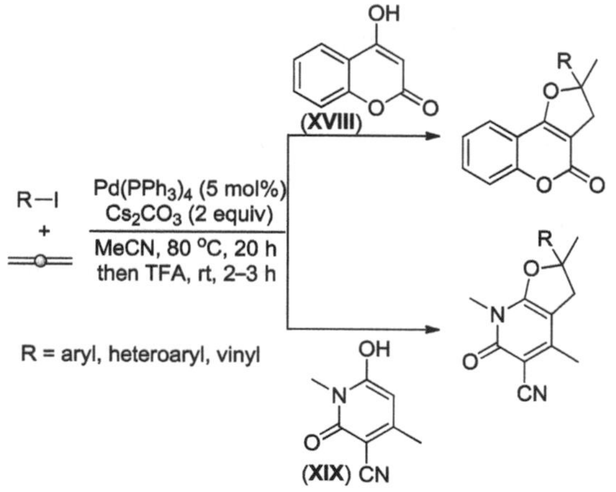

SCHEME 6.17 Synthesis of fused dihydrofurans using π-allyl Pd complex.

an initial intramolecular Heck-type reaction of arenes with alkynes/alkenes (**XVI**) or (**XVII**) followed by allene insertion forming a π-allyl palladium intermediate. Trapping of π-allyl palladium complex with an external amine nucleophile afforded the desired heterocycle (Scheme 6.16).[27b]

This *in situ* generated π-allyl Pd-complex brought about via the reaction of an allene with an organopalladium species was successfully applied to the synthesis of annelated dihydrofurans by the reaction of heterocyclic enols (**XVIII**) or (**XIX**) in the presence of a base and Pd(0) catalyst. This was then followed by a trifluoroacetic acid (TFA)-mediated cyclization of the allylation product, leading to desired dihydrofuran derivatives (Scheme 6.17).[27c]

6.2.4.2.2 *Cyclization via π-Allyl Palladium Complex Containing Pendant Nucleophile*

Among the various transition metal–allyl complexes, π-allyl palladium complexes are commonly used in multicomponent cyclization processes for the synthesis of various heterocycles. The insertion of allenes to organopalladium species generates

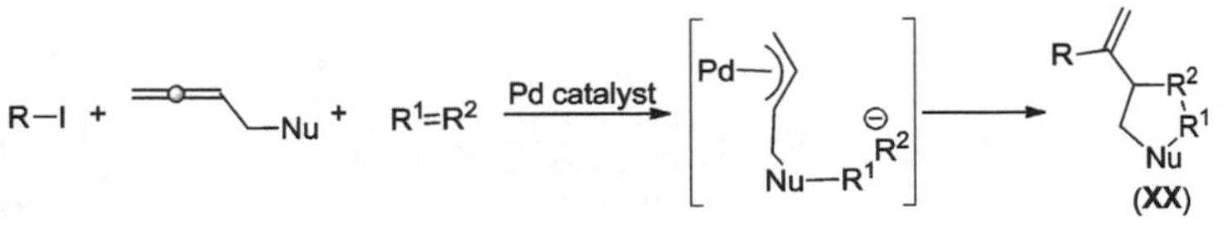

SCHEME 6.18 Nucleophile tethered allenes in multicomponent cyclizations.

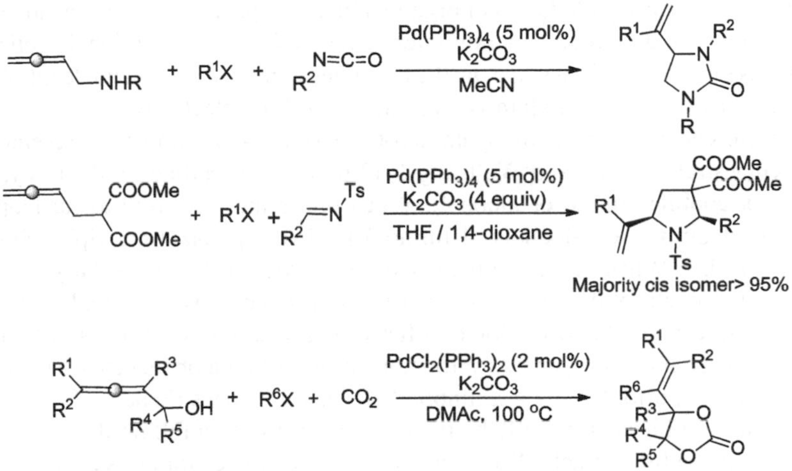

SCHEME 6.19 Synthesis of heterocycles via cyclization of π-allyl palladium species.

π-allyl palladium complexes. However, if the allene contains a pendant nucleophile, which can couple with another unsaturated component, then final intramolecular nucleophilic attack to the cationic π-allyl palladium species generates the desired heterocycles (**XX**) (Scheme 6.18).

The Ma group utilized these nucleophile-tethered allenes for the synthesis of heterocycles via multicomponent reactions. They utilized amine-tethered allenes, isocyanates, and aryl iodides for the synthesis of imidazolones (Scheme 6.19).[28a] The formation of π-allyl palladium species is believed to be the important intermediate that further undergoes coupling with isocyanates and subsequent cyclization to generate imidazolone derivatives. The selective formation of pyrrolidines has been demonstrated by Ma and co-workers via a three-component reaction of imine, aryl halide, and allenyl malonates under analogous reaction conditions (Scheme 6.19). By varying the organohalide and imine, several polysubstituted *cis*-pyrrolidine derivatives can be obtained.[28b] Further modification has been demonstrated by Inoue et al. in which one of the components was CO_2, leading to the formation of cyclic carbonates (Scheme 6.19).[28c]

6.2.4.3 Use of Isocyanides

Similarly to conjugated dienes and allenes, unsaturated moieties like isocyanides can also be employed for the synthesis of important heterocycles via insertion of organopalladium species followed by cyclization. Lang et al. developed a protocol for the synthesis of oxazolines using ethanolamine as the nucleophile (Scheme 6.20).

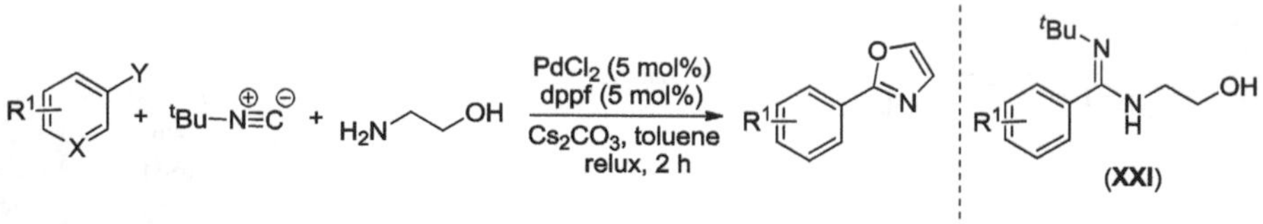

SCHEME 6.20 Synthesis of oxazoline via Pd catalyzed isocyanide insertion.

In this reaction carbopalladation of organopalladium species across isocyanides and further attack of an amine generate intermediate (**XXI**). An additional cyclization step takes place in the final stage, as these amines contain a second nucleophile, and final elimination of *tert*-butylamine provides oxazoline selectively.[29a]

The efficient procedure for the synthesis of quinazolin-4(3*H*)-imines[29b] (Scheme 6.21) was demonstrated by Pu and Wu using *N*-(2-bromoaryl) carbodiimide, isocyanide, and a nucleophile (amine or alcohol). When phosphites are used as nucleophiles in place of amines or alcohols, 4-imino-3,4-dihydroquinazolin-2-ylphosphonates are obtained.[29c] When *o,o'*-Diiodo carbodiimides were reacted similarly with isocyanide and nucleophile under identical reaction conditions, a second cyclization occurred, leading to the formation of a fused ring heterocycle.[29d] In a similar palladium catalyzed multicomponent cyclization, the use of 2-halobenzoates, isocyanides and hydrazines, afforded the 4-aminophthalazin-1(2*H*)-ones (Scheme 6.21).[29e] This reaction also initiates identically by the isocyanide insertion followed by lactamization via nucleophilic attack. The synthesized product had limited substrate scope; however, dealkylation of the product can be used as synthon for further organic transformations.

Jiang et al. reported the synthesis of seven-membered heterocycle 4-amine-benzo[*b*][1,4]oxazepines via palladium catalyzed, three-component reactions of *o*-aminophenols, bromoalkynes, and isocyanides (Scheme 6.22).[29f] Herein the reaction is proposed to initiate via the nucleophilic attack of *o*-aminophenols to bromoalkynes, which, after the oxidative addition in the presence of Pd(0), generates a vinyl palladium species (**XXII**). This intermediate (**XXII**), after insertion across the isocyanide moiety, forms (**XXIII**). Intermediate (**XXII**) finally leads to final product 4-amine-benzo[*b*][1,4]oxazepines after base mediated cyclization and subsequent rearrangement.

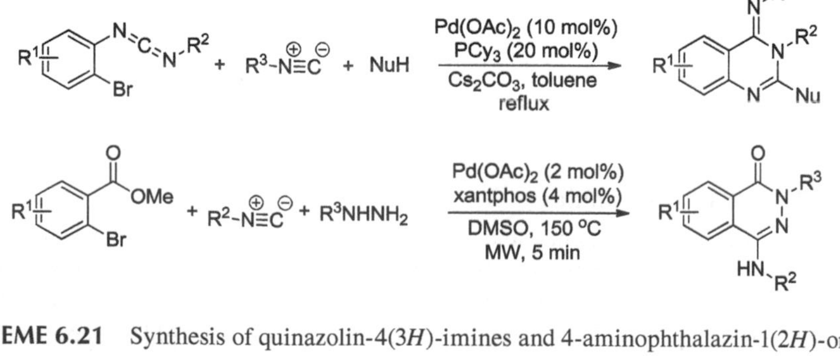

SCHEME 6.21 Synthesis of quinazolin-4(3*H*)-imines and 4-aminophthalazin-1(2*H*)-ones.

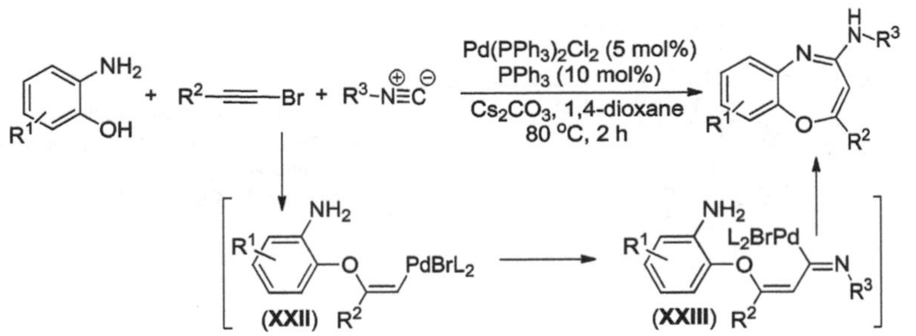

SCHEME 6.22 Synthesis of 4-amine-benzo[*b*][1,4]oxazepines via Pd catalyzed isocyanide insertion.

1,4-Benzoxazepines have been further synthesized by Cai et al. via a palladium catalyzed, three-component reaction of *N*-tosylaziridines, 2-iodophenols, and isocyanides. Ring opening of the aziridine moiety with 2-iodophenol followed by a Pd catalyzed isocyanide insertion affords the desired product in moderate to high yields (Scheme 6.23).[29g]

Takahashi and co-workers reported the use of *o*-alkenyl arylisonitrile for the synthesis of 2,3-disubstituted indoles (Scheme 6.24). This three-component reaction also proceeds in the presence of an *in situ* generated Pd(0) catalyst. The organopalladium species generated via the reacion of aryl iodide and Pd(0) catalyst after insertion into an isonitrile unit of an *o*-alkenyl arylisonitrile generates species (**XXIV**). After subsequent cyclization and 1,3-migration of hydrogen, this (**XXIV**) forms π-allyl palladium complex (**XXV**). Trapping of the nucleophilic amine component to this complex (**XXV**) leads to the selective formation of 2,3-disubstituted indoles.[29h]

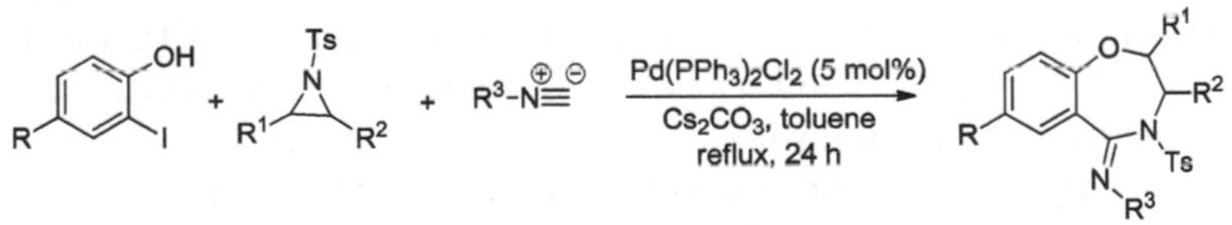

SCHEME 6.23 Pd catalyzed isocyanide insertion via aziridine ring opening.

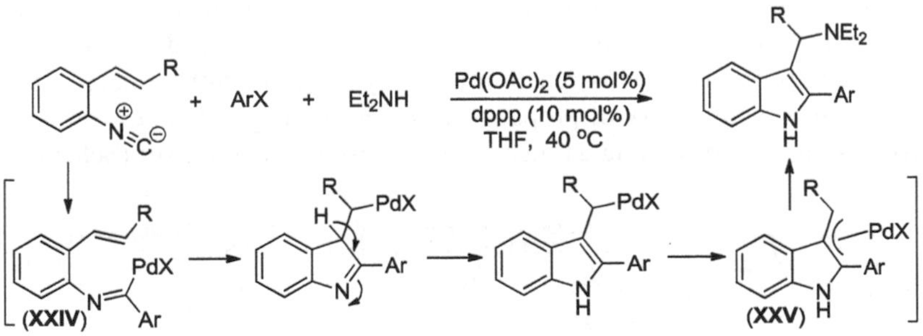

SCHEME 6.24 Synthesis of 2,3-disubstituted indoles from *o*-alkenyl arylisonitrile.

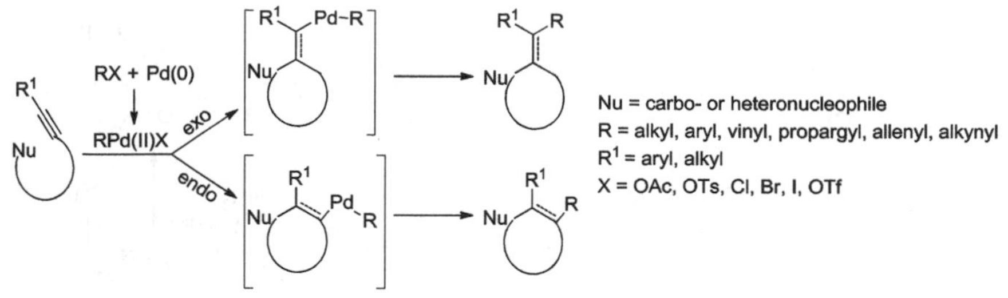

SCHEME 6.25 Cyclization with alkenes or alkynes containing pendant nucleophiles.

6.2.5 Cyclization of Unsaturated C–C π–Components Containing Tethered Nucleophiles

One rational way to generate heterocyclic moieties via transition metal catalyzed multicomponent reactions is by the activation of unsaturated substrates in the presence of organopalladium species and subsequent attack by the pendant nucleophile. This route gives access to the synthesis of five- or six-membered heterocycles (Scheme 6.25).

Designing an alkene or alkyne substrate containing a suitably positioned proximal nucleophile (carbo- or heteronucleophile) is crucial for the success of this process. The first report of such reactions was demonstrated by Inoue and co-workers in 1990 with limited substrate scope.[30]

6.2.5.1 Cyclization via Carbonucleophiles

Balme and co-workers reported the synthesis of highly substituted tetrahydrofurans via the palladium catalyzed multicomponent cyclization of *in situ* generated carbonucleophiles.[31a] In this process initial Michael addition of allylic alcohol to the conjugate acceptor generates enolate carbonucleophile (**XXVI**), which, after activation with organopalladium species and subsequent intramolecular cyclization, provides the substituted tetrahydrofuran moiety (Scheme 6.26). This reaction demonstrates an easy access to substituted tetrahydrofurans from simple allylic alcohol with moderate yields; however, the use of secondary and tertiary allylic alcohols gave lower yields. The crucial factor of this reaction is the slow addition technique of the allylic alkoxide component, which prevents the possible side reactions.

The use of more reactive propargyl alkoxide during the process allows addition of all three reaction components simultaneously to afford 3-arylidene (and alkenylidene) tetrahydrofurans.[31b] This reaction is highly versatile with respect to starting materials—that is, a variety of propargyl alcohols and organohalides can be effective for this process, providing a range of tetrahydrofuran derivatives (Scheme 6.27).

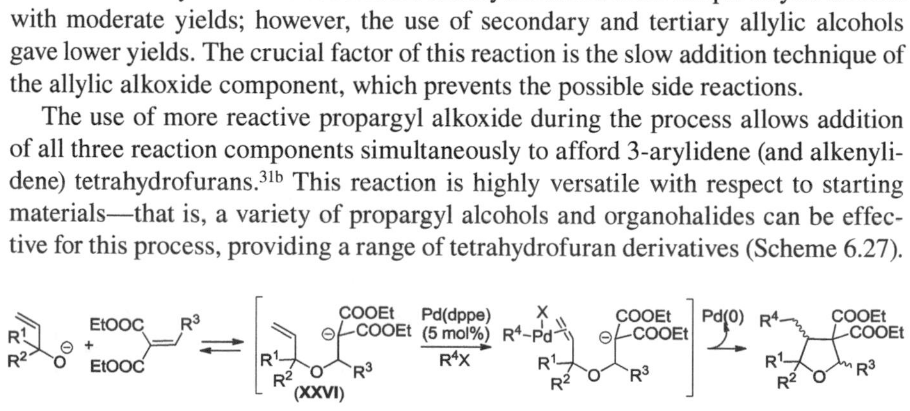

SCHEME 6.26 Synthesis of substituted tetrahydrofurans via carbonucleophiles.

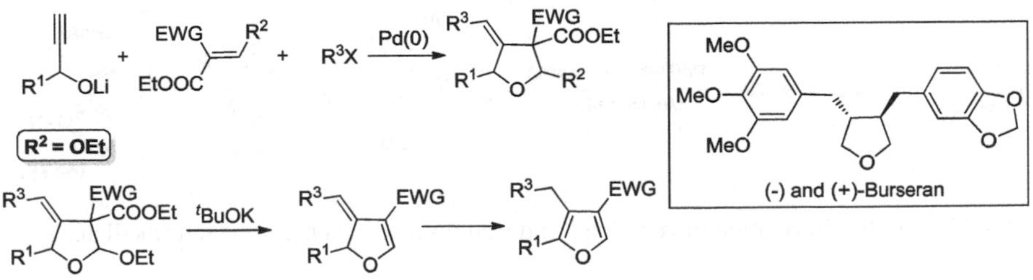

SCHEME 6.27 Formation of 3-arylidene tetrahydrofurans via carbocyclization.

This methodology was further extended for the one-pot synthesis of furan derivatives using diethyl ethoxymethylenemalonate as a Michael acceptor. Base potassium *t*-butoxide was used in excess in this case for converting tetrahydrofuran derivatives to furan moiety via an elimination process.[31c] The reaction in sum involves sequential conjugate addition, palladium catalyzed cyclization followed by an alkoxide-mediate decarboxylative elimination, and subsequent double bond isomerization. This reaction was further applied for the formal total synthesis of lignan antitumor Burseran (Scheme 6.27).

Recently, Lu and Liu reported a similar protocol for the synthesis of stereodefined allylidene tetrahydrofurans utilizing lithium propargylic alkoxides and alkylidene malonates. Contrary to the previous reactions, the catalytic amount of Pd(II) salt is effective during the cyclization process rather than the Pd(0) catalyst.[31d] In a similar fashion, the use of propergyl amine tethered activated olefins and aryl iodine under similar Pd catalyzed conditions leads to the formation of pyrrolidines selectively.[31e,f]

6.2.5.2 Cyclization via Heteronucleophiles

This type of multicomponent reaction is based on a sequential Pd mediated Sonogashira coupling of terminal alkynes with heteronucleophile tethered *o*-haloarenes followed by a cyclization process in the presence of organohalides. The majority of these fused ring heterocycle synthesis processes have already been discussed previously. A few additional examples are discussed here.

Synthesis of benzannulated nitrogen heterocycles has been reported by Wu et al. using 2-alknynylbenzonitriles, sodium methoxide, and iodoarenes.[32a] This reaction is based on the addition of sodium methoxide to 2-alkynylbenzonitriles forming intermediate (**XXVII**). It was further followed by Pd(0) catalyzed heteroannulation of ketimine intermediate (**XXVII**) with an organopalladium-activated alkynyl system, leading to the formation of isoindoles or isoquinolines depending on 5-exo or 5-endo type cyclization. During this cyclization process, endo/exo ratio is dependent on steric environment between the substituents of terminal alkynes as well as on the ketimine unit (Scheme 6.28).

This multicomponent heteronucleophilic cyclization is also employed for the direct access to a late intermediate in the synthesis of marine sesquiterpenoid frondosin B. This unusual transformation proceeds via initial Sonogashira coupling followed by cyclization and further 1,7-hydrogen shift with subsequent π-electrocyclization (Scheme 6.29).[32b]

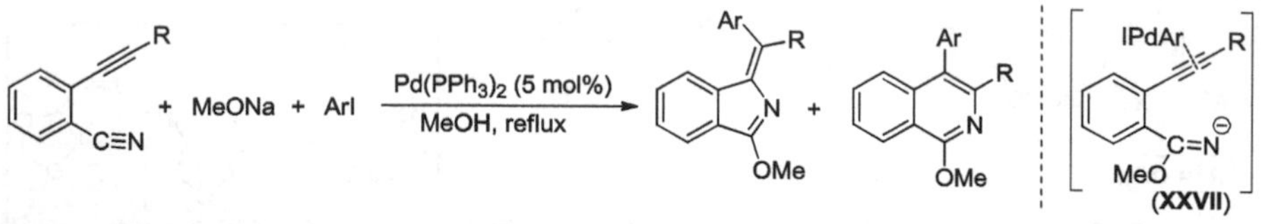

SCHEME 6.28 Formation of isoindoles and isoquinolines via ketimine intermediate.

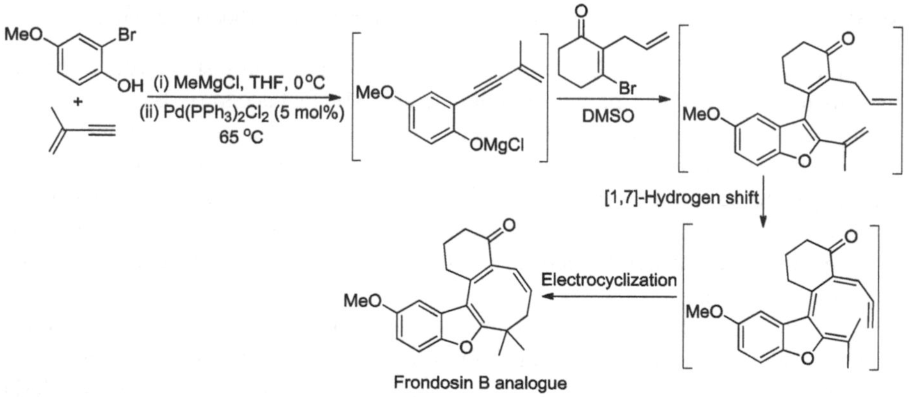

SCHEME 6.29 Multicomponent coupling strategy to benzofurans and further ring expansion.

6.2.6 Cyclization via A³-Type Coupling Involving Amines

A^3 coupling reactions are the three-component reactions between aldehydes, amines, and alkynes to yield propargylamine derivatives under various catalytic conditions (Scheme 6.30).

Among various multicomponent cyclizations using amines, A^3-type coupling with tandem cyclization represents the most efficient and promising approach for the synthesis of bioactive heterocycles. By taking advantage of versatile reactivity of propergylamines, tandem reactions have been developed that are especially useful for the synthesis of heterocyclic compounds. These A^3-type coupling reactions are classified depending on the location of the second functional group that initiates the tandem annulation after A^3 coupling.

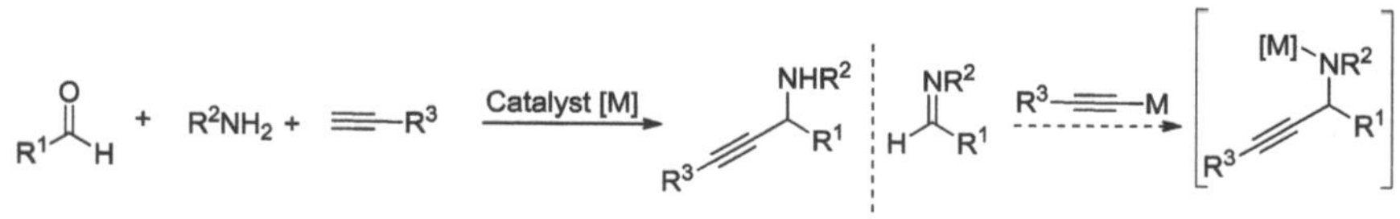

SCHEME 6.30 General strategy for A^3 type couplings.

6.2.6.1 Reactions Based on Functional Amine-Participated Tandem A³ Coupling Transformation

Amine, the most versatile and reactive component, can be found in many forms (viz. primary amines, secondary amines, or even ammonia or in the form of aromatic amines, aliphatic amines). Thus, the most convenient approach to access heterocyclic scaffolds via A^3 coupling based tandem reaction is by using functionalized amines. In their pioneering study, Iqbal et al. reported a convenient Cu catalyzed A^3 coupling using aromatic amines, alkynes, and aldehydes for the synthesis of quinolines.[33a] Formation of quinolines proceeds by the formation of A^3 coupling product (**XXVIII**) via the reaction of *in situ* generated imine with terminal alkyne. After tautomerization and, further, in the presence of a copper catalyst, intermediate (**XXVIII**) leads to a transition state (**XXIX**). Final cyclization of (**XXIX**) via a nucleophilic carbon site in amines leads to the formation of desired quinolines (Scheme 6.31). A little excess (30 mol%) of catalyst CuCl and THF solvent is the optimum to provide quinoline derivatives in moderate yields.

A modified protocol has been reported by Wang et al. using a mixed catalytic system of $AuCl_3$ and CuBr in MeOH solvent to afford the same quinoline derivatives with enhanced product yields (Scheme 6.32).[33b] Using $FeCl_3$ as the catalyst, Tu and co-workers subsequently developed an interesting protocol for the synthesis of quinolines (Scheme 6.32).[33c] The same quinoline moiety was further synthesized by many others using various cheap catalysts, including $Fe(CF_3COO)_3$ and $Fe(OTf)_3$.[33d,e] Many important transition metal catalyzed syntheses of heterocyclic scaffolds via functional amine participated A^3 coupling are cited here.[33f–j]

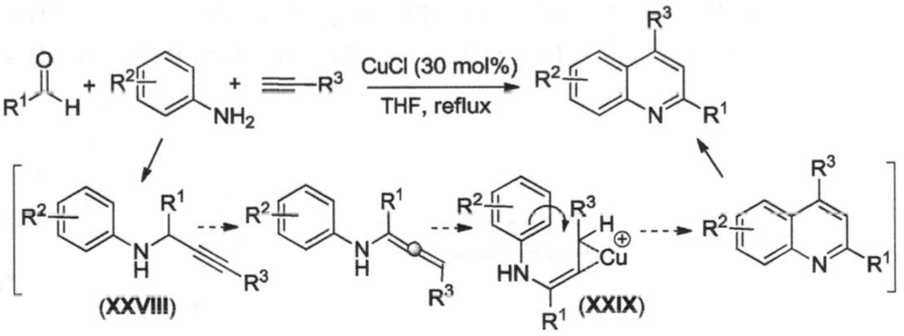

SCHEME 6.31 Synthesis of quinolines via A^3 couplings using functionalized amines.

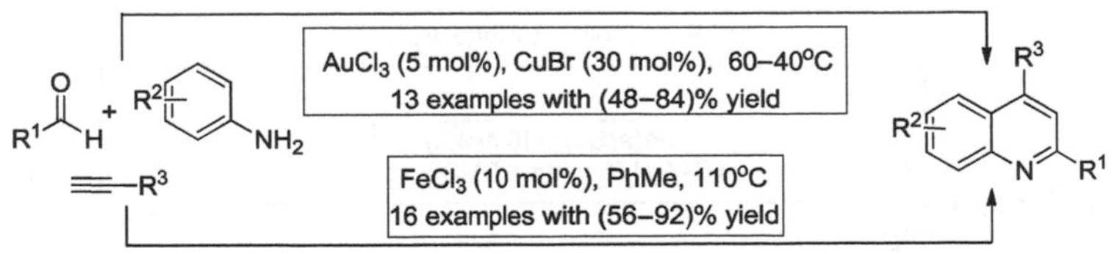

SCHEME 6.32 Various routes to quinolines using transition metals.

6.2.6.2 Reactions Based on Functional Aldehyde Participated Tandem A³ Coupling Transformation

As the electrophilic component aldehydes contain many other reactive functional groups that are usually easily accessible, employing functional aldehydes for the design of A³ coupling-based tandem reactions for heterocycle synthesis is also important. Liu and co-workers developed a three-component synthesis of indolizine by reacting functionalized aldehydes (pyridinly-2-aldehydes), terminal alkynes, and amines.[34a] This process efficiently promoted the synthesis of different indolizines under solvent-free operation in the presence of 1 mol% $NaAuCl_4{\cdot}2H_2O$ catalyst at 60 °C. The key transformation of the reaction was the A³ coupling of three components yielding corresponding propargylamine intermediates (**XXX**), and the insertion of the Au catalyst to the C–C triple bond on (**XXX**) initiated further tandem annulation to provide final indolizine products with moderate to good yields (Scheme 6.33).

Successful synthesis of indolizine derivatives via the A³ coupling triggered considerable research efforts in improving the reactions through a number of other catalytic methods. For example, the $AgBF_4$/toluene,[34b] $Fe(acac)_3$/TBAOH,[34c] and the CuI/PEG[34d] system, etc. have all been found to be good methods for this three-component synthesis of indolizines (Scheme 6.34). Transition metal catalyzed syntheses of various heterocyclic scaffolds via functional aldehyde-participated A³ coupling are cited herein.[34e–j]

6.2.6.3 Reactions Based on Functional Alkyne Participated Tandem A³ Coupling Transformation

Owing to the high reactivity of alkynes, these compounds, especially terminal alkynes, are usually unstable, and corresponding alkynes bearing other reactive functional groups are accordingly of less availability compared with amines or

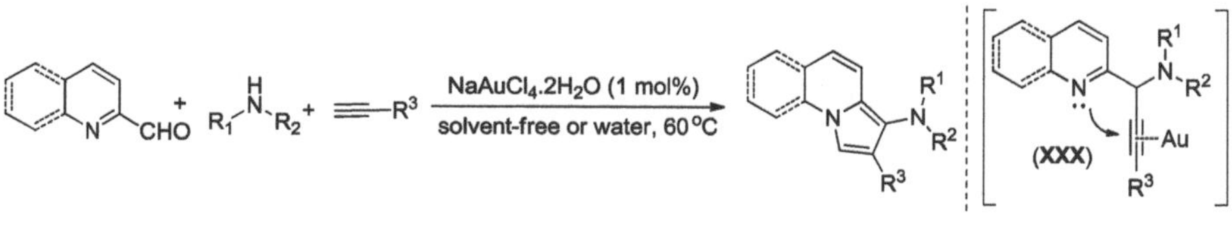

NaAuCl4.2H2O (1 mol%)
solvent-free or water, 60 °C
(XXX)

SCHEME 6.33 Synthesis of indolizines via A³ couplings using functionalized aldehydes.

AgBF4 (5 mol%), toluene, 85 °C
Fe(acac)3 (10 mol%) TBAOH (5 mol%), DMSO, rt
CuCl (10 mol%), PEG, 90 °C, N2

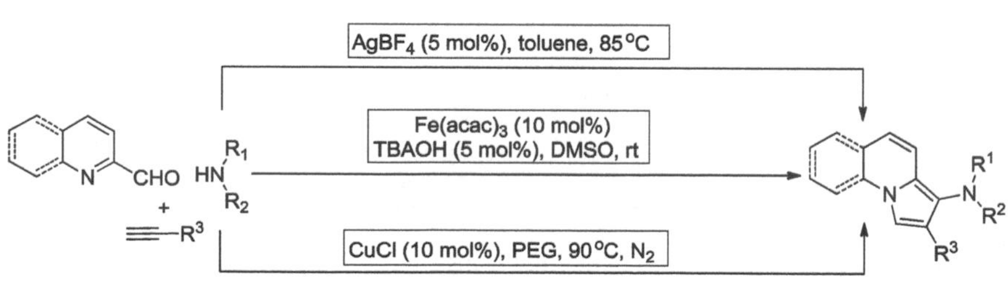

SCHEME 6.34 Various other catalytic routes to indolizines using A³ coupling.

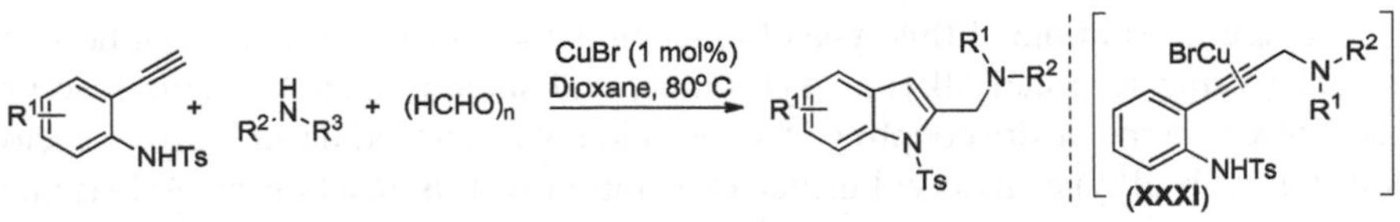

SCHEME 6.35 Synthesis of indolizines via A^3 couplings using functionalized amines.

aldehydes. Therefore, fewer examples of functional alkyne-participated A^3 coupling-based tandem reactions are available in the literature. A typical example based on functional alkynes was the three-component synthesis of indole derivatives using *o*-aminophenylacetylene, formaldehyde, and secondary amines. In the presence of 1 mol% CuBr catalyst, the three-component reactions performed in dioxane at 80 °C and provided indoles directly. This reaction proceeds via the formation of CuBr catalyzed A^3 coupling, which gives intermediates (**XXXI**) and further copper catalyzed intramolecular annulation leading to the formation of 2-substituted indoles (Scheme 6.35).[35]

6.2.6.4 Reactions of More Than Three Components Based on A^3 Coupling Tandem Transformation

Based on the preceding three classes, it could be easily concluded that those heterocycle syntheses were all designed by making use of any of these three reactants containing additional functional groups (i.e., these are intramolecular tandem transformations of propargylamine intermediates). However, the present case is an intermolecular transformation of A^3 coupled propargylamine products as this multicomponent cyclization process employs more than three components. For example, Li and co-workers developed a useful method for the synthesis of isoindolines by employing amine (1 equiv.), formaldehyde (2 equiv.), and terminal alkynes (3 equiv.). During the reactions, intermediates (**XXXII**) generated via double A^3 coupling reactions underwent [2+2+2] cycloaddition in the presence of additional terminal alkynes to afford target products (Scheme 6.36).[36]

6.3 MULTICOMPONENT CYCLIZATION VIA DIPOLAR CYCLOADDITION REACTIONS

Transition metal catalysts are often employed for the formation of 1,3-dipolar units that, upon further cycloaddition with another component, generate various heterocyclic scaffolds via an overall multicomponent process.

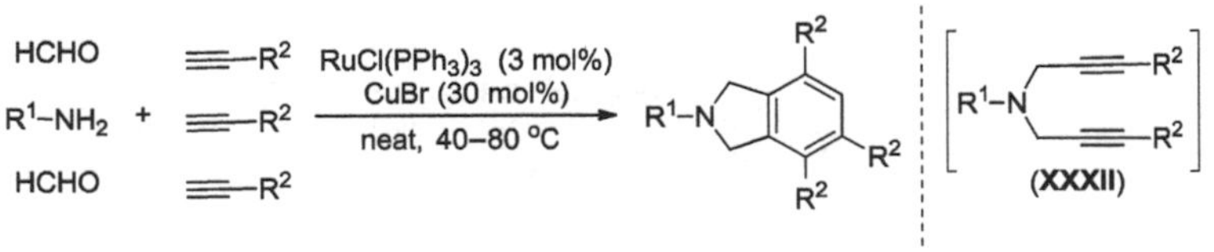

SCHEME 6.36 Synthesis of heterocycles via A^3 couplings with more than three components.

One useful example of this type of reaction is the copper catalyzed synthesis of triazoles using terminal alkyne, aryl iodide, and sodium azide.[37a,b] This reaction takes place via the *in situ* coupling of azide anion with aryl iodide and further cycloaddition with alkyne moiety. Further exploration of this reaction by Ackermann et al. involved the four-component coupling of terminal alkynes, sodium azide, and two different aryl iodides (Scheme 6.37).[37c] This reaction proceeds via analogous Cu catalyzed coupling of azides with aryl iodide and further copper catalyzed 1,3-dipolar cycloaddition generating triazole moiety (**XXXIII**). Functionalization of sp^2 C–H present in the triazole moiety with a second aryl iodide and in the presence of LiO^tBu resulted in the formation of 1,4,5-trisubstituted triazole products in good yields.

Palladium catalysts have also been introduced for the formation of triazole moiety by Yamamoto et al. using trimethylsilyl azide, alkynes, and allylic reagents. This reaction proceeds via the formation of π-allyl palladium azide complex followed by [3+2] cycloaddition with terminal alkynes generating *N*-allylated triazole via final reductive elimination (Scheme 6.38).[38a,b] Suitable modification of this chemistry has also been developed for the formation of tetrazole derivatives via cycloaddition with nitrile substrates (Scheme 6.38).[38c]

Palladium catalysts have also been used for multicomponent cyclization via the use of a 1,3-oxazolium-5-oxides (Münchnones) intermediate followed by concurrent cycloaddition reaction (Scheme 6.39).[39a,b] A Pd catalyzed multicomponent reaction of imines, acid chlorides, and carbon monoxide (CO) generates 1,3-dipolar 1,3-oxazolium-5-oxides (**XXXIV**). Further, cycloaddition of this Münchnones (**XXXIV**) with a variety of unsaturated systems generates a range of heterocyclic products. For example, cycloaddition of (**XXXIV**) with internal alkynes generates polysubstituted pyrroles with loss of CO_2 (Scheme 6.39).[39c,d] Use of electron-poor *N*-tosyl substituted imines in lieu of alkynes provides imidazole derivatives[39e] The addition of imines to

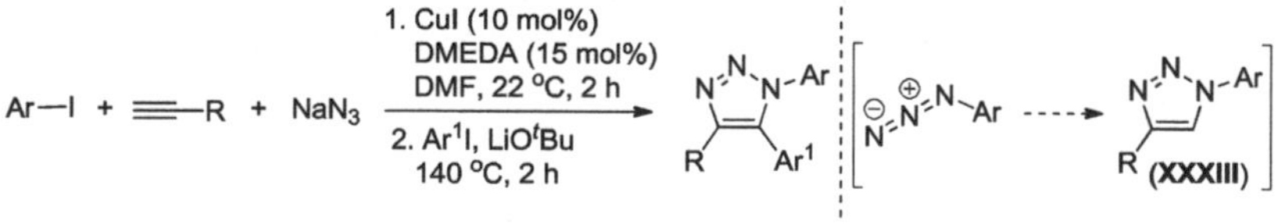

SCHEME 6.37 Synthesis of trisubstituted triazoles using Cu catalyzed MCR.

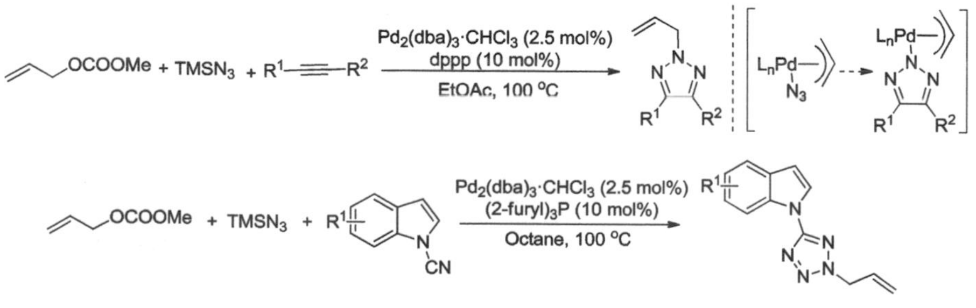

SCHEME 6.38 Synthesis of triazoles and tetrazoles using Pd catalyzed MCR.

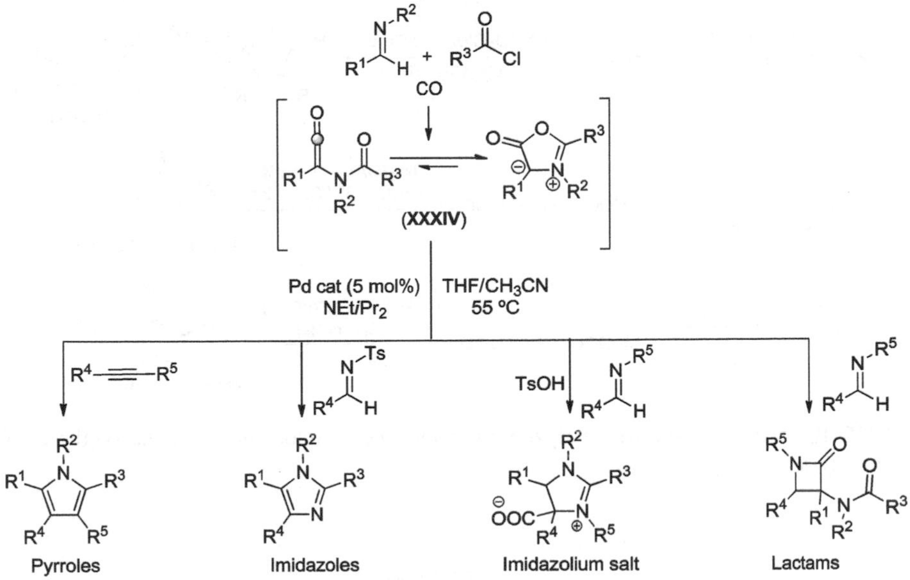

SCHEME 6.39 Pd catalyzed multicomponent cycloaddition via Münchnone intermediate.

Münchnones (**XXXIV**) in the presence of an acid catalyst leads to the formation of imidazolium carboxylate salts (Scheme 6.39).[39f,g] Amide substituted β-lactams are found to be the sole product via the [2+2] cycloaddition to the ketene tautomer of Münchnones with imines (Scheme 6.39).[39h] Recently, aryl halides have been used as a greener alternative for acid chlorides because Pd catalyzed carbonylation of aryl halides generates an analogous intermediate, thus giving access to these heterocycles via an overall five-component reaction.[39i]

1,3-Dipolar cycloaddition of azomethine and carbonyl ylides is also demonstrated for the synthesis of bioactive heterocycles in the presence of suitable transition metals. Although Cu catalysts are initially employed for this type of multicomponent cyclization process for the synthesis of heterocycles like tetrahydrofurans,[40a] recent studies demonstrate that the use of rhodium and ruthenium catalysts is more reactive for these processes. For example, $Rh_2\left(CO_2^tBu\right)_4$ was found to be an efficient catalyst for the formation of tetrahydropyrans with high diastereoselectivity with no side reactions (Scheme 6.40).[40b] This reaction proceeds via rhodium carbenoid, which reacts with aldehyde generating carbonyl ylides (**XXXV**) for subsequent alkyne cycloaddition. A similar approach is also applied for the formation of azomethine ylides that, after cycloaddition, provide efficient synthesis to pyrroles[40c] and pyrrolines (Scheme 6.40).[40d]

In a separate approach Lu and Wang recently reported synthesis of substituted pyrrole derivatives via Cu catalyzed coupling of α-diazokeones, amines, and nitrostyrene (Scheme 6.41).[41a] Here, the formation of azomethine ylide (**XXXVI**) via the N–H insertion of the *in situ* generated carbene, followed by dehydrogenation and further [3+2] cycloaddition with nitrostyrene, provides the desired pyrrole.

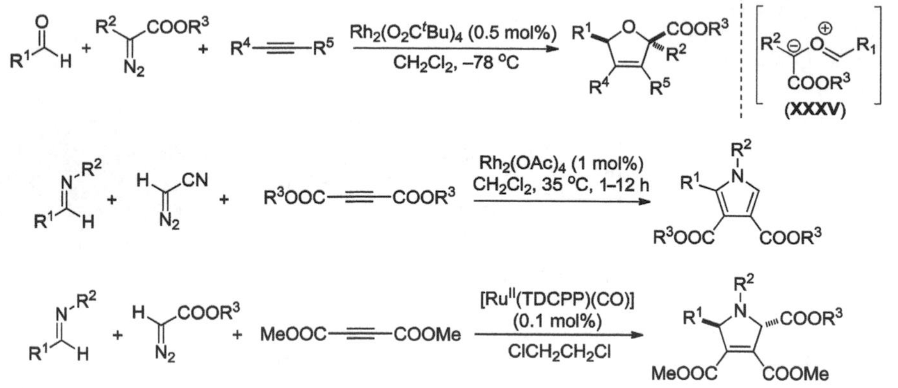

SCHEME 6.40 Rh- and Ru catalyzed multicomponent cycloaddition for the synthesis of heterocycles.

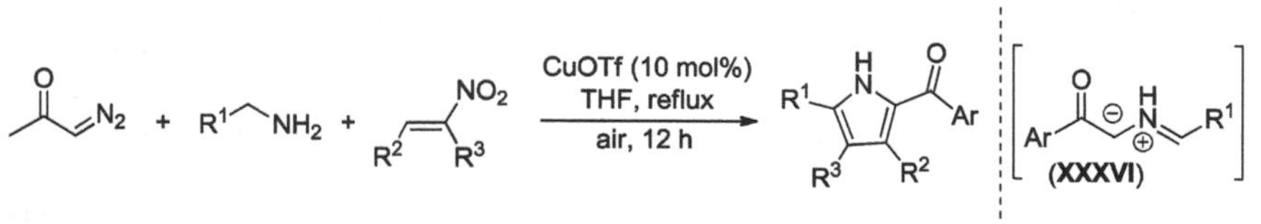

SCHEME 6.41 Cu catalyzed synthesis of pyrroles using dipolar cycloaddition.

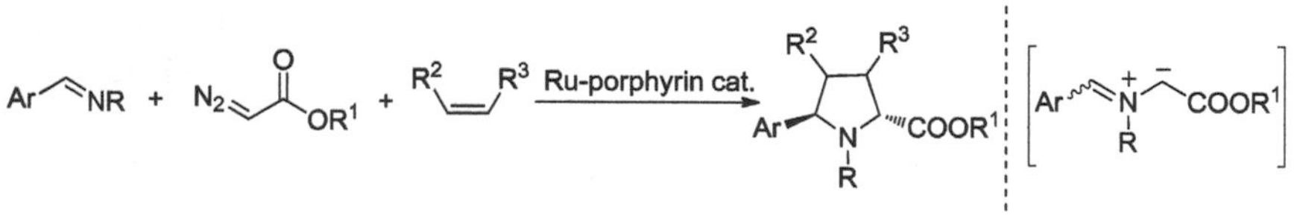

SCHEME 6.42 Synthesis of pyrrolidines using a Ru–porphyrin catalyst.

In 2003, Scheidt et al. reported a Cu(I) catalyzed, three-component reaction of α-diazo esters, amines, and various alkenes and alkynes to form substituted pyrrolidines with good diastereo selectivities with high yields. The generated azomethine ylide by the reaction of an imine from α-diazo ester and copper catalyst further reacted with an external dipolarophile, affording desired pyrrolidine derivatives.[41b] Che and co-workers reported synthesis of analogous pyrrolidines using a Ru-porphyrin catalyst (Scheme 6.42).[41c]

6.4 USE OF METALLACYCLE IN MULTICOMPONENT CYCLIZATION

Cycloaddition of two unsaturated fragments to generate a five-membered ring metallacycle is an integral step in a number of multicomponent reactions for the synthesis of a variety of bioactive heterocycles and carbocycles. In this context, the commonly employed metal for the formation of metallacycle is titanium.[42a–e] Due to some interesting properties of early electropositive transition metal catalysts, they are employed in these reactions as compared to the late transition metal counterparts because the former transition metals are easily able to form metallacycles that are not so prone to reductive elimination and thus can readily incorporate heteroatoms. Odom and co-workers developed various modifications for the reactivity of these metallacycles. The example given here demonstrated the formation of α,β-unsaturated β-iminoamines by the selective coupling of primary amines, alkynes, and isocyanides via the *in situ* formation of titanium–imido complexes (Scheme 6.43).[42f] Now the trapping of 1,3-unsaturated intermediate (**XXXVII**) with various nucleophiles can generate a variety of important heterocycles. This unsaturated intermediate (**XXXVII**), upon reaction with hydrazine, provided 1,2,3-trisubstituted pyrazoles,[42g] while the use of hydroxylamine hydrochlorides at room temperature afforded isoxazoles[42h] and using amidines afforded trisubstituted pyrimidines as the exclusive product (Scheme 6.43).[42i] The use of unsymmetrical internal alkynes in lieu of its symmetrical components afforded desired products with good regio- and chemoselectivity.

Polysubstituted quinolines are also synthesized in this process via the generation of α,β-unsaturated imine using aromatic amine followed by the insertion of alkyne and nitrile with subsequent acid catalyzed thermal cyclization in the final step afforded the desired quinolines (Scheme 6.44).[43] This reaction is also applicable to heteroaromatic amines for the synthesis of fused pyrroles, thiophenes, and indoles.

Suitable modification of the catalytic conditions and using two equivalents of the isocyanite substrate, this reaction can be used to generate 2,3-diaminopyrroles (Scheme 6.45).[44] The use of twofold ${}^tBuNC$ generates a six-membered titanacycle (**XXXVIII**) via a second ${}^tBuNC$ addition. Protonation of this ligand from the titanium complex (**XXXVIII**) and subsequent intramolecular cyclization provide the pyrrole product. The use of a bulkier *tert*-butyl group in the isonitrile moiety is crucial for product selectivity as the smaller alkyl isonitriles lead to simple imines.

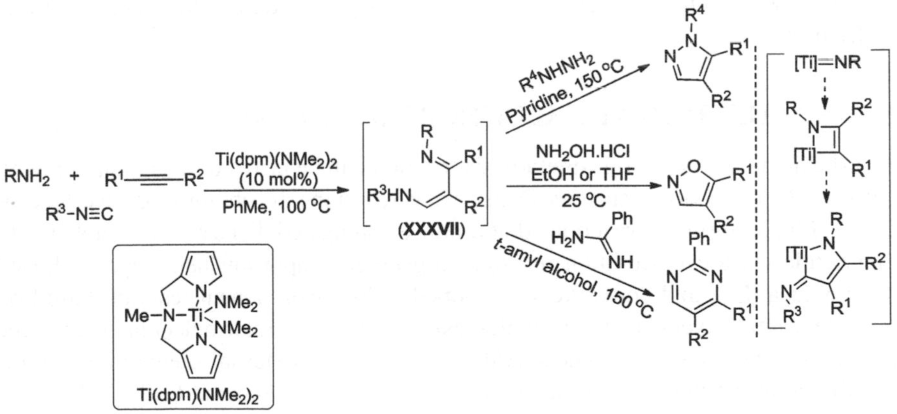

SCHEME 6.43 Use of titenacycle for the synthesis of pyrazoles, isoxazoles, and pyrimidines.

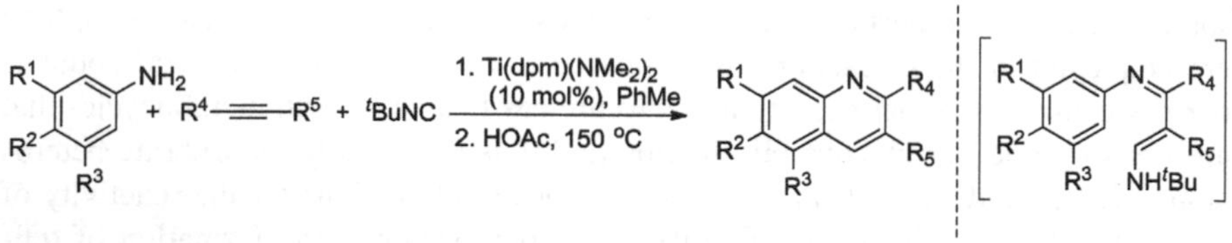

SCHEME 6.44 Three-component synthesis of quinolines using a Ti catalyst.

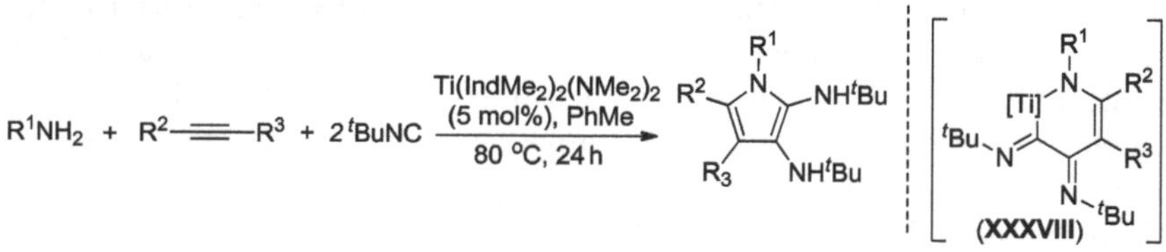

SCHEME 6.45 Ti catalyzed synthesis of 2,3-disubstituted pyrroles.

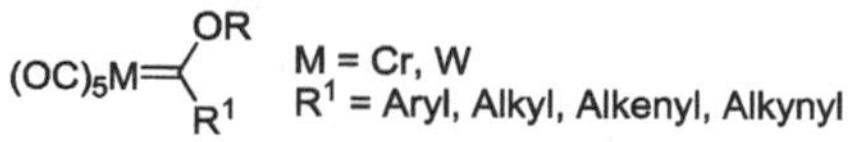

SCHEME 6.46 General structure of Fischer-type carbenes.

6.5 USE OF METALLOCARBENE IN MULTICOMPONENT CYCLIZATION

Various metallocarbenes have been found to be effective for the synthesis of bioactive heterocycles via metal catalyzed multicomponent reactions. The metalocarbenes generated *in situ* by the reaction of α-diazoesters and metal catalysts have already been discussed in the cycloaddition section.

Fischer-type carbenes are also an efficient path to synthesize a variety of bioactive complex heterocycles using multicomponent reactions.[45] Generally, these Fischer-type carbenes are generated by using early transition metals such as Cr and W (Scheme 6.46).

6.6 CYCLIZATION VIA METATHESIS REACTIONS

A ruthenium catalyzed cross-metathesis reaction can be utilized in conjugation with various other reactions, such as Diels–Alder reactions, to obtain bioactive heterocycles. Multicomponent reactions that produce conjugated diene can be coupled with dienophile via a cycloaddition reaction to generate important heterocycles. Based on this idea, Lee and co-workers developed a three-component reaction based on a tandem intramolecular enyne metathesis followed by intermolecular diene–ene metathesis and subsequent Diels–Alder reaction with external dienophile, affording the desired heterocycle in a stereoselective manner (Scheme 6.47).[46a] To avoid undesired side reactions, dienophile was added only after the tandem metathesis was completed.

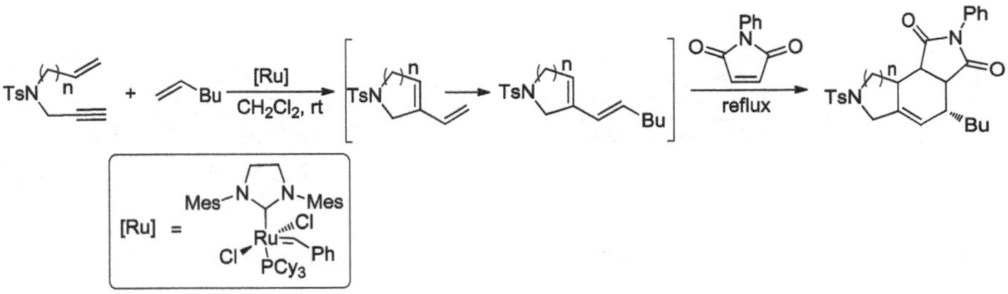

SCHEME 6.47 One-pot, multicomponent cyclization via tandem metathesis/Diels–Alder reaction.

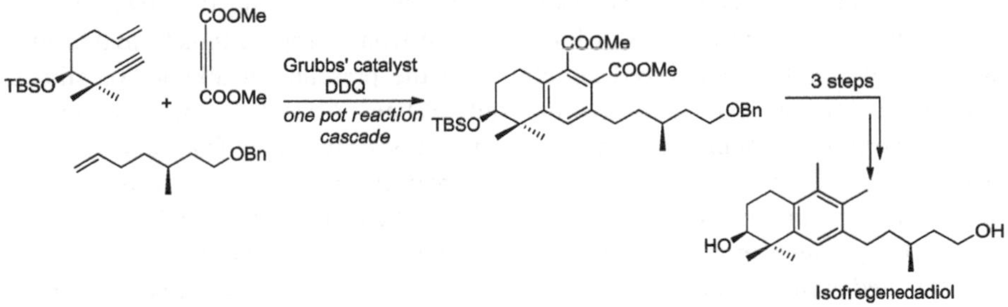

SCHEME 6.48 Total synthesis of isofregenedadiol via ring-closing metathesis/cross metathesis/Diels–Alder.

This idea was further utilized for the total synthesis of a bioactive diterpene isofregenedadiol. This reaction proceeds via a one-pot reaction sequence comprising an enyne ring-closing metathesis/cross-metathesis/Diels–Alder reaction/aromatization leading to the formation of the target skeleton (Scheme 6.48).[46b]

6.7 MULTICOMPONENT CYCLIZATION USING C–H FUNCTIONALIZATION

Transition metal catalyzed C–H functionalization represents the ideal step: a redox and atom economic bond forming process for the synthesis of complex organic frameworks.[47] Along with many sp C–H and sp^2 C–H, various sp^3 C–H, such as benzylic, allylic, and adjacent, heteroatoms and even inert alkanes can easily be functionalized without using preactivated starting material and be further merged with another tandem cyclization step. Lautens and co-workers demonstrated an early example of multicomponent cyclization via sp^2 C–H functionalization for the synthesis of fused oxygen heterocycles.[48] This reaction is initiated via the reversible insertion of norbornene (a strained alkene) to the organopalladium species, and subsequent two *ortho*-C–H activation affords the fused ring oxygen heterocycles via an overall three-component reaction (Scheme 6.49).[49]

Formation of oxindoles has been recently reported by Zhu et al. following a Pd catalyzed, three-component reaction. Starting from alkyne tethered amide and two different

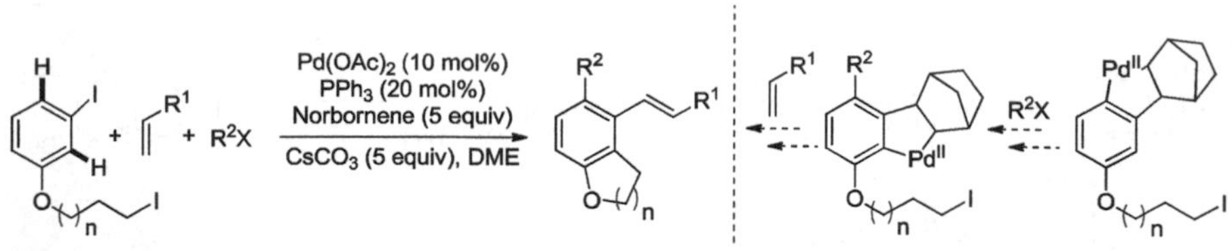

SCHEME 6.49 Fused oxygen heterocycles via norbornene mediated C–H functionalization.

aryl iodides, this process provides desired oxyindoles following Sonogashira coupling, carbopalladation, and subsequent *ortho*-C–H functionalization (Scheme 6.50).[50a,b]

Imines are found to be useful building blocks for the synthesis of a variety of heterocycles using multicomponent reactions via tandem C–H functionalization. The Cheng group reported the *in situ* formed imine from carbonyl moiety and amine, which, after rhodium catalyzed annulation in the presence of alkynes or alkene, leads to the formation of isoquinolinium salts (Scheme 6.51).[51a] A similar system for the synthesis of isoquinolines and heterocycle-fused pyridines from aryl ketones, hydroxylamine, and alkynes has been demonstrated by Hua and co-workers via similar *in situ* imine directed C–H annulation (Scheme 6.51).[51b] α,β-Unsaturated ketocarbonyls and ammonium acetate in combination with symmetrical alkynes reacting under rhodium catalyzed reaction conditions afforded pyridine derivatives (Scheme 6.51).[51c] Jiang et al. reported a Pd catalyzed protocol for the synthesis of quinolones from aldehydes, anilines, and electron-deficient alkenes (Scheme 6.51).[51d]

A multitasking oxyacetamide directing group has been developed by Zhao and Huang for the direct and selective double C–H annulation for the synthesis of polycyclic heterocycles in one-pot sequence using two different alkynes. This Rh catalyzed reaction proceeds via the reaction of, first, alkyne with oxyacetamide directing a substrate-generating intermediate (**XXXIX**), and, further, enamine-directed C–H activation of (**XXXIX**) with subsequent annulation, with another alkyne afforded the desired heterocycle (Scheme 6.52).[52]

A Pd catalyzed, ligand-promoted triple C–H activation sequence was demonstrated by Yu and co-workers from 1 equiv. of propionamide and 2 equiv. of aryliodide for the synthesis of 4-aryl-2-quinolinone derivatives. This interesting reaction occurs via initial amide-directed sp^3 C–H arylation followed by dehydrogenation generating intermediate (**XL**). This intermediate (**XL**), after subsequent Heck reaction with a second molecule of iodoarene and a final sp^2 C–H amination, afforded the desired heterocycles (Scheme 6.53).[53]

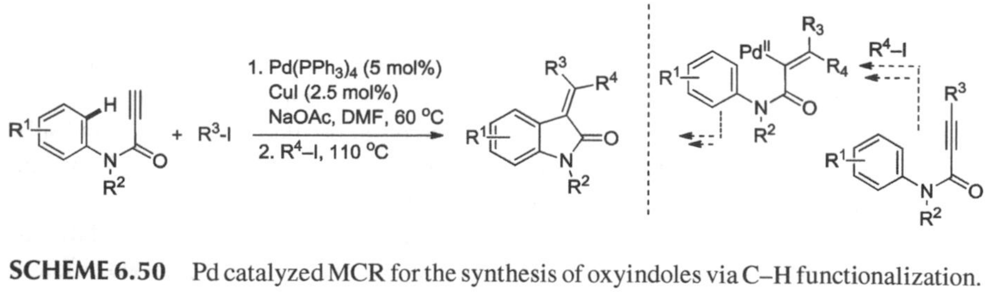

SCHEME 6.50 Pd catalyzed MCR for the synthesis of oxyindoles via C–H functionalization.

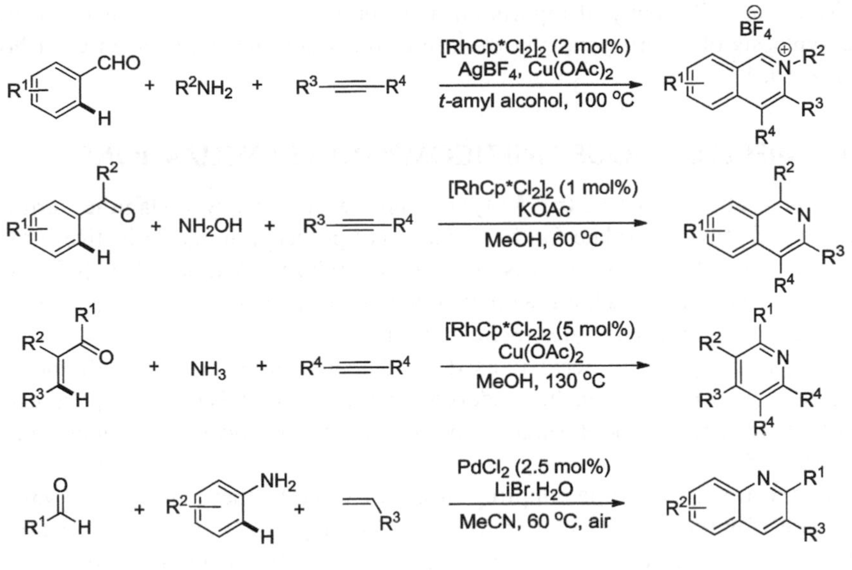

SCHEME 6.51 Transition metal catalyzed synthesis of heterocycles via multicomponent cyclization.

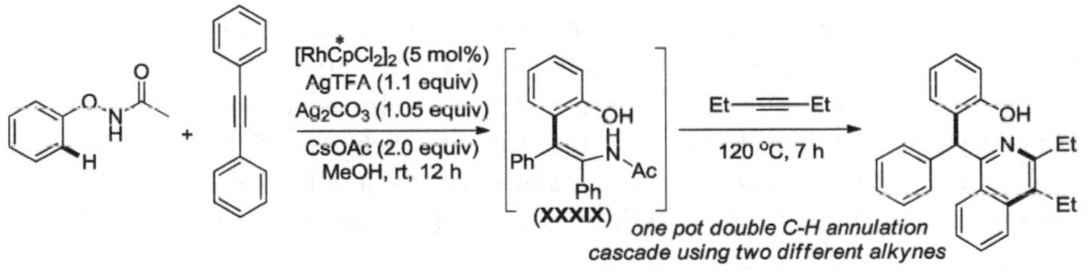

SCHEME 6.52 Synthesis of polycyclic heterocycles via cascade annulation.

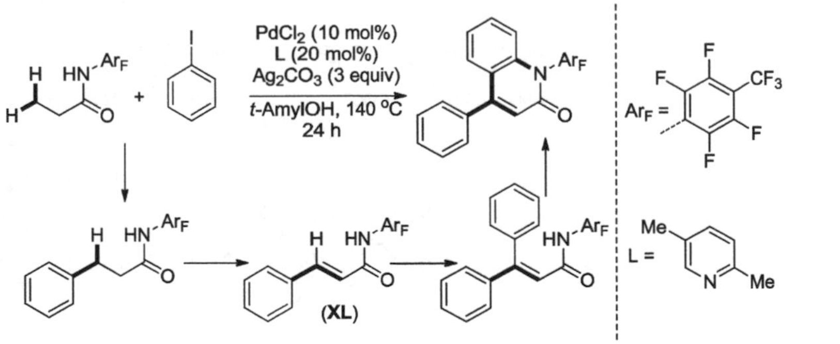

SCHEME 6.53 Synthesis of 4-aryl-2-quinolinone derivatives via triple C–H activation.

There are still plenty of reports on multicomponent cyclizations that deal with the synthesis of heterocycles via the use of transition metal catalyzed C–H bond activation.[54]

6.8 MISCELLANEOUS MULTICOMPONENT CYCLIZATIONS

Yamamoto et al. reported the A^3-type coupling of *N*-benzylallylamine, terminal alkyne, and ethyl glyoxalate for the formation of propergylamine derivatives (**XLI**). This intermediate cycloisomerizes under an Ir catalyzed condition. This is followed by a Diels–Alder cycloaddition with dienophiles, and a final dehydrogenation affords polycyclic pyrrole carboxylates (Scheme 6.54).[55]

The propergylamine derivatives derived from amine, terminal alkyne, and *o*-arylated benzaldehyde further undergo intramolecular Huisgen 1,3-dipolar cycloaddition, leading to the formation of aza-analogs of bis-benzocyclooctadienes (Scheme 6.55).[56]

A Cu catalyzed A^3 type coupling product can be further merged with various processes for the synthesis of imidazole-type rings with proper modifications. The propergyl amine generated in this A^3-type coupling is modified further to a propergylcyanamide intermediate (**XLII**). A lanthanide mediated hydroamination/isomerization of this intermediate (**XLII**) provides dialkylamino imidazole derivatives (Scheme 6.56).[57a] The use of thiol or alcohol as the nucleophilic source affords 2-thio- and 2-oxo-imidazoles as the sole product (Scheme 6.56).[57b]

Beller and co-workers developed a Ru catalyzed, three-component reaction between ketones, amines, and diols for the synthesis of various pyrrole derivatives. The use of *t*-amyl alcohol as the solvent, $[Ru_3(CO)_{12}]$/xantphos (3:1) as the catalytic combination, and K_2CO_3 as the base afforded polysubstituted pyrroles in moderate to good yields with high levels of regioselectivity (Scheme 6.57).[58a] As this methodology is limited to benzylic ketones and sterically less hindered vicinal diols, a suitable modification of the catalytic system ($[Ru(p\text{-cymene})Cl_2]_2$/xantphos/^{t}BuOK) was further developed to overcome these limitations.[58b]

Dhawan and Arndtsen developed an intriguing protocol for the Pd catalyzed synthesis of pyrrole moiety starting from internal alkynes, acid chlorides, and imines. The mechanism of this reaction is proposed to proceed via initial Pd catalyzed oxidative addition of an imine–acid chloride adduct followed by carbonylation generating

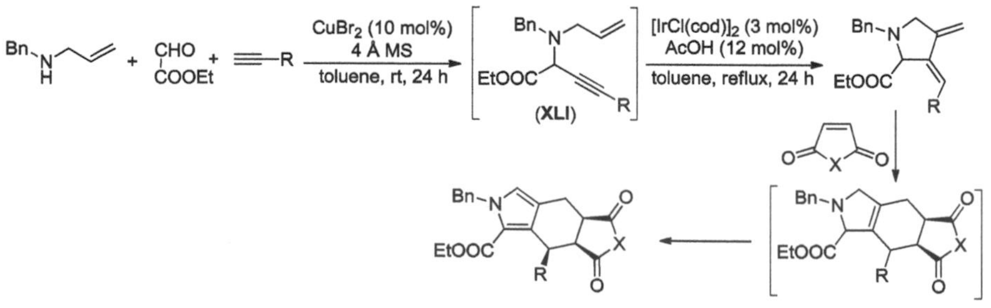

SCHEME 6.54 Synthesis of polycyclic pyrrole carboxylates via A^3-type coupling and further cyclization.

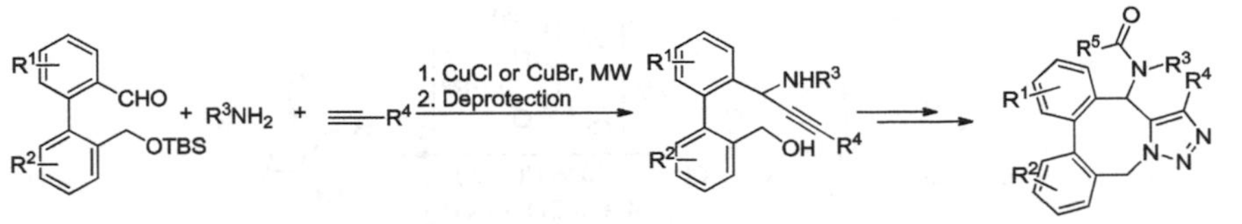

SCHEME 6.55 Synthesis of bisbenzocyclooctadienes via A^3-type coupling/cycloaddition.

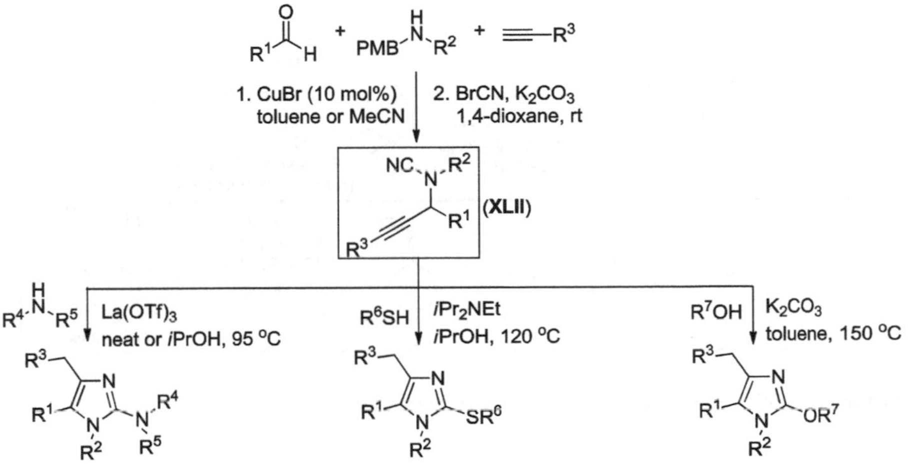

SCHEME 6.56 Synthesis of imidazole type rings via A^3-type coupling/cyclization.

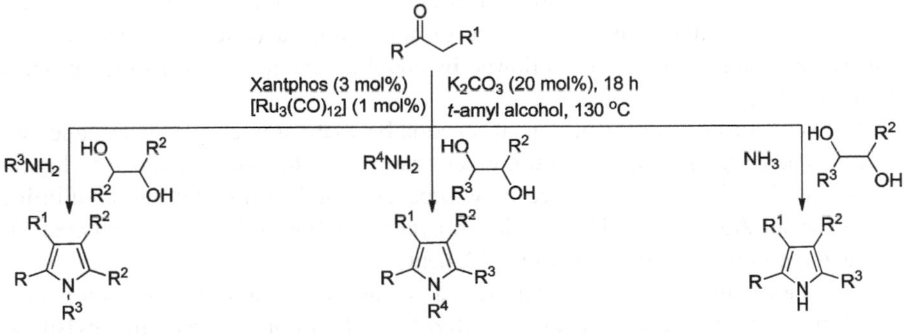

SCHEME 6.57 Synthesis of various pyrrole derivatives via Ru catalyzed MCRs.

Münchnones intermediate. This intermediate via cycloaddition with internal alkynes affords pyrrole derivatives via loss of CO_2 (Scheme 6.58).[59]

A subsequent illustration of this reaction is the incorporation of a further level of product complexity into the pyrrole structure with minimal steps, as demonstrated in Scheme 6.59. The members of this class of multicyclic pyrroles are of utility as potential therapeutics such as retinoic acid regulators, which can be generated in three steps from aldehyde, alkyne, amine, and acid chloride following this strategy.

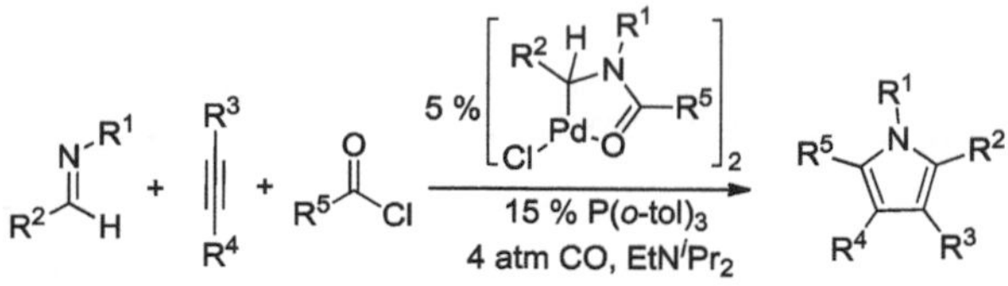

SCHEME 6.58 Three-component synthesis of pyrroles via münchnones intermediate/cycloaddition.

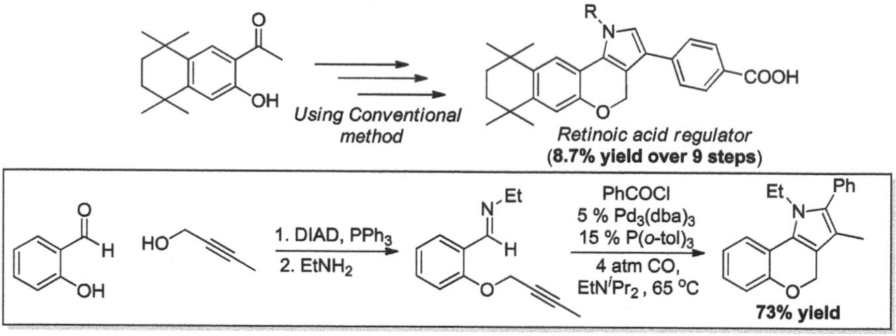

SCHEME 6.59 Total synthesis of retinoic acid regulator analogs.

However, its conventional long synthetic route (nine steps) suffers from very low product yield (Scheme 6.59).

A multicomponent reaction of imine, chloroformate, organotin reagent, and carbon monoxide leads to the formation of imidazolones via the condensation of *in situ* formed ketocarbamates in the presence of ammonium acetate. This process consists of an initial carbonylation followed by coupling and a final cyclization process (Scheme 6.60).[60]

A silver catalyzed multicomponent reaction between amines, aldehydes, and isocyanides bearing an acidic α-proton gives easy access to a diverse range of highly substituted 2-imidazolines. This reaction proceeds via the initial formation of imine and a further Aldol-type addition with isocyanide, followed by subsequent cyclization afforded the desired imidazoline. (Scheme 6.61).[61]

Yamamoto and co-workers reported a palladium catalyzed three-component coupling reaction of aryl isocyanides, allyl methyl carbonate, and trimethylsilylazide (Scheme 6.62).[62] This reaction proceeds via the generation of π-allyl palladium carbodiimide complex (**XLIII**) via the carbopalladation of initial π-allyl palladium azide complex to *o*-alkynyl isonitrile. The isomerization of π-allyl palladium

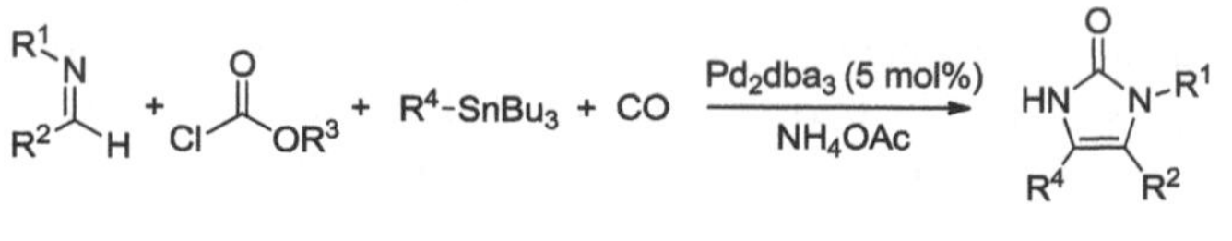

SCHEME 6.60 Pd catalyzed, four-component synthesis of imidazoles.

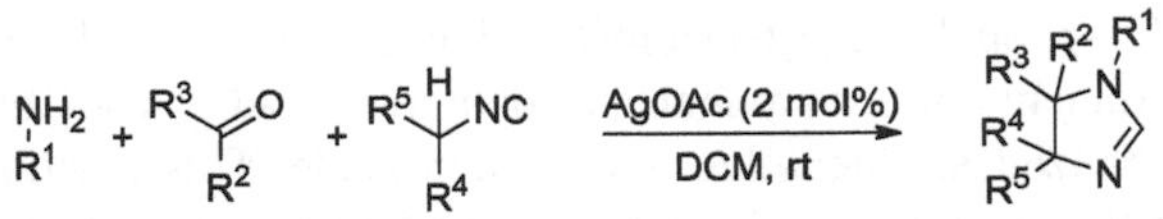

SCHEME 6.61 Silver catalyzed synthesis of highly substituted 2-imidazolines.

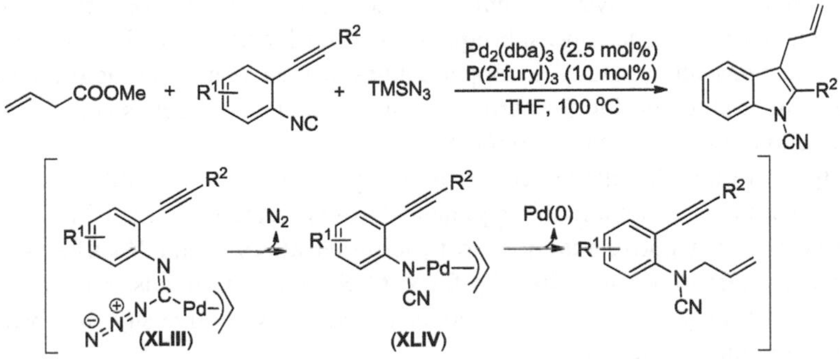

SCHEME 6.62 Pd catalyzed synthesis of *N*-cyanoindoles via isocyanide insertion/cyclization.

carbodiimide complex (**XLIII**) to π-allyl palladium cyanamide complex (**XLIV**) is the key step in this process that further cyclizes generating *N*-cyanoindoles.

Andreana et al. demonstrated four-component Ugi condensation for the synthesis of polysubstituted 1,4-benzodiazepin-3-ones (**XLV**) and (**XLVI**). This reaction proceeds via an initial isocyanide based reaction followed by an aza-Michael condensation, leading to the synthesis of polysubstituted 1,4-benzodiazepin-3-ones (Scheme 6.63).[63] By using protic solvents, controlled microwave irradiation, and a reducing agent Fe(0)/NH_4Cl, these derivatives can be prepared efficiently through a one-pot, two-step process. Interestingly, *o*-nitro benzaldehyde and *o*-nitrobenyzlamine not only act as bifunctional substrates, but also selectively direct a substitution pattern for benzodiazepin-3-ones of type (**XLV**) and (**XLVI**).

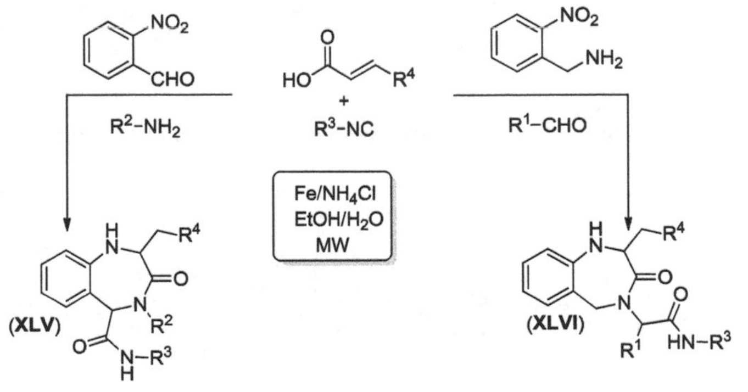

SCHEME 6.63 One-pot synthesis of 1,4-benzodiazepin-3-ones using MCR.

Synthesis of 5-phenyl-[1,2,3]triazolo[1,5-*c*]quinazolines has been demonstrated using a highly efficient multicomponent protocol from readily available (*E*)-1-bromo-2-(2-nitrovinyl)benzenes, aldehydes, and sodium azide. This reaction proceeds via consecutive [3+2] cycloaddition, Cu catalyzed S_NAr reaction, reduction, cyclization, and oxidation sequences (Scheme 6.64).[64]

An interesting protocol for the synthesis of bicyclic isoxazolidines was developed using a Cu(II) catalyst in a one-pot process. This reaction occurred via an *in situ* formation of nitrones and subsequent [3+2] cycloaddition with dienophiles. In this efficient cascade strategy, a copper catalyst serves as a Lewis acid for the ene reaction, an organometallic for aerobic oxidation, and also a Lewis acid for a [3+2] cycloaddition reaction. (Scheme 6.65).[65]

Fully substituted thiophene derivatives are synthesized from readily available triethylammonium 1-(2-oxoindolin-3-ylidene)-2-aroylethanethiolates using copper(I) catalyzed denitrogenative reactions with terminal alkynes and *N*-sulfonyl azides. This process simultaneously installs C–N, C–S, and C–C bonds, allowing direct formation of highly functionalized thiophenes with a wide diversity in substituents in a one-pot manner (Scheme 6.66).[66]

A Cu catalyzed, cascade, three-component reaction for the synthesis of 2-*N*-substituted benzothiazoles has been reported from 2-haloanilines, carbon disulfide, and *N* nucleophiles (Scheme 6.67). Various *N* nucleophiles, such as primary or secondary aliphatic/aromatic amines, cyclic secondary amines, and aromatic *N*-containing heterocycles such as pyrrole, indole, imidazole, were found to be effective for this process (Scheme 6.67).[67]

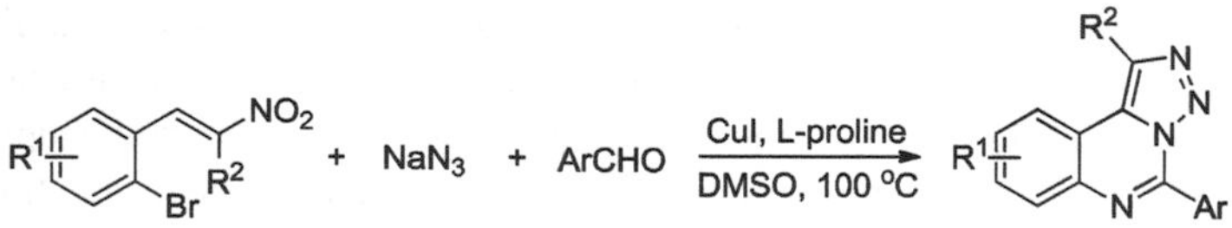

SCHEME 6.64 Multicomponent synthesis of triazolo-quinazolines using Cu(I) catalyst.

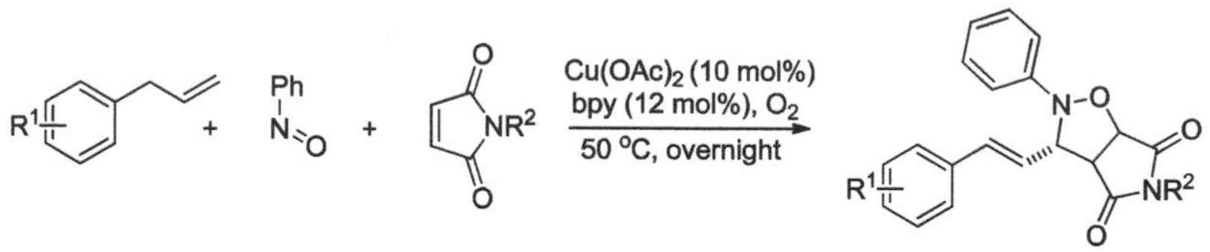

SCHEME 6.65 Synthesis of bicyclic isoxazolidines via dipolar cycloaddition.

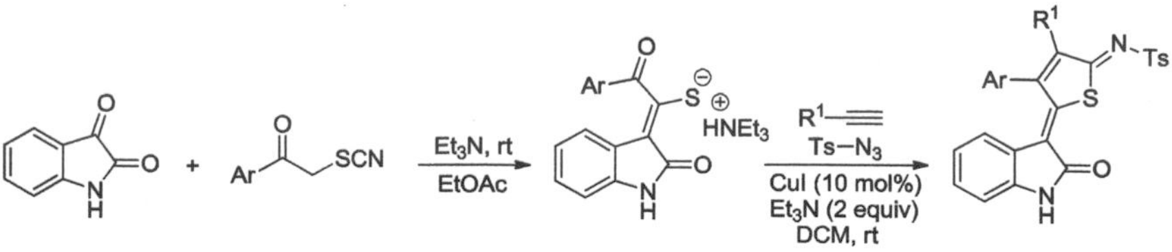

SCHEME 6.66 One-pot synthesis of fully substituted thiophenes.

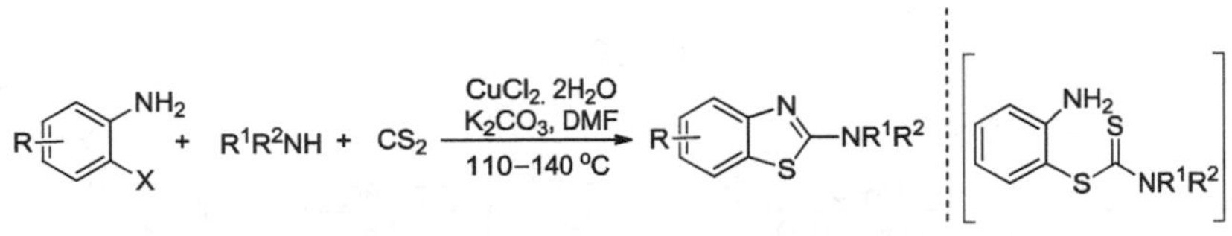

SCHEME 6.67 One-pot, three-component synthesis of 2-substituted benzothiazoles.

In 2011, a copper catalyzed, one-pot, three-component synthesis of 2*H*-indazoles was reported using 2-bromobenzaldehydes, primary amines, and sodium azide (Scheme 6.68).[68] 2-Bromobenzaldehydes are reacted with primary amines and NaN_3 to afford good to excellent yields of a variety of 2*H*-indazoles via the formation of one C–N and one N–N bond.

Eilbracht and Schmidt have reported the synthesis of indole derivatives starting from allylamine, phenyl hydrazine with CO/H_2. In this process, the initial hydroformylation of allylamines with CO/H_2 generates aldehydes followed by hydrazone formation; after Fischer indole synthesis this generates biologically important tryptamide derivatives (Scheme 6.69).[69]

The Ugi–Smiles reaction in combination with a Heck cyclization has also been reported for the synthesis of indole scaffolds.[70] 2-Iodo-4-nitrophenol, allylamine, aldehydes, and isocyanides are combined in an Ugi–Smiles coupling to afford the intermediate (**XLVII**), which is converted in one pot to indoles under Heck coupling conditions (Scheme 6.70). A one-pot reaction was possible if the residual isocyanide was neutralized prior to the addition of the palladium catalyst.

Barluenga and co-workers reported the synthesis of spiroacetals by a Pd(II) catalyzed reaction of an alkynol, an aldehyde, and a primary amine.[71] This reaction proceeds via the initial attack of the hydroxyl group to the activated triple bond by a Pd(II) catalyst generating intermediate (**XLVIII**). Subsequent protodemetalation provides methylidenefuran (**XLIX**), which reacts further with imine in the presence of a Pd(II) catalyst generating intermediate (**L**). Final attack of the hydroxyl group to

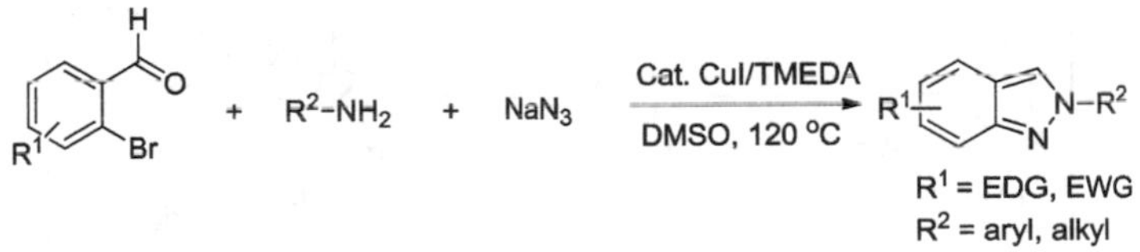

SCHEME 6.68 Synthesis of 2*H*-indazoles via Cu catalyzed MCR.

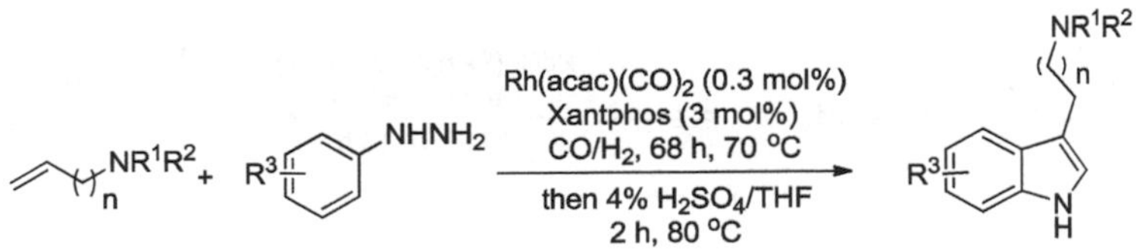

SCHEME 6.69 Rh catalyzed synthesis of 3-substituted indoles using CO/H_2.

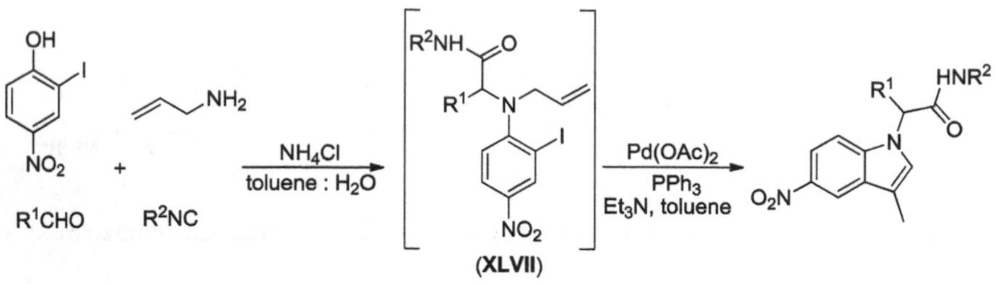

SCHEME 6.70 Four-component synthesis of substituted indoles via Ugi–Smiles/Heck combination.

the oxonium species (**L**) leads to spiroacetal. Further treatment of the crude mixture with $Mg(ClO_4)_2$ and $HClO_4$ in CH_2Cl_2/CH_3CN at room temperature is necessary to avoid the formation of an equimolar amount of two diastereomers (Scheme 6.71).

A palladium catalyzed multicomponent reaction of primary amine in combination with diiodobutenoic acid and a terminal alkyne afforded 1,3-disubstituted pyrroles (Scheme 6.72).[72] This reaction proceeds via a one-pot allylic amination followed by Sonogashira cross coupling utilizing terminal alkyne, and further intramolecular hydroamination and final isomerization lead to the desired 1,3-disubstituted pyrole. The delicate choice of amine is important for the successful transformation as aryl, alkyl, and benzylamines provide desired pyrrole derivatives smoothly; however, the use of nitrogen nucleophiles such as tosylamine and benzylcarbamate was unsuccessful in giving the desired product.

Catellani et al. have worked out a simple procedure that allows the synthesis of compounds belonging to the important class of dihydrodibenzoazepine and

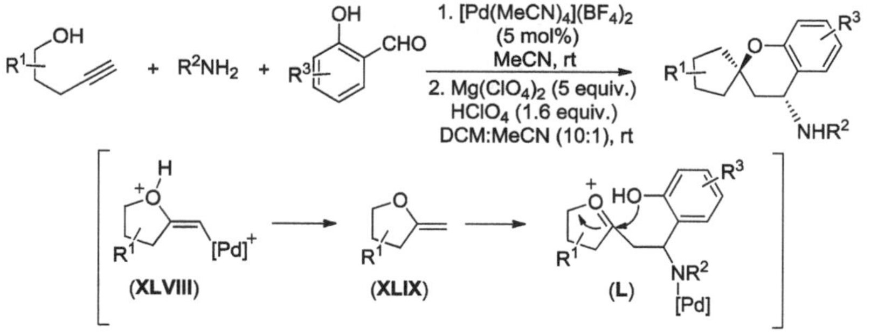

SCHEME 6.71 Pd catalyzed synthesis of spiroacetals via intramolecular double cyclization.

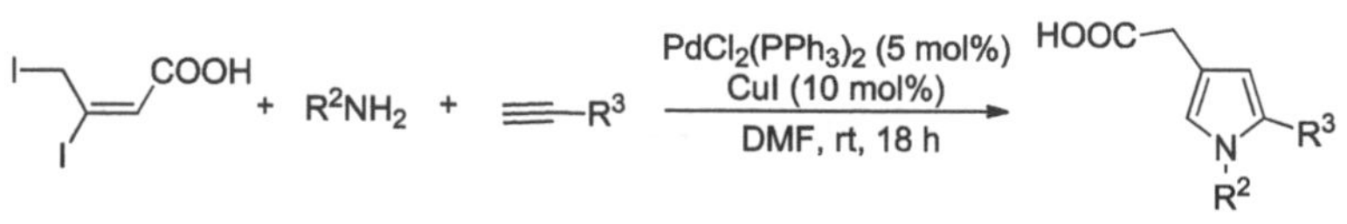

SCHEME 6.72 Synthesis of pyrrole derivatives via allylic amination/Sonogashira/heterocyclization.

dibenzoazepine derivatives via a sequential three-component reaction. Starting from simple and commercially available substrates such as aryl iodide, bromoaniline, and either norbornene or norbornadiene and palladium as catalyst, the use of norbornene leads to dihydrodibenzoazepine derivatives; however, the use of norbornadiene allows an additional step consisting of a retro Diels–Alder reaction, thus leading to the dibenzoazepines. Chelation of the amino group to palladium plays the key role, causing a deviation from the usual C_{sp2}–C_{sp2} bond formation to a C_{sp2}–C_{sp3} bond formation, which leads to the selective formation of a dibenzoazepine product (Scheme 6.73).[73]

An intriguing multicomponent reaction was developed using 2-iodophenol or *N*-tosyl iodoarene, bicyclopropylidene, and carbon monoxide for the synthesis of spirocycle products (**LII**) or annelated lactone derivatives (**LIII**) (Scheme 6.74). This reaction is believed to proceed via initial carbonylative carbopalladation with the insertion of two carbon monoxides generating a common intermediate (**LI**). Acylcarbopalladation of intermediate (**LI**), followed by intramolecular carbopalladation and final reductive elimination, provides spirocycle (**LII**). In an alternate pathway, the intermediate (**LI**) undergoes an intramolecular Michael addition and final cyclization, furnishing annelated product (**LIII**) (Scheme 6.74).[74]

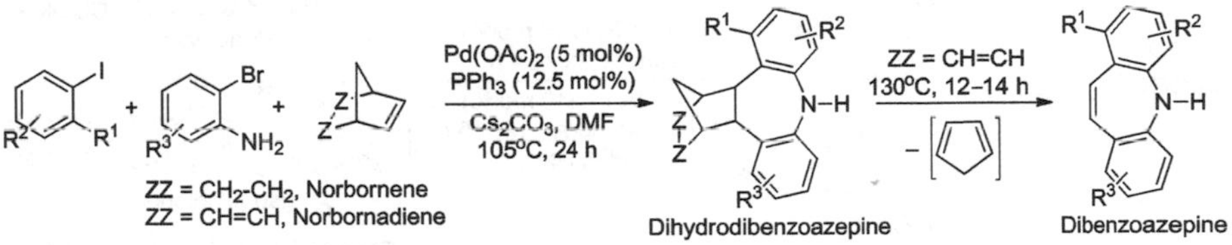

SCHEME 6.73 Dihydrodibenzoazepine and dibenzoazepine derivatives via Pd catalyzed MCR.

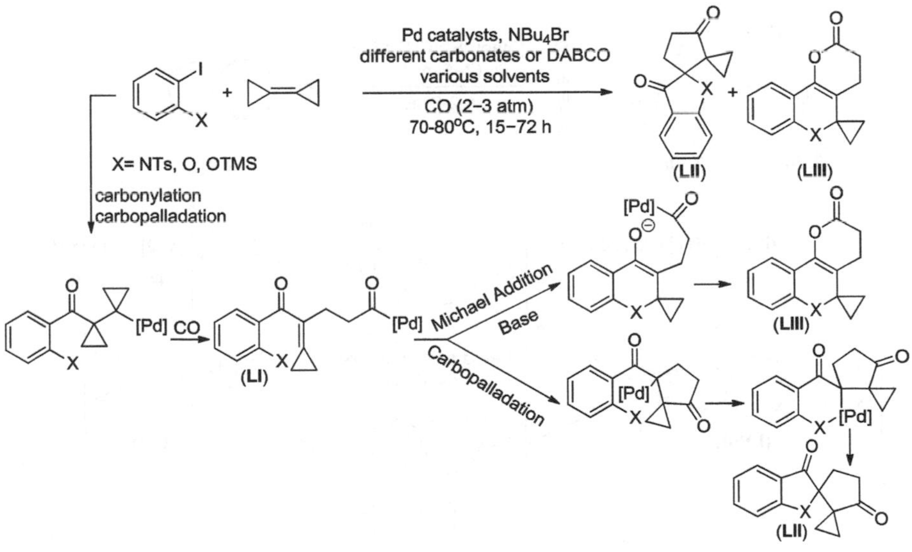

SCHEME 6.74 Synthesis of spirocycles or annelated lactones via multicomponent cascade.

An interesting [2+2+2] cyclotrimerization protocol was developed by Grigg et al. using a bimetallic catalytic combination (Rh and Pd) for the synthesis of indolone or isoquinolone derivatives. Reaction of alkynyl enamides (**LIV**) or (**LV**), diynes, and organostannane or organoboron compounds under the present bimetallic catalytic conditions (Wilkinson's catalyst and Pd(0)) generates indolones or isoquinolones in moderate yields (Scheme 6.75).[75]

Wender et al. reported a remarkable multicomponent process in which a novel 1,2,3-butatriene equivalent [TMSBO: (**LVI**), 4-(trimethylsilyl)but-2-yn-1-ol) engages chemospecifically as a two-carbon alkyne component in Rh catalyzed [5+2] cycloaddition with a vinylcyclopropane (**LVII**) to produce an intermediate cycloadduct (**LVIII**). This intermediate under the reaction conditions further undergoes a rapid 1,4-Peterson elimination, producing a reactive four-carbon diene intermediate (**LIX**), which reacted further by either a metal catalyzed or a thermal [4+2] cycloaddition, leading to the complex polycycles (**LX**) (Scheme 6.76).[76]

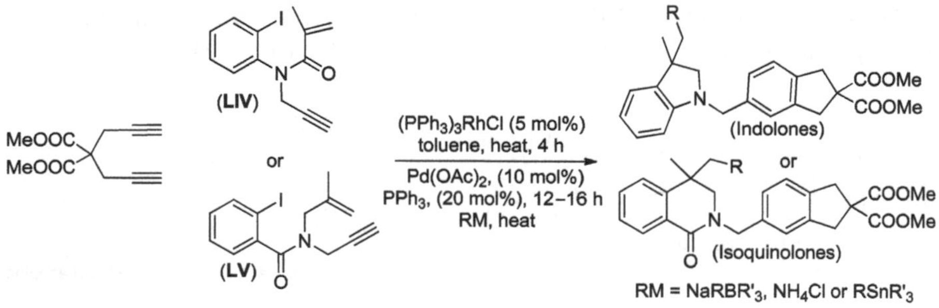

SCHEME 6.75 Cyclotrimerization cascade for the synthesis of indolones/isoquinolones using Rh/Pd.

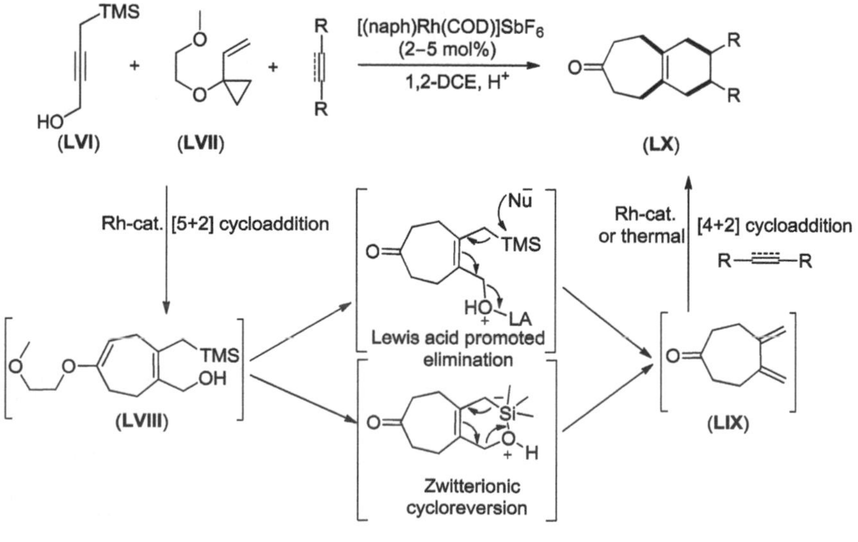

SCHEME 6.76 Synthesis of polycycles using Rh catalyzed cycloaddition cascade.

6.9 CONCLUSION

In conclusion, this chapter describes many examples of bioactive heterocycles using transition metal catalyzed multicomponent reactions. It is clear that transition metal catalysis has proven itself as a powerful tool for MCR based heterocyclic scaffold generation. Nowadays, transition metal catalysis is an emerging area with many new strategies, such as multimetallic catalysis, cooperative metal catalysis, multifunctional catalysis, and C–H bond activations; combining these processes with other transformations will provide the platform for various newer multicomponent routes for the generation of complex heterocycles.

REFERENCES

1. *Synthesis of heterocycles via multicomponent reactions* (eds. R. V. A. Orru and E. Ruijter), Springer, **2010**.
2. (a) D'Souza, D. M.; Müller, T. J. J. *Chem. Soc. Rev.* **2007**, *36*, 1095; (b) Balme, G.; Bouyssi, D.; Mont, N. *Multicomponent reactions* (eds. J. Zhu and H. Bienaymé), Wiley-VCH Verlag GmbH, Weinheim, **2005**, p. 224; (c) Balme, G.; Bossharth, E.; Monteiro, N. *Eur. J. Org. Chem.* **2003**, 4101.
3. (a) Schoenberg, A.; Bartoletti, I.; Heck, R. F. *J. Org. Chem.* **1974**, 39, 3318; (b) Schoenberg, A.; Heck, R. F. *J. Org. Chem.* **1974**, 39, 3327; (c) Schoenberg, A.; Heck, R. F. *J. Am. Chem. Soc.* **1974**, *96*, 7761
4. (a) Torii, S.; Okumoto, H.; Xu, L. H. *Tetrahedron Lett.* **1991**, *32*, 237; (b) Kalinin, V. N.; Shostakovsky, M. V.; Ponomaryov, A. B. *Tetrahedron Lett.* **1990**, *31*, 4073; (c) Yang, Q.; Alper, H. *J. Org. Chem.* **2010**, *75*, 948; (d) Ye, F.; Alper, H. *J. Org. Chem.* **2007**, *72*, 3218; (e) Kadnikov, D. V.; Larock, R. C. *J. Org. Chem.* **2004**, *69*, 6772; (f) Kadnikov, D. V.; Larock, R. C. *J. Org. Chem.* **2003**, *68*, 9423; (g) Grigg, R.; Liu, A.; Shaw, D.; Suganthan, S.; Woodall, D. E.; Yoganathan, G. *Tetrahedron Lett.* **2000**, *41*, 7125; (h) Xiao, W.-J.; Alper, H. *J. Org. Chem.* **1999**, *64*, 9646; (i) Larksarp, C.; Alper, H. *J. Org. Chem.* **1999**, *64*, 9194; (j) Okuro, K.; Alper, H. *J. Org. Chem.* **1997**, *62*, 1566.
5. Cao, H.; McNamee, L.; Alper, H. *Org. Lett.* **2008**, *10*, 5281.
6. Abbiati, G.; Arcadi, A.; Canevari, V.;Capezzuto, L.; Rossi, *E. J. Org. Chem.* **2005**, *70*, 6454.
7. (a) Majumdar, K. C.; Samanta, S.; Sinha, B. *Synthesis* **2012**, *44*, 817; (b) Cacchi, S.; Fabrizi, G. *Chem. Rev.* **2011**, *111*, PR215; (c) Zeni, G.; Larock, R. C. *Chem. Rev.* **2006**, *106*, 4644.
8. (a) Arcadi, A.; Cacchi, S.; Carnicelli, V.; Marinelli, F. *Tetrahedron* **1994**, *50*, 437; (b) Arcadi, A.; Cacchi, S.; Marinelli, F. *Tetrahedron Lett.*, **1992**, *33*, 3915.
9. Chaplin, J. H.; Flynn, B. L. *Chem. Commun.* **2001**, 1594.
10. Markina, N. A.; Chen, Y.; Larock, R. C. *Tetrahedron* **2013**, *69*, 2701;
11. (a) Ackermann, L. *Org. Lett.* **2005**, *7*, 439; (b) Kaspar, L. T.; Ackermann, L. *Tetrahedron* **2005**, *61*, 11311.
12. Knapp, J. M.; Zhu, J. S.; Tantillo, D. J.; Kurth, M. J. *Angew. Chem. Int. Ed.* **2012**, *51*, 10588.
13. Liang, Y.; Meng, T.; Zhang, H.-J.; Xi, Z. *Synlett* **2011**, 911.
14. (a) Dell'Acqua, M.; Facoetti, D.; Abbiati, G.; Rossi, E. *Tetrahedron* **2011**, *67*, 1552; (b) Dell'Acqua, M.; Abbiati, G.; Rossi, E. *Synlett* **2010**, *17*, 2672; (c) Bossharth, E.; Desbordes, P.; Monteiro, N.; Balme, G. *Org. Lett.* **2003**, *5*, 2441; (d) Shekarrao, K.; Nath, D.; Kaishap, P. P.; Gogoi, S.; Boruah, R. C. *Steroids* **2013**, *78*, 1126.
15. Pilgrim, B. S.; Gatland, A. E.; McTernan, C. T.; Procopiou, P. A.; Donohoe, T. J. *Org. Lett.* **2013**, *15*, 6190.

16. (a) Grigg, R.; Sridharan, V.; Terrier, C. *Tetrahedron Lett.* **1996**, *37*, 4221; (b) Grigg, R.; Savic, V. *Tetrahedron Lett.* **1996**, *37*, 6565.
17. Grigg, R.; Mariani, E.; Sridharan, V. *Tetrahedron Lett.* **2001**, *42*, 8677.
18. (a) Dahl, T.; Tornøe, C. W.; Bang-Andersen, B.; Nielsen, P.; Jørgensen, M. *Angew. Chem. Int. Ed.* **2008**, *47*, 1726; (b) Tsoung, J.; Panteleev, J.; Tesch, M.; Lautens, M. *Org. Lett.* **2014**, *16*, 110.
19. Zhang, L.; Sonaglia, L.; Stacey, J.; Lautens, M. *Org. Lett.* **2013**, *15*, 2128.
20. (a) Karpov, A. S.; Merkul, E.; Rominger, F.; Müller, T. J. J. *Angew. Chem. Int. Ed.* **2005**, *44*, 6951; (b) Karpov, A. S.; Müller, T. J. J. *Org. Lett.* **2003**, *5*, 3451; (c) Karpov, A. S.; Müller, T. J. J. *Synthesis* **2003**, 2815.
21. Mohamed Ahmed, M. S.; Kobayashi, K.; Mori, A. *Org. Lett.* **2005**, *7*, 4487.
22. (a) Willy, B.; Rominger, F.; Müller, T. J. J. *Synthesis* **2008**, 293; (b) Rotaru, A. V., Druta, I. D.; Oeser, T.; Müller, T. J. J. *Helv. Chim. Acta* **2005**, *88*, 1798; (c) Merkul, E.; Boersch, C.; Frank, W.; Müller, T. J. J. *Org. Lett.* **2009**, *11*, 2269; (d) Karpov, A. S.; Merkul, E.; Oeser, T.; Müller, T. J. J. *Eur. J. Org. Chem.* **2006**, 2991.
23. (a) Willy, B.; Dallos, T.; Rominger, F.; Schönhaber, J.; Müller, T. J. J. *Eur. J. Org. Chem.* **2008**, 4796; (b) Palimkar, S. S.; Lahoti, R. J.; Srinivasan, K. V. *Green Chem.* **2007**, *9*, 146.
24. (a) Karpov, A. S.; Rominger, F.; Müller, T. J. *J. Org. Biomol. Chem.* **2005**, *3*, 4382; (b) Karpov, A. S.; Oeser, T.; Müller, T. J. *J. Chem. Commun.* **2004**, 1502.
25. (a) Staben, S. T.; Blaquiere, N. *Angew. Chem. Int. Ed.* **2010**, *49*, 325; (b) Zhang, L.; Malinakova, H. C. *J. Org. Chem.* **2007**, *72*, 1484.
26. Xie, X.; Lu, X. *Tetrahedron Lett.* **1999**, *40*, 8415.
27. (a) Aftab, T.; Grigg, R.; Ladlow, M.; Sridharan, V.; Thornton-Pett, M. *Chem. Commun.* **2002**, 1754; (b) Grigg, R.; Savic, V.; Sridharan, V. *Tetrahedron* **2002**, *58*, 8613; (c) Grigg, R.; Nurnabi, M.; Sarkar, M. R. A. *Tetrahedron* **2004**, *60*, 3359.
28. (a) Shu, W.; Yu, Q.; Jia, G.; Ma, S. *Chem. Eur. J.* **2011**, *17*, 4720; (b) Ma, S.; Jiao, N. *Angew. Chem. Int. Ed.* **2002**, *41*, 4737; (c) Uemura, K.; Shiraishi, D.; Noziri, M.; Inoue, Y. *Bull. Chem. Soc. Jpn.* **1999**, *72*, 1063.
29. (a) Boissarie, P. J.; Hamilton, Z. E.; Lang, S.; Murphy, J. A.; Suckling, C. J. *Org. Lett.* **2011**, *13*, 6256; (b) Qiu, G.; Liu, G.; Pu, S.; Wu, J. *Chem. Commun.* **2012**, *48*, 2903; (c) Qiu, G.; Lu, Y.; Wu, J. *Org. Biomol. Chem.* **2013**, *11*, 798; (d) Qiu, G.; He, Y.; Wu, J. *Chem. Commun.* **2012**, *48*, 3836; (e) Vlaar, T.; Ruijter, E.; Znabet, A.; Janssen, E.; de Kanter, F. J. J.; Maes, B. U. W.; Orru, R. V. A. *Org. Lett.* **2011**, *13*, 6496; (f) Liu, B.; Li, Y.; Yin, M.; Wu, W.; Jiang, H. *Chem. Commun.* **2012**, *48*, 11446; (g) Ji, F.; Lv, M.-F.; Yi, W.-B.; Cai, C. *Adv. Synth. Catal.* **2013**, *355*, 3401; (h) Onitsuka, K.; Suzuki, S.; Takahashi, S. *Tetrahedron Lett.* **2002**, *43*, 6197.
30. Inoue, Y.; Itoh, Y.; Yen, I. F.; Imaizumi, S. *J. Mol. Cat.* **1990**, *60*, L1–L3.
31. (a) Cavicchioli, M.; Sixdenier, E.; Derrey, A.; Bouyssi, D.; Balme, G. *Tetrahedron Lett.* **1997**, *38*, 1763; (b) Bottex, M.; Cavicchioli, M.; Hartmann, B.; Monteiro, N.; Balme, G. *J. Org. Chem.* **2001**, *66*, 175; (c) Garcon, S.; Vassiliou, S.; Cavicchioli, M.; Hartmann, B.; Monteiro, N.; Balme, G. *J. Org. Chem.* **2001**, *66*, 4069; (d) Liu, G.; Lu, X. *Tetrahedron* **2003**, *44*, 467; (e) Azoulay, S.; Monteiro, N.; Balme, G. *Tetrahedron Lett.* **2002**, *43*, 9311; (f) Clique, B.; Vassiliou, S.; Monteiro, N.; Balme, G. *Eur. J. Org. Chem.* **2002**, 1493.
32. (a) Wei, L.-M.; Lin, C.-F.; Wu, M. J. *Tetrahedron Lett.* **2000**, *41*, 1215; (b) Kerr, D. J.; Willis, A. C.; Flynn, B. L. *Org. Lett.* **2004**, *6*, 457.
33. (a) Huma, H. Z. S.; Halder, R.; Kalra, S. S.; Das, J.; Iqbal, J. *Tetrahedron Lett.* **2002**, *43*, 6485; (b) Xiao, F.; Chen, Y.; Liu, Y.; Wang, J. *Tetrahedron* **2008**, *64*, 2755; (c) Cao, K.; Zhang, F.-M.; Tu, Y.-Q.; Zhang, X.-T.; Fan, C.-A. *Chem. Eur. J.* **2009**, *15*, 6332; (d) Zhang, M.; Wang, T.; Xiong, B.; Yan, F.; Wang, X.; Ding, Y.; Song, Q. *Heterocycles* **2012**, *85*, 639; (e) Yao, C.; Qin, B.; Zhang, H.; Lu, J.; Wang, D.; Tu, S. *RSC Adv.* **2012**, *2*,

3759; (f) Gaddam, V.; Ramesh, S.; Nagarajan, R. *Tetrahedron* **2010**, *66*, 4218; (g) Chen, X.-L.; Zhang, J.-M.; Shang, W.-L.; Lu, B.-Q.; Jin, J.-A. *J. Fluor. Chem.* **2012**, *133*, 139; (h) Liu, P.; Pan, Y.-M.; Xu, Y.-L.; Wang, H.-S. *Org. Biomol. Chem.* **2012**, *10*, 4696; (i) Mishra, S.; Ghosh, R. *Synthesis* **2011**, 3463; (j) Guchhait, S. K.; Chandgude, A. L.; Priyadarshani, G. *J. Org. Chem.* **2012**, *77*, 4438.

34. (a) Yan, B.; Liu, Y. *Org. Lett.* **2007**, *9*, 4323; (b) Bai, Y.; Zeng, J.; Ma, J.; Gorityala, B. K.; Liu, X.-W. *J. Comb. Chem.* **2010**, *12*, 696; (c) Patil, S. S.; Patil, S. V.; Bobade, V. D. *Synlett* **2011**, 2379; (d) Mishra, S.; Naskar, B.; Ghosh, R. *Tetrahedron Lett.* **2012**, *53*, 5483; (e) Dighe, S. U.; Hutait, S.; Batra, S. *ACS Comb. Sci.* **2012**, *14*, 665; (f) Nguyen, R.-V.; Li, C.-J. *Synlett* **2008**, 1897; (g) Sakai, N.; Uchida, N.; Konakahara, T. *Tetrahedron Lett.* **2008**, *49*, 3437; (h) Li, H.; Liu, J.; Yan, B.; Li, Y. *Tetrahedron Lett.* **2009**, *50*, 2353; (i) Yu, M.; Wang, Y.; Li, C.-J.; Yao, H. *Tetrahedron Lett.* **2009**, *50*, 6791; (j) Chernyak, D.; Chernyak, N.; Gevorgyan, V. *Adv. Synth. Catal.* **2010**, *352*, 961.
35. Ohno, H.; Ohta, Y.; Oishi, S.; Fujii, N. *Angew. Chem. Int. Ed.* **2007**, *46*, 2295.
36. Bonfield, E. R.; Li, C.-J. *Adv. Synth. Catal.* **2008**, *350*, 370.
37. (a) Fokin, V. V.; Matyjaszewski, K. *Organic chemistry—Breakthroughs and perspectives.* Wiley-VCH Verlag GmbH, Weinheim, **2012**, p. 247; (b) Liang, L.; Astruc, D. *Coord. Chem. Rev.* **2011**, *255*, 2933; (c) Ackermann, L.; Potukuchi, H. K.; Landsberg, D.; Vicente, R. *Org. Lett.* **2008**, *10*, 3081.
38. (a) Kamijo, S.; Jin, T.; Huo, Z.; Yamamoto, Y. *Tetrahedron Lett.* **2002**, *43*, 9707; (b) Kamijo, S.; Jin, T.; Huo, Z.; Yamamoto, Y. *J. Am. Chem. Soc.* **2003**, *125*, 7786; (c) Kamijo, S.; Jin, T.; Yamamoto, Y. *J. Org. Chem.* **2002**, *67*, 7413.
39. (a) Arndtsen, B. A. *Chem. Eur. J.* **2009**, *15*, 302; (b) Dhawan, R.; Dghaym, R. D.; Arndtsen, B. A. *J. Am. Chem. Soc.* **2003**, *125*, 1474; (c) Dhawan, R.; Arndtsen, B. A. *J. Am. Chem. Soc.* **2004**, *126*, 468; (d) Lu, Y.; Arndtsen, B. A. *Angew. Chem. Int. Ed.* **2008**, *47*, 5430; (e) Siamaki, A. R.; Arndtsen, B. A. *J. Am. Chem. Soc.* **2006**, *128*, 6050; (f) Worrall, K.; Xu, B.; Bontemps, S.; Arndtsen, B. A. *J. Org. Chem.* **2011**, *76*, 170; (g) Dghaym, R. D.; Dhawan, R.; Arndtsen, B. A. *Angew. Chem. Int. Ed.* **2001**, *40*, 3228; (h) Dhawan, R.; Dghaym, R. D.; St. Cyr, D. J.; Arndtsen, B. A. *Org. Lett.* **2006**, *8*, 3927; (i) Bontemps, S.; Quesnel, J. S.; Worrall, K.; Arndtsen, B. A. *Angew. Chem. Int. Ed.* **2011**, *50*, 8948.
40. (a) Huisgen, R.; De March, P. *J. Am. Chem. Soc.* **1982**, *104*, 4953; (b) DeAngelis, A.; Taylor, M. T.; Fox, J. M. *J. Am. Chem. Soc.* **2009**, *131*, 1101; (c) Galliford, C. V.; Scheidt, K. A. *J. Org. Chem.* **2007**, *72*, 1811; (d) Li, G.-Y.; Chen, J.; Yu, W.-Y.; Hong, W.; Che, C.-M. *Org. Lett.* **2003**, *5*, 2153.
41. (a) Hong, D.; Zhu, Y.; Li, Y.; Lin, X.; Lu, P.; Wang, Y. *Org. Lett.* **2011**, *13*, 4668; (b) Galliford, C. V.; Beenen, M. A.; Nguyen, S. T.; Scheidt, K. A. *Org. Lett.* **2003**, *5*, 3487; (c) Li, G.-Y.; Chen, J.; Yu, W.-Y.; Hong, W.; Che, C.-M. *Org. Lett.* **2003**, *5*, 2153.
42. (a) Lee, H.-W.; Kwong, F.-Y. *Eur. J. Org. Chem.* **2010**, 789; (b) Chopade, P. R.; Louie, J. *Adv. Synth. Catal.* **2006**, *348*, 2307; (c) Varela, J. A.; Saá, C. *Chem. Rev.* **2003**, *103*, 3787; (d) Brummond, K. M.; Kent, J. L. *Tetrahedron* **2000**, *56*, 3263; (e) Odom, A. L.; McDaniel, T. J. *Acc. Chem. Res.* **2015**, *48*, 2822; (f) Banerjee, S.; Shi, Y.; Cao, C.; Odom, A. L. *J. Organomet. Chem.* **2005**, *690*, 5066; (g) Majumder, S.; Gipson, K. R.; Staples, R. J.; Odom, A. L. *Adv. Synth. Catal.* **2009**, *351*, 2013; (h) Dissanayake, A. A.; Odom, A. L. *Tetrahedron* **2012**, *68*, 807; (i) Majumder, S.; Odom, A. L. *Tetrahedron* **2010**, *66*, 3152.
43. Majumder, S.; Gipson, K. R.; Odom, A. L. *Org. Lett.* **2009**, *11*, 4720.
44. Barnea, E.; Majumder, S.; Staples, R. J.; Odom, A. L. *Organometallics* **2009**, *28*, 3876.
45. (a) Fernández-Rodríguez, M. A.; García-García, P.; Aguilar, E. *Chem. Commun.* **2010**, *46*, 7670; (b) Barluenga, J.; Perez-Sanchez, I.; Rubio, E.; Florez, J. *Angew. Chem. Int. Ed.* **2003**, *42*, 5860; (c) Barluenga, J.; Alonso, J.; Fananas, F. J. *J. Am. Chem. Soc.* **2003**, *125*, 2610; (d) Barluenga, J.; Fernandez-Rodriguez, M. A.; Andina, F.; Aguilar, E.

J. Am. Chem. Soc. **2002**, *124*, 10978; (e) Barluenga, J.; Dieguez, A.; Rodriguez, F.; Florez, J.; Fananas, F. J. *J. Am. Chem. Soc.* **2002**, *124*, 9056; (f) Wu, H.-P.; Aumann, R., Fröhlich, R.; Wibbeling, B.; Kataeva, O. *Chem. Eur. J.* **2001**, 7, 5084; (g) Wu, H.-P.; Aumann, R., Fröhlich, R.; Wibbeling, B. *Chem. Eur. J.* **2002**, *8*, 910.

46. (a) Lee, H.-Y; Kim, H. Y.; Tae, H.; Kim, B. G.; Lee, J. *Org. Lett.* **2003**, *5*, 3439; (b) Kurhade, S. E.; Sanchawala, A. I.; Ravikumar, V.; Bhuniya, D.; Reddy, D. S. *Org. Lett.* **2011**, *13*, 3690.
47. (a) Wencel-Delord, J.; Droege, T.; Liu, F.; Glorius, F. *Chem. Soc. Rev.* **2011**, *40*, 4740; (b) Cho, S. H.; Kim, J. Y.; Kwak, J.; Chang, S. *Chem. Soc. Rev.* **2011**, *40*, 5068; (c) Alberico, D.; Scott, M. E.; Lautens, M. *Chem. Rev.* **2007**, *107*, 174.
48. Pache, S.; Lautens, M. *Org. Lett.* **2003**, *5*, 4827.
49. Martins, A.; Mariampillai, B.; Lautens, M. *C–H activation*, vol. 292 (eds. J.-Q. Yu and Z. Shi), Springer, Berlin, **2010**, p. 1.
50. (a) Pinto, A.; Neuville, L.; Zhu, J. *Tetrahedron Lett.* **2009**, *50*, 3602; (b) Pinto, A.; Neuville, L.; Zhu, J. *Angew. Chem. Int. Ed.* **2007**, *46*, 3291.
51. (a) Jayakumar, J.; Parthasarathy, K.; Cheng, C.-H. *Angew. Chem. Int. Ed.* **2012**, *51*, 197; (b) Zheng, L.; Ju, J.; Bin, Y.; Hua, R. *J. Org. Chem.* **2012**, *77*, 5794; (c) Sim, Y.-K.; Lee, H.; Park, J.-W.; Kim, D.-S.; Jun, C.-H. *Chem. Commun.* **2012**, *48*, 11787; (d) Ji, X.; Huang, H.; Li, Y.; Chen, H.; Jiang, H. *Angew. Chem. Int. Ed.* **2012**, *51*, 7292.
52. Chen, Y.; Wang, D.; Duan, P.; Ben, R.; Dai, L.; Shao, X.; Hong, M.; Zhao, J.; Huang, Y. *Nat. Commun.* **2014**, *5*, 4610.
53. Deng, Y.; Gong, W.; He, J.; Yu, J.-Q. *Angew. Chem. Int. Ed.* **2014**, *53*, 6692.
54. (a) Wang, Y.; Chen, C.; Peng, J.; Li, M. *Angew. Chem. Int. Ed.* **2013**, *52*, 5323; (b) Su, X.; Chen, C.; Wang, Y.; Chen, J.; Lou, Z.; Li, M. *Chem. Commun.* **2013**, *49*, 6752; (c) Rohlmann, R.; Stopka, T.; Richter, H.; Garcia Mancheño, O. *J. Org. Chem.* **2013**, *78*, 6050.
55. Yamamoto, Y.; Hayashi, H.; Saigoku, T.; Nishiyama, H. *J. Am. Chem. Soc.* **2005**, *127*, 10804.
56. Mont, N.; Mehta, V. P.; Appukkuttan, P.; Beryozkina, T.; Toppet, S.; Van Hecke, K.; Meervelt, L. V.; Voet, A.; DeMaeyer, M.; Eycken., E. V. *J. Org. Chem.* **2008**, *73*, 7509.
57. (a) Giles, R. L.; Sullivan, J. D.; Steiner, A. M.; Looper, R. E.; *Angew. Chem. Int. Ed.* **2009**, *48*, 3116; (b) Giles, R. L.; Nkansah, R. A.; Looper, R. E. *J. Org. Chem.* **2010**, *75*, 261.
58. (a) Zhang, M.; Neumann, H.; Beller, M. *Angew. Chem. Int. Ed.* **2013**, *52*, 597; (b) Zhang, M.; Fang, X.; Neumann, H.; Beller, M. *J. Am. Chem. Soc.* **2013**, *135*, 11384.
59. Dhawan, R.; Arndtsen, B. A. *J. Am. Chem. Soc.* **2004**, *126*, 468.
60. Siamaki, A. R.; Black, D. A.; Arndtsen, B. A. *J. Org. Chem.* **2008**, *73*, 1135.
61. (a) Bon, R. S.; Vliet, B. V. Sprenkels, N. E.; Schmitz, R. F.; de Kanter, F. J. J.; Stevens, C. V. Swart, M.; Bickelhaupt, F. M.; Groen, M. B.; Orru, R. V. A. *J. Org. Chem.* **2005**, *70*, 3542; (b) Elders, N.; Schmitz, R. F.; de Kanter, F. J. J.; Ruijter, E.; Groen, M. B.; Orru, R. V. A. *J. Org. Chem.*, **2007**, *72*, 6135.
62. Kamijo, S.; Yamamoto, Y. *J. Am. Chem. Soc.* **2002**, *124*, 11940.
63. De Silva, R. A.; Santra, S.; Andreana, P. R. *Org. Lett.* **2008**, *10*, 4541.
64. Jia, F.-C.; Xu, C.; Zhou, Z.-W.; Cai, Q.; Li, D.-K.; Wu, A.-X. *Org. Lett.* **2015**, *17*, 2820.
65. Chen, H.; Wang, Z.; Zhang, Y.; Huang, Y. *J. Org. Chem.* **2013**, *78*, 3503.
66. Jiang, B.; Tu, X.-J.; Wang, X.; Tu, S.-J.; Li, G. *Org Lett.* **2014**, *16*, 3656.
67. Ma, D.; Lu, X.; Shi, L.; Zhang, H.; Jiang, Y.; Liu, X. *Angew. Chem. Int. Ed.* **2011**, *50*, 1118.
68. Kumar, M. R.; Park, A.; Park, N.; Lee, S. *Org. Lett.* **2011**, *13*, 3542.
69. Schmidt, A. M.; Eilbracht, P. *Org. Biomol. Chem.* **2005**, *3*, 2333.
70. El Kaim, L.; Gizzi, M.; Grimaud, L. *Org. Lett.* **2008**, *10*, 3417.

71. Barluenga, J.; Mendoza, A.; Rodríguez, F.; Fañanás, F. J. *Angew. Chem. Int. Ed.* **2009**, *48*, 1644.
72. Lamande-Langle, S.; Abarbri, M.; Thibonnet, J.; Duchêne, A.; Parrain, J.-L. *Chem. Commun.* **2010**, *46*, 5157.
73. Della Cá, N.; Maestri, G.; Malacria, M.; Derat, E.; Catellani, M. *Angew. Chem. Int. Ed.* **2011**, *50*, 12257.
74. von Seebach, M.; Grigg, R.; De Meijere, A. *Eur. J. Org. Chem.* **2002**, 3268.
75. Grigg, R.; Sridharan, V.; Zhang, J. *Tetrahedron Lett.* **1999**, *40*, 8277.
76. Wender, P. A.; Fournogerakis, D. N.; Jeffreys, M. S.; Quiroz, R. V.; Inagaki, F.; Pfaffenbach, M. *Nature Chem.* **2014**, *6*, 448.

7 Synthesis of Fluorinated Heterocycles via Multicomponent Reactions

Sandip S. Shinde, S. N. Thore, Rajendra P. Pawar,* K. L. Ameta, and Sunil N. Patil*

CONTENTS

7.1 INTRODUCTION

Fluorine is one of the most important elements in the periodic table because it is the most electronegative element, makes the strongest C–F bond, and is more polarized in organic synthesis. These unique and distinctive properties of fluorine make it very useful and widely applicable in the pharmaceutical and agrochemical industries. Hydrogen replaced with the fluorine of an organic compound can tune the pharmacokinetic and pharmacological properties of the molecule, which improve solubility, membrane permeability,

* Corresponding authors: ss.shinde@ncl.res.in and rppawar@yahoo.com.

metabolic stability, and receptor-binding properties compared with nonfluorinated analogs.[1] Due to these unique properties of fluorine, organofluorine compounds are very applicative in medicinal, pharmaceutical, agrochemical, and material chemistry. This wide applicability of fluorine and organofluorine compounds attracts the attention of researchers and chemists. Labeled fluorine-18 is one of the most commonly used positron emitting radioisotopes in positron emission tomography (PET) imaging technology, which can detect presymptomatic biochemical changes in body tissues. In addition to its use in PET-CT, fluorine is used in preparation of chemically resistant polymer materials such as polytetrafluoroethylene (Teflon) or polarity to piezoelectric material such as polyvinylidene fluoride and organic liquid crystals for displays. Because of these many advantages of fluorine, organic chemists have developed various methods to introduce fluorine in organic molecules. During the last two decades, numerous organofluorine heterocyclics have been developed; in this chapter, we have summarized synthesis of multicomponent reactions (MCRs) in application of fluorine compound synthesis.[2]

Multicomponent reactions are the processes in which more than two substrates can react to produce simple and flexible building blocks, with advantages of practical, time-saving, one-pot reaction conditions. Thus, MCRs have attracted considerable interest in combinatorial libraries of compounds, drug discovery, and automated synthesis.[3] The MCRs are a process in which there is an inherent formation of various bonds in one procedure, without isolating the intermediates, changing the reaction parameters, or adding additional reagents. Numerous heterocyclic compounds can be synthesized by using MCR strategies under basic or acidic conditions. For a decade, the application of MCRs in synthesis of fluorinated heterocycles was applied in agrochemical and pharmaceutical fields, including the radiopharamceutical industries, to achieve value-added products and drugs.[4]

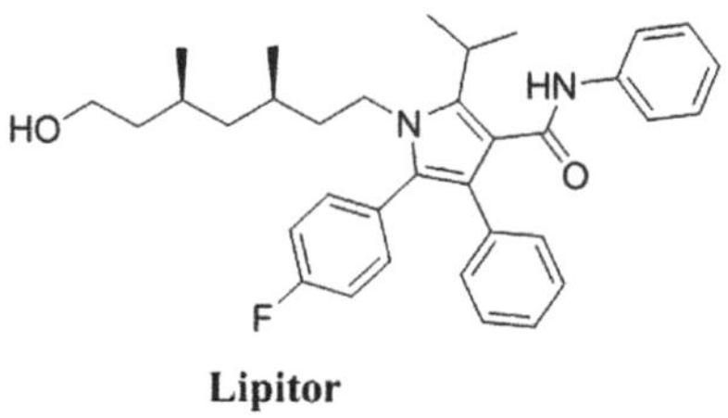

Lipitor

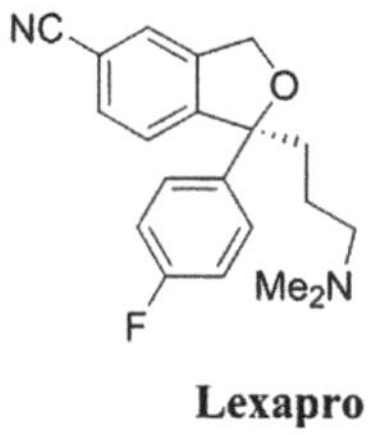

Lexapro

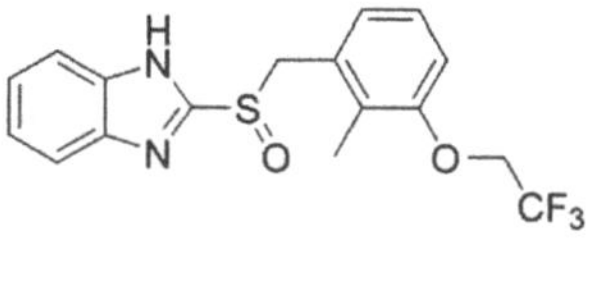

Prevacid

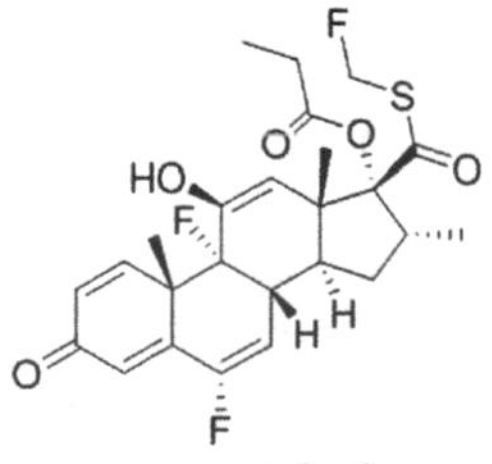

Advair

FIGURE 7.1 Fluorinated drug molecules.

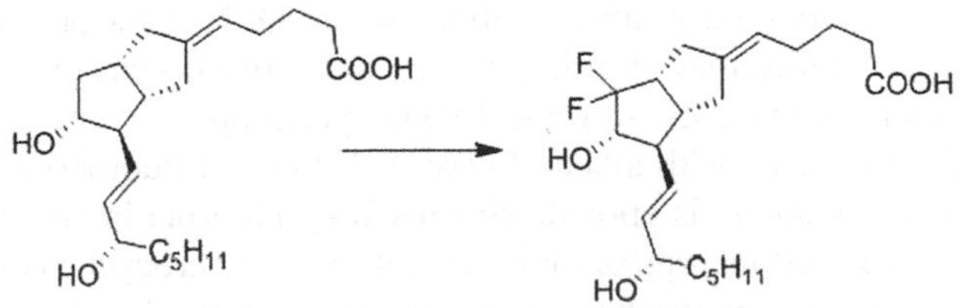

$t_{1/2}$ = 10 min
Prostacyclin

$t_{1/2}$ = 24 hr
10,10-Difluoro-13-dehydroprostacyclin

FIGURE 7.2 Stability and biological activity of nonfluorinated and fluorinated analogs.

	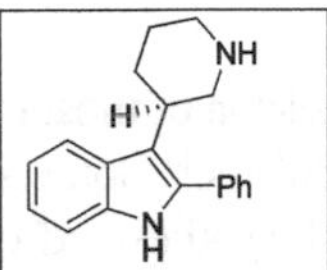	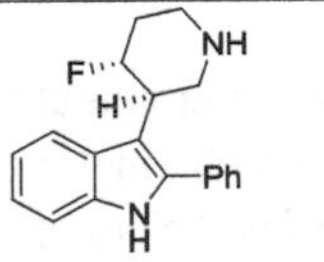	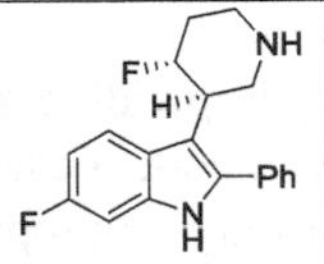
Binding affinity	0.99	0.43	0.06
Bioavailability	1%	18%	80%

FIGURE 7.3 Comparative study of biological activity of fluorinated and nonfluorinated compounds.

Currently, 25% of pharmaceutical[5] and 30% of agrochemical areas are occupied by fluorinated compounds. A few of the commercial product structures are shown in Figure 7.1.

Another important use of fluorinated molecules is their enhanced stability of molecules and that they can be radioactive.[6] Incorporating a fluorine atom in a molecule increases its half-life period tremendously. It is used in medicinal chemistry (Figure 7.2).

The relation between binding affinity and bioavailability of fluorinated heterocycles is noticeable in pharmacokinetics[6] (Figure 7.3).

Bioavailability is defined as the proportion of a drug or other substance that enters the circulation when introduced into the body and is able to have an active effect. The high binding affinity of a poor drug or receptor is enhanced by introducing a fluorine atom into the drug molecule that also affects the bioavailability of the drug. A difluorinated compound has more bioavailability, so it is considered as a more efficient drug candidate. By looking at these unique and applicable properties of fluorine, various research groups have developed a lot of methodologies and ways to incorporate fluorine in organic molecule. One of the efficient and versatile routes of synthesis of such fluorinated molecules is the multicomponent approach.

7.2 SYNTHESIS OF FLUORINATED PYRIMIDINE

Like fluorinated molecules, fluorinated heterocycles show unique and differential activity and thus the need to access a new class of compounds is justified. Elena B. Averina et al. synthesized fluoropyridine-*N*-oxide by a multicomponent approach from

1-bromo-1-fluoro-cyclopropanes upon treatment with $NOBF_4$ in the presence of nitriles as a solvent; unexpectedly, they afforded pyridine derivative instead of isoxazole derivative.[7] As shown in Scheme 7.1, 1-bromo-1-fluorocyclopropanes were prepared. They first treated dihalocarbene with alkene to access 1-bromo-1-fluorocyclopropanes followed by heterocyclization via epoxide ring opening. Fluorine in epoxide is responsible for this unexpected incorporation of solvent in the heterocyclic ring.[8–10] Thus, a multicomponent reaction approach is useful to synthesis of novel, unknown fluorinated heterocycles that can show immense biological activity.

In 2014, Averina et al. expanded the scope of this MCR strategy and synthesized fluoropyridine-*N*-oxide followed by fluoropyridine, using PCl_3/chloroform as a reducing agent.[11] They developed a mild method of conversion of pyridine *N*-oxide to be very reactive in aromatic nucleophilic substitution to obtain fluorinated pyridine (Scheme 7.2).

1,4-Dihydropyridine is a fundamental skeleton of most important classes of drugs.[12,13] It is a very important calcium channel modulator.[14] Some structural modification alters its pharmacological properties.[15] Song et al. synthesized one-pot, four-component fluorinated derivatives of DHP (dihydropyridine) using aromatic aldehyde, Meldrum's acid, ethyl-4,4,4-trifluoro-1,3-dioxobutanoate, and ammonium acetate to afford ethyl 2-hydroxy-6-oxo-4-aryl-2-(trifluoromethyl)piperidine-3carboxylate.[16] The first step was the Michael addition, followed by intermolecular cyclization (Scheme 7.3).

They developed simple multicomponent reaction conditions that showed that the one-pot, four-component reaction provides a simple and convenient approach to ethyl 2-hydroxy-6-oxo-4-aryl-2-(trifluoromethyl)-piperidine-3-carboxylate and ethyl 6-oxo-4-aryl-2-(trifluoromethyl)-1,4,5,6-tetrahydropyridine-3-carboxylate from

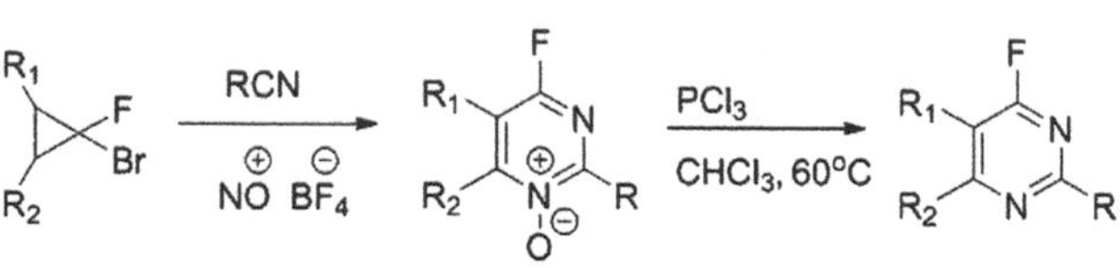

SCHEME 7.1 Synthesis of fluorinated *N*-oxide-pyrimidine.

SCHEME 7.2 Synthesis of fluorinated pyrimidine.

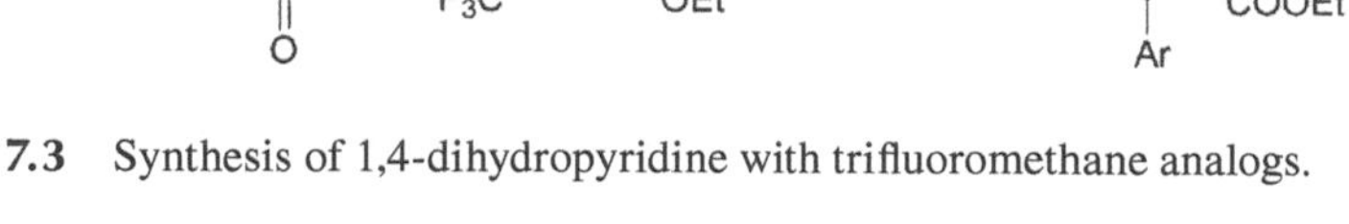

SCHEME 7.3 Synthesis of 1,4-dihydropyridine with trifluoromethane analogs.

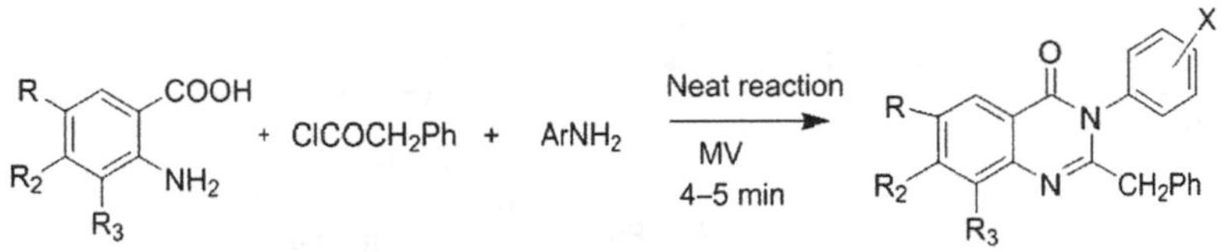

SCHEME 7.4 Fluorinated derivatives of quinazolinone by MCRs in a microwave.

readily available starting materials by dehydration of hemi-aminal moiety under mild conditions. These may be considered as useful CF_3-containing substrates for the synthesis of a wide variety of heterocyclic compounds with potential.

7.3 SYNTHESIS OF QUINAZOLINONE

Fluorinated quinazoline attracts the attention of scientists and chemists due to its biological activity. In 2004, by a neat reaction process, Dandia et al. synthesized one-pot fluorinated 2,3-disubstituted quinazolin-4-(3*H*)-ones that have antifungal activity.[17] They used anthranilic acid and phenyl acetyl chloride and substituted anilines for three components under microwave irradiation and neat reaction conditions. The reaction is very fast under microwave conditions and is one of the green procedures due to absence of solvent in reaction.

As shown in Scheme 7.4, amines such as electron-withdrawing as well as electron-donating *ortho*- or *para*-substituted amines give the desired product. The synthesized compound showed antifungal activity with different pathogenic fungi, like *Rhizoctonia solani* (cause of root rot of okra), *Fusarium oxysprorum* (cause of wilt of mustard), and *Colletotrichum capsici* (cause of leaf spot and fruit rot of chili). Heterocycles containing fluorine enhance its biological activity and therefore such an approach is useful to build a molecular library, which is useful for antifungal, herbicidal, pesticidal, CNS depressant, and AMPA inhibitors.

7.4 SYNTHESIS OF FLUORINDATED CYCLIC AMIDINES

In 2012, Prof. R. Lavilla explored a multicomponent reaction like the Mannich–Ritter transformation and formation of fluorinated heterocycles, using benzo- and pehyl-nitirle, which were suitable substrates to obtaining the corresponding cyclic amidine products as depicted in Scheme 7.5. Also, functionalized nitrile, such as methyl cyanoacetate, afforded the corresponding MCR adducts.[18]

7.5 SYNTHESIS OF THIAZOLIDINE-2,4-(1*H*)-DIONE

Like fluorine and nitrogen/oxygen heterocycles, fluorine and sulfur/phosphorus play an important role in biological activity enhancement. Such types of compounds have biological antiproliferative,[19] antitumor,[20] antimicrobial,[21] anti-inflammatory,[22] antihistamic,[23] and anticonvulsant activity. Coupling of sulfur/phosphorus heterocycles with organofluorinated compounds became a more interesting topic for researchers. In 2014 Ali et al. prepared a 2,3-disubstituted-4-thiazolidinones type of compound

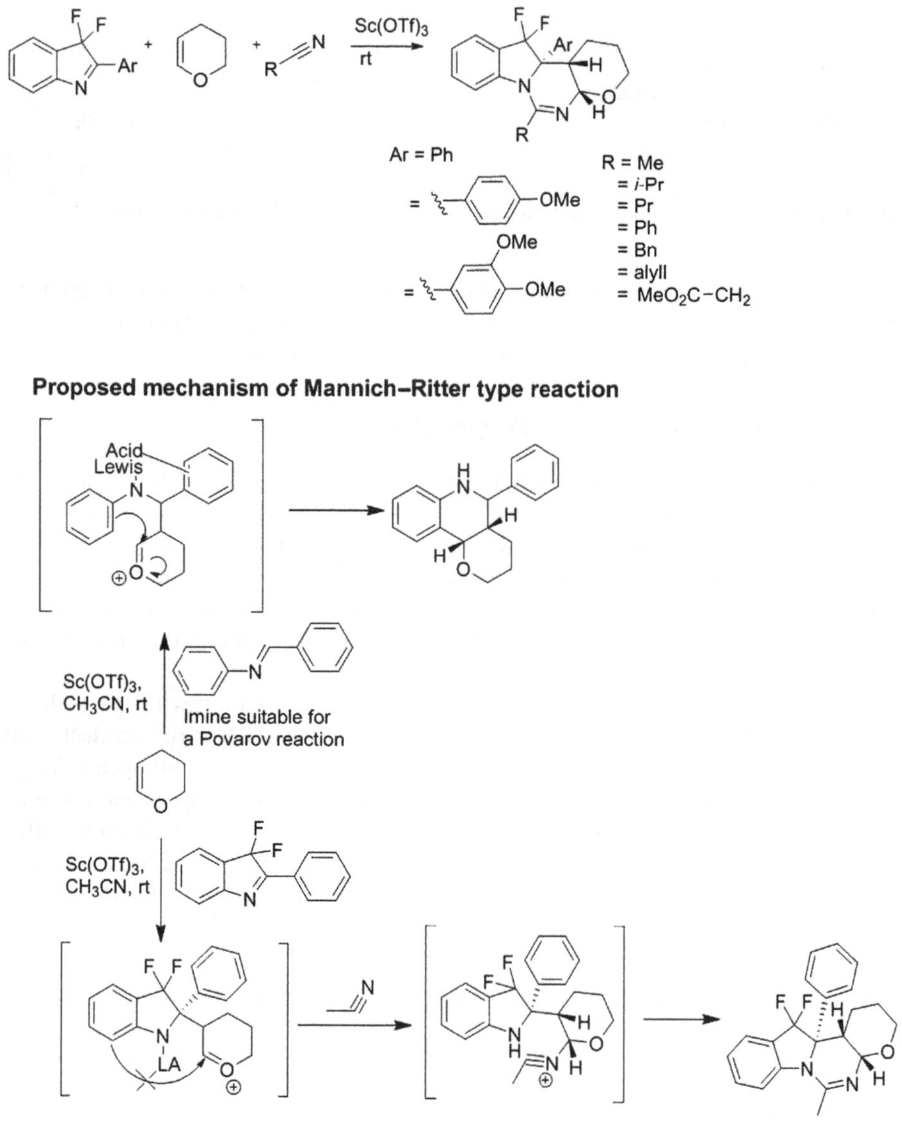

$Sc(OTf)_3$, rt

Ar = Ph

R = Me
= *i*-Pr
= Pr
= Ph
= Bn
= alyll
= MeO_2C-CH_2

Proposed mechanism of Mannich–Ritter type reaction

$Sc(OTf)_3$, CH_3CN, rt

Imine suitable for a Povarov reaction

$Sc(OTf)_3$, CH_3CN, rt

SCHEME 7.5 Amidine synthesis from reaction of dihydropyrane, various fluorinated cyclic imines, and alkyl nitriles.

like 5-fluoro-3-(4-fluorophenyl)-4-spiro[indole-3,2-thiazolidine]-2,4-(1*H*)-dione using 5-fluoroisatin, 4-fluoroaniline, and thioglycolic acid reaction as an MCR approach (Scheme 7.6).[24]

Other fluorinated derivatives of this building block are achieved by condensation with trifluoroacetamide, 2-chloro-6-fluorobenzaldehyde and hydrazine hydrate as follows.

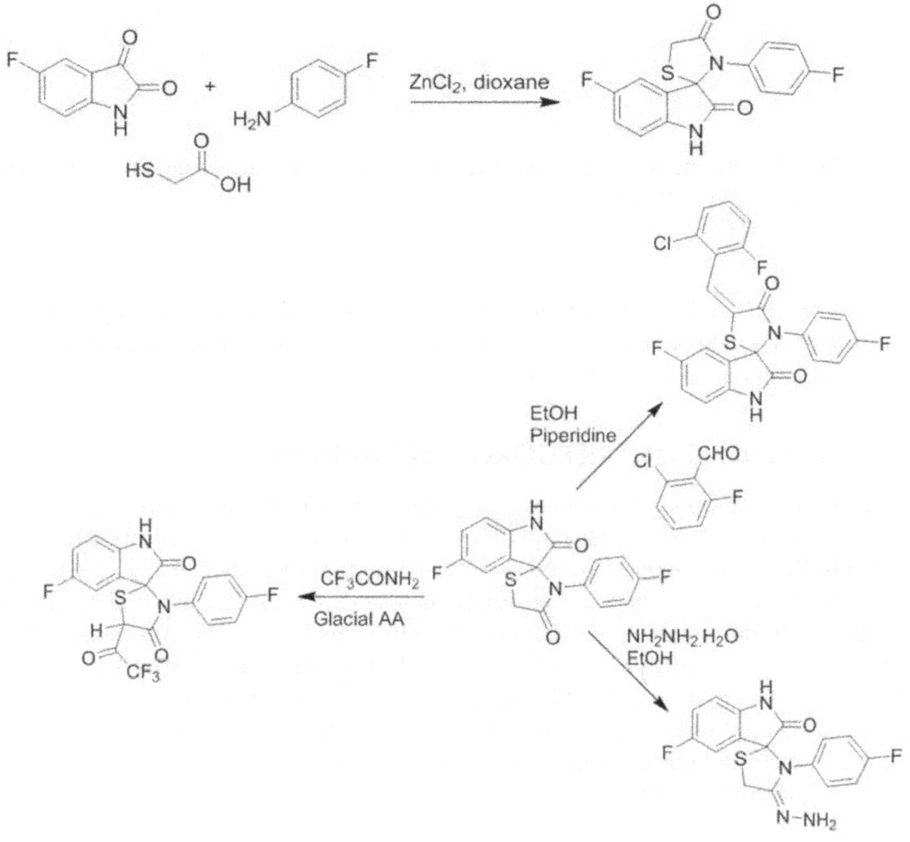

SCHEME 7.6 Synthesis of fluorinated thiazolidinedione analogs using a $ZnCl_2$ catalyst.

Here, formed hydrazones could also be precursors of several bioactive compounds. All synthesized compounds showing antioxidant activity are checked by the 1,1-diphenyl-2-picrylhydrazyl (DPPH) method.[25,26] They had prepared novel compounds, which might be used in medicinal chemistry, and developed the shortest synthetic route for such compounds.

7.6 SYNTHESIS OF FLUORINATED TETRAHEDRONPYRIMIDO [1,2-β] BENZOTHIAZOLE

Shaabani et al. reported the facile synthesis of fluorinated tetrahedropyrimido [1,2-β] benzothiazole by three-component MCRs of aryl aldhyde, 2-aminobenzothiazole, and trifluoromethyldicarbonyl under solvent-free conditions at 90 °C as depicted in Scheme 7.7.[27] Demonstrated reactions afforded stereoselective fluorinated tetrahydropyrimido[1,2-β] benzothiazoles in excellent yields from simple and

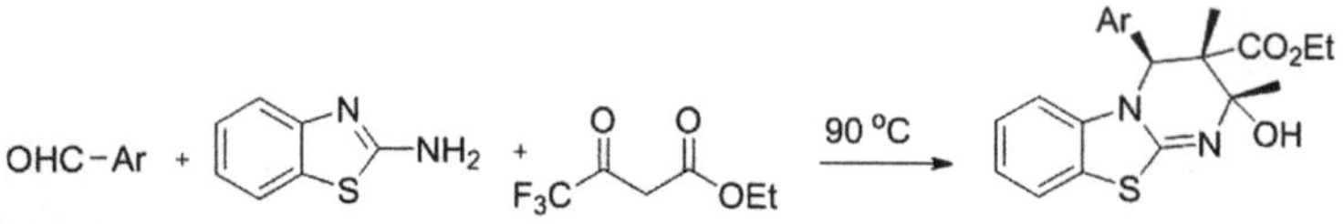

SCHEME 7.7 Synthesis of stereoselective three-component reaction fluorinated tetrahedronpyrimido [1,2-β] benzothiazole.

commercially available precursors under neutral conditions without any catalyst. Applying this method, various substituted aryl aldheyde tetrahydropyrimidobenzothiazole compounds were synthesized.

7.7 SYNTHESIS OF TRIFLUOROPYRUVAMIDES

Multifluorinated organic compounds are very useful in agrochemicals so there is a need to search for the easiest modalities to prepare multifluorinated organic compounds. By looking at the importance of fluorinated compounds, scientists are trying to build up such compounds using simple starting materials. El Kaïm and his group prepared trifluoropyruvamides using electrophilic addition of trifluoroacetic anhydride to isocynides as shown in Scheme 7.8.[28] They further reacted these trifluoropyruvamides with diazo compounds to access epoxide[29] and Friedel–Crafts reactions.[30]

Further, they investigated reactivity of trifluoromethane intermediate with different ketones and made a library of multifluorinated compounds, which is difficult with any other processes.[31]

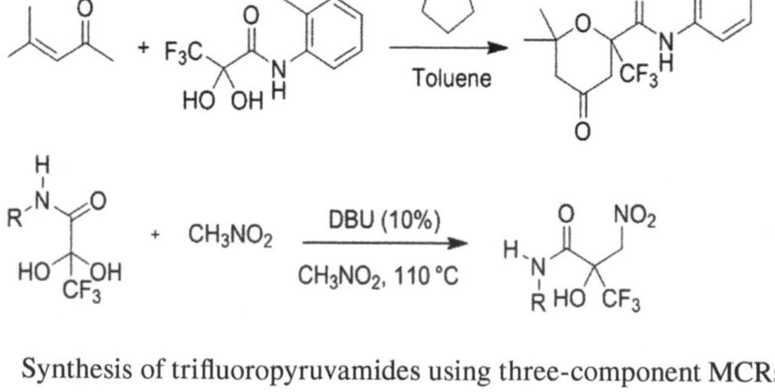

SCHEME 7.8 Synthesis of trifluoropyruvamides using three-component MCRs.

7.8 SYNTHESIS OF FLUORINATED PIPERAZINES BY MULTIPLE UGI REACTIONS

Due to the hydrogen bond, donor and acceptor sites of tetrazole-diketopiperazines family compounds are bioactive and can enhance bioactivity by incorporating fluorine atom in tetrazole-diketopiperazines analogs (Scheme 7.9). Such a class of compounds shows platelet-activating factor inhibitor, and Ugi reactions are very useful in the construction of α-amino tetrazoles.[32] In 2015 Alexander Dömling et al. synthesized such types of compounds, which show bioactivity. They prepared valuable fluorinated heterocycles by sequential consecutive two steps of Ugi reaction conditions as shown in Scheme 7.9.[33]

With this approach of a twofold multicomponent Ugi reaction, they prepared a library of randomly generated tetrazole fused ketopiperazines; all are very important from a pharmacological point of view.

7.9 MULTICOMPONENT REACTIONS IN APPLICATION IN FLUORINE RADIOLABELING

The availability of radiolabeled probes is important for *in vivo* studies by PET. Among the frontier challenges in 18F-radiochemistry are the interconnected goals of increasing synthetic efficiency and diversity in the construction of 18F-labeled radiotracers. 18F-radioretrosynthetic strategies implemented to date are typically linear sequences of transformations designed with the aim of introducing the 18F-label ideally in the last step or at least as late as possible. The convergent 18F-radiochemistry allows for the rapid assembly of functionalized 18F-radiotracers from readily accessible 18F-labeled

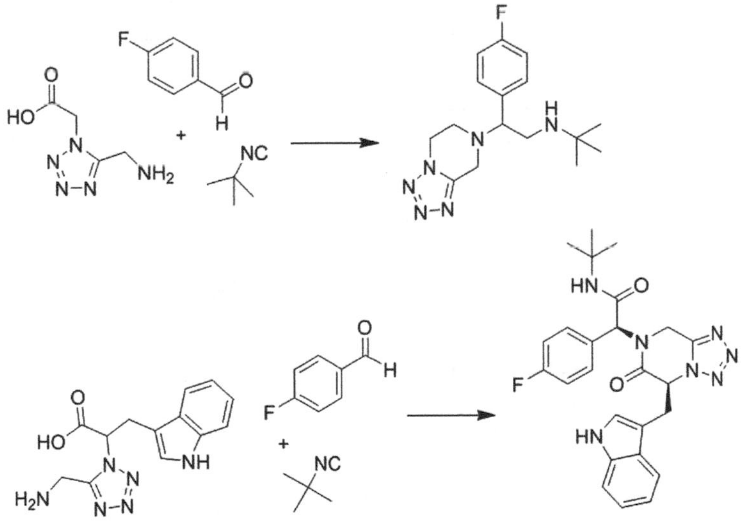

SCHEME 7.9 Synthesis of fluorinated piperazines by multiple Ugi reactions.

prosthetic groups. Scheme 7.10 shows the use of MCRs for 18F-labeling concept, Ugi-4CR, Passerini-3CR, Biginelli-3CR, and Groebke-3CR were performed successfully using 18F-benzaldehydes. These highly convergent reactions delivered, in high radiochemical yield (RCY), structurally complex 18F-radiotracers with the label positioned on an aryl motif not responsive to direct nucleophilic fluorination.[34]

Using an MCR as the key transformation, an extremely concise two-step synthetic sequence for the synthesis of various 18F-labeled 3,4-dihydropyrimidin-2-(1*H*)-ones, imidazo[1,2-*a*]pyridines, α-acyloxyamides, and peptide-type products has been validated. Since the development of novel MCRs inclusive of asymmetric transformations,

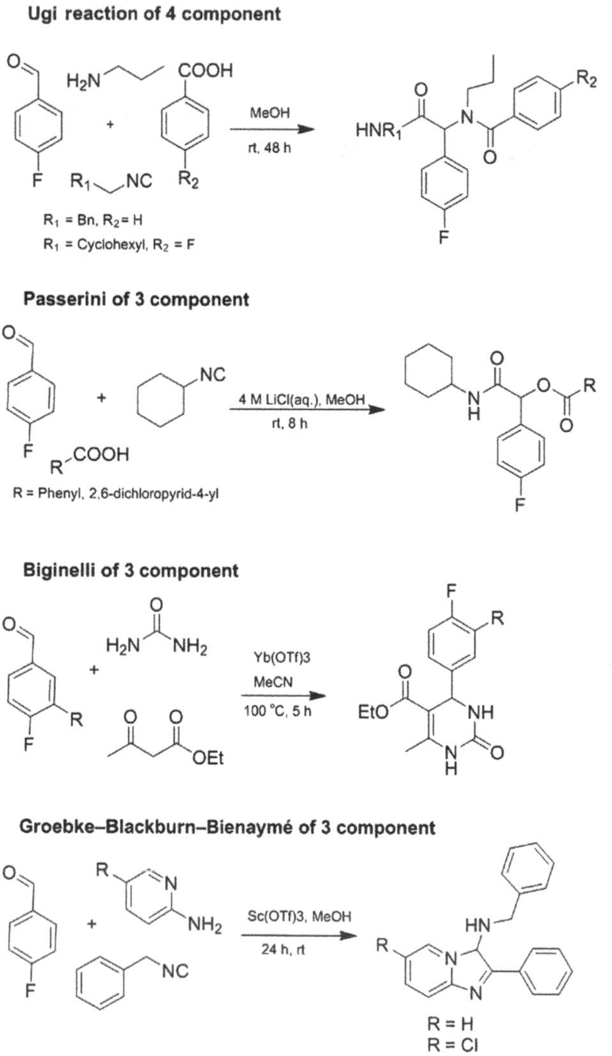

Ugi reaction of 4 component

MeOH, rt, 48 h

R_1 = Bn, R_2 = H

R_1 = Cyclohexyl, R_2 = F

Passerini of 3 component

4 M LiCl(aq.), MeOH, rt, 8 h

R = Phenyl, 2,6-dichloropyrid-4-yl

Biginelli of 3 component

Yb(OTf)3, MeCN, 100 °C, 5 h

Groebke–Blackburn–Bienaymé of 3 component

Sc(OTf)3, MeOH, 24 h, rt

R = H

R = Cl

SCHEME 7.10 Various MCRs with radiolabeled 4-(F-18) benzaldehyde.

the range of opportunities emerges for the synthesis of structurally complex 18F-labeled PET tracers but also to access biomarkers for other imaging modalities.

7.10 SYNTHESIS OF FLUORINATED PIPERAZINES FROM CHIRAL AMINO ACID

Another new approach toward diketopiperazines was published using dipeptides, aldehydes, and isocyanides. For example, bifunctional L-Ala-L-Pro reacts with 4-fluorobenzaldehyde and benzyl isocyanide to form product in 38% yield in a ratio of 3:2 (Scheme 7.11).[35]

Scheme 7.11 showed that the intramolecular reaction is assumed to proceed via the nine-membered intermediate R-adduct, which collapses to the six-membered diketopiperazine, by an intramolecular transacylation.

A Korean group described the stereoselective IMCR via an efficient Ugi four-center three-component reaction leading to 2,5-diketopiperazine derivatives. This process was a one-pot reaction involving commercially available dipeptides containing an amine and a carboxylic acid functionality in one component, aldehydes and isocyanides. The advantage of this protocol is that a wide variety in the substitution pattern of 2,5-diketopiperazines can be easily achieved by changing the components to generate a library of interesting pharmacophores.

7.11 SYNTHESIS OF FLUORINATED QUINAZOLINONES

The quinolin-2-(1*H*)-one scaffold series is very important biologically.[36] In 2010, McCluskey et al. synthesized quinolin-2-(1*H*)-one scaffold by rapid sequential four-component Ugi–Knoevenagel condensation of aminophenylketone, aldehyde, cyanoacetic acid, and isocynide as shown in Scheme 7.12.[37] But in this approach, the requirement is aromatic aldehyde as well as an electron donating group, and

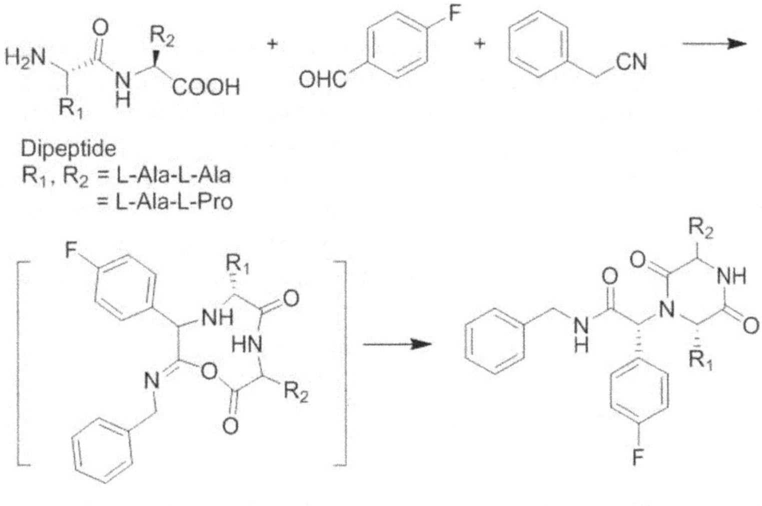

SCHEME 7.11 Synthesis of fluorinated piperazines.

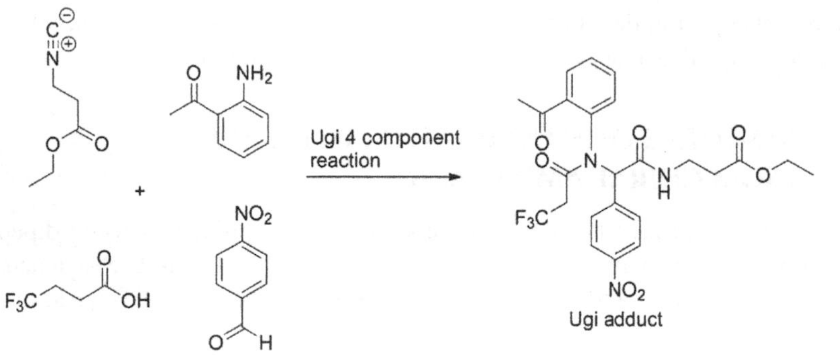

SCHEME 7.12 Ugi adduct from four different components.

aliphatic isocyanide, if an aromatic aldehyde-containing electron-withdrawing group or isocynide with aromatic ester unit, mixture of a quinoline-2(1*H*)-one and amino amide (Ugi component adduct) is formed. The research group used an Ugi four-component reaction approaches to synthesis of multifunctional compounds contain at least one fluorine atom in the final product.

Fluorinated quinazolinone compounds using simple multicomponent approach by replacing any component by fluorinated component give access to new, complex, multifunctional compounds.

7.12 SYNTHESIS OF FLUORINE MOIETY WITH GLUCOSE

The asymmetric multicomponent reaction (AMCR) tool is very important in synthesis of products that contain at least a chiral center.[38] By this approach we can synthesize a compound in either an enantioselective or a distereoselective manner. In 2006 Dondoni and his group synthesized a new class of heterocyclic C-glycoconjugates and carbon-linked sugar and heterocyclic amino acids by AMCRs as shown in Scheme 7.13.

They proposed to prepare DHPM (dihydropyrimidone), which is used for treatment of prostatic hyperplasia. Another biologically relevant drug candidate, Cerebrocast, is a novel drug with antidiabetic and neuroprotectant activity (Scheme 7.14).

Here, the fluorine substituted C2-glycosylate-dihydropyridine (DHP), which was achieved by one-pot two step via Hantzsch and Reformatsky reactions conditions as shown in Scheme 7.14. In this target-oriented synthesis of library of fluorinated-glycosylated-DHPs molecules showed much higher biological activity compared with nonfluorinated glycosylated-DHPs derivatives; thus, the fluoride exchange

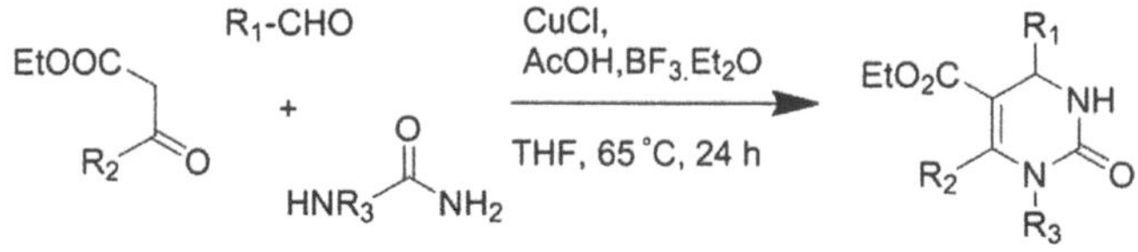

SCHEME 7.13 Dihydropyrimidone glycoconjugates prepared via Biginelli three-component reactions.

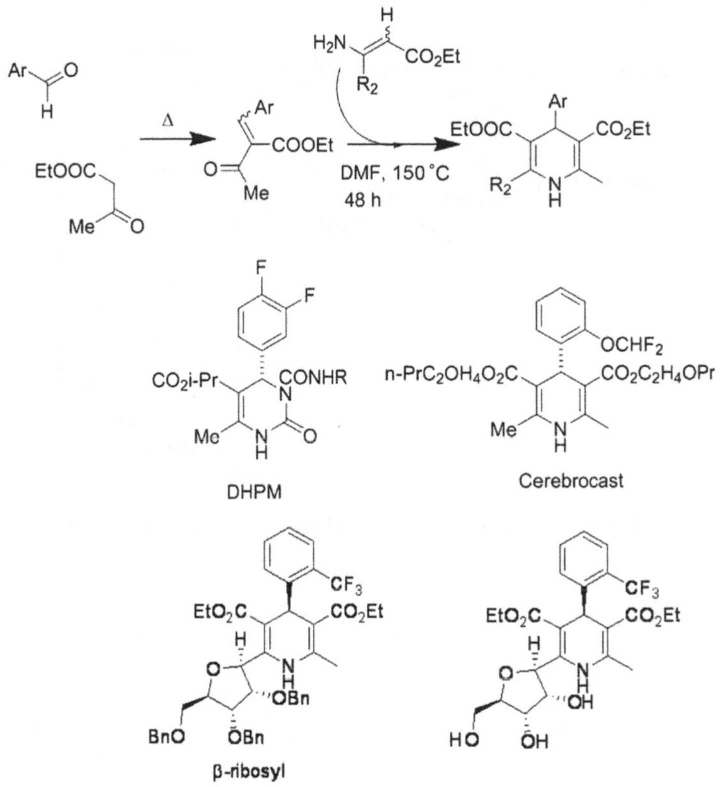

SCHEME 7.14 Synthesis of C2-glycosylated DHPs(dihydropyridine) via two-step, one-pot Hantzsch 3 CRs.

with target molecule leads to alter physicochemical properties and pharmacological profiles.

The targeted *C*-glycosyl α,α-difluoro β-amino esters were achieved by the reaction of imine derivative from *p*-methoxyphenylamine (PMP) and carbohydrate aldehyde reactions, and with Reformatsky reagent BrZnCF2CO2Et reaction condition as shown in Scheme 7.15. The approach to obtain biologically active β-amino ester via multicomponent procedure was useful to achieve chiral drug molecules with atom economy, low labor, and material cost with chiral building blocks.[38]

7.13 SYNTHESIS OF ACRIDINE

Polyhydroacridine derivatives show high fluorescence efficiency; they also have similar importance with nicotinamide adenine dinucleotide (NADH) and nicotinamide adenine dinucleotide phosphate (NADPH). By looking at such properties of a group of compounds, Shizheng and his group synthesized fluorinated polyhydrobenzoacridine-1-one derivatives under microwave irradiation as well as solvent-free conditions in one pot as shown in Scheme 7.16.[39]

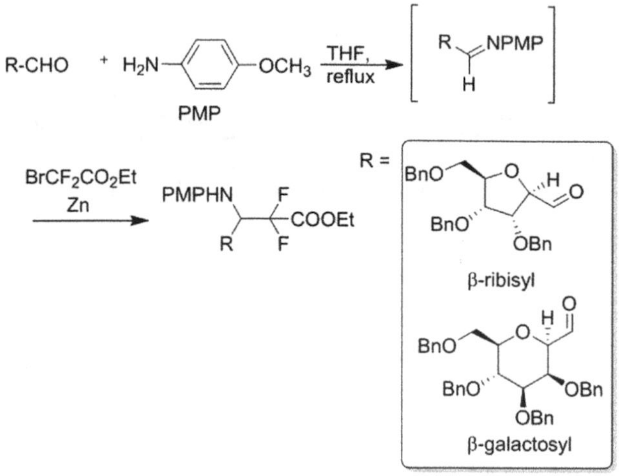

SCHEME 7.15 Stepwise synthesis of *C*-glycosyl α,α-difluoro β-amino esters.

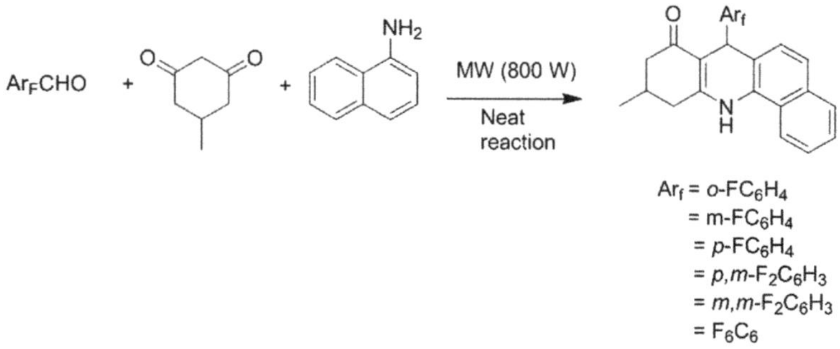

SCHEME 7.16 Synthesis of fluorinated acridine heterocycle.

They prepared a molecular library of poly- and perfluorinated polyhydrobenzoacridines under microwave, solvent-free condition, which is environment friendly. This process is simple, short, and an easy workup; it is efficient to synthesize such types of compounds.

7.14 SYNTHESIS OF QUINOLINE

Scheme 7.17 shows the microwave-promoted one-pot preparation of fluorinated propargylamines and their chemical transformation method by Jian-Min Zhang. A series of fluorinated propargylamines have been synthesized from the one-pot, three-component reaction of fluorobenzaldehyde, aniline, and phenylacetylene under solvent-free and microwave irradiation. The fluorinated propargylamines were then further transformed to chalcones or quinoline derivatives, respectively, depending on the different structures of the propargylamines.

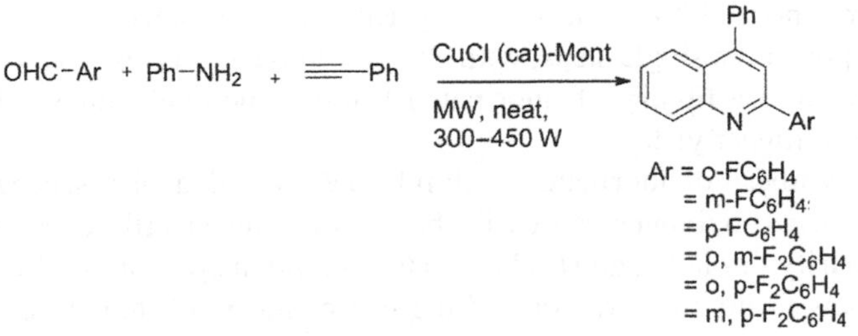

SCHEME 7.17 Synthesis of fluorinated quinoline analogs.

Employing fluorobenzaldehyde, aniline, and phenylacetylene as the reagents and CuCl as the catalyst in a N2 atmosphere, a series of fluorinated propargylamines were prepared under solvent-free microwave irradiation conditions. After further microwave irradiation in the presence of the CuCl catalyst under atmospheric conditions, the propargylamines transformed to quinoline derivatives or decomposed to chalcones and anilines, depending on the substituent on the aniline ring. In the meantime, quinoline derivatives were obtained from the one-pot, three-component reaction in the presence of montmorillonite doped with CuCl under microwave irradiation and solvent-free conditions.[40]

7.15 OTHER HETEROCYCLES

Like other heterocycles, benzothiazolines, benzoxazolines, and dihydrobenzoxazoxazinones are important and pharmaceutically applicable classes of compounds. Incorporation of fluorine in such classes of compounds increases its applicability in pharmaceuticals. The need to incorporate fluorine is a challenge. In 2006, Olah et al. synthesized one-pot fluorinated benzimidazolines, benzothiazolines, benzoxazolines, and dihydrobenzoxazinones using gallium(III) triflate as a catalyst (Scheme 7.18).[41]

They used gallium(III) triflate as a Lewis acid, moisture-stable, recoverable, reusable, and environmentally friendly compound. This catalyst can be used even in

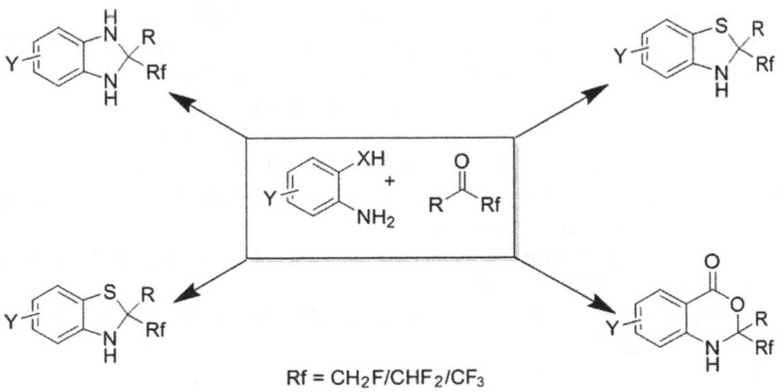

SCHEME 7.18 Synthesis of fluorinated quinoline analogs.

catalytic amounts and it is nontoxic. They accessed fluorinated heterocycles, which are very important in single steps, with easy workup and easy purification.

They prepared a library of fluorinated heterocyclic compounds using such an approach with higher yield.

One-pot synthesis of fluorinated benimidazoles, benzthiazolines, benzoxazolines, and dihydrobenzoxazinones can easily be achieved under mild conditions in high yields and purity through Ga(OTf)3 mediated condition cyclizations. Introduction of fluorine atoms favors the formation of the five-membered heterocycles over seven-membered heterocycles.

7.16 CONCLUDING REMARKS

In this chapter we covered some valuable MCRs such as the Ugi, Mannich–Ritter, Passerini, and Benigalli reactions. These reactions were applied in the synthesis of fluorinated derivatives of various heterocyclic scaffolds such as pyrimidine, pyridine, quinazolinones, triazines, benzimidazolines, spiro[piperidine-4,4′-pyrano[3,2-c] quinolines, benzimidazolines, benzothiazolines, benzoxazolines, and dihydrobenzoxazinones. The fluorine atom is intriguing in the development of novel active compounds in pharmaceuticals and agriculture research. The synthesis of fluorinated molecules by MCRs has attracted much interest due to its presence, increased metabolic stability of molecule, liphophilicity, and receptor-binding properties.

REFERENCES

1. (a) Wang, J.; Rosello, M. S.; Acena, J. L.; Pozo, C. D.; Sorochinsky, A. E.; Fustero, S.; Soloshonok, V. A.; Liu, H., *Chem. Rev.* **2014**, 114, 2432–2506; (b) Huiban, M.; Tredwell, M.; Mizuta, S.; Wan, Z.; Zhang, X.; Collier, T. L.; Gouverneur, V.; Passchier, J., *Nature Chem.* **2013**, 5, 941–944.
2. (a) Yang, X.; Wu, T.; Phipps, R. J.; Toste, F. D., *Chem. Rev.* **2015**, 115, 826–870; (b) Ojima, I., *J. Org. Chem.* **2013**, 78, 6358–6383.
3. Domling, A., *Chem Rev.* **2006**, 106, 17–89.
4. (a) Sunderhaus, J. D.; Martin, S. F., *Chem. Eur. J.* **2009**, 15, 1300–1308; (b) Dondoni, A.; Massi, A., *Acc. Chem. Res.* **2006**, 39, 451.
5. (a) Muller, K.; Faeh, C.; Diederich, F., *Science* **2007**, 317, 1881; (b) McGrath, N.; Brichacek, M.; Njardarson, J., *J. Chem. Ed.* **2010**, 87, 1348.
6. Fried, D.; Mitra, D.; Nagarajan, M.; Mehrotra, M., *J. Med. Chem.* **1980**, 23, 234.
7. Sedenkova, K. N.; Averina, E. B.; Grishin, Y. K.; Kutateladze, A. G.; Rybakov, V. B.; Kuznetsova, T. S.; Zefirov, N. S., *J. Org. Chem.* **2012**, 77, 9893.
8. Lin, S.-T.; Lin, L.-H.; Yao, Y.-F., *Tetrahedron Lett.* **1992**, 33, 3155.
9. Lin, S.-T.; Kuo, S.-H.; Yang, F.-M., *J. Org. Chem.* **1997**, 62, 5229.
10. Kadzaeva, A. Z.; Trofimova, E. V.; Gazzaeva, R. A.; Fedotov, A. N.; Mochalov, S. S., *Moscow Univ. Chem. Bull.* **2009**, 64, 28.
11. Sedenkova, K. N.; Averina, E. B.; Grishin, Y. K.; Kuznetsova, T. S.; Zefirov, N. S., *Tetrahedron Lett.* **2014**, 55, 483.
12. (a) Lavilla, R., *J. Chem. Soc., Perkin Trans.* 1, **2002**, 9, 1141; (b) Sausins, A.; Duburs, G., *Heterocycles* **1988**, 27, 269.
13. (a) Meguro, K.; Aizawa, M.; Sohda, T.; Kawamatsu, Y.; Nagaoka, A., *Chem. Pharm. Bull.* **1985**, 33, 3787; (b) Triggle, D., *J. Cell. Mol. Neurobiol.* **2003**, 23, 293; (c) De Simone, R. W.; Currie, K. S.; Mitchell, S. A.; Darrow, J. W.; Pippin, D. A., *Comb.*

Chem. **2004**, 7, 473; (d) Loev, B.; Ehrreich, S. J.; Tedeschi, R. E., *J. Pharm. Pharmacol.* **1972**, 24, 917; (e) Mojarrad, J. S.; Vo, D.; Velazquez, C.; Knaus, E. E., *Bioorg. Med. Chem.* **2005**, 13, 4085.

14. (a) Bossert, F.; Meyer, H.; Wehinger, E., *Angew. Chem. Int. Ed.* **1981**, 20, 762; (b) Stout, D. M.; Meyers, A. I., *Chem. Rev.* **1982**, 82, 223, (c) Bossert, F.; Vater, W., *Med. Res. Rev.* **1989**, 9, 291; (d) Marchalin, S.; Chudik, M.; Mastihuba, V.; Decroix, B., *Heterocycles* **1998**, 48, 1943.
15. (a) Schramm, M.; Thomas, G.; Towan, I. L.; Franckowiak, G., *Nature* **1983**, 303, 535; (b) Brown, A. M.; Kunz, D. L.; Yatani, A., *Nature* **1984**, 311, 570; (c) Chorvat, R. J.; Rorig, K. J., *J. Org. Chem.* **1988**, 53, 5779; (d) Kappe, C. O., *Tetrahedron* **1993**, 49, 6937; (e) Kappe, C. O.; Fabian, W. M. F., *Tetrahedron* **1997**, 53, 2803.
16. Wang, P.; Song, L.; Yi, H.; Zhang, M.; Zhu, S.; Deng, H.; Shao, M., *Tetrahedron Lett.* **2010**, 51, 3975.
17. Dandia, A.; Singh, R.; Sarawgi, P., *J. Fluorine Chem.* **2005**, 126, 307.
18. Preciado, S.; Vicente-Garcia, E.; Llabres, S.; Luque, L. F.; Lavilla, R., *Angew. Chem. Int. Ed.*, **2012**, 51, 6874.
19. Ramshid, P. K.; Agadeeshan, J. S.; Krishnan, A.; Mathew, M.; Nair, S. A.; Pillai, M. R., *Med. Chem.* **2010**, 6, 306.
20. Wang, S.; Zhao, Y.; Zhang, G.; Lv, Y.; Zhang, N.; Gong, P., *Eur. J. Med. Chem.* **2011**, 46, 3509.
21. Sakhuja, R.; Panda, S. S.; Khanna, L.; Khurana, S.; Jain, S. C., *Bioorg. Med. Chem. Lett.* **2011**, 21, 5465.
22. Tripathi, A. C.; Gupta, S. J.; Fatima, G. N.; Sonar, P. K.; Verma, A.; Saraf, S. K., *Eur. J. Med. Chem.* **2014**, 72, 52.
23. Arya, K.; Rawat, D. S.; Dandia, A.; Sasai, H., *J. Fluorine Chem.* **2012**, 137, 117.
24. Tarik, E.; Ali; Reda, M.; Abdel-R., *J. Sulfur Chem.* **2014**, 35, 399.
25. Kato, K.; Terao, S.; Shimamoto, N.; Hirata, M., *J. Med. Chem.* **1988**, 31, 793.
26. Siddhuraju, P.; Becker, K., *Food Chem.* **2007**, 101, 10.
27. Shaabani, A.; Rahmati, A; Rezayan, A. H.; Khavasi, H. R., *J. Iran. Chem. Soc.* **2011**, 8, 24.
28. El Kaïm, L., *Tetrahedron Lett.* **1994**, 35, 6669.
29. El Kaïm, L.; Guyoton, S.; Meyer, C., *Tetrahedron Lett.* **1996**, 37, 375.
30. El Kaïm, L.; Ollivier, C., *Synlett* **1997**, 7, 797.
31. Colin, T.; Kaïm, El. L.; Gaultier, L.; Grimaud, L.; Gatay, L.; Michaut, V., *Tetrahedron Lett.* **2004**, 45, 5611.
32. (a) Ugi, I.; Steinbrückner, C., *Chem. Ber.* **1961**, 94, 734; (b) Ugi I., *Angew. Chem. Int. Ed.*, **1962**, 1, 8.
33. Zarganes-Tzitzikas, T.; Patil, P.; Khoury, K.; Herdtweck, K.; Dömling, A., *Eur. J. Org. Chem.* **2015**, 2015, 51.
34. Li, L.; Hopkinson, M. N; Yona, R. L.; Bejot, R.; Gee, A. D.; Gouverneur, V., *Chem. Sci.* **2011**, 2, 123.
35. Cho, S.; Keum, G.; Kang, S. B.; Han, S.-Y.; Kim, Y., *Mol. Diversity* **2003**, 6, 283.
36. Pan, C. S.; List, B., *Angew. Chem. Int. Ed.*, **2008**, 47, 3622.
37. Gordon, C. P.; Young, K. A.; Hizartzidis, L.; Deane, F. M.; McCluskey, A., *Org. Biomol. Chem.* **2011**, 9, 1419.
38. Dondoni, A.; Massi, A., *Acc. Chem. Res.* **2006**, 39, 451.
39. Jianan, J.; Jianmin, Z.; Fengrui, L.; Wenli, S.; Yong, X.; Shizheng, Z., *Chin. J. Chem.* **2010**, 28, 1217.
40. Chen, X.-L.; Zhang, J.-M.; Shang, W.-L.; Lu, B. Q.; Jin, J.-A., *J. Fluorine Chem.* **2012**, 133, 139.
41. Surya Prakash, G. K.; Mathew, T.; Panja, C.; Vaghoo, H.; Venkataraman, K.; Olah, G. A., *Org. Lett.* **2007**, 9, 179.

8 One-Pot Total Synthesis of Bioactive Natural Products

*Jiachen Xiang, Miao Wang, Yu Sun, and Anxin Wu**

CONTENTS

8.1 INTRODUCTION

The evolution of molecules on Earth for nearly 3.8 billion years gives us a great opportunity to read incalculable chemical structures with diversity, versatility, and functionality. Over the past centuries, scientists have decoded the valuable chemical genome of our planet through continuously innovated analytical tools.[1] In the meanwhile, synthetic chemists dedicated their lifelong careers with never fading passions to natural products' total synthesis, which was determined as the Mt. Qomolangma (Mt. Everest) of organic synthesis.[2] Natural products provide rich sources of information and inspiration; moreover, as gifts of Mother Nature, the natural products library offers us endless possibilities to hunt and create drugs for relieving humans' pain from disease: Around half of the medicines in stock are derived from natural products.[3] As a result, natural products' total synthesis has drawn attention from not only synthetic and analytical chemists, but also biological, pharmaceutical, and industrial scientists.[4] There is no doubt that total synthesis defines and also promotes the frontier of chemistry. Top chemists can now claim that we can conquer molecules with structural complexity in more or less any degree—for example, the distinguished synthesis of longifolene by E. J. Corey,[5] esteemed synthesis of palytoxin by Kishi,[6] the breathtaking synthesis of brevetoxin B in impressive style by K. C. Nicolaou.[7] These are just a glimpse of the milestones in total synthesis history.[8]

* Corresponding author: chwuax@mail.ccnu.edu.cn.

Since the twentieth century, the tendency to pursue the limits of structural complexity has changed by means of the motivation behind and the standards for total synthesis. A new generation of chemists holding novel concepts has surged up tidal waves in this research field. For instance, P. A. Wender reemphasized "ideal synthesis," which was defined by Hendrickson in 1975, to call the economies of synthesis design.[9] P. S. Baran, with singular talent, enriched this concept with his short and protecting group-free total syntheses.[10] Suggested by Clarke recently, "pot economy" proposed higher evaluation standards of a total synthesis, which aims to "complete an entire multi-step, multi-reaction synthesis in a single pot."[11] As an ultimate goal of "pot economy," one-pot total synthesis protocol has emerged in recent years. "Similar to cook in the kitchen, synthetic chemists strive to save time and resources by avoiding purifications between individual steps within a multistep synthesis, thus minimizing the transfer of material between vessels."[12] Either using one-pot, multistages or a direct one-pot protocol, this synthetic strategy holds promises to afford the "last total synthesis" for specific natural products. For example, historically, the first total synthesis of Tropinone was achieved via 20 artificial steps, 0.75% overall yield in 1896 via several functional group transformations.[13] However, tedious operations and harsh conditions were definitely unfavorable to its subsequent application. In 1917, Robinson et al. reported a three-compound reaction, employing commercially available stating materials, they obtained Tropinone artistically in above 42% yield with a one-pot, one-stage protocol (Scheme 8.1).[14]

From at least three points of view, the horizon of one-pot total synthesis will be spectacular, to provide concise synthetic routes toward certain natural products with the minimum steps, to develop mechanically novel methodologies for producing natural product-derived congeners, and, moreover, to establish molecule libraries with higher efficiency to accelerate drug discovery.

To the best of our knowledge, there are only a few reports on one-pot total syntheses; many of them relied on multicomponent domino reactions (MDRs).[15] It is well understood that employing multicomponents could obviously increase the structural diversity of products, which means some logical changes of substrates will lead to opportunities for constructing a focused library of a natural product family (referring to this point hereinafter). Additionally, domino reactions[16] play a decisive role in improving the efficiency of the entire total synthesis.

The main focus of the present chapter covers representative literature on the topic of one-pot total synthesis and highlights the process that relates to MDRs of them. We also added mono- or bimolecular reaction, which successfully culminated in one-pot total synthesis to reveal the entire landscape of this research field. The one-pot preparation of natural biopolymers such as the synthesis of oligosaccharides[17] and proteins[18] was not described. The organization of this chapter is categorized in the following order: (a) one-pot, multistage total synthesis, (b) microwave-assisted one-pot total synthesis, and (c) direct one-pot total synthesis. Brief descriptions of backgrounds and operation details are attached at the beginning of each part, respectively. A traditional synthetic method (using a multiple-steps approach) is described to make a comparison with the current one-pot process toward the same target molecule while necessary. Mechanistic

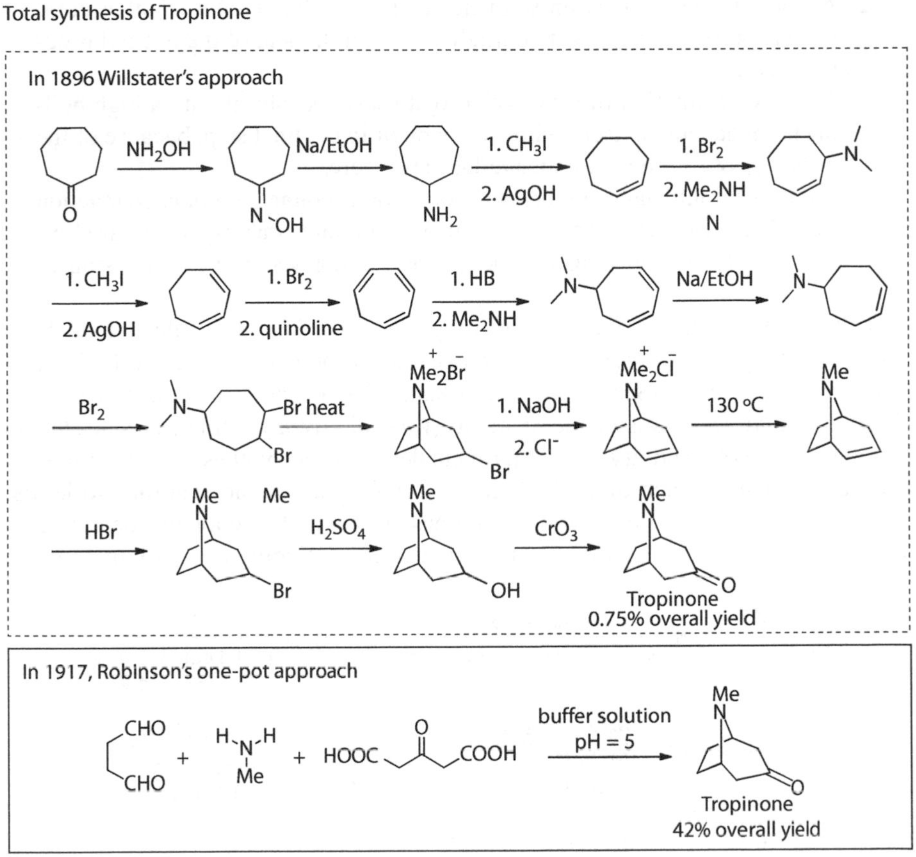

Total synthesis of Tropinone

In 1896 Willstater's approach

NH_2OH → Na/EtOH → 1. CH_3I 2. AgOH → 1. Br_2 2. Me_2NH

1. CH_3I 2. AgOH → 1. Br_2 2. quinoline → 1. HB 2. Me_2NH → Na/EtOH

Br_2 → Br heat → 1. NaOH 2. Cl^- → 130 °C

HBr → H_2SO_4 → CrO_3 → Tropinone 0.75% overall yield

In 1917, Robinson's one-pot approach

buffer solution pH = 5 → Tropinone 42% overall yield

SCHEME 8.1 Total synthesis of tropinone.

discussions are also given to provide information about possible reaction pathways in some cases.

8.2 ONE-POT, MULTISTAGE TOTAL SYNTHESIS

In a one-pot, multistage total synthesis, reagents are added in sequence to the reaction flask; workup and purification operation should be taken once as the final step.[19]

In 2011, Hayashi[20] enumerated the challenges to be overcome in a one-pot, multistage total synthesis as followed:

1. Some transformations are not suitable for a one-pot reaction. When substrate **A** reacts with substrate **B** to give **C** as target intermediate, along with **D** as a by-product, the subsequent reaction has to be carried out in the presence of **D**, which might cause problems—for example, poor chemoselectivity or incompatible chemical environment.

2. As the number of reaction sequence increases, the amount of other compounds present increases. Eventually, the overall yield of the desired product decreases.
3. There is a limitation in terms of solvent usage. A solvent with a high boiling point cannot be used, with exception of in the final step, because of the difficulty of its removal under reduced pressure.
4. A one-pot, multistage synthesis is not a simple connection of each reaction step. The reagents that can be used in one-pot, multistage synthesis are limited. Highly reactive reagents should not be employed in the design stage.

Loh et al.[21] reported a one-pot, multistage approach from aldehyde and aldehyde equivalent to give two THP-ring backbone natural products: (±)-centrolobine and (±)-civet cat secretion (Schemes 8.2 and 8.3). (±)-Civet cat secretion was isolated from glandular secretions of the civet cat (*Veverra civetta*), while (±)-centrolobine exhibited antibiotic activity.[22] The main challenge of the synthesis is screening out the best solvent for each individual step and finding a suitable condition to let the hydroxy groups survive during the whole transformation. Protection and deprotection process can be avoided by this means. It is worth mentioning that indium complex

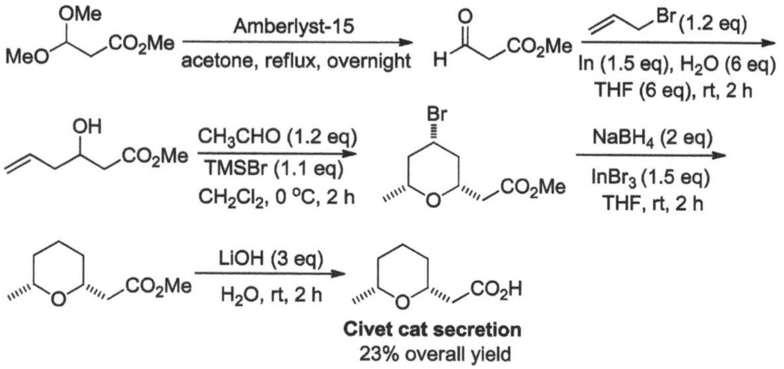

SCHEME 8.2 Total synthesis of civet cat secretion.

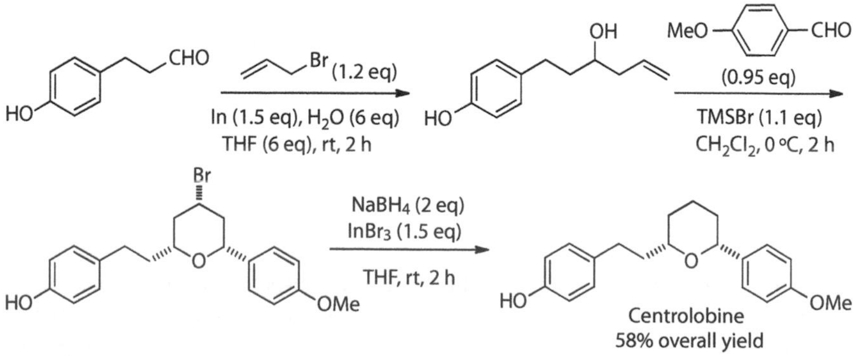

SCHEME 8.3 Total synthesis of centrolobine.

by-product,[23] which generated from the indium-mediated allylation of aldehydes with allyl bromide, would also catalyze the downstream Prins cyclization and dehalogenation. Eight THP-ring backbone natural products were obtained to provide a small molecular library using this one-pot protocol.

Horne et al.[24] described a biomimetic synthesis of grossularines-1 with a one-pot, multistage strategy (Scheme 8.4).

The grossularine family represents the first examples of α-carboline metabolites. Grossularine-1 was isolated from the Britannia marine tunicate *Dendrodoa grossularia*, which exhibit pronounced effects against solid human tumor cell lines.[25] Historically, only one total synthesis was achieved by the Hibino group in 1995.[26] They used 12–13 linear steps to construct grossularines-1 and metabolically related sibling grossularines-2 suffering from some harsh experimental conditions. In Horne's endeavor, they employed oxotryptamine to prepare **1**, and then react it with MeOH and NH_3 for 1 day for a dimerization to occur. Subsequently, 6πe-electrocyclic closure formed an α-carboline ring as the key step. Automatic oxidation and aminolysis led to **2** with satisfactory yields. Then grossularines-1 can be formed under hydrolysis condition with quantitative yield in one pot. This work supports a plausible biogenetic pathway of the grossularine family, and it also provides a novel synthetic methodology to α-carbolines.

Luotonin A has been used in Chinese medicine for the treatment of rheumatism and inflammation in the early stage. Isolated from the plant *Peganum nigellastrum*,[27] this quinazoline-containing alkaloid demonstrated as a human DNA topoisomerase I

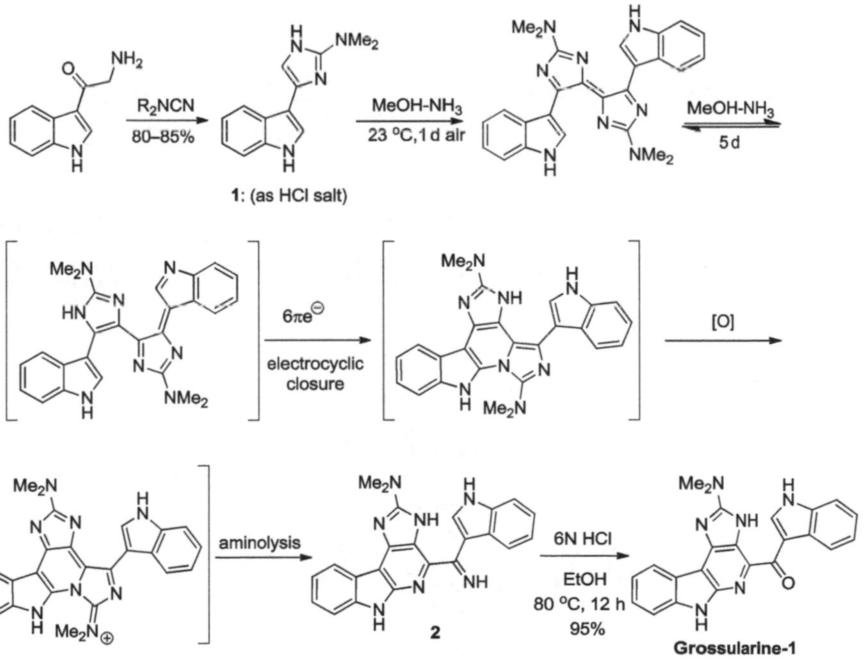

SCHEME 8.4 Total synthesis of grossularine-1.

(hTopI) poison to stabilize a DNA/enzyme binary complex and presented potent cytotoxicity against P-388 cells.[28] The syntheses of the luotonin family of alkaloids have attracted worldwide interest. According to previous literature, luotonin A can be formed through five single steps in transformation from anthranilamide (Scheme 8.5a).[29]

Chu et al. recently reported a one-pot, multistage protocol that begins with isatoic anhydride and propargylamine (Scheme 8.6).

2-Aminobenzamide **3** can be obtained in a nearly quantitative yield.[30] After removing the solvent, a domino reaction, including condensation twice and a $Yb(OTf)_3$ catalyzed aza-Diels–Alder [4⁺+2] cycloaddition reaction, gave the final product luotonin A in one pot with 35% yield (Scheme 8.5c). Literarily, this is the shortest synthesis for luotonin A since the reaction requires only one purifying operation and no more special precautions. With this facile protocol in hand, the author prepared some analogs with diversity of substituents on ring E and some B-ring-expanded analogs. Three compounds (**2c**, **2d**, and **3b**) exhibited more potent inhibitory activities than luotonin A (Scheme 8.7). This is an excellent example to show advantages of one-pot total synthesis from both synthetic and biological screening views.

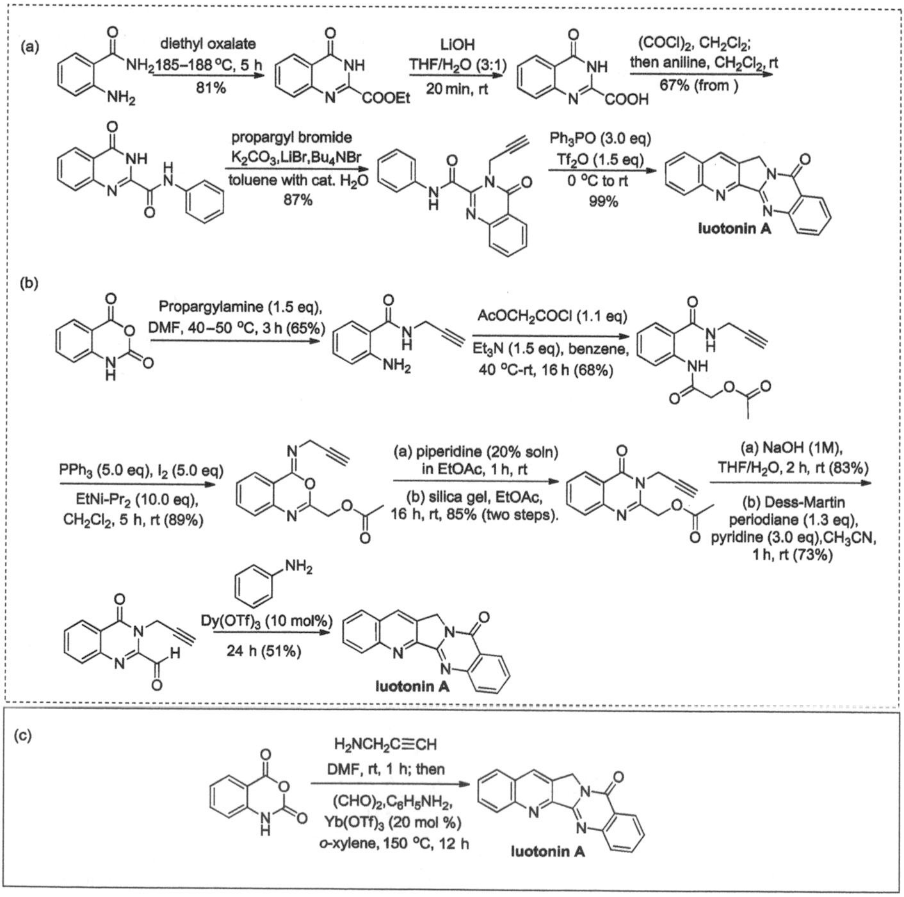

SCHEME 8.5 Synthesis approaches of luotonin A.

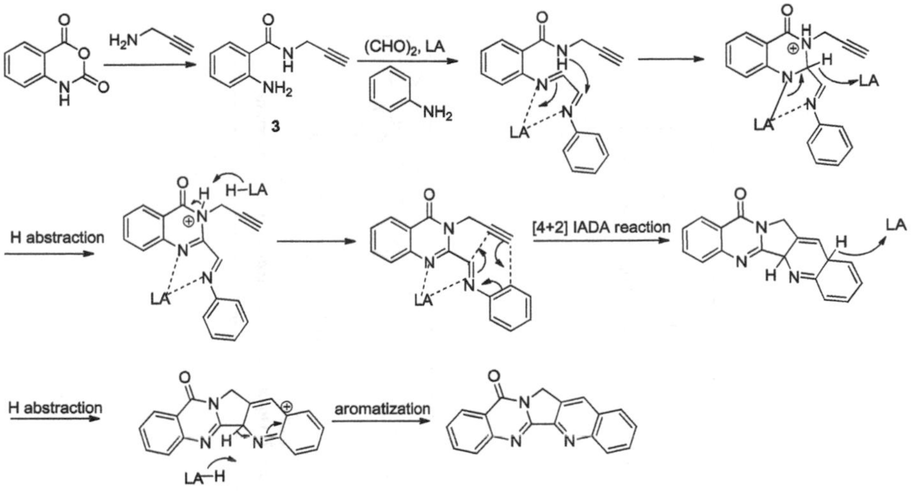

SCHEME 8.6 Possible mechanism.

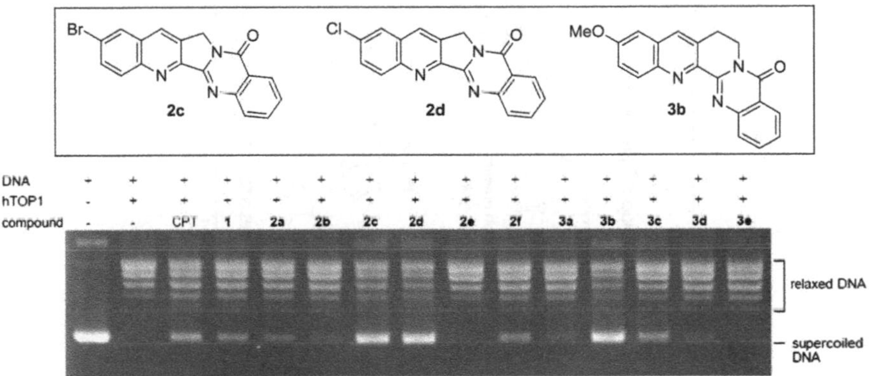

SCHEME 8.7 Unnatural congeners and biochemical tests.

Wu et al. used a family of amino acids to synthesize a family of 5-(3-indolyl) oxazoles alkaloids in the test tube. This included natural congeners pimprinine, pimprinethine, pimprinaphine, WS-30581A, WS-30581B, laboradorin 1, and natural congeners uguenenazole, balsoxine, and texamine (Scheme 8.8).[31]

Isolated from various bacterial sources or marine ascidians, naturally occurring 5-(3-indolyl)oxazoles alkaloids present a broad spectrum of biological activity.[32] For example, pimprinine demonstrated weak antifungal activity and turned out to inhibit monoamine oxidase (MAO). In this one-pot total synthesis protocol, the author used 1-(1H-indol-3-yl)ethanone with iodine to offer oxidative carbonylative product aromatic α-keto aldehyde **1ab** as the common precursor (Scheme 8.9).

After thin layer chromatography monitoring, amino acids were added to the flask to trigger a subsequent domino reaction. There is no necessity to change solvent or temperature; sequentially, condensation, annulation, decarboxylation, and oxidation

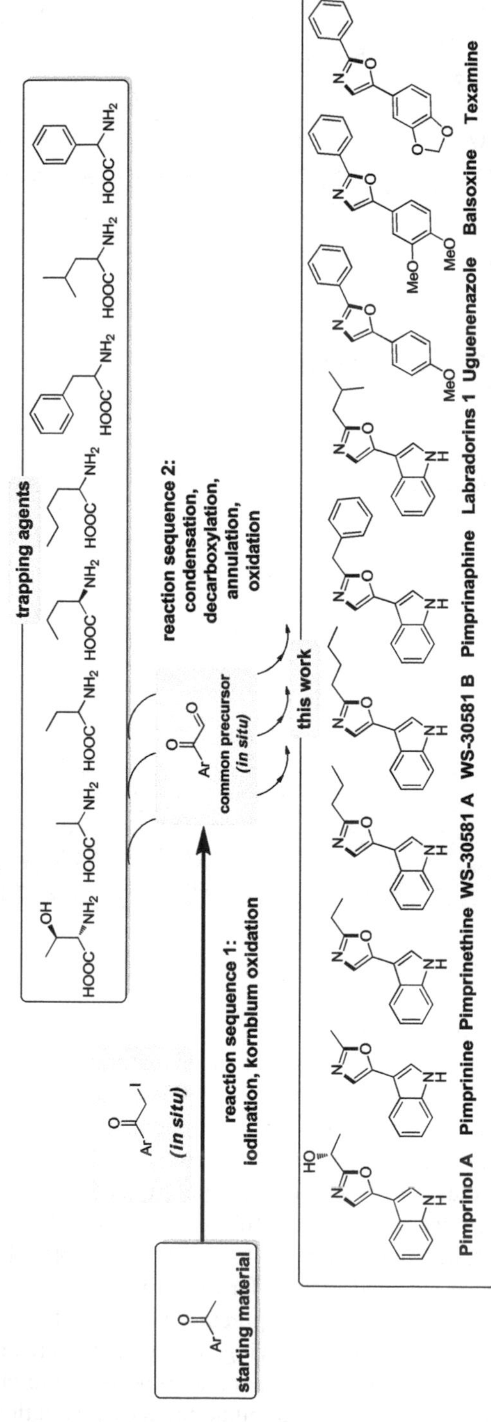

SCHEME 8.8 Total synthesis of 5-(3-indolyl)oxazoles alkaloids.

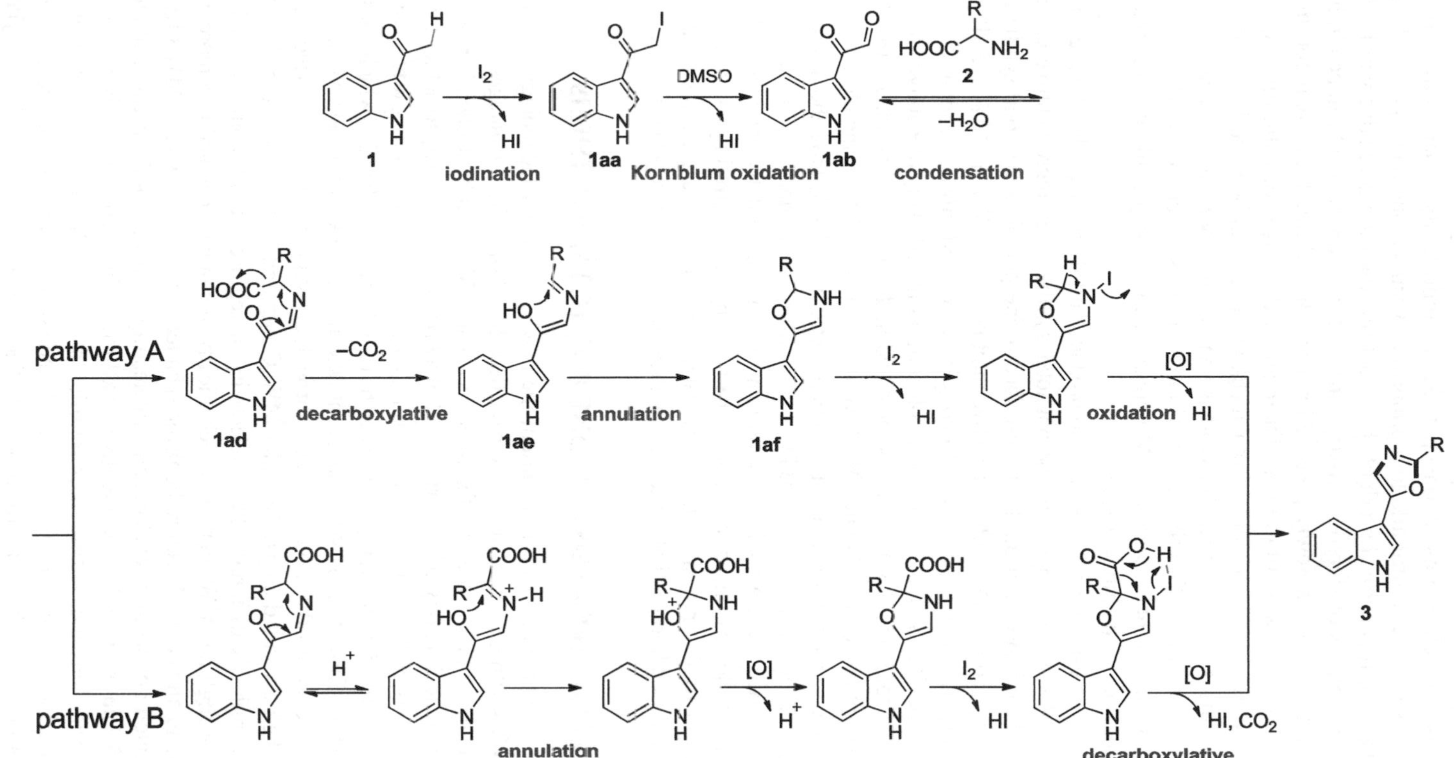

SCHEME 8.9 Possible mechanism.

reaction take places and eventually constructs 5-(3-indolyl)oxazoles alkaloids with amino acids' branch chains in one pot. More importantly, when L-threonine is employed as substrate, chiral alkaloid pimprinol A can be obtained. This is the first total synthesis of this natural product. The reaction condition is relatively mild. As a consequence, centers of stereochemistry inherited from amino acid can be retained. The optical purity of the newly prepared pimprinol A through this one-pot, multi-stage total synthesis was up to 96%. There are wide ranges of naturally occurring alkaloids containing amino acids' branch chains. They can be regarded as metabolite from amino acids. From the biosynthesis view, employing amino acids to build those natural products directly via domino reaction is inherently biomimetic and holds significant synthetic potential. Moreover, such strategy could provide reasonable information for its biosynthesis route.

In addition to using chiral substrates to produce chiral products, the chiral catalytic process can also be employed in one-pot total synthesis. A representative example was reported by Hayashi et al. They described a one-pot, six-stage total synthesis of the dipeptidylpeptidase IV (DPP4) selective inhibitor.[33] ABT-341 in 63% overall yield through an organocatalytic process and an $i\text{Pr}_2\text{EtN}$ promoted chiral inversion isomerization.[34] To be specific (Scheme 8.10), the whole process included organocatalytic Michael addition of acetaldehyde to the nitroalkene **4** and addition of the resulting nitroalkane and vinylphosphonate via a Homer–Wadsworth–Emmons (HWE) ring closure to give **5**. After epimerization, **6** treated with trifluoroacetic acid (TFA) gave the desired carboxylic acid **7**, then underwent a TBTU-mediated amide formation to build amide **8**. The final step was a Zn/AcOH promoted reduction and eventually obtained ABT-341 through purification by chromatography. The building and controlling of chiral center in the one-pot total synthesis is highly valuable, since it gives a great opportunity to establish a focused library of pharmaceutical ingredients with stereochemistry diversity.

8.3 MICROWAVE-ASSISTED, ONE-POT TOTAL SYNTHESIS

Instead of conventional heating reaction, the newly developed technology of microwave (MW) synthetic chemistry in recent decades shows the advantages of facile, mild, economic, and eco-friendly features. As a result, MW synthetic chemistry is well known as the "technology of tomorrow."[35] It may be an available way to achieve green chemistry, or so-called sustainable chemistry. One advantage of a microwave-assisted reaction is that the reaction temperature can be rapidly raised in 10 °C/s, which does not require the transformation of MW energy into thermal energy as a detour. More importantly, some reactions that do not execute through conventional heating or leading low yields can be performed elegantly under MW. Compared with a conventional heating reaction, completely different chemoselectivity or regioselectivity could be achieved relatively through the MW method. Mechanically, it can be attributed to a more polar transient state and could therefore be favored under MW,[36] which means MW gives us unprecedented opportunities to get some interesting results beyond our expectations.

In 2005, Liu et al. had a continuous breakthrough in microwave-assisted, one-pot total synthesis. Ten kinds within three families of heterocyclic alkaloids were available to be obtained via the one-pot approach. Each endeavor provided the simplest synthetic method for corresponding natural products. The reaction process involves acylation

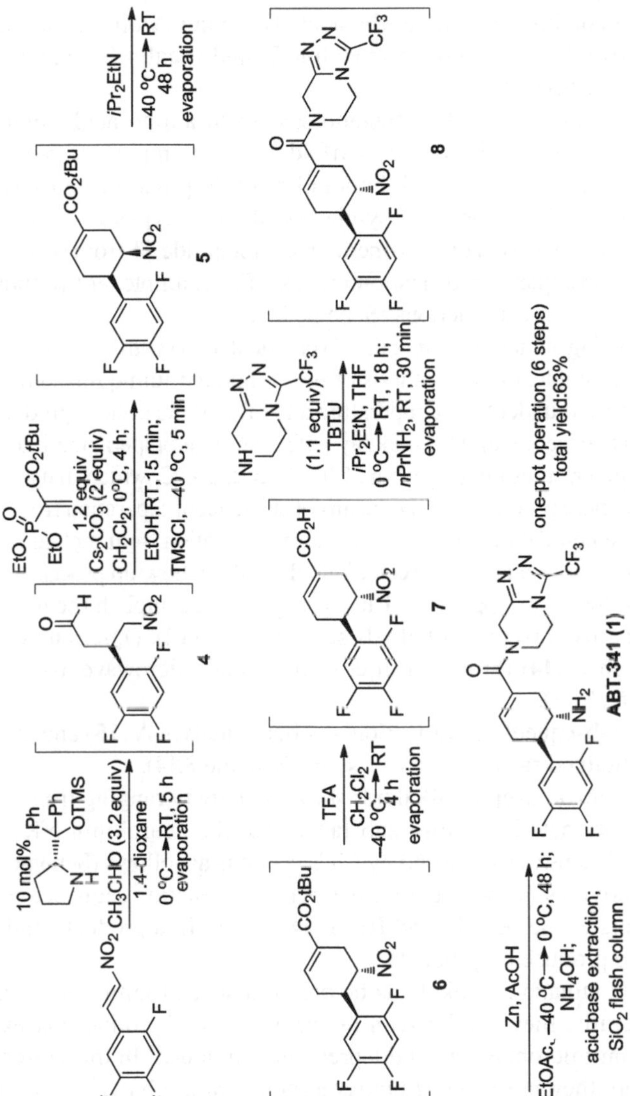

SCHEME 8.10 Total synthesis of ABT-341.

reaction through the condensation of carboxylic acid and amine, which is a similar way to the peptide bond formation process. With the assistance of microwave irradiation, this committed step can be easily reached. There is no need to use condensing agents such as DCC, HBTU, or HOBt compared to conventional heating reactions.

As the first attempt in this series of work, the authors added *N*-protected a-amino acid **9** and tryptophan methyl ester hydrochloride **10** in sequence under the microwave irradiation condition of anthranilic acid. Three quinazolin-4-one fungal natural products—glyantrypine, funmiquinazoline F, and fiscalin B—can be furnished in one pot (Scheme 8.11).[37]

This kind of alkaloid, which is metabolized from amino acid, contains a core pyrazino[2,1-*b*]quinazoline-3,6-dione scaffold holding important bioactivity. For example, they are potent known inhibitors of multidrug resistance (MDR) to antitumor compounds.[38] Previous total syntheses of these types of natural products depended on prolix formation of a six-member cyclopeptide **11** from two amino acids, followed with a subsequent modification process. For example, *ent*-fumiquinazoline G required nine-step linear reactions (Scheme 8.12).[39]

However, applying a microwave reaction to practical total synthesis will also bring disadvantages. For example, in the synthesis of the compound fumiquinazoline F, epimerization at C4 occurred under high temperature and offered the desired product with only 70% enantiomeric excess (ee). Unavoidably, effective diketopiperazine-like cyclization only works at temperatures ranging from >210 °C to <230 °C, which make it difficult to produce optically pure products. By the means of subsequent separation from preparative high-performance liquid chromatography or recrystallization from methanol, the author obtained three kinds of optically correct alkaloids in the follow-up process.

A similar strategy was applied by Liu's group[40] to accomplish the total syntheses of quinazolinobenzodiazepine alkaloids sclerotigenin (**12**), (±)-circumdatin F (**13**), and (±)-asperlicin C (**14**) through a three-component, microwave-assisted, one-pot protocol (Scheme 8.13).

With simple subsequent derivatization, (±)-benzomalvin A (**15**) and (±)-asperlicin E (**16**) can be obtained rapidly with two steps (Scheme 8.14).

Quinazolinobenzodiazepine alkaloids show bioactivity ranging from CCK antagonists to inhibitory activity against substance P at the neurokinin NK1 receptor.[41] The retrosynthetic analysis proposed by Liu's group is as follows (Scheme 8.15): Due to different reactivities of the coupling partners, the reaction sequence between the three components would be I, II, and III to give the desired products, rather than II, I, and then III to produce by-products.

Pyrrolo[2,1-*b*]quinazoline alkaloids belong to a special family of quinazolinone alkaloids. Generally, they can be seen as metabolites of proline that exhibit anti-inflammatory, antimicrobial, and antidepressant activities.[42] In the subsequent work of the Liu group, they used 4-(*tert*-butoxycarbonylamino) butyric acid and anthranilic acid derivatives under the condition of microwave irradiation around 200 °C for 10 min. Deoxyvasicinone (**17**), mackinazolinone (**18**), and 8-hydroxydeoxyvasicinone (**19**) (Scheme 8.16) can be obtained in a one-pot, one-stage protocol.[43]

A possible reaction pathway is depicted in Scheme 8.17. It suggests that a complex nonacyclic lactam could be a key intermediate **20** of the transformation.

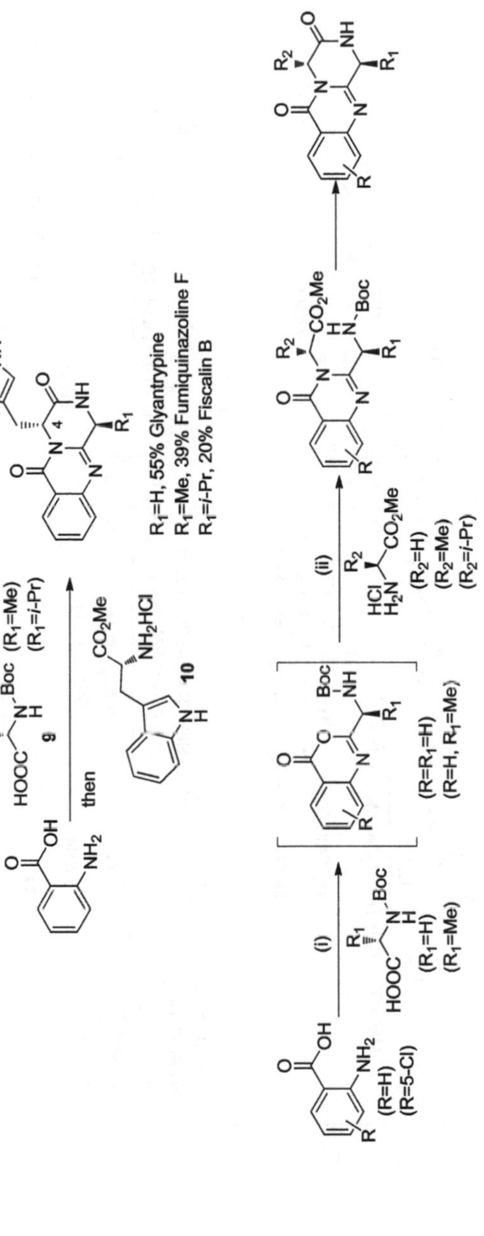
HOOC
9
(R1=H)
(R1=Me)
(R1=i-Pr)
then
10
R1=H, 55% Glyantrypine
R1=Me, 39% Fumiquinazoline F
R1=i-Pr, 20% Fiscalin B
(R=H)
(R=5-Cl)
(i)
(R1=H)
(R1=Me)
(R=R1=H)
(R=H, R1=Me)
(ii)
(R2=H)
(R2=Me)
(R2=i-Pr)

SCHEME 8.11 Total synthesis of quinazolin-4-one fungal natural products.

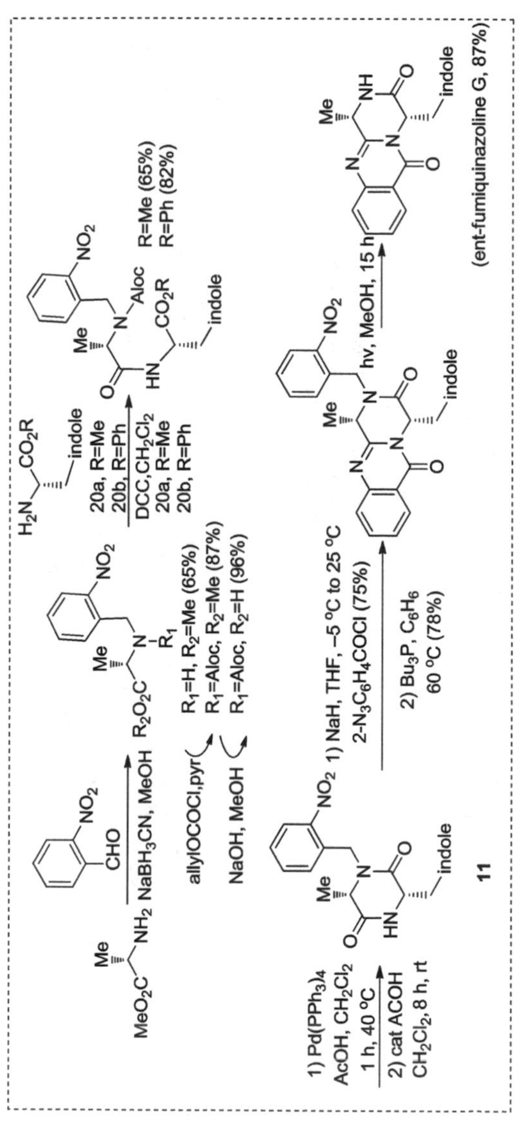

SCHEME 8.12 Total synthesis of ent-fumiquinazoline G.

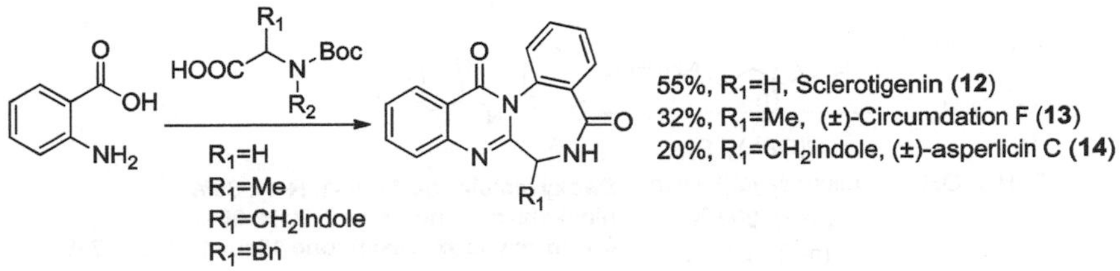

SCHEME 8.13 Total synthesis of quinazolinobenzodiazepine alkaloids.

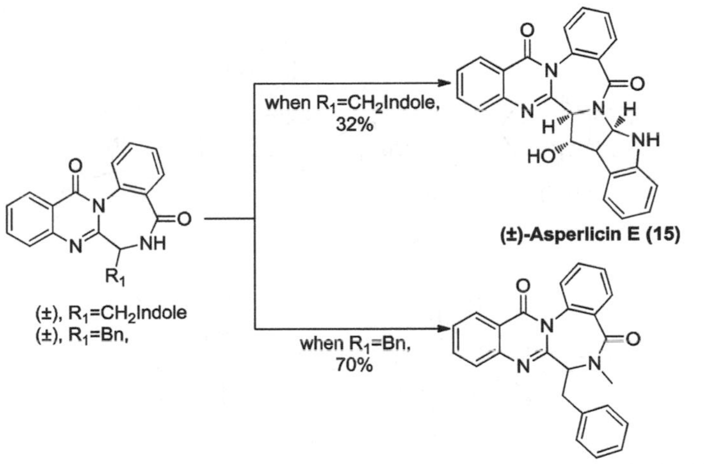

SCHEME 8.14 Synthesis of (±)-benzomalvin A and (±)-asperlicin E.

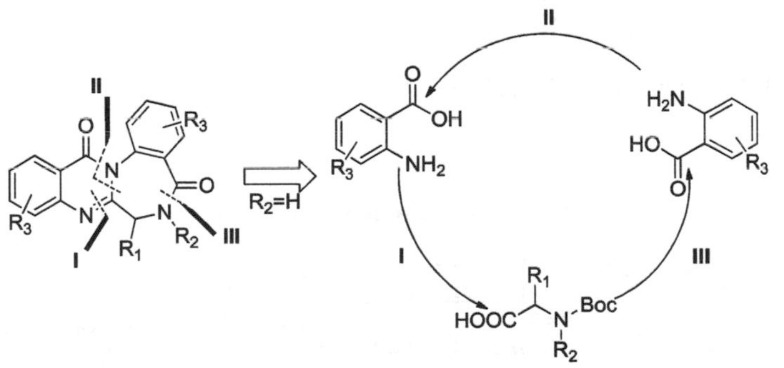

SCHEME 8.15 Retrosynthetic analysis.

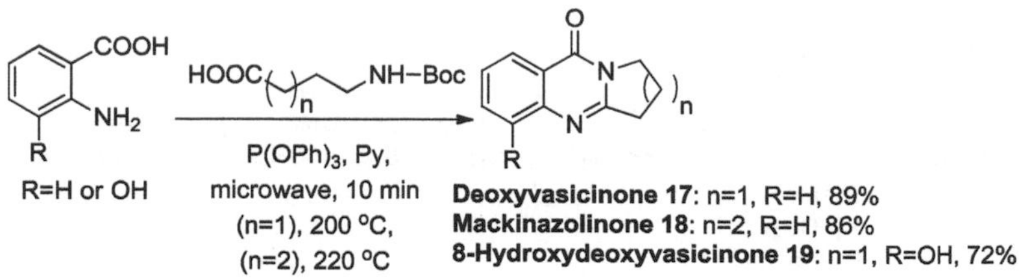

COOH
NH$_2$
R
R=H or OH
HOOC($)_n$NH–Boc
P(OPh)$_3$, Py,
microwave, 10 min
(n=1), 200 °C,
(n=2), 220 °C
O
N
N
R
n
Deoxyvasicinone 17: n=1, R=H, 89%
Mackinazolinone 18: n=2, R=H, 86%
8-Hydroxydeoxyvasicinone 19: n=1, R=OH, 72%

SCHEME 8.16 Total synthesis of pyrrolo[2,1-*b*]quinazoline alkaloids.

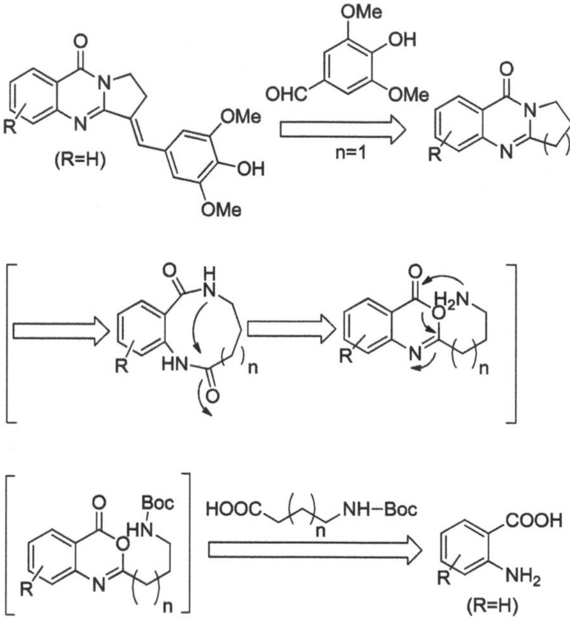

O
N
N
R
(R=H)
OMe
OH
OMe
OMe
OH
OHC
OMe
n=1
O
N
N
R
n
O
H
N
R
HN
O
n
O
H$_2$N
O
N
R
n
O
Boc
HN
O
N
R
n
HOOC($)_n$NH–Boc
COOH
NH$_2$
R
(R=H)

SCHEME 8.17 Retrosynthetic analysis.

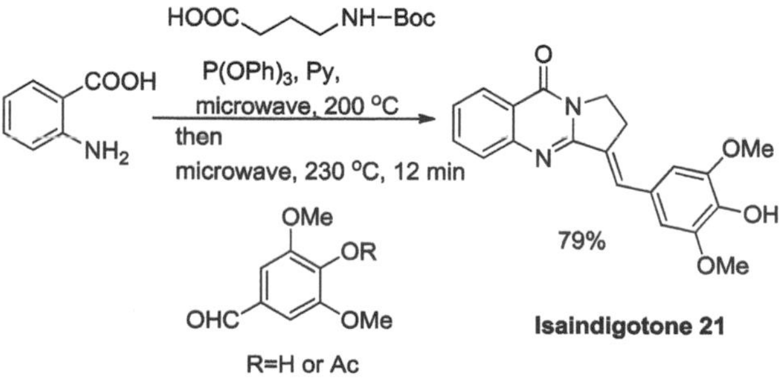

HOOC NH–Boc
COOH
NH$_2$
P(OPh)$_3$, Py,
microwave, 200 °C
then
microwave, 230 °C, 12 min
OMe
OR
OHC
OMe
R=H or Ac
O
N
N
OMe
OH
OMe
79%
Isaindigotone 21

SCHEME 8.18 Total synthesis of isaindigotone 21.

Subsequent studies of the synthesis show that it is possible to transfer a five-membered ring into a six-membered ring. Liu and associates synthesized unnatural congeners for biological activity scanning. When they added aldehyde **11a** as the third component into the microwave reaction system and continued heating for 12 minutes at 230 °C, isaindigotone **21** could be formed with high regioselectivity and chemoselectivity (Scheme 8.18).

As a consequence, the present total synthesis of this natural product belongs to a two-step process in one pot, which is much shorter than the previously reported route with six linear steps.[44]

8.4 DIRECT ONE-POT TOTAL SYNTHESIS

The most facile one-pot total synthesis way we can imagine is direct one-pot total synthesis in which natural products are formed in only one pot with only one stage. Compared with the other two types, direct one-pot total synthesis requires more ingenious design and more arduous attempts. Furthermore, the concept turns out to be more abstract. In a conventional retrosynthetic analysis, people usually search a set of reactions for synthesizing a target by step-by-step chemical synthesis, which was devised by E. J. Corey.[45] However, in the retrosynthetic analysis of direct one-pot total synthesis, we approach this goal by identifying appropriate reaction routes whose steps can be integrated compatibly under a common condition such that the entire sequence proceeds in one pot (Scheme 8.19).[46]

Direct one-pot total synthesis should focus not only on a specific natural product but also on a collective congeners library, which is similar to a diverted total synthesis (DTS) presented by Danishefsky.[47] DTS is highly valued in the synthesis of natural products and analogs because it achieves both structural complexity and diversity (Scheme 8.20a).[48] Scheme 8.20(b) conceptualizes our complementary corresponding direct one-pot total synthesis model.

This improved strategy can be viewed as a one-pot version of a DTS that relies on precise reaction conditions in which a set of reaction sequences can be simultaneously integrated. Advantageously, this approach does not require the purification or even the stability of any of the intermediates and common precursors generated *in situ* from available starting materials. Trapping agents originally present in the

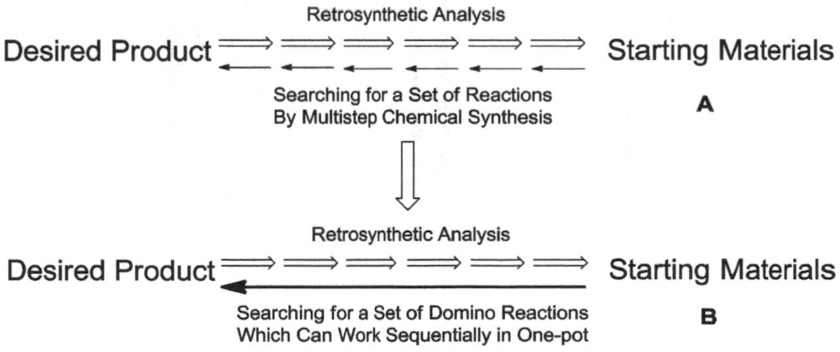

SCHEME 8.19 Retrosynthetic analysis of a direct one-pot total synthesis.

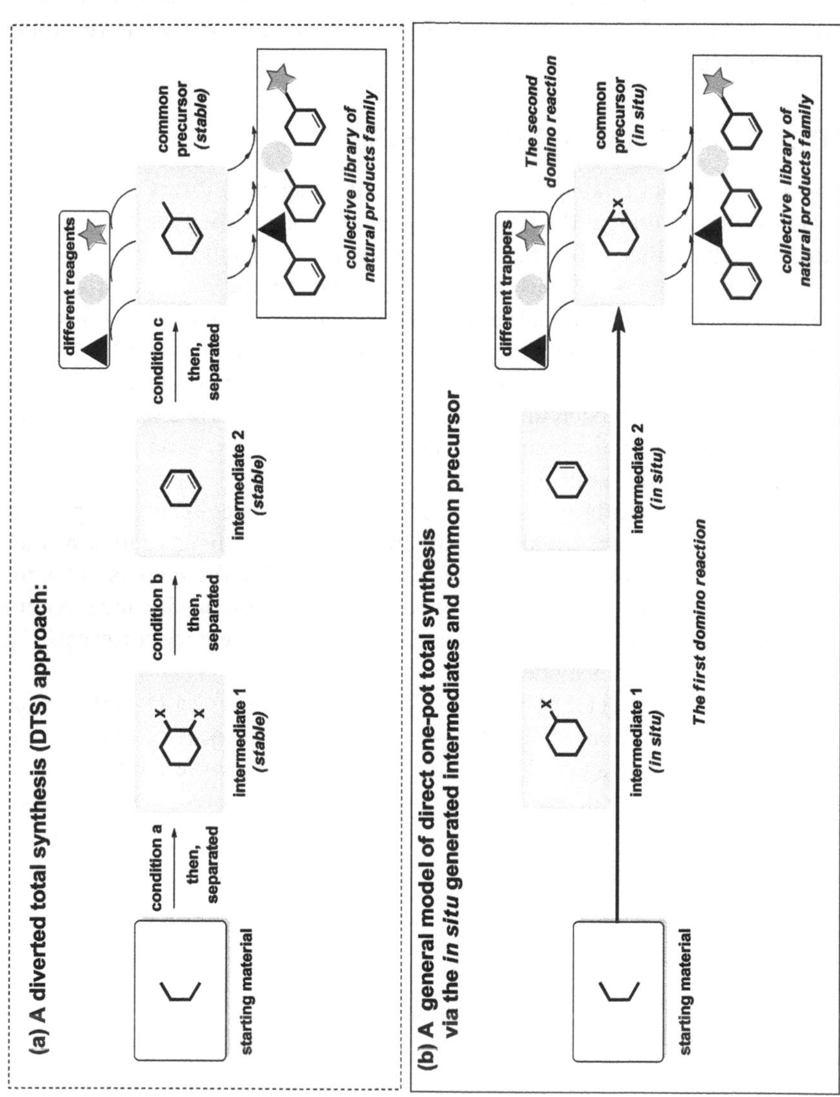

SCHEME 8.20 Direct one-pot total synthesis model.

system can skillfully react with the common precursor with high priority/selectivity as compared to other intermediates. In this manner, a focused collection of natural product analogs with substituent diversity could be directly obtained in one pot.

Guided by this strategy, Wu et al. continuously focused on direct one-pot total synthesis of naturally occurring alkaloids from 2012. The first try traced back to luotonin F.[49] Isolated from *Peganum nigellastrum* Bunge, the luotonin family was used in Chinese medicine for the treatment of rheumatism, inflammation, abscesses, and other maladies.[50] Retrosynthetically, luotonin F could be obtained from a sequential condensation and aromatization reaction of 2-oxo-2-(quinolin-3-yl)-acetaldehyde and 2-aminobenzamide. The common precursor **22** could be formed *in situ* from **24** through I_2/DMSO promoted oxidative carbonylative process. With careful optimization of reaction conditions, it was determined that treating 3-acetylquinoline **24** with 2-aminobenzamide **21** in the presence of I_2 in DMSO at 110 °C for 1 h afforded the desired luotonin F in 72% yield. This endeavor was valued as our first example to prove multiple reactions would self-sequentially take place in one pot to accomplish natural products directly (Scheme 8.21).

A similar strategy was taken into the one-pot total synthesis of β-carbolines and isoquinoline corresponding natural products[51] to prove synthetic universality by the same group (Scheme 8.22).[52]

Isolated from various natural sources, β-carboline alkaloids with a pyrido[3,4-*b*] indoles core structure have a broad range of biological activities, including antimicrobial, antifungal, antitumor, cytotoxic, and antiplasmodial properties. Based on a logical retrosynthetic analysis, employing aryl ethyl ketone **25**, tryptamine derivatives, and stoichiometric molecular iodine, eight β-carboline alkaloids, including the eudistomins Y1–Y6 family, papaveraldine, and pityiacitrin, can be formed directly through a one-pot protocol. This is the shortest route we have ever known compared to what can be found in the literature. The mechanism study (Scheme 8.23) revealed that the key step is a Pictet–Spengler condensation of tryptamine with the *in situ* generated common precursor phenylglyoxal **27**. Additionally, H_2O_2 presents as an assist to oxidation in the

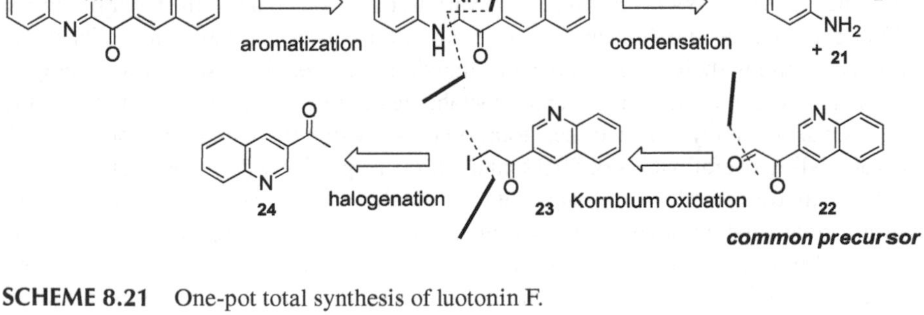

SCHEME 8.21 One-pot total synthesis of luotonin F.

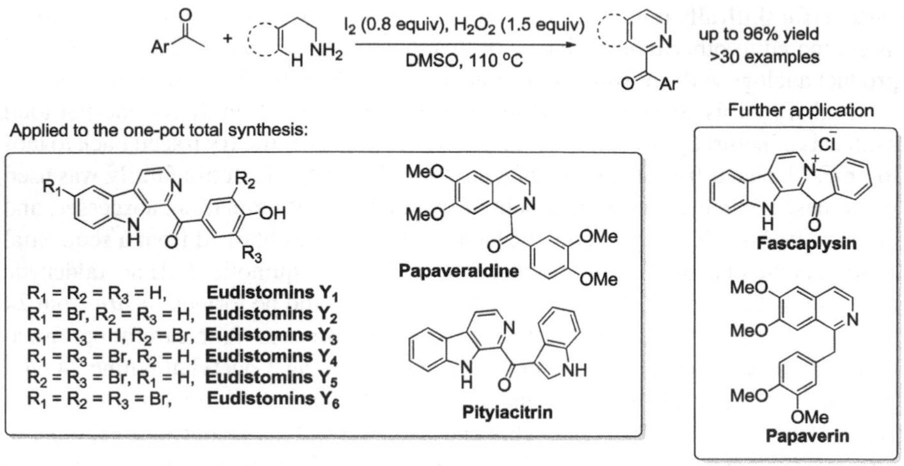

SCHEME 8.22 One-pot total synthesis of β-carbolines and isoquinoline.

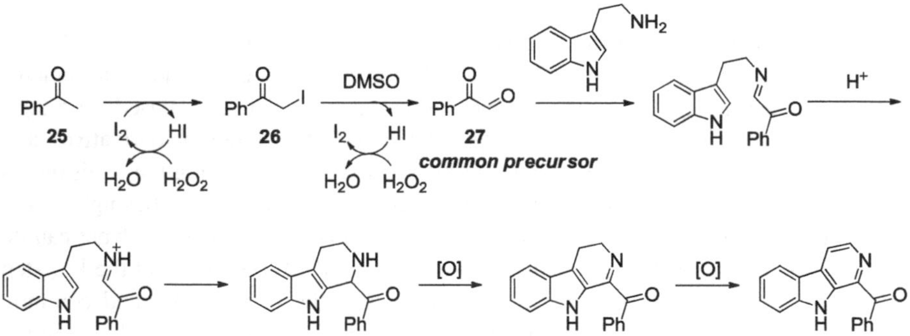

SCHEME 8.23 Possible mechanism.

transformation. More importantly, this reaction can be used as the crucial step to afford fascaplysin and papaverin with simple derivatization according to the procedure reported before.

Another kind of indole alkaloid that draws a lot synthetic interest is 3,3′-bis(indolyl) methanes (BIMs).[53] Widely isolated from various terrestrial and marine natural sources, BIMs exhibit potential bioactivities ranging from cancer inhibition to antibacterial activity.[54] Among these, streptindole and arsindoline are relatively structurally complex members that can be seen as potential metabolites of serine. The Sambri and Ishikura groups independently reported their total synthesis within three and seven steps,[55] with corresponding overall yields of 59.9% and 24.6%, respectively (Scheme 8.24).

Wu et al. described a direct one-pot total synthesis of streptindole and arsindoline via a formal tandem decarboxylative deaminative dual coupling reaction of amino acids and indoles with yields of 82% and 84%, respectively (Scheme 8.25).[56] Their congeners—tris(3-indolyl)methane, 3,3′-bisindolylphenylmethane, vibrindole A,

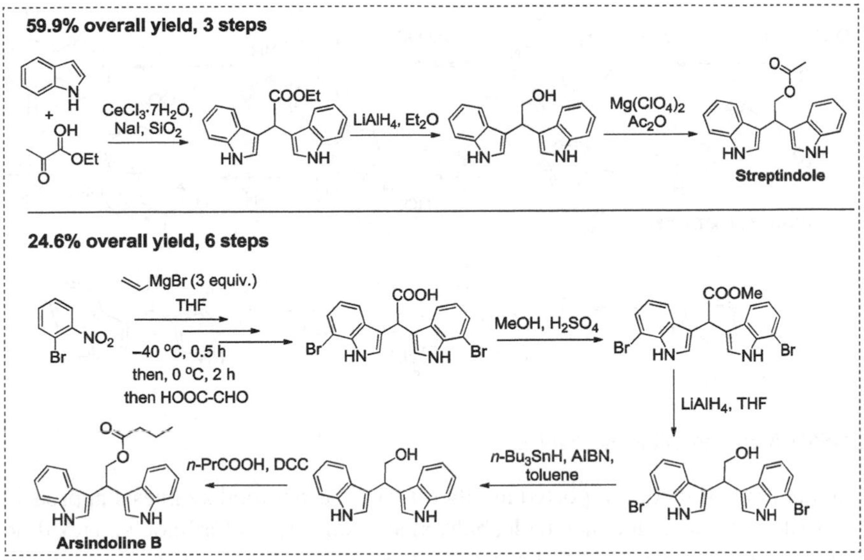

SCHEME 8.24 Traditional total synthesis of 3,3′-bis(indolyl)methanes.

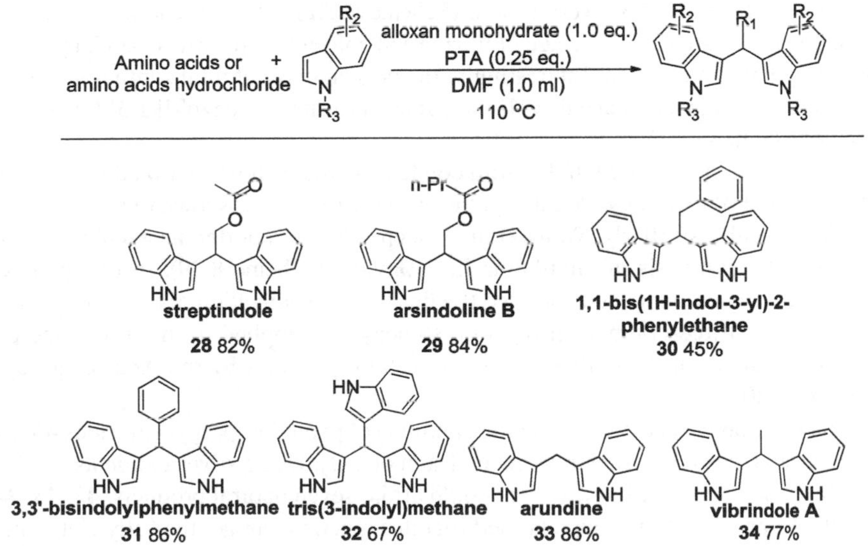

SCHEME 8.25 One-pot total synthesis of 3,3′-bis(indolyl)methanes.

arundine, and 1,1-(3,3′-diindolyl)-2-phenylethane—can also be obtained from this transformation (Scheme 8.25).

The reaction pathway undergoes an alloxan-promoted condensation; a dehydration reaction sequence, leading amino acids to generate to the corresponding aldehyde species as common precursors; and are finally captured by two indole molecules (Scheme 8.26). The entire conversion holds biomimetic value.

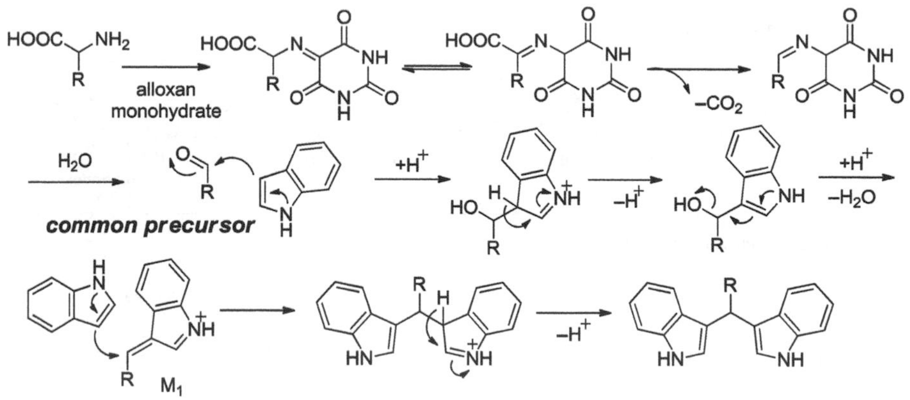

SCHEME 8.26 Possible mechanism.

In addition, Wu et al. reported another direct one-pot total synthesis approach in 2013 that can furnish uguenenazole, balsoxine, texamine, halfordinol, and texaline.[57] Those naturally occurring 2,5-disubstituted oxazoles have remarkable impact on the treatment of tuberculosis, according to research contributions.[58] In 2005, Giddens et al. described a total synthesis of texaline from 1-(benzo[*d*][1,3]dioxol-5-yl)ethanone with five steps and 3.6% overall yield (Scheme 8.27a).[59] Harsh reaction conditions were employed, such as the usage of nitrine and concentrated sulfuric acid, hydrogenation by Pd/C, and so on. Interestingly, the present one-pot total synthesis of texaline and its congeners started with the same substrate 1-(benzo[*d*][1,3]dioxol-5-yl) ethanone (Scheme 8.27b).

Only one step is needed under mild conditions, and it shows reasonable synthetic advantages of our strategy. Possible pathway of the reaction is shown in Scheme 8.28.

Mechanically similarly, Wang et al.[60] independently reported an excellent direct one-pot total synthesis of annuloline in 75% yield (Scheme 8.29). In this process, TBHP acted as oxidant and promoted the final *in situ* cyclization of the oxazole ring.

In 2010, a three-component reaction strategy was applied to the direct one-pot total synthesis of arylnaphthalene lactone natural products by the Anastas group[61] (Scheme 8.30).

The reaction was performed on the condition of phenylpropargyl chlorides **40** and phenylacetylenes **41** as substrate under 1 atm of CO_2. Using silver iodide as catalyst and 18-crown-6 as auxiliary reagent could afford seven natural products: **42**, **43**, **44**, **45**, **46**, **47**, and **48**. The group claimed that this reaction can get the fully elaborated arylnaphthalene lactone family in a rapid methodology. This means that this methodology could be applied to obtain a broad range of arylnaphthalene lactones and the closely related analogs, both naturally occurring and otherwise, in a parallel or high-throughput way, thereby allowing for rapid exploration of the properties of this class of compounds.

Tryptanthrin is a kind of tryptophan-derived alkaloid that is an important member of indoloquinazoline natural products. Tryptanthrin was identified as a molecule useful in pharmaceutical activity, such as excellent cytotoxicity against human breast carcinoma (MCF-7), lung carcinoma (NCI-H460), and central nervous system

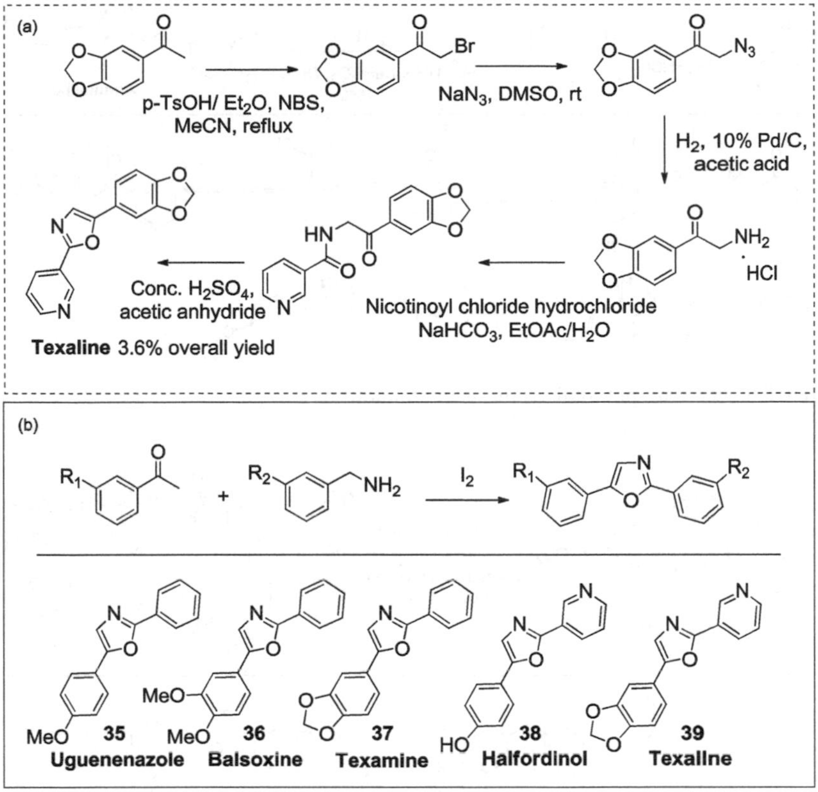

(a)

p-TsOH/ Et_2O, NBS, MeCN, reflux

Br

NaN_3, DMSO, rt

N_3

H_2, 10% Pd/C, acetic acid

NH_2 · HCl

Nicotinoyl chloride hydrochloride $NaHCO_3$, EtOAc/H_2O

HN

Conc. H_2SO_4, acetic anhydride

Texaline 3.6% overall yield

(b)

R_1 + R_2 NH_2 I_2 R_1 R_2

MeO **35** **Uguenenazole**

MeO MeO **36** **Balsoxine**

37 **Texamine**

HO **38** **Halfordinol**

39 **Texallne**

SCHEME 8.27 Onc-pot total synthesis of texaline and its congeners.

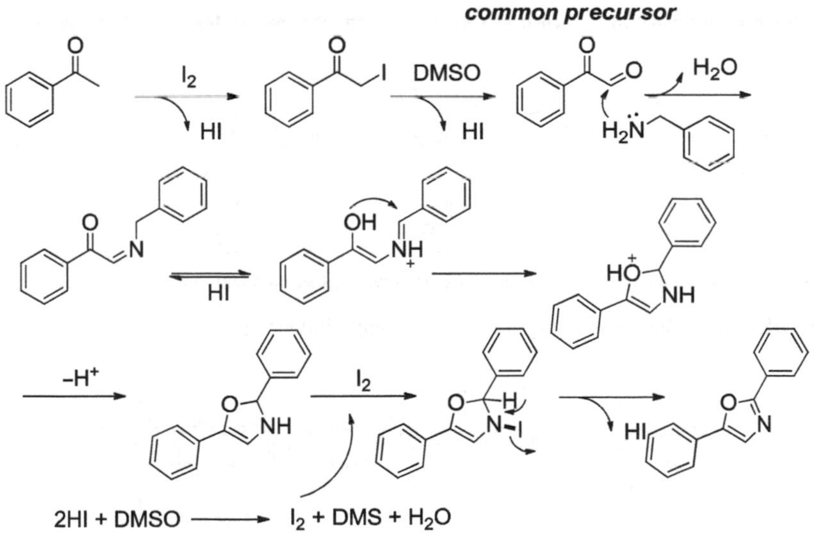

common precursor

I_2 HI

DMSO HI

H_2O

H_2N

HI

OH NH

HO NH

$-H^+$

O NH

I_2

N–I H

HI

2HI + DMSO ⟶ I_2 + DMS + H_2O

SCHEME 8.28 Possible mechanism.

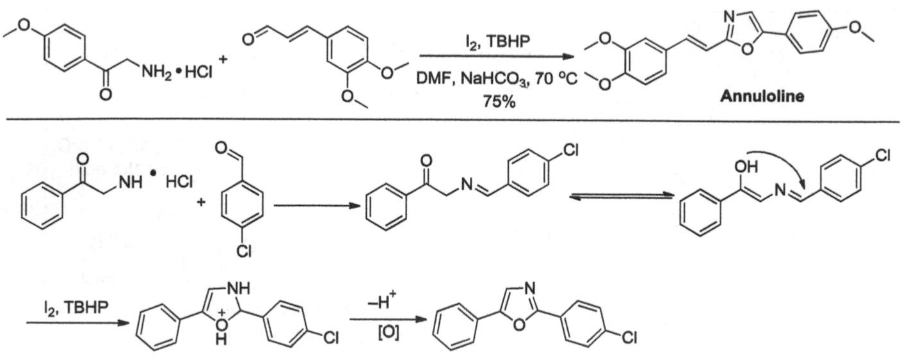

SCHEME 8.29 One-pot total synthesis of annuloline.

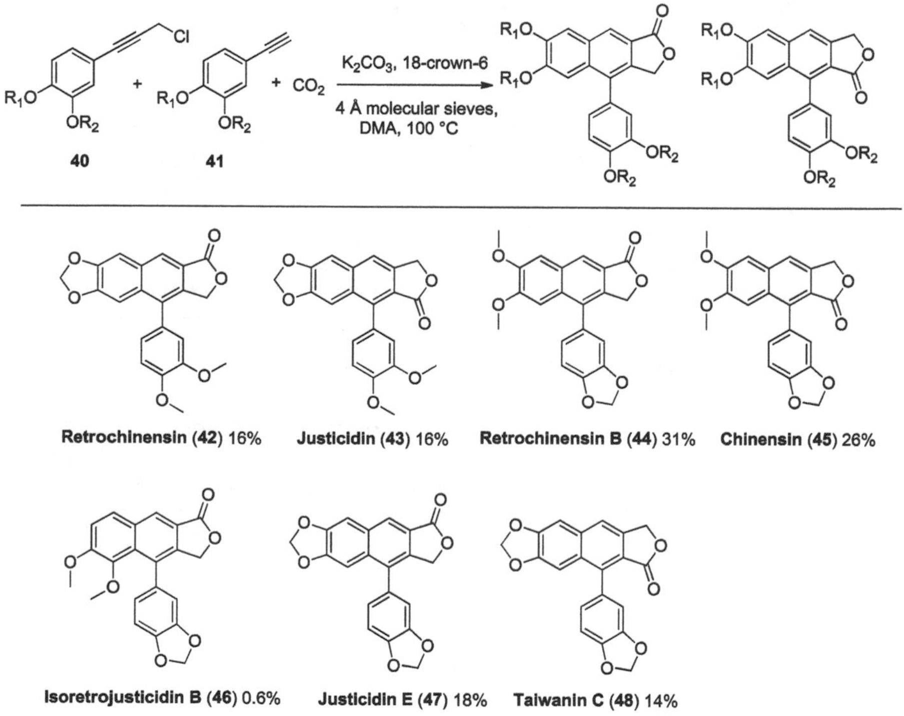

SCHEME 8.30 One-pot total synthesis of arylnaphthalene lactone.

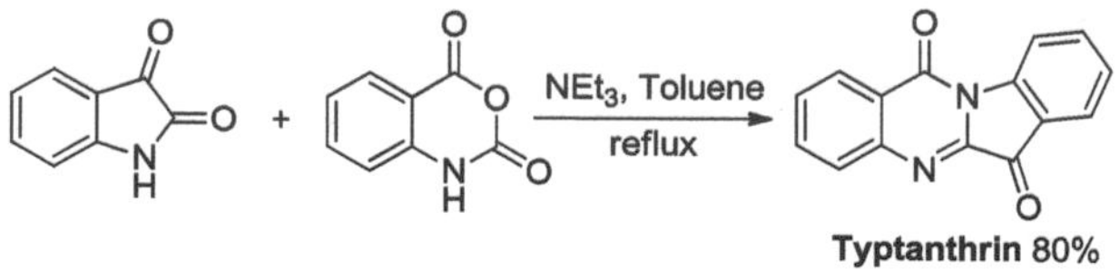

SCHEME 8.31 One-pot total synthesis of tryptanthrin.

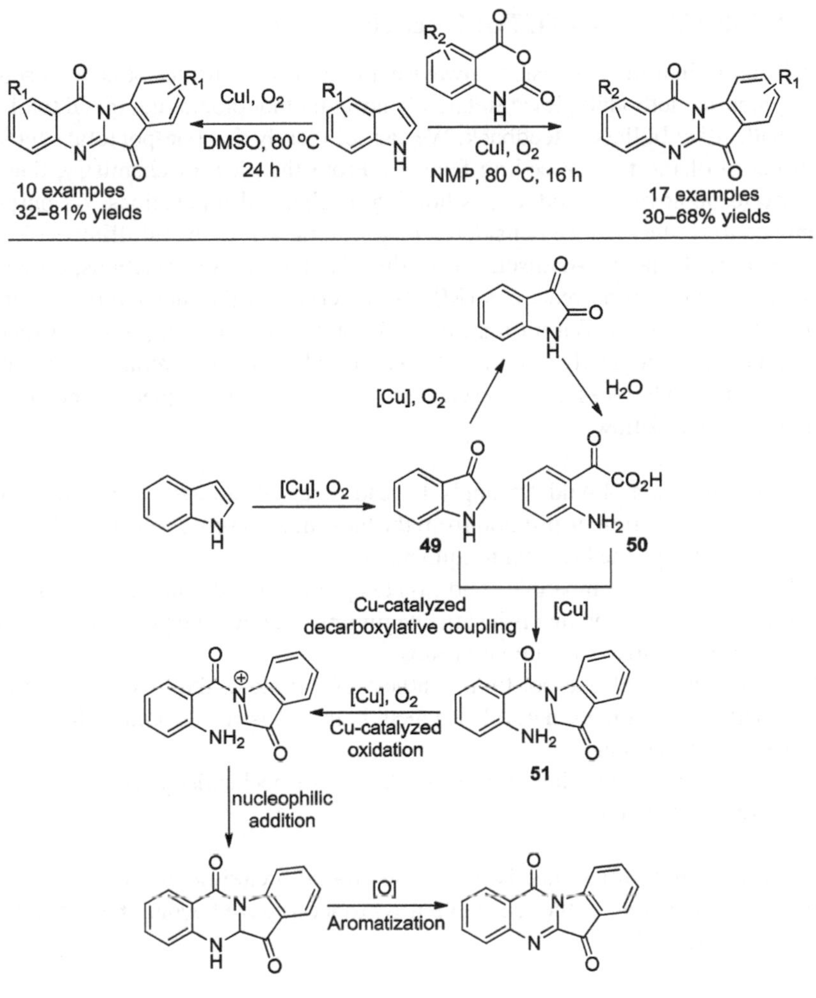

SCHEME 8.32 Possible mechanism.

carcinoma (SF-268) cell lines.[62] Chen et al. reported a two-component reaction of isatin and isatoic anhydride with triethylamine as base and in the reflex condition of toluene to generate tryptanthrin in 80% yield (Scheme 8.31). However, the scope of isatin anhydride is relatively limited.

Recently, Wang et al.[63] reported a mechanically different but more universal method from indole and isatin or only indole to afford tryptanthrin (Scheme 8.32).

The reaction involving a CuI catalyzes a process to generate intermediates **49** and **50** in suit; subsequently, a decarboxylation coupling of **49** and **50** could give intermediate **51**. Finally, a sequential oxidation, cyclization, and aromatization process gave the indoloquinazolin alkaloid and relevant derivatives. This reaction provided a very efficient method for the synthesis of the asymmetric alkaloids of this family.

8.5 SUMMARY AND FUTURE PERSPECTIVES

The adventure in total synthesis is always enriched with a surplus of novel discoveries in synthetic methodologies.[64] Relatively novel methodologies equipped total syntheses with silver bullets as feedback. We can see clearly that one-pot total syntheses contain both of the two preceding features. From the view of chemistry, they are supposed to be novel methodologies holding mechanical innovations. As purposeful applications, they can be considered elegant total syntheses fulfilling ecological and economic demands. As discussed in this chapter, domino reactions, especially multicomponent domino reactions (MDRs), emerged as the major force to realize one-pot total synthesis. Within that, one-pot total synthesis is no longer a random attempt but can be logically designed. However, to be honest, several scientific challenges in this newly blossoming research topic need to be conquered, and some of them are listed as follow:

1. What principle should we apply to lead a retrosynthesis of one-pot total synthesis? Getting inspiration from the biosynthesis process is one solution, but not always the best and unique method.
2. Searching for a compatible condition to optimize the domino reaction itself is tedious work, and that is just the beginning of achieving product diversity carrying sensitive substituent groups.
3. The examples of one-pot total synthesis are limited; there is no report of structurally complex natural products being constructed under the guidance of this strategy.
4. Building and controlling multiple centers of stereochemistry simultaneously are rather difficult.

Scientific communities eagerly anticipate ground-breaking progress in this area, since that might be one of the reachable ways to let us gaze beyond the limitation of chemical synthesis.[65]

REFERENCES

1. Wender, P. A.; Miller, B. L. *Nature* **2009**, *460*, 197.
2. Hoffmann, R. W. *Angew. Chem. Int. Ed.* **2013**, *52*, 123.
3. Paterson, I.; Anderson, E. A. *Science* **2005**, *21*, 451.
4. Jesse, W. H.; Li, J.; Vederas, C. *Science* **2009**, *325*, 161.
5. Corey, E. J.; Ohno, M.; Mitra, R. B.; Vatakencherry, P. A. *J. Am. Chem. Soc.* **1964**, *86*, 478.
6. Suh, E. M.; Kishi. Y. *J. Am. Chem. Soc.* **1994**, *116*, 11205.
7. (a) Nicolaou, K. C.; Theodorakis, E. A.; Rutjes, F. P. J. T.; Tiebes, J.; Sato, M.; Untersteller, E.; Xiao, X. Y. *J. Am. Chem. Soc.* **1995**, *117*, 1171; (b) Nicolaou, K. C.; Rutjes, F. P. J. T.; Theodorakis, E. A.; Tiebes, J.; Sato, M.; Untersteller, E. *J. Am. Chem. Soc.* **1995**, *117*, 1173.
8. Nicolaou, K. C.; Vourloumis, D.; Winssinger, N.; Baran. P. S. *Angew. Chem. Int. Ed.* **2000**, *39*, 44.
9. (a) Wender, P. A.; Miller. B. L. *Nature* **2009**, *460*, 197; (b) Wender, P. A. *Tetrahedron* **2013**, *69*, 7529.

10. Gaich, T.; Baran, P. S. *J. Org. Chem.* **2010**, *75*, 4657.
11. Clarke, P. A.; Santos, S.; Martin, H. C. *Green Chem.* **2007**, *9*, 438.
12. (a) Vaxelaire, C.; Winter, P.; Christmann. M. *Angew. Chem. Int. Ed.* **2011**, *50*, 3605; (b) Walji, A. M.; MacMillan, D. W. C. *Synlett.* **2007**, *10*, 1477.
13. Willstater, R. B. *Dtsch. Chem. Ges.* **1896**, *29*, 393.
14. Robinson, R. *J. Chem. Soc.* **1917**, *111*, 762.
15. For a selected review, please see Barry, B.; Toure, D.; Hall, G. *Chem. Rev.* **2009**, *109*, 4439.
16. Tietze, L. F. *Chem. Rev.* **1996**, *96*, 115.
17. Mong, T. K. K.; Lee, H. K.; Duron, S. G.; Wong. C. H. *PNAS* **2003**, *100*, 797.
18. (a) Bang, D.; Kent, S. B. H. *Angew. Chem. Int. Ed.* **2004**, *43*, 2534; (b) Fauvet, B.; Butterfield, S. M.; Fuks, J.; Brikb, A.; Lashuel, H. A. *Chem. Commun.* **2013**, *49*, 9254.
19. Vaxelaire, C.; Winter, P.; Christmann, M. *Angew. Chem. Int. Ed.* **2011**, *50*, 3605.
20. Ishikawa, H.; Honma, M.; Hayashi. Y.; *Angew. Chem. Int. Ed.* **2011**, *50*, 2824.
21. Zhou, H; Loh, T. P. *Tetrahedron Lett.* **2009**, *50*, 4368.
22. (a) Evans, P. A.; Cui, J.; Gharpure, S. J. *Org. Lett.* **2003**, *5*, 3883; (b) Colobert, F.; Des, M. R.; Solladie, G.; Carreno, M. C. *Org. Lett.* **2002**, *4*, 1723.
23. (a) Loh, T. P.; Zhou, J. R. *Tetrahedron Lett.* **1999**, *40*, 9115; (b) Li, C. J.; Chan, T. H. *Tetrahedron Lett.* **1991**, *32*, 7017; (c) Auge, J.; Germain, N. L.; Uziel, J. *Synthesis* **2007**, *12*, 1739.
24. Miyake, F. Y.; Yakushijin, K.; Horne, D. A. *Angew. Chem. Int. Ed.* **2005**, *44*, 3280.
25. (a) Pattey, C. M.; Guyot, M. *Tetrahedron* **1989**, *45*, 3445; (b) Loukaci, A.; Guyot, M. *Magn. Reson. Chem.* **1996**, *34*, 143; (c) Helbecque, N.; Moquin, C.; Bernier, J. L.; Morel, E.; Guyot, M.; Henichart, J. P. *Cancer Biochem. Biophys.* **1987**, *9*, 271.
26. Choshi, T.; Yamada, S.; Sugino, E.; Kuwada, T.; Hibino, S. *J. Org. Chem.* **1995**, *60*, 5899.
27. Ma, Z. Z.; Hano, Y.; Nomura, T.; Chen, Y. J. *Heterocycles* **1997**, *46*, 541.
28. Cagir, A.; Jones, S. H.; Gao, R.; Eisenhauer, B. M.; Hecht, S. M. *J. Am. Chem. Soc.* **2003**, *125*, 13628.
29. (a) Zhou, H. B.; Liu, G. S.; Yao, Z. J. *J. Org. Chem.* **2007**, *72*, 6270; (b) Twin, H.; Batey, R. A. *Org. Lett.* **2004**, *6*, 4913.
30. Tseng, M. C.; Chu, Y. W.; Tsai, H. P.; Lin, C. M.; Hwang, J.; Chu, Y. H. *Org. Lett.* **2011**, *13*, 920.
31. Xiang, J. C.; Wang, J. G.; Wang, M.; Meng, X. G.; Wu, A. X. *Tetrahedron* **2014**, *70*, 7470.
32. (a) Takeuchi, T.; Ogawa, K.; Iinuma, H.; Suda, H.; Ukita, K. *J. Antibiot.* **1973**, *26*, 162; (b) Naik, S. R.; Harindran, J.; Varde, A. B. *J. Biotechnol.* **2001**, *88*, 1.
33. (a) Weber, A. E. *J. Med. Chem.* **2004**, *47*, 4135; (b) Gwaltney II, S. L.; Stafford, J. A. *Annu. Rep. Med. Chem.* **2005**, *40*, 149.
34. Ishikawa, H.; Honma, M.; Hayashi. Y. *Angew. Chem. Int. Ed.* **2011**, *50*, 2824.
35. Ameta, K. L.; Dandia A. *Green chemistry: Synthesis of bioactive heterocycles.* Springer, New York, chap. 6, **2014**.
36. Loupy, A.; Maurel, F.; Gogová, A. S. *Tetrahedron* **2004**, *60*, 1683.
37. Liu, J. F.; Ye, P.; Zhang, B.; Bi, G.; Sargent, K.; Yu, L.; Yohannes, D.; Baldino. C. M. *J. Org. Chem.* **2005**, *70*, 6339.
38. (a) Roninson, I. B., ed. *Molecular and cellular biology of multidrug resistance in tumor cells*, Plenum Press: New York, **1991**; (b) Kane, S. E. *Adv. Drug Res.* **1996**, *28*, 181.
39. Snider, B. B.; Busuyek, M. V. *Tetrahedron* **2001**, *57*, 3301.
40. Liu, J. F.; Kaselj, M.; Isome, Y.; Chapnick, J.; Zhang, B.; Bi, G.; Yohannes, D.; Yu, L.; Baldino, C. M. *J. Org. Chem.* **2005**, *70*, 10488.
41. Sun, H. H.; Barrow, C. J.; Sedlock, D. M.; Gillum, A. M.; Cooper, R. *J. Antibiot.* **1994**, *47*, 515.

42. For recent reviews on quinazoline alkaloids, see (a) Michael, J. P. *Nat. Prod. Rep.* **2004**, *21*, 650; (b) Johne, S. In *Supplements to the 2nd edition of Rodd's chemistry of carbon compounds*, Ansell, M. F., ed.; Elsevier: Amsterdam, vol. IV I/J, p. 223, **1995**.
43. Liu, J. F.; Ye, P.; Sprague, K.; Sargent, K.; Yohannes, D.; Baldino, C. M.; Wilson, C. J.; Ng, S. C. *Org. Lett.* **2005**, *7*, 3363.
44. Molina, P.; Tarraga, A.; Tejero, A. G. *Synthesis* **2000**, *11*, 1523.
45. Corey, E. J.; Cheng, X. M. *The logic of chemical synthesis*, Wiley: New York, **1989**.
46. Xue, W. J.; Li, H. Z.; Gao, F. F.; Wu, A. X. *Tetrahedron* **2014**, *70*, 239.
47. Njardarson, J. T.; Gaul, C.; Shan, D.; Huang, X. Y.; Danishefsky, S. J. *J. Am. Chem. Soc.* **2004**, *126*, 1038.
48. Dai, W. M. *Diversity Oriented Synthesis* **2012**, *1*, 11.
49. Zhu, Y. P.; Fei, Z.; Liu, M. C.; Jia, F. C.; Wu, A. X. *Org. Lett.* **2013**, *15*, 378.
51. (a) Xiao, X. H.; Qiu, G. L.; Wang, H. L.; Liu, L. S.; Zheng, R. L.; Jia, Z. J.; Deng, Z. B. *Chin. J. Pharmacol. Toxicol.* **1988**, *2*, 232; (b) Ma, Z. Z.; Hano, Y.; Nomura, T.; Chen, Y. *J. Heterocycles* **1997**, *46*, 541.
52. Zhu, Y. P.; Liu, M. C.; Cai, Q.; Jia, F. C.; Wu, A. X. *Chem.-Eur. J.* **2013**, *19*, 10132.
53. (a) Fahy, E.; Potts, B. C. M.; Faulkner, D. J.; Smith, K. *J. Nat. Prod.* **1991**, *54*, 564; (b) Bell, R.; Carmeli, S.; Sar, N. *J. Nat. Prod.* **1994**, *57*, 1587; (c) Bifulco, G.; Bruno, I.; Riccio, R.; Lavayre, J.; Bourdy, G. *J. Nat. Prod.* **2000**, *63*, 596.
54. (a) Ichite, N.; Chougule, M. B.; Jackson, T.; Fulzele, S. V.; Safe, S.; Singh, M. *Clin. Cancer Res.* **2009**, *15*, 543; (b) Bell, R.; Carmeli, S.; Sar, N. *J. Nat. Prod.* **1994**, *57*, 1587.
55. (a) Bartoli, G.; Bosco, M.; Foglia, G.; Giuliani, A.; Marcantoni, E.; Sambri, L. *Synthesis* **2004**, *06*, 895; (b) Abe, T.; Nakamura, S.; Yanada, R.; Choshi, T.; Hibino, S.; Ishikura, M. *Org. Lett.* **2013**, *15*, 3622.
56. Xiang, J. C.; Wang, J. G.; Wang, M.; Meng, X. G.; Wu, A. X. *Org. Biomol. Chem.* **2015**, *13*, 4240.
57. Gao, Q. H.; Fei, Z.; Zhu, Y. P.; Lian, M.; Jia, F. C.; Liu, M. C.; She, N. F.; Wu, A. X. *Tetrahedron* **2013**, *69*, 22.
58. Cheung, C. W.; Buchwald, S. L. *J. Org. Chem.* **2012**, *77*, 7526, and references cited therein.
59. Giddens, A. C.; Boshoff, H. I. M.; Franzblau, S. G.; Barry III, C. E.; Copp, B. R. *Tetrahedron Lett.* **2005**, *46*, 7355.
60. Wan, C. F.; Gao, L. F.; Wang, Q.; Zhang, J. T.; Wang. Z. Y. *Org. Lett.* **2010**, *12*, 3902.
61. Foley, P.; Eghbali, N.; Anastas. P. T. *J. Nat. Prod.* **2010**, *73*, 811.
62. (a) Jao, C. W.; Lin, W. C.; Wu, Y. T.; Wu, P. L. *J. Nat. Prod.* **2008**, *71*, 1275. (b) Yu, S. T.; Chern, J. W.; Chen, T. M.; Chiu, Y. F.; Chen, H. T.; Chen, Y. H. *Acta Pharmacol. Sin.* **2010**, *31*, 259.
63. Wang, C.; Zhang, L. P.; Ren, A. N.; Lu, P.; Wang. Y. G. *Org. Lett.* **2013**, *15*, 2982.
64. For selected examples, please see Nicolaou, K. C.; Baran, P. S. *Angew. Chem. Int. Ed.* **2002**, *41*, 2678.
65. Kennedy, D.; Norman, C. *Science* **2005**, *309,* 75.

9 Modern Synthesis of Bioactive Heterocycles via IMCR Modification

Anshu Dandia, Shahnawaz Khan, Vijay Parewa, Amit Sharma, and Begraj Kumawat*

CONTENTS

9.1 INTRODUCTION

Isocyanides, formally known as isonitriles, are organic compounds with the functional group –NC. The organic part is associated to the isocyanide group via nitrogen atom, not via carbon. Isocyanides are described by two resonance structures: first, one with a triple bond between the nitrogen and the carbon with a positively charged nitrogen and a negatively charged carbon and, second, one with a double bond between nitrogen and carbon. Surprisingly, isocyanides are the class of stable organic compounds with a formally divalent carbon. The exceptional properties of the isocyano group, which may be utilized both as an electrophile and as a nucleophile, differ from isocyanides from other functional groups in organic chemistry.[1] Only carbene and carbon monoxide share this property with isocyanides. The first isocyanide synthesized by the chemist Lieke in 1859, from the reaction of allyl iodide and silver cyanide, was allyl cyanide; Lieke did not recognize them as such and first believed them to be nitriles.[2] Furthermore, it was Gautier who first discovered the isomeric nature of the relationship between the isocyanide and nitriles.[3] At the same time, Hofmann synthesized an isocyanide by the reaction of aniline with chloroform in the presence of potassium hydroxide.[4] The synthetic method for these compounds has not been much explored due to the unpleasant odor of the simplest and the most

* Corresponding author: dranshudandia@yahoo.co.in.

volatile isocyanides, which limited chemists to the development of efficient strategies for the synthesis of isocyanides. People who have inhaled volatile isocyanides such as allyl, benzyl, cyclohexyl, or *tert*-butyl isocyanide over a longer period of time report the sensory perception of the smell of hay. There are more than a hundred isocyanides containing natural products that have been isolated from marine species. Various naturally occurring isocyanides showed antibiotic, fungicidal, and antineoplastic effects. However, more than a dozen methods for the syntheses of isocyanides have been reported.[4] The reaction of *N*-formamide with phosgene or phosgene surrogate such as di- and tri-phosgene and a matching base is the method of choice with regard to cost, yield, and execution in most cases. Another convenient method is the dehydration with $POCl_3$.[5]

On the other hand, *N*-heterocyclic compounds have always been at the forefront of attention due to their numerous uses in pharmaceutical applications. *N*-heterocyclic moieties also serve as an integral part of a broad variety of biologically active natural products and synthetic compounds.[6] The overwhelming majority of commercially available synthetic drugs (up to 80%) have a *N*-heterocyclic structural component.[7] Due to the widespread interest in *N*-heterocycles, the synthesis of these compounds has always been among the most important research areas in synthetic chemistry. In the past decades, many conventional methods for the preparation of *N*-heterocycles have been improved to better meet the demands of modern combinatorial synthesis and medicinal chemistry.[8]

This review covers the synthesis of *N*-fused heterocycles using (1) isocyanide-based multicomponent reactions, (2) transition metal catalyzed multicomponent reactions using isocyanides as a building block, and (3) transition metal catalyzed insertion reactions using isocyanides—specifically those that appeared in the last 4 years.

9.2 ISOCYANIDE-BASED MCR

Multicomponent reactions (MCRs) are generally defined as reactions where more than two starting materials react to form a product, incorporating essentially all of the atoms of the educts (Figure 9.1).[9] Ideal multicomponent synthesis allows the simultaneous addition of all reactants, reagents, and catalysts at the onset of the reaction, which requires that all reactants combine in a uniquely ordered manner under the same reaction conditions. Thus, an MCR is a sequence of mono- and bimolecular events that proceeds sequentially until an irreversible final step traps the product. All these processes are highly efficient, for they create molecular complexity by

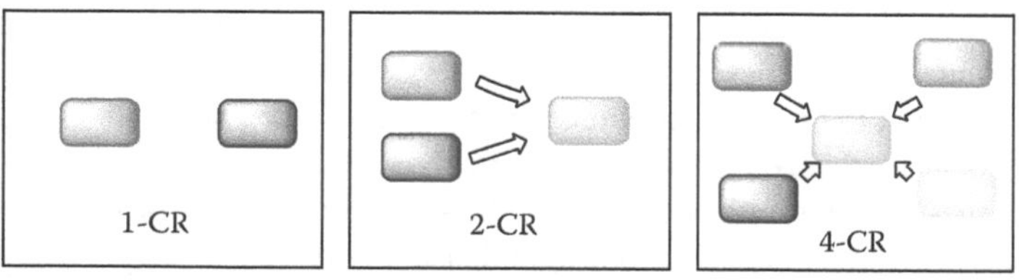

FIGURE 9.1 A divergent one-component reaction and convergent two- and four-component reactions.

generating more than two chemical bonds per operation. Hundreds of MCRs have been described over the years.[10] Perhaps the earliest MCR described is Hantzsch's dihydropyrimidine synthesis dated more than 150 years ago. However, the discovery of novel MCRs is rather a theme of the past decade. With the emergence of high-throughput screening in the pharmaceutical industry over a decade ago, synthetic chemists were faced with the challenge of preparing large collections of molecules to satisfy the demand for new screening compounds. By virtue of its inherently high exploratory power, research on MCRs has naturally become a rapidly evolving field and since 1995 has attracted attention from both academic and industrial researchers. It is therefore not surprising that many efforts are currently being devoted to this new area of research.

Generally, there are different classification schemes of MCRs possible (e.g., according to the reaction mechanisms, the components involved, or their intrinsic variability). For example, recently Sonoda et al. described a three-component reaction (3-CR) of epoxides **1**, elemental sulfur **2**, and carbon monoxide **3**, yielding, under basic conditions, 1,3-oxathiolan-2-ones **4** (Scheme 9.1).[11] From the viewpoint of simple operation, mild reaction conditions, and good yields, the present reaction provides a useful method for synthesis of 1,3-oxathiolan-2-ones. However, this reaction is not very useful to prepare large combinatorial libraries of compounds, since there is only one variable starting material, the epoxide, whereas the other two starting materials are fixed in all reactions. This 3-CR constitutes an MCR of low variability.

On the other hand, a recent publication introduced the union of two highly variable MCRs: the Petasis and Ugi reactions. Both reactions use starting materials that are commercially available in very large quantities. Theoretically, this combination of MCRs spans a chemical space of greater than $1{,}000 \times 200 \times 500 \times 1{,}000 \times 1{,}000 = 10^{14}$ small molecules. This constitutes a combination of MCRs of very high variability, covering a large chemical space.[12]

Another way to describe the usefulness of a multicomponent reaction is correlated to numerous factors: the number of bonds formed in one sequence, which Tietze[13] has referred to as the bond-forming efficiency (BFE, or bond-forming economy); the increase in structural complexity (structure economy); and finally, its suitability for general application. Multicomponent reactions have attracted considerable interest owing to their exceptional synthetic efficiency. The BFE is an important measure to determine the quality of a multicomponent reaction (seen later in Scheme 9.3).[14,15]

Unlike the usual stepwise formation of individual bonds in the target molecule, the crucial attribute of MCRs is the inherent formation of several bonds in one operation without isolating the intermediates, changing the reaction conditions, or adding

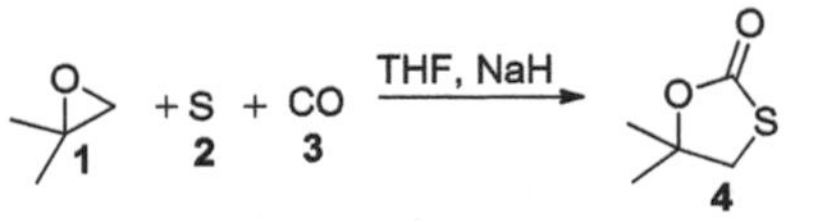

1 + S (2) + CO (3) —THF, NaH→ 4

SCHEME 9.1 Sonoda's 3-CR of epoxides **1**, sulfur **2**, and carbon monoxide **3**, yielding 1,3-oxathiolan-2-ones **4**.

further reagents. It is obvious that adopting such strategies would allow the minimization of both waste production and expenditure of human labor. The products are formed simply by mixing the corresponding set of starting materials. Since the structures of the products carry portions of all the reactants employed, MCRs that have a high attendant BFE assure a marked increase in molecular complexity and diversity. A wide variation among these starting materials opens up versatile opportunities for the synthesis of compound libraries. The generalization to as many available starting materials as possible is an indispensable characteristic for the most general application. Multicomponent reactions thus address the requirements for efficient high-throughput synthesis of compounds in a cost- and time-effective manner. Reactions that build up carbon–carbon, carbon–nitrogen, and other carbon–heteroatom bonds and at the same time introduce heteroatom-containing functionality into the structural framework are especially attractive for the rapid construction of organic molecules. Special subclasses are isocyanide based MCRs (IMCRs). They are particularly interesting because they are more versatile and diverse than the remaining MCRs. The enormous potential of isocyanides for the development of multicomponent reactions lies in the diversity of bond-forming processes available, their functional group tolerance, and the high levels of chemo-, regio-, and stereoselectivity often observed. Also, IMCRs among MCRs are at the minority overall, at the moment; nevertheless, they provide the largest chemical space.

9.2.1 Isocyanide-Based Ugi and Passerini Reactions

Today most MCR chemistry performed with isocyanides relates to the classical reactions of Passerini and Ugi (Figure 9.2).[16,17] Indeed, the large number of different scaffolds now available mostly builds on these two MCRs and their combination with other types of reactions. Passerini reactions involve an oxo component, an isocyanide, and a nucleophile. Ugi reactions are defined as the reaction of a Schiff base or an enamine with a nucleophile and an isocyanide, followed by a (Mumm) rearrangement reaction. Interestingly, Ugi reactions as compared to Passerini reactions are much more versatile not only in terms of library size but also in terms of scaffolds. This can be attributed to the many different acid components or nucleophiles and amine components that have been described to date for the Ugi reaction. In the past decades, several reviews on the Ugi and Passerini reactions toward the synthesis

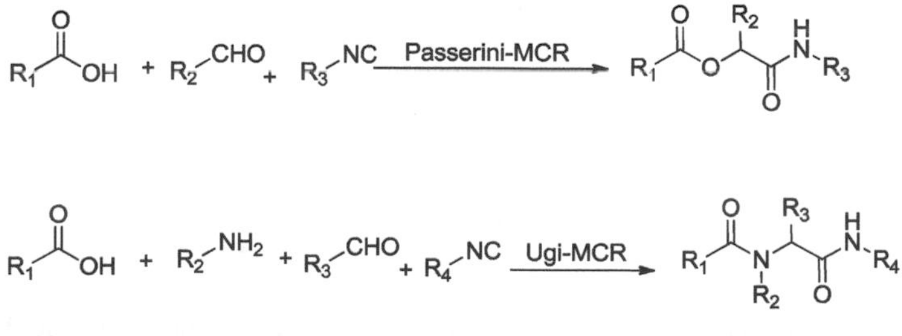

$$R_1COOH + R_2CHO + R_3NC \xrightarrow{\text{Passerini-MCR}} R_1C(=O)OCH(R_2)C(=O)NHR_3$$

$$R_1COOH + R_2NH_2 + R_3CHO + R_4NC \xrightarrow{\text{Ugi-MCR}} R_1C(=O)N(R_2)CH(R_3)C(=O)NHR_4$$

FIGURE 9.2 Isocyanide based Passerini and Ugi multicomponent reaction.

N-heterocycles have appeared.[18] Some recent updates on these reactions are presented in the current review.

Thus, Wanner et al. reported the synthesis of *N*-silyl-4,4-disubstituted 1, 4-dihydropyridine **7** via Ugi type MCR (Scheme 9.2).[19] This reaction was typically performed under acidic reaction conditions, necessary to generate iminium ions as an integral part of the reaction sequence. When applied to 1,4-dihydropyridines **5** under acidic reaction conditions, in addition to transforming into an iminium salt, they are likely to also cause a desilylation reaction. Accordingly, as reactive intermediate, the *N*-desilylated cyclic iminium ion should form. But with the iminium ion as intermediate, the Ugi reaction can proceed along the common pathway leading to Ugi products provided with a *N*-acyl group resulting from the carboxylic acid **6** used in the reaction. Thus, now depending on the carboxylic acid selected for the Ugi reaction, final products with different *N*-acyl groups can be obtained. This reaction has been realized for a series of different carboxylic acids and isocyanides and was found to proceed with high distereoselectivities and with good yields.

The one-pot synthesis of uracil polyoxin C analogs **12** using Ugi multicomponent reaction has been described by Williams and co-workers (Scheme 9.3).[20] The four components employed in the Ugi reaction are 2′,3′-isopropylidine-protected uridine-5′-aldehyde **8**, 2,4 dimethoxybenzylamine **10**, an isoxazolecarboxylic acid **11**, and the convertible isonitrile *N*-(2-{[(*tert*-butyldimethylsilyl)oxy]methyl}phenyl) carbonitrile **9**. Following the Ugi reaction, treatment with HCl in MeOH achieves deportation of the isopropylidene group and the *N*-benzyl group and conversion of the isonitrile-derived amide (the Ugi product) into the corresponding methyl ester.

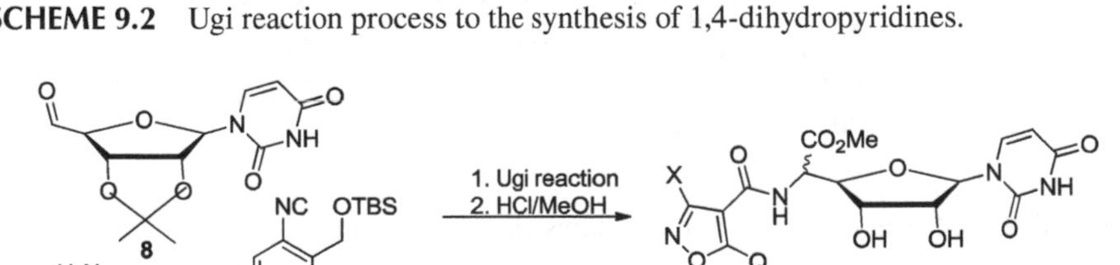

SCHEME 9.2 Ugi reaction process to the synthesis of 1,4-dihydropyridines.

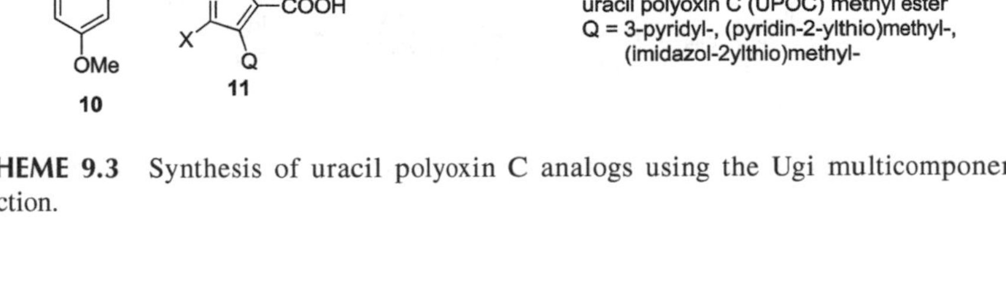

SCHEME 9.3 Synthesis of uracil polyoxin C analogs using the Ugi multicomponent reaction.

The procedure is amenable to automated multiparallel synthesis of novel compounds related to the polyoxin and nikkomycin nucleoside-peptide antibiotics.

Tron et al. reported the synthesis of Passerini–Ugi hybrids **17** by a four-component reaction (Scheme 9.4).[21] In this reaction, the imino anhydride intermediate, generated using a secondary amine in the Ugi reaction, can be intramolecularly attacked by other nucleophiles before the well-known synthetic fates can take place allowing the formation of new scaffolds.

The proline-like β-turn mimics **21** accessed via Ugi reaction involving monoprotected hydrazines **19**. In this work, a new IMCR using various aliphatic keto carboxylic acids **18** in combination with a Boc- or Cbz-protected hydrazine **19** as a surrogate for the amine component, which was expected to yield substituted *N*-aminolactams, has been described (Scheme 9.5).[22]

The synthesis of five- and six-membered lactams **25** via solvent-free microwave Ugi reaction has been reported by Deprez et al. (Scheme 9.6).[23] Five- and six-membered lactams were synthesized via a four-center, three-component Ugi reaction by combining amines **24**, isocyanides **23**, and keto acids **22** under solvent-free microwave conditions. The reaction was carried out in much shorter times and the yields were improved in comparison to classical conditions.

Westermann et al. described seven-component reactions by sequential chemoselective Ugi–Mumm/Ugi–Smiles reaction leading to highly diverse products (Scheme 9.7).[24] After adding a first set of Ugi starting materials (formaldehyde,

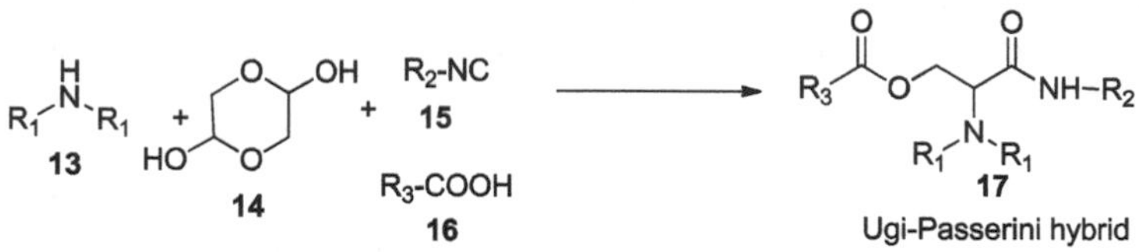

SCHEME 9.4 Ugi–Passerini hybrids by a four-component reaction.

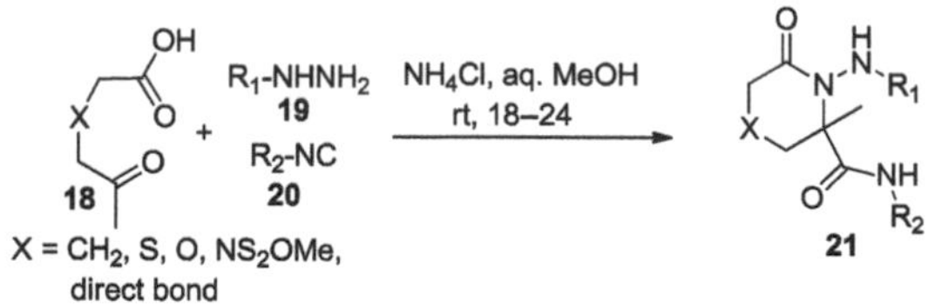

SCHEME 9.5 Preparation of *N*-aminolactams via Ugi reaction.

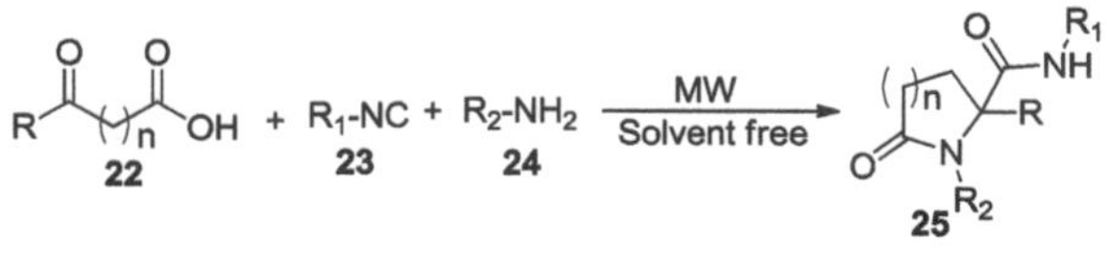

SCHEME 9.6 Synthesis of five- and six-membered lactams via solvent-free microwave Ugi reaction.

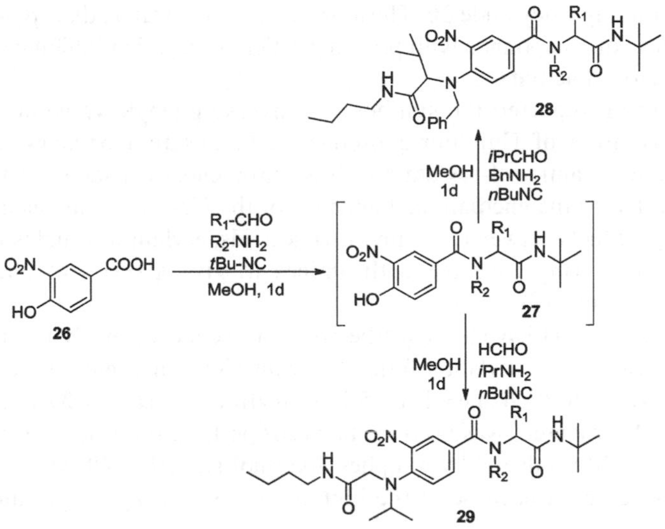

SCHEME 9.7 Seven-component reactions leading to highly diverse products.

isopropyl amine, and *tert*-butyl isonitrile), a second set of Ugi components (isobutyric aldehyde, benzyl amine, and *n*-butyl isonitrile) is added after 24 h, upon which the mixed Ugi–Mumm/Ugi–Smiles product can be isolated.

The synthesis of tetrazolo[1,5-*a*][1,4]benzodiazepines **34** via a novel multicomponent reaction has been disclosed (Scheme 9.8).[25] The tetrazolo[1,5-*a*][1,4] benzodiazepines were obtained by a facile azide Ugi five-center, four-component reaction (U-5C-4CR) using ketones **32**, sodium azide **31**, ammonium chloride **33**,

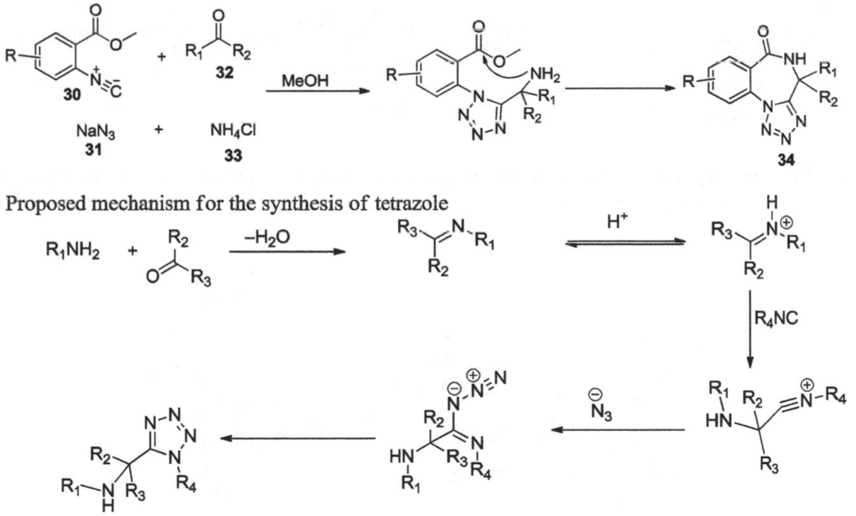

SCHEME 9.8 Synthesis of tetrazolo[1,5-*a*][1,4]benzodiazepines via a novel multicomponent isocyanide based condensation.

and corresponding isocyanide **30**. The aforementioned tetrazolodiazepines represent a notable class of compounds with proven platelet aggregation inhibitory and cholecystokinin agonist activities.

Menchi et al. reported the synthesis of diverse phenylglycine derivatives **37** via transformation of Ugi four-component condensation primary adducts **36**. 3-(*N*-substituted) amino-4-arylamino-1*H*-isochromenones (isocoumarins)—which can be regarded as the enediamine tautomers of the Ugi four-component condensation primary adducts between 2-formylbenzoic acids, arylamines, and isocyanides—undergo a facile ring cleavage with amines to give a series of phenylglycine derivatives (Scheme 9.9).[26]

Chiral γ-lactams **41** have been synthesized via a sequence involving an enantioselective Friedel–Crafts alkylation followed by an Ugi four-center, three-component reaction (U-4C-3CR) (Scheme 9.10).[27] 5-Hydroxyfuran-2(5*H*)-one **39**, a readily available renewable resource, was used as an electrophile in the Friedel–Crafts alkylation of indoles **38** catalyzed by a diphenylprolinol silylether **40**. Moderate catalyst loading was achieved because of the high reactivity of 5-hydroxyfuran-2(5*H*)-one in this process.

Charton et al. reported the synthesis of bivalent compounds **43** via 4C-Ugi reaction using squaric acid **42** as a suitable building block (Scheme 9.11).[28]

The synthesis of *N*-cyclohexyl-2-(2-hydroxyphenylamino) acetamide derivatives **45** via zinc chloride catalyzed three-component Ugi reaction has been disclosed by Shaabani and co-workers (Scheme 9.12).[29]

Ghandi et al. reported a three-component, intramolecular Ugi reaction for the synthesis of indoloketopiperazines **50** (Scheme 9.13).[30] 2-(3-Chloro-2-formyl-1*H*-indol-1-yl) acetic acid **48**, as a bifunctional formyl acid, is prepared in three steps.

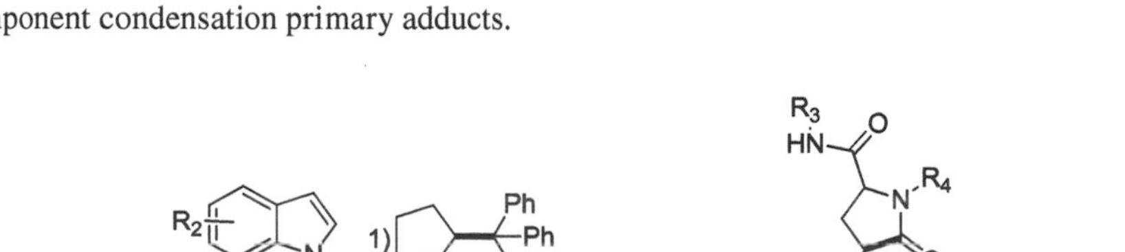

SCHEME 9.9 Synthesis of diverse phenylglycine derivatives via transformation of Ugi four-component condensation primary adducts.

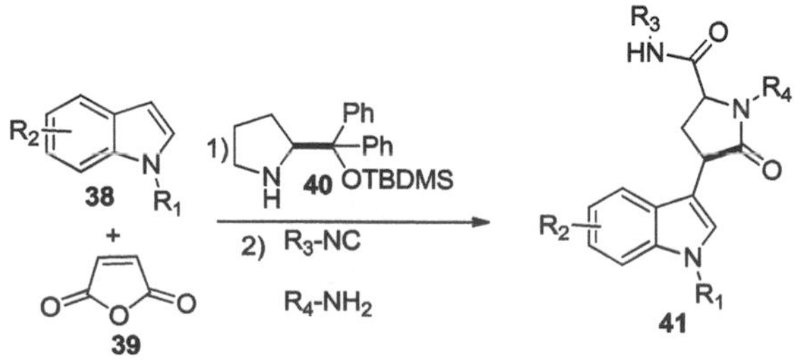

SCHEME 9.10 Synthesis of chiral γ-lactams via a sequence involving an enantioselective Friedel–Crafts alkylation followed by an Ugi four-center, three-component reaction.

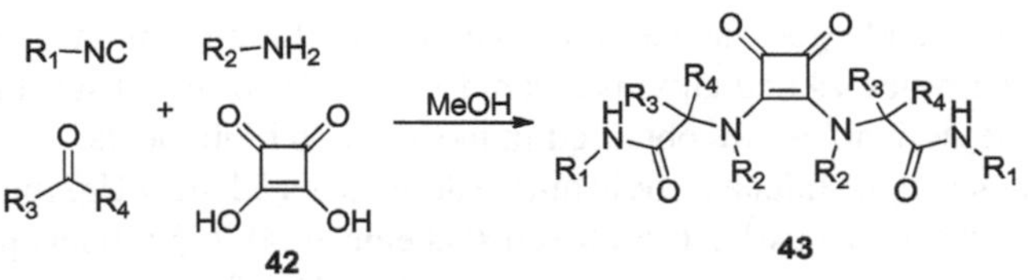

SCHEME 9.11 Synthesis of bivalent compounds via double 4C-Ugi reaction.

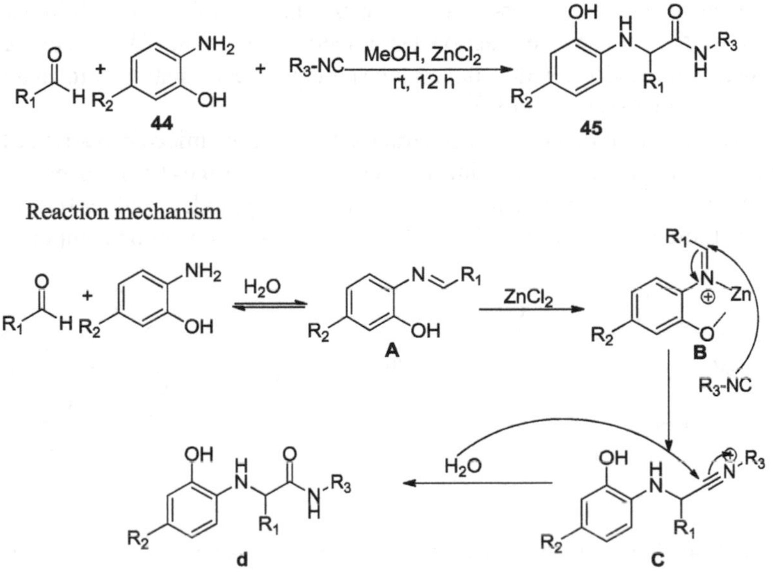

Reaction mechanism

SCHEME 9.12 Synthesis of *N*-cyclohexyl-2-(2-hydroxyphenylamino) acetamide derivatives.

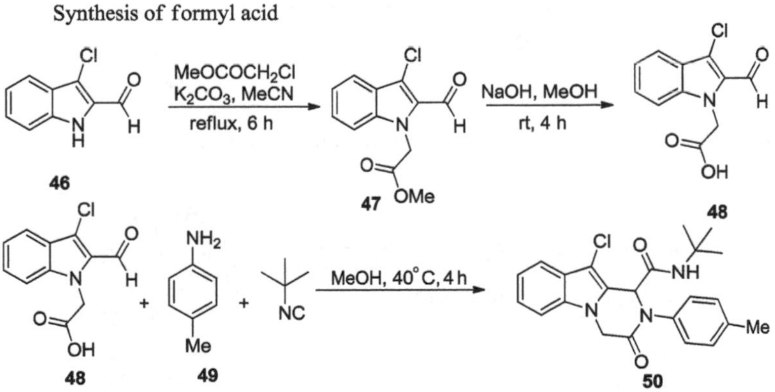

Synthesis of formyl acid

SCHEME 9.13 Synthesis of indoloketopiperazines via a three-component, intramolecular Ugi reaction.

This compound undergoes a one-pot, four-center, three-component Ugi reaction with primary amines **49** and alkyl isocyanides. A series of novel substituted indoloketopiperazine derivatives are obtained in moderate to high yields.

The synthesis of arylidenepyruvic amide derivatives **52** via a Ugi four-component condensation reaction has been archived (Scheme 9.14).[31] Arylidenepyruvic acids **51** (APAs) have been successfully employed in an Ugi four-component condensation reaction that yielded polyfunctional amides. Condensation of various APAs, aldehydes, amines, and isocyanides at room temperature in 96% ethanol as a green solvent proceeded in good yields.

Hulme et al. reported a one-pot, two-step synthesis of bis-pyrrolidinone tetrazoles **54** via the Ugi–Azide reaction using methyl levulinate **53**, primary amines, isocyanides, and azidotrimethylsilane with subsequent acid treatment to catalyze the lactam formation (Scheme 9.15).[32]

The synthesis of morpholin- or piperazine-ketocarboxamide derivatives **56** using Ugi three-component reaction induced by chiral cyclic imines **55** has been described (Scheme 9.16).[33] The Ugi reaction of the imines, isocyanides, and carboxylic acids opens an efficient access to novel morpholin-2-one-3-carboxamide compounds. The

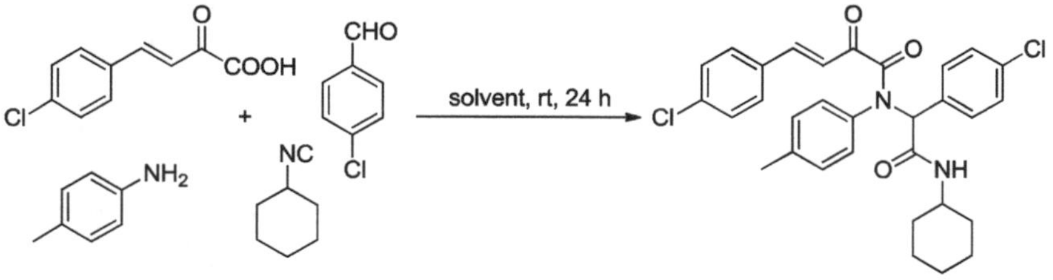

SCHEME 9.14 Synthesis of arylidenepyruvic amide derivatives via Ugi four-component condensation.

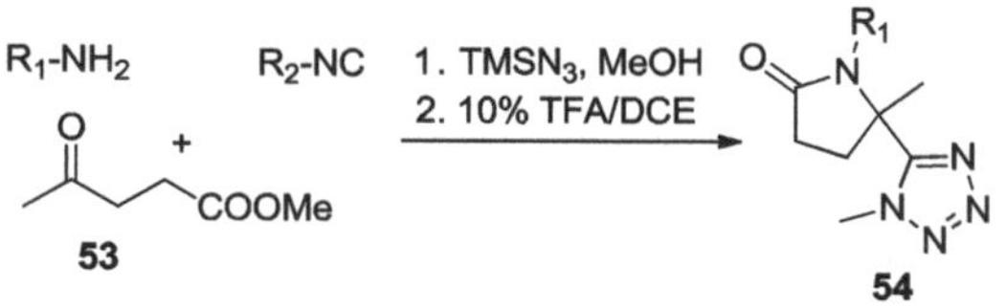

SCHEME 9.15 Two-step synthesis of bis-pyrrolidinone tetrazole.

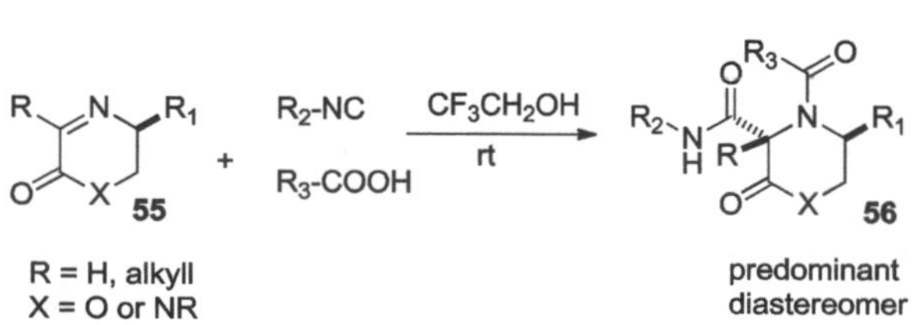

SCHEME 9.16 Synthesis of morpholin- or piperazine-ketocarboxamide derivatives via Ugi three-component reaction.

chiral imines showed promising stereoinduction for the new chiral center of the Ugi products, and predominant *trans*-isomers were obtained in most cases. Addition of some Lewis acids or proton acids could improve the diastereoselectivity further but usually led to a drop in total yield.

A library of symmetrical **62** and unsymmetrical bis-(β-aminoamides) **61** has been prepared starting from symmetrical secondary diamines **57** by using a double Ugi four-component reaction. The use of 2-hydroxymethyl benzoic acid **58** is necessary to suppress the competing split Ugi reaction, increasing the yield and simplifying the purification step (Scheme 9.17).[34]

Orru et al. represent an efficient combination of MAO-N catalyzed desymmetrization of cyclic mesoamines **63** with Ugi–Smiles multicomponent chemistry that produced optically pure *N*-aryl proline amides **65** (Scheme 9.18).[35]

The detailed investigation of the Paserini reaction with CF_3-carbonyl compounds **66** has been reported. The reaction provides a new approach to trifluorolactic acid derivatives and CF_3 substituted depsipeptides **69** (Scheme 9.19).[36] The method is promising for the synthesis of chiral trifluoromethyl depsipeptides—that is, orthogonally protected building blocks for incorporation into naturally occurring depsipeptides.

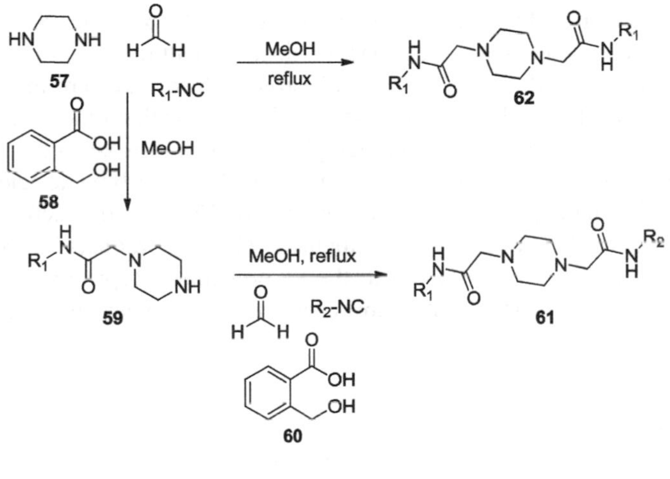

SCHEME 9.17 Synthesis of symmetrical and unsymmetrical bis-(β-aminoamides).

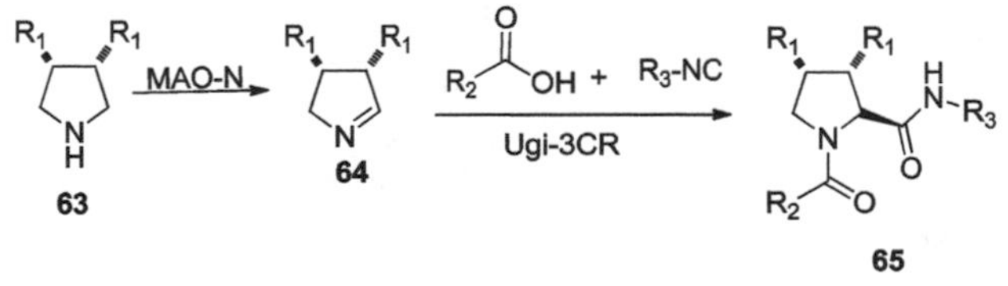

SCHEME 9.18 Asymmetric synthesis of prolyl peptides by MAO-N oxidation and Ugi-type 3CR.

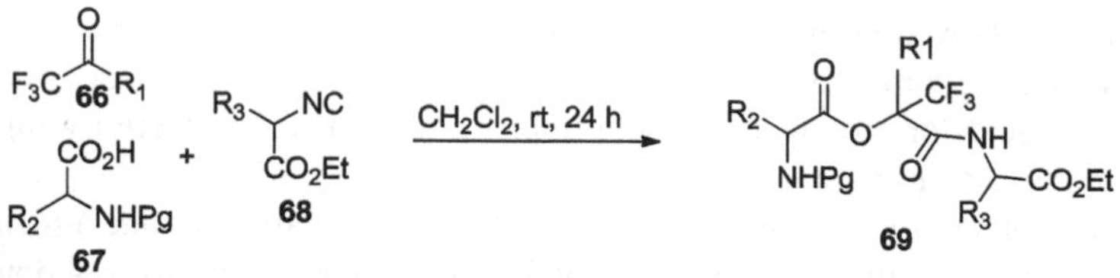

SCHEME 9.19 Synthesis of CF_3-depsipeptides.

9.2.2 Post-Ugi and -Paserini Transformations

The MCR with other synthetic transformation has been proven to be a powerful tool for the construction of complex molecules in a few steps and, generally, from readily available starting materials.[37] Among these MCRs, the isocyanide based Ugi multicomponent reaction is especially useful for the synthesis of complex molecules due to its convergent character, atom economy, operational simplicity, and high variability, and it provides the possibility of further transformations including condensation, an intramolecular SNAr reaction, ring-closure metathesis, cycloaddition, macrolacton, etc.[38] Currently, much attention has been paid to the arrangement of Ugi MCR and Pictet–Spengler reactions for the construction of natural product-like compounds.

In this context, Domling et al. reported the synthesis of polycyclic indole derivative **74** via Ugi MCR followed by Pictet–Spengler (PS) reaction (Scheme 9.20).[39] Polycyclic indole moieties are often part of bioactive natural or synthetic products; however, traditionally they have to be synthesized over several steps involving time-consuming sequential multistep syntheses. Theirs was the first report on the Ugi–PS combination where electron-rich indolethylamine-derived isocyanides **70** react in the Ugi–PS reactions with a diversity of bifunctional keto carboxylic acid **71** derivatives and orthogonally protected aminoacetaldehyde **72**, to yield structurally intriguing polycyclic indole alkaloid-type compounds. The Ugi products **73** were isolated and purified before subjecting them to the PS procedure. Typically, the Ugi products were formed as mixtures of enantiomers or diastereomers in equal or almost

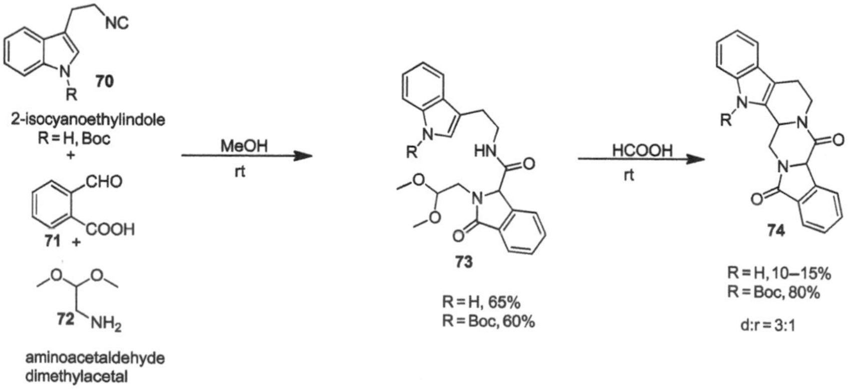

SCHEME 9.20 Synthesis sequence involving an intramolecular U-4CR and a subsequent PS-2CR.

equal ratio, respectively. Interestingly, protection of the indole proton by a Boc group increased the yield of the PS reaction to 80% upon treatment of the Ugi product with formic acid at room temperature. Significantly, aliphatic acyclic and cyclic as well as aromatic oxo acids are substrates for this reaction sequence.

This group also reported the synthesis of newly discovered natural product bacillamide C and several derivatives for the first time and in only three steps. The key transformation constitutes a thiazole Ugi multicomponent reaction (Scheme 9.21).[40] These compounds will serve to elucidate chemical biology and structure activity relationship of this potent antialgae natural product and show the synthetic pathway to related natural products. Thus, reaction of the four starting materials affords the Ugi intermediate **78** in 60% yields under ambient conditions. The next step, the cleavage of the 2,4-dimethoxybenzyl protecting group, is usually performed with trifluoroacetic acid (TFA) at ambient temperature. However, attempts to analogously deprotect under these conditions did not show any conversion. Only prolonged treatment with TFA at 50 °C yielded the deprotected product **79** in 58% yields. A recently described one-pot procedure for the above transformation overcomes the reported drawbacks, i.e., multistep saponification, activation, and coupling sequences to transform the methyl ester into the amide. Thus, 1,5,7-triazabicyclo[4.4.0]dec-5-ene (TBD) catalyzed direct amidation of with tryptamine yielded racemic target natural product bacillamide C **80**. Overall, racemic bacillamide C is thus accessible from commercially available starting materials in three simple to perform steps in overall 30% yield.

The synthesis of highly diverse indole derivatives via isocyanide based Ugi MCR followed by Pictet–Spengler reaction has also been described by this group (Scheme 9.22).[41] They have synthesized tryptophan- and tryptamine-derived isocyanides **81** and reacted them in the Ugi 4-CR with aldehydes, primary amines, and carboxylic acids.

Aminoacetaldehyde dimethylacetal was used as bifunctional amine components that, in a subsequent step, would undergo the Pictet–Spengler reaction to afford

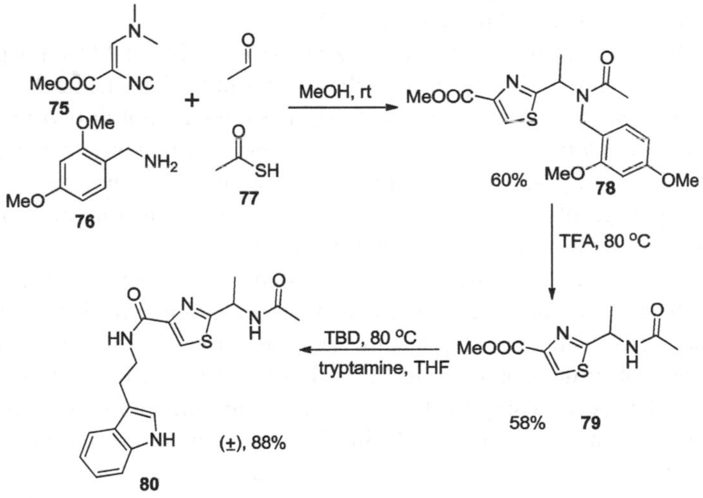

SCHEME 9.21 Racemic four-component bacillamide C synthesis.

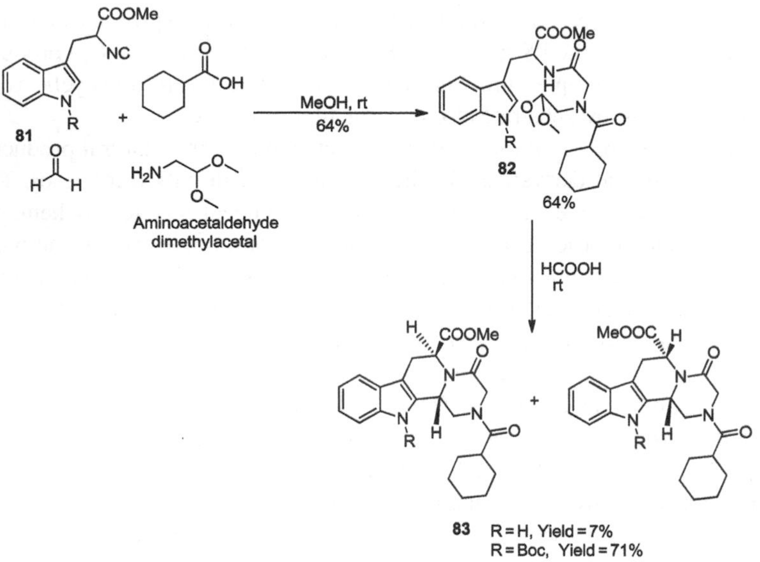

SCHEME 9.22 Synthesis of indole derivatives via Ugi-4CR followed by Pictet–Spengler reaction.

highly substituted indole derivatives **83**. Such a two-step process could have advantages over currently sequential processes, since the product diversity should be much larger and the effort to synthesize the compound should be drastically reduced.

Orru et al. described the biocatalytic desymmetrization of 3,4-*cis* substituted mesopyrrolidines using monoamine oxidase N (MAO-N) from *Aspergillus niger* with an Ugi-type multicomponent reaction followed *in situ* by a Pictet–Spengler-type cyclization reaction sequence for the rapid asymmetric synthesis of alkaloid-like polycyclic compounds **86** (Scheme 9.23).[42]

The synthesis of benzotriazole and benzimidazole scaffolds **90** from Ugi–Smiles couplings of isocyanides has been described by Kaim et al. (Scheme 9.24).[43] When allylamine **88** was used as the amino input in Ugi–Smiles a coupling of *o*-nitrophenols **87**, the resulting adducts **89** can be deallylated by a palladium catalyzed process leading to a formal Ugi–Smiles coupling with ammonia. This new sequence, combined with hydrogenolysis of the nitro group, offers an interesting multicomponent entry to benzotriazole and benzimidazole scaffolds.

Laconde et al. reported the synthesis of 4-aminopiperidine-4-carboxylic acid derivatives **92** and their application to the synthesis of carfentanil **93** and remifentanil **94** (Scheme 9.25).[44] The first step of this synthesis is the Ugi MCR using propionic acid, aniline, substituted piperidone **91**, and 1-cyclohexenyl isocyanide in methanol for 24 h. The second step for the synthesis of carfentanil and remifentanil was achieved by methanolysis in 10% AcCl in MeOH at room temperature for 24 h. The mechanism for the Ugi reaction with piperidone has been described. The 4-anilino group at the piperidine quaternary carbon is particularly unreactive. However, in this particular situation, the acylated isoamide **A** readily undergoes cyclization to

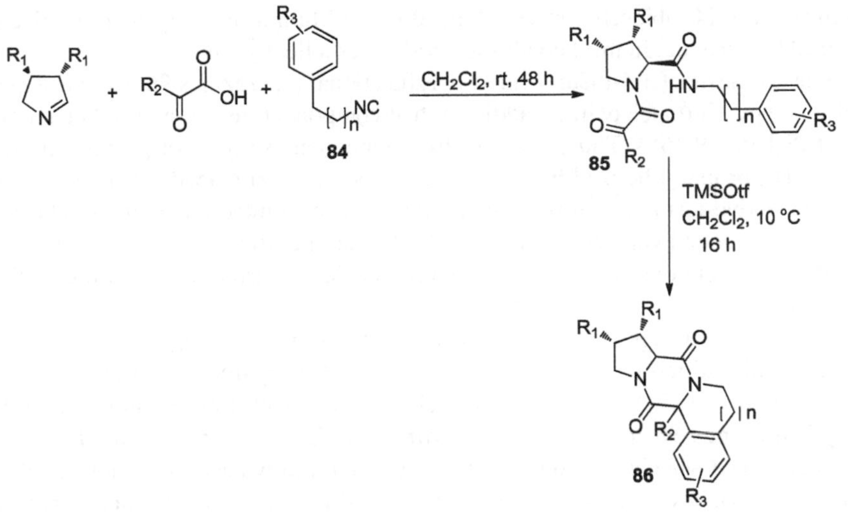

SCHEME 9.23 MAO-N oxidation/Ugi MCR/Pictet–Spengler type cyclization (MUPS) sequence.

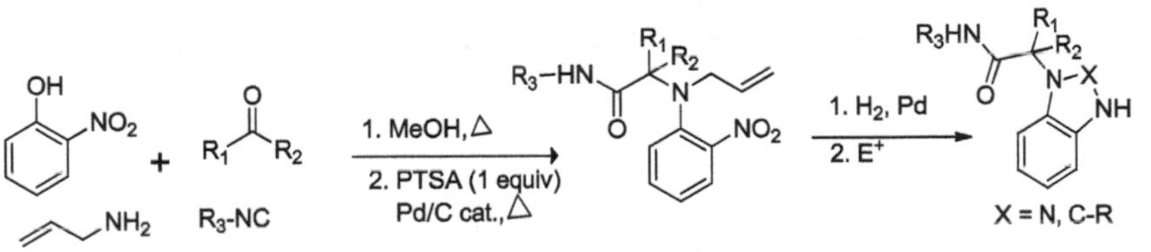

SCHEME 9.24 Synthesis of benzotriazole and benzimidazole scaffolds.

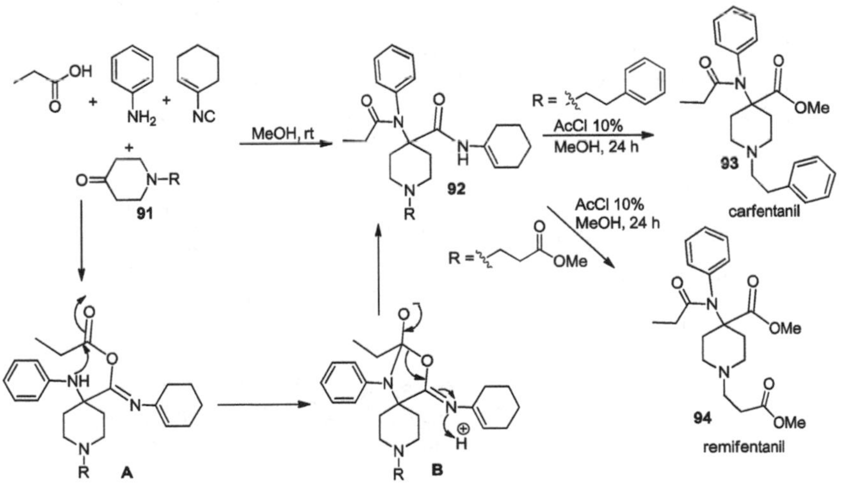

SCHEME 9.25 Synthesis of 4-aminopiperidine-4-carboxylic acid derivatives and their application to the synthesis of carfentanil and remifentanil.

form the spiro[4.5]bicyclic intermediate **B** of the Mumm rearrangement to give the desired Ugi product in mild conditions with an excellent yield.

The synthesis of 1,4-disubstituted polyfunctional piperazines **96** using a sequential one-pot Ugi/nucleophilic addition five-component reaction has been developed (Scheme 9.26).[45] Thus, one-pot, five-component reaction of primary amines, aldehydes, propargylic acid **95**, isocyanides, and piperazine leads to the regio- and stereoselective formation of polyfunctional 1,4-disubstituted piperazines. The reaction could proceed via formation of an *N*-substituted-2-alkynamide as an intermediate **B**, which contains an active triple bond suitable for further nucleophilic addition reactions.

Krasavin et al. reported the diversity-oriented synthesis of pyrazol-3-one based on hydrazinodipeptide-like units **99** and **100** prepared via the Ugi reaction (Scheme 9.27).[46] In this approach, N-cyanoacetyl-N^{1}-trifluoroacetyl-N^{11}-alkylhydrazines **98**, prepared via a hydrazino-Ugi reaction, provided different pyrazol-3-ones when exposed to mildly acidic and mildly basic conditions at 60 °C. These approaches offer easy access to two different pyrazol-3-one-containing chemotypes in a diversity-oriented fashion, in only two chemical operations from simple precursors. Parchinsky et al. described the synthesis of 3-oxoisoindolines **103** and **106**, which includes preparation of Ugi adducts containing thiophene and fumaric acid residues **102** and **105**. When treated with excess *m*-CPBA at room temperature, these precursors undergo a simple oxidative cycloaddition/aromatization

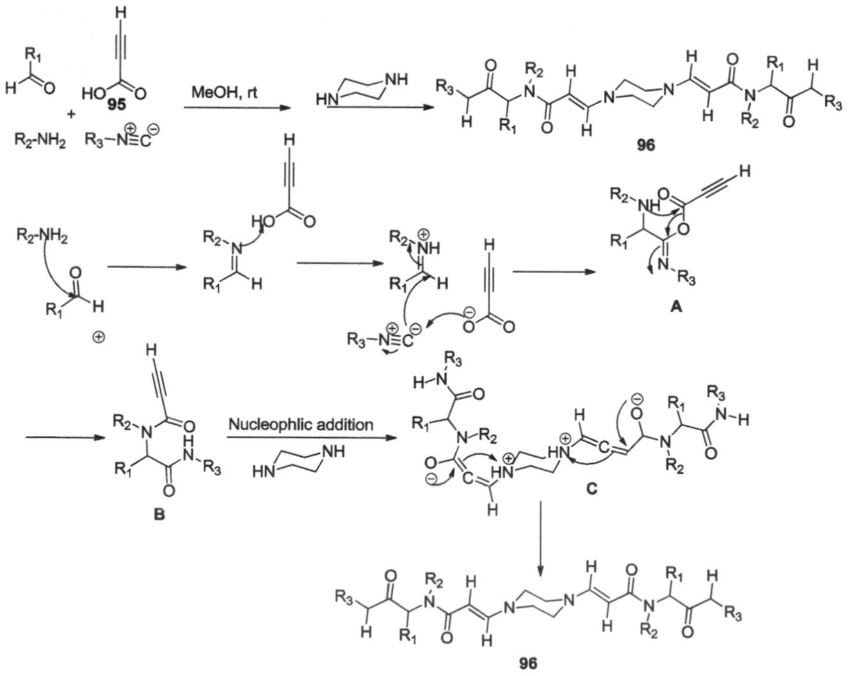

SCHEME 9.26 Synthesis of 1,4-disubstituted polyfunctional piperazines via a sequential one-pot Ugi/nucleophilic addition five-component reaction.

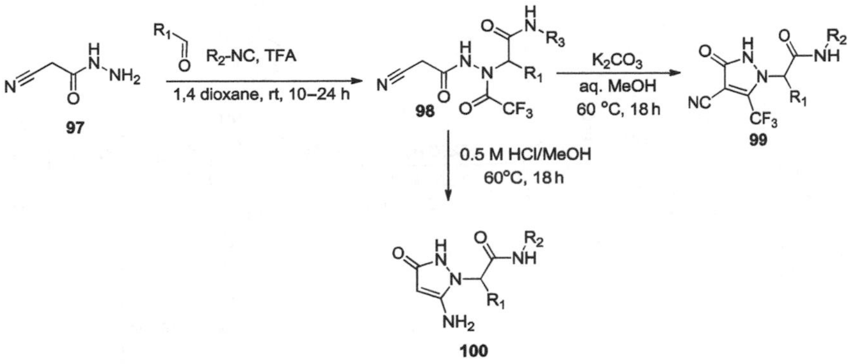

SCHEME 9.27 Synthesis of pyrazol-3-one based on hydrazinodipeptide-like units prepared via the Ugi reaction.

transformation and the corresponding 3-oxoisoindoline products are isolated in fair chemical yield over two steps (Scheme 9.28).[47]

The Gámez-Montaño group described the synthesis of tetrahydroisoquinolin-pyrrolopyridinones **113** by Ugi-3CR–aza Diels–Alder/S-oxidation/Pummerer (Scheme 9.29).[48] A series of tetrahydroisoquinolin-pyrrolopyridinones were prepared from an easily accessible aldehyde, a commercially available amine, a

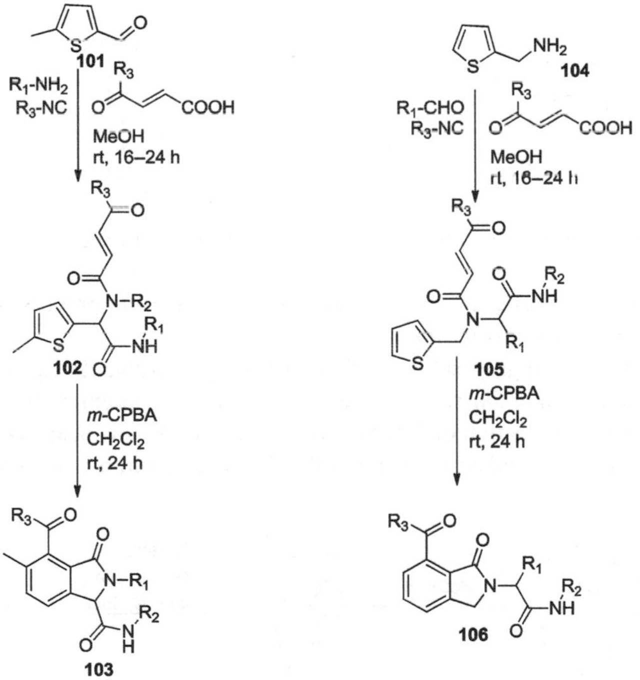

SCHEME 9.28 Synthesis of thiophene-containing precursors via the Ugi reaction and their transformation into 3-oxoisoindolines by treatment with *m*-CPBA.

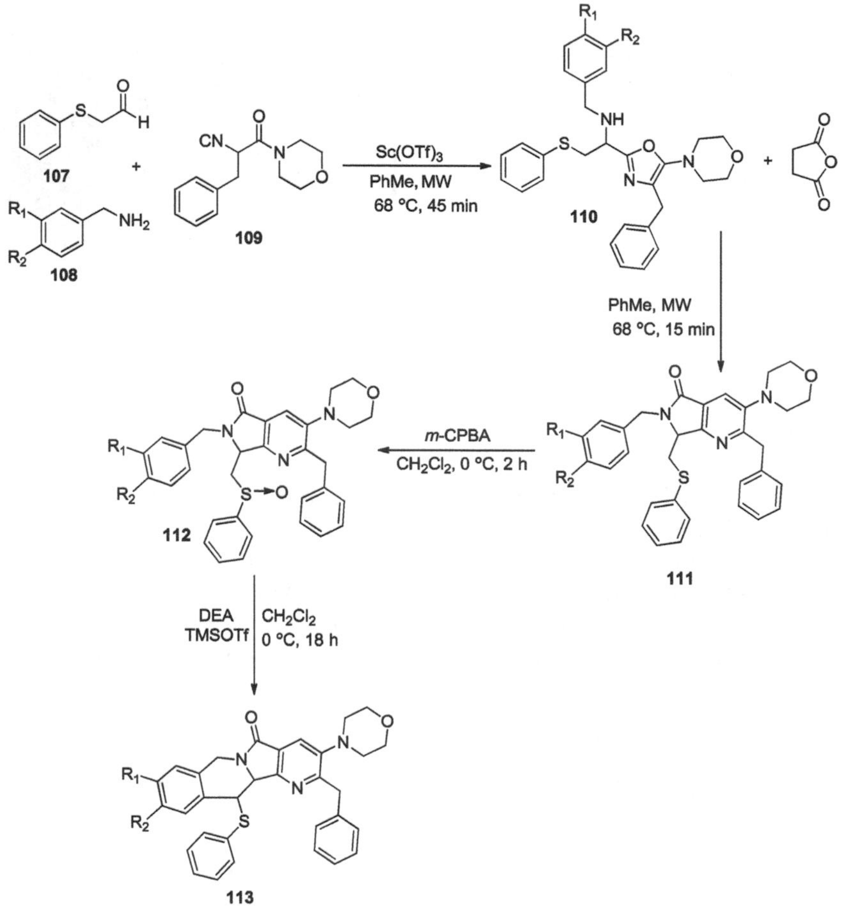

SCHEME 9.29 Synthesis of tetrahydroisoquinolinpyrrolopyridinones.

readily isolable isonitrile, and maleic anhydride via a triple process: Ugi-3CR-aza Diels–Alder/S-oxidation/Pummerer.

The synthesis of highly substituted tetrahydroepoxyisoindole carboxamides **117** using Ugi intramolecular Dielse–Alder reactions has been reported (Scheme 9.30).[49] The four component Ugi reaction of 2-furaldehyde **115**, an alkenoic acid **114**, an isonitrile and an amine affords rapid access to a family of acetylenic furan analogs **116**, which on

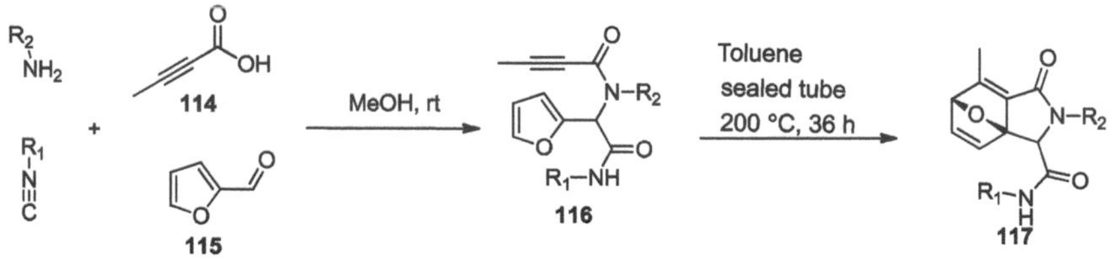

SCHEME 9.30 Synthesis of highly substituted tetrahydroepoxyisoindole carboxamides.

heating undergo an intramolecular Dielse–Alder (IMDA) reaction yielding highly substituted tricyclic lactams in good to excellent yields. This Ugi–IMDA reaction proved to be highly substituent tolerant across both the isonitriles and amines examined.

The synthesis of 3-hydroxypyrazoles **123** via hydrazine-mediated cyclization of Ugi products **121** has been reported (Scheme 9.31).[50] The reactions proceed via a tandem Ugi/debenzylation/hydrazine-mediated cyclization sequence. Herein, *n*-butyl isocyanide was utilized as an alternative to classical convertible isocyanides, enabling high-yielding hydrazine-mediated cyclization.

Chattopadhyay et al. reported a two-step reaction protocol for the synthesis of eight-member 1,5-benzodiazocine-2-ones **127** by Ugi four-center, three-component coupling reaction (U-4C-3CR) and subsequent reductive cyclization using Fe/NH_4Cl in protic solvent (Scheme 9.32).[51]

Spatz et al. used Ugi–Dieckmann reaction sequence to the synthesis of tetramic acid derivatives (Scheme 9.33).[52] In this reaction 1,1,-dimethyl-2-isocyanoethyl-methylcarbonate **128** was used as cleavable isocyanide for the Ugi-4CR.

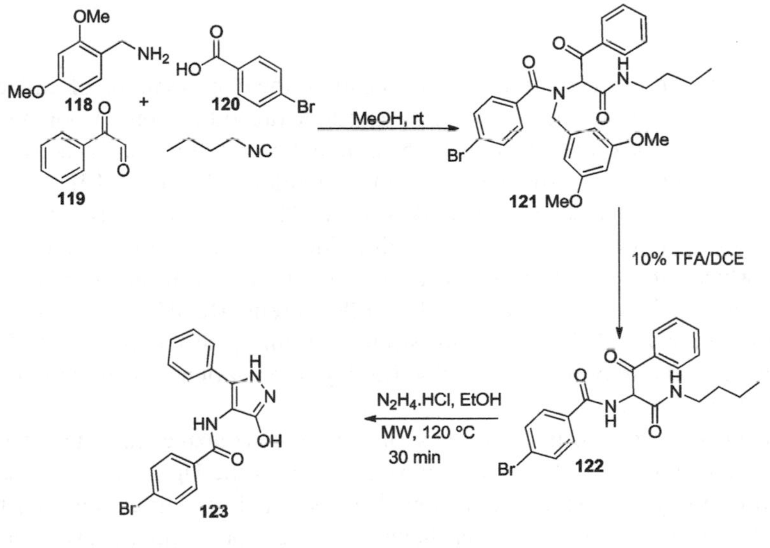

SCHEME 9.31 Synthesis of 3-hydroxypyrazoles via hydrazine mediated cyclization of Ugi products.

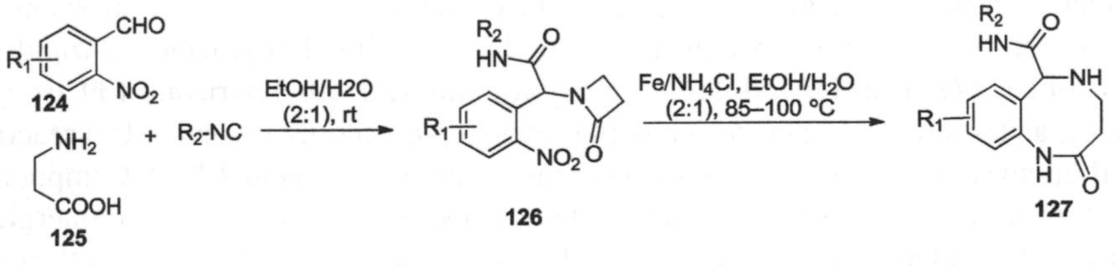

SCHEME 9.32 Synthesis of eight-membered 1,5-benzodiazocine-2-ones by Ugi four-center, three-component reaction.

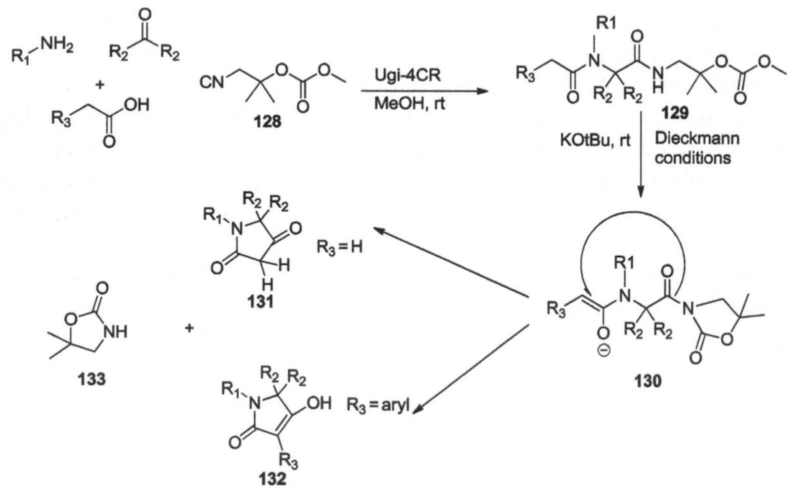

SCHEME 9.33 Synthesis of tetramic acids by an Ugi–Dieckmann condensation.

In the following postcondensation modification, the deprotonation of amide initiates the cyclization to the *N*-acyl-5, 5-dimethyloxazolidin-2-one. Upon attack by the enolized carboxylic acid moiety, 5-dimethyloxazolidin-2-one acts as leaving group and a Dieckmann-like cyclization to pyrrolidine-2,4-dione **131** or hydroxy-dihydropyrrolidone **132** structures takes place. The Ugi reaction is generally initiated by the condensation of amine with aldehyde, leading to an intermediate imine, which subsequently reacts with a CH-acidic carboxylic acid/acetic acid and isocyanide to afford the desired product **129**. Herein, MeOH turned out to be the best solvent for the MCR step. In this strategy, amines, carbonyls, and a CH-acidic carboxylic acid can be varied broadly, leading to compounds with three potential points of diversity.

The microwave assisted synthesis of triazadibenzoazulenones **137** has been reported (Scheme 9.34).[53] The methodology employs the Ugi reaction to assemble desired diversity, and acid treatment enables two tandem ring closing transformations. The order of ring closure is shown to be key for optimal conversion to the desired tetracyclic product and initially proceeds through a benzimidazole intermediate, followed by a second ring closure to give the desired fused benzodiazepine. The two-step protocol was further facilitated by microwave irradiation. Prudent selection of the isonitrile reagent enables the correct order of ring-forming events.

An efficient method to construct drug-like 2,3-dihydropyrazino[1,2-*a*]indole-1,4-diones **141** from 1*H*-indole-2-carboxylic acids **138**, ethyl pyruvate **139**, isocyanides, and primary amines via a one-pot, two-step procedure involving Ugi reaction and microwave assisted cyclization has been described (Scheme 9.35).[54] Compounds containing such motifs have been reported to possess cytotoxic,[22] melatoninergic,[23] antiviral,[24] and antifungal[25] activities. This broad medicinal relevance and the constrained peptidomimetic character make these new compounds attractive additions to any diversity set for biological screening.

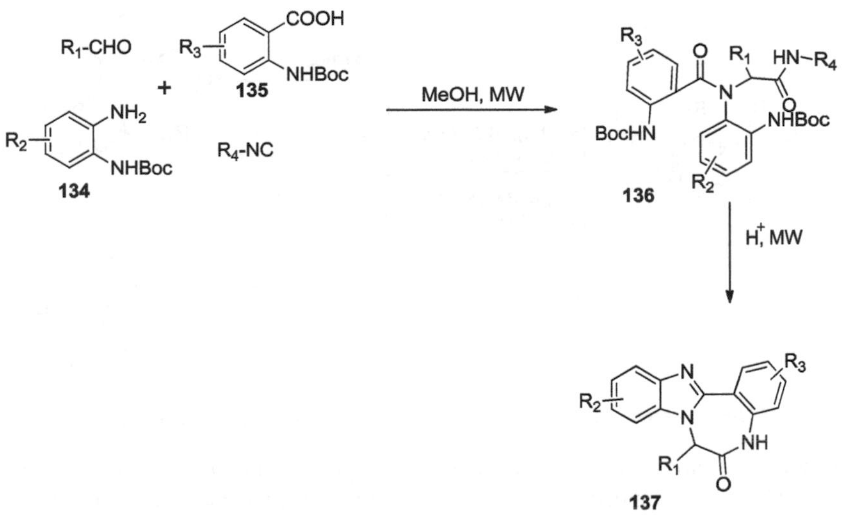

SCHEME 9.34 Synthesis of triazadibenzoazulenones via post-Ugi modification.

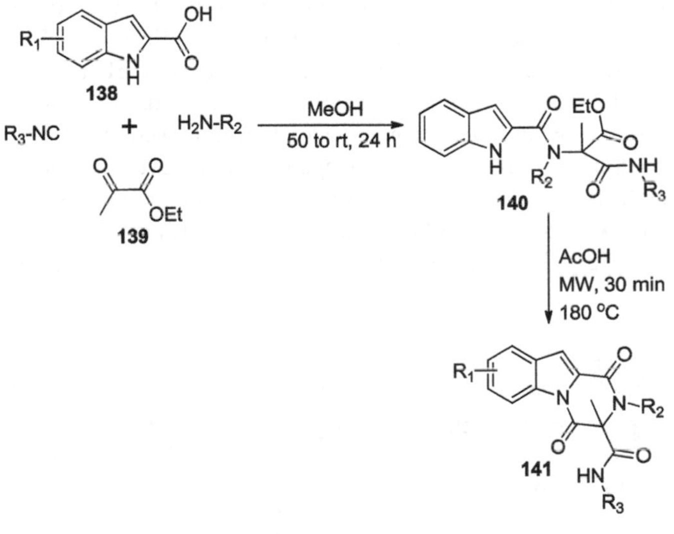

SCHEME 9.35 Synthesis of 2,3-dihydropyrazino [1,2-*a*]indole-1,4-diones via a one-pot, two-step procedure involving Ugi reaction and microwave-assisted cyclization.

Zhu et al. reported the synthesis of two distinct heterocycles via Ugi post-functionalization, from a single set of Ugi adducts **142** (Scheme 9.36).[55] Linear amides, prepared in one step by the Ugi four-component reaction, were converted to 3,4-dihydroquinoxalin-3-ones **143** or to 2-(2-oxoindolin-1-yl)acetamides **144**, dependent on the catalytic conditions. While microwave irradiation was found to be determinant on the reaction efficiency, the choice of ligand diverged the reaction pathways. Heating a solution of **142** in dioxane/MeCN (v/v) (85/15) under microwave

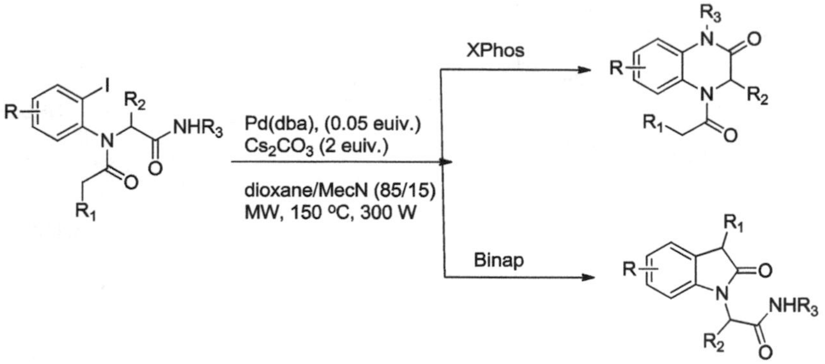

SCHEME 9.36 Synthesis of two distinct heterocycles via Ugi postfunctionalization.

irradiation conditions in the presence of $Pd(dba)_2$ (0.05 equiv.) and Cs_2CO_3 (2 equiv.), using XPhos as a supporting ligand, afforded the 3,4-dihydroquinoxalin-3-ones **143** via an intramolecular *N*-arylation of the secondary amide. On the other hand, using BINAP as ligand under otherwise identical conditions, intramolecular R–CH arylation of tertiary amide occurred to furnish the oxindoles **144**.

The synthesis of pseudopeptidic (S)-6-amino-5-oxo-1,4-diazepines and (S)-3-benzyl-2-oxo-1,4-benzodiazepines by an Ugi-4CC Staudinger/aza-Wittig sequence has been disclosed (Scheme 9.37).[56] The sequential Ugi reaction between *p*-substituted arylglyoxals **146**, alkylamines **147**, cyclohexyl isocyanide **152**, and 3-azido-(S)-2-(*tert*-butoxycarbonylamino)propanoic acid **145**, followed by a Staudinger/aza-Wittig cyclization in the presence of triphenylphosphine, gave rise to enantiomerically pure *N*-cyclohexyl 4 alkyl-2-aryl-5-oxo-(S)-6-(*tert*-butoxycarbonylamino)-4,5,6,7-tetrahydro-1*H*-1,4-diazepine-3-carboxamides **149**. By the same sequence, *p*-substituted benzaldehydes **150**, 2-aminobenzophenone **153**, cyclohexyl isocyanide **152**, and

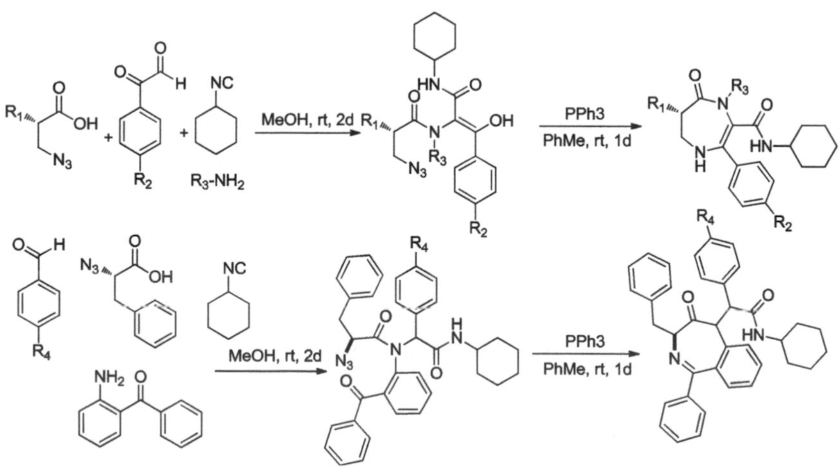

SCHEME 9.37 5-Oxo-1,4-diazepine-3-carboxamides by the Ugi 4CC/Staudinger/aza-Wittig sequence.

(S)-3-phenyl-2-azidopropionic acid gave rise to *N*-cyclohexyl 2-((S)-3-benzyl-2-oxo-5-phenyl-2,3-dihydro-1*H*-benzo[*e*][1,4]diazepin-1-yl)-(R/S)-2-arylacetamides **155**.

The tetrazolo fused diazepinones **161** have been synthesized via post-Ugi cyclization using isonitriles from the Baylis–Hillman adducts of acrylates (Scheme 9.38).[57] First, the substituted allyl isonitriles from the primary allyl amines afforded from the Baylis–Hillman adducts of acrylates is synthesized. Further, the Ugi reaction of these isonitriles with $TMSN_3$, aliphatic amines, and aldehydes or ketone affords 1-substituted tetrazoles, which have been demonstrated to be suitable substrates for producing tetrazolo fused diazepinones.

We have developed a new strategy for the syntheses of skeletal, diverse, *N*-fused polycyclic compounds via an Ugi-type MCR followed by a CuI catalyzed coupling reaction or tandem Pictet–Spengler reaction. This two-step sequence provides eight distinct skeletons of fused {6-5-5-6}, {5-5-5-6}, {6-5-6-6}, and {5-5-6-6} ring systems (Scheme 9.39).[58] The synthesis of Ugi-type product **164** was achieved by the condensation of aromatic heterocyclic 2-aminoazines, aldehydes, and isocyanides in methanol catalyzed by PTSA. Ugi product **164** was used for further elaboration (i.e., Cu catalyzed C–N coupling and Pictet–Spengler reaction). The optimal condition for the synthesis of product **166** CuI as catalyst, Cs_2CO_3 as base 1, 10-phenonthroline as ligand, and DMF as the solvent. The Pictet–Spengler reaction went properly in 50% TFA in dichloroethane (DCE) for the synthesis of product **165**.

We have also reported a novel ligand-free palladium catalyzed cascade reaction for the synthesis of highly diverse isoquinolin-1(2*H*)-one derivatives **168** from isocyanide and amide precursors synthesized by Ugi MCR **167** (Scheme 9.40).[59] A broad variety of acids, amines, and isocyanides were used as starting materials for

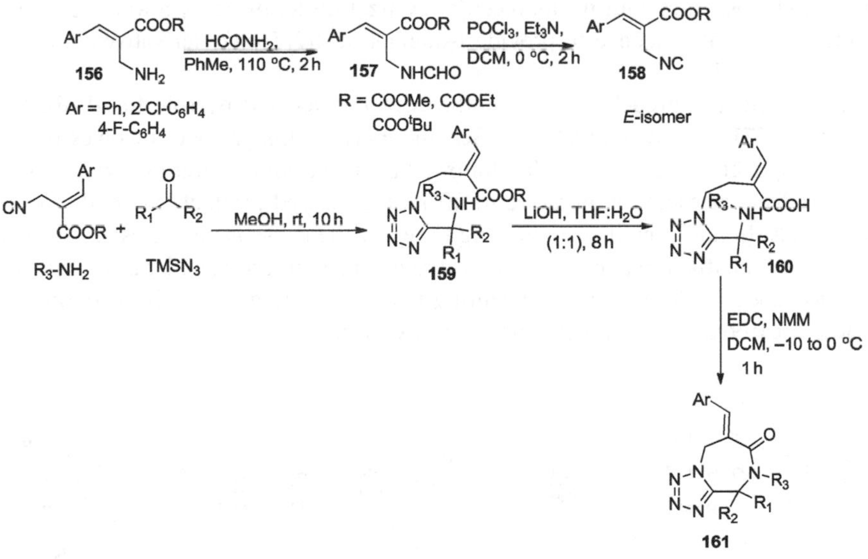

Ar, COOR, 156, NH2, HCONH2, PhMe, 110 °C, 2h, Ar, COOR, 157, NHCHO, POCl3, Et3N, DCM, 0 °C, 2h, Ar, COOR, 158, NC
Ar = Ph, 2-Cl-C6H4, 4-F-C6H4
R = COOMe, COOEt, COOtBu
E-isomer
CN, Ar, COOR, +, R1, O, R2, R3-NH2, TMSN3, MeOH, rt, 10h, 159, LiOH, THF:H2O (1:1), 8h, 160, EDC, NMM, DCM, −10 to 0 °C, 1h, 161

SCHEME 9.38 Synthesis of tetrazolo fused diazepinones via post-Ugi cyclization using isonitriles from the Baylis–Hillman adducts of acrylates.

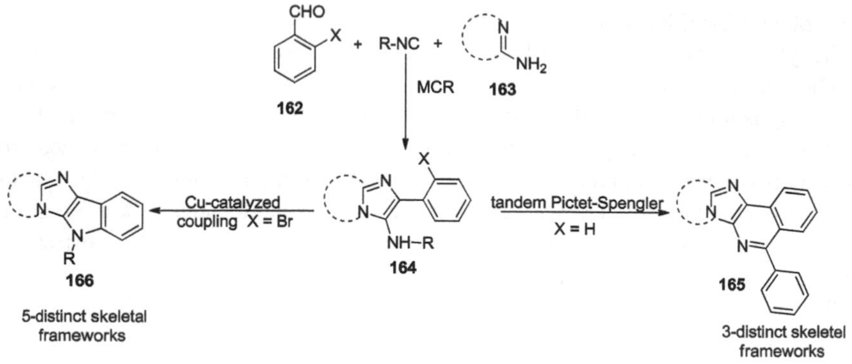

SCHEME 9.39 Two-step synthesis of *N*-fused polycyclic heterocycles.

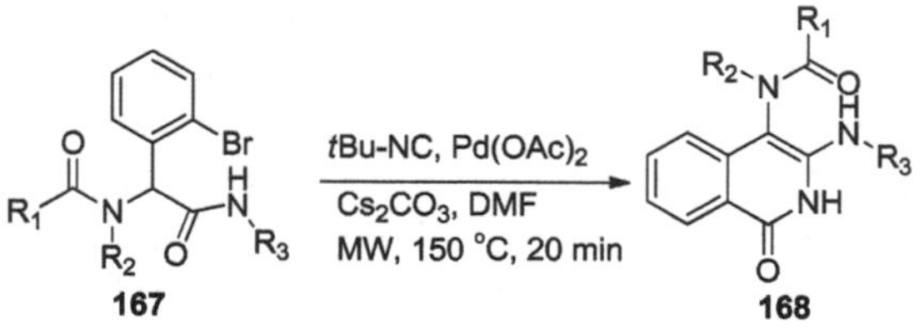

SCHEME 9.40 Synthesis of substituted isoquinolin-1(2*H*)-one via a Pd catalyzed coupling reaction of amide and isocyanide.

Ugi MCR, leading to various amide precursors, which in turn provided entry into diverse isoquinolin-1(2*H*)-one derivatives. The reaction proceeds through tandem isocyanide insertion with intramolecular cyclization followed by a Mazurciewitcz–Ganesan type of sequence to provide isoquinoline-1(2*H*)-one derivatives in moderate to good yield.

An efficient approach for the synthesis of indole- and pyrrole fused diketopiperazines **171** has been developed (Scheme 9.41).[60] This protocol involves the Ugi four-component reaction (U-4CR) followed by an intramolecular cyclization of the Ugi products at room temperature to afford the desired products in good to excellent yields. In addition, it is interesting to report the subsequent regioselective ring-opening of the diketopiperazine unit occurring via an intermolecular transamidation reaction under mild conditions, resulting in the formation of highly functionalized indole-2-carboxamides and pyrrole-2-carboxamides **172**.

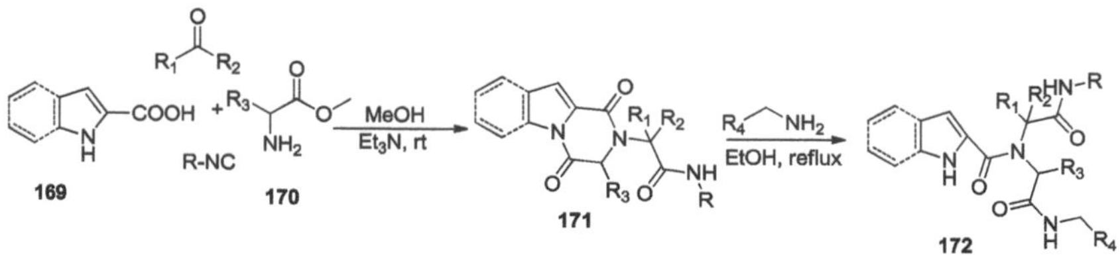

SCHEME 9.41 Synthesis of indole fused DKPs by utilizing Ugi-4CR.

The synthesis of tricyclic *N*-heterocycles via Ugi reaction followed by tandem SN^2–Heck double cyclization was reported by Riva et al. (Scheme 9.42).[61]

Ramazani et al. reported the synthesis of 2,5-disubstituted 1,3,4-oxadiazolederivatives **182** by a Ugi-4CR/aza-Wittig sequence using (*N*-isocyanimino)triphenylphosphorane **181**, a secondary amine **179**, a carboxylic acid **180**, and an aromatic aldehyde **181** in CH_2Cl_2 at ambient temperature in high yields without using any catalyst or activation (Scheme 9.43).[62]

The synthesis of indolo[1,2-*a*]quinoxalinones **186** by sequential Ugi/Ullmann type reaction catalyzed by CuI/L-proline has been described (Scheme 9.44).[63] The procedures combine the Ugi four-component reaction of aldehydes, 2-iodoaniline, 2-indole carboxylic acid, and isocyanides followed by the copper catalyzed intramolecular *N*-arylation of the Ugi product in a one-pot procedure, which afforded the desired products in good to excellent yields.

The Balalaic group described an efficient method for the stereoselective synthesis of 3-(diarylmethylene)-2-oxindoles and 3-(arylmethylene)-2-oxindoles **189** via a carbopalladation reaction. In this approach, an Ugi four-component reaction (4-CR), adduct **188** was used as the starting material. A one-pot sequence involving intermolecular carbopalladation C–H activation/C–C bond formation efficiently afforded the oxindole derivatives (Scheme 9.45).[64]

A method for the chemoselective reduction of Ugi-type lactam amides **191** at the lactam carbonyl functionality with borane complexes has been developed (Scheme 9.46).[65] The novel reduction products can be further manipulated synthetically to yield various novel N and C terminally active unnatural amino acid **192** building blocks.

Wessjohann and co-workers described a Pd II/IV catalyzed oxidative cyclization of 1,6-enynes **195** derived by the Ugi four-component reaction to produce *N*-substituted 3-aza-bicyclo[3.1.0]hexan-2-ones **196** (Scheme 9.47).[66] Different

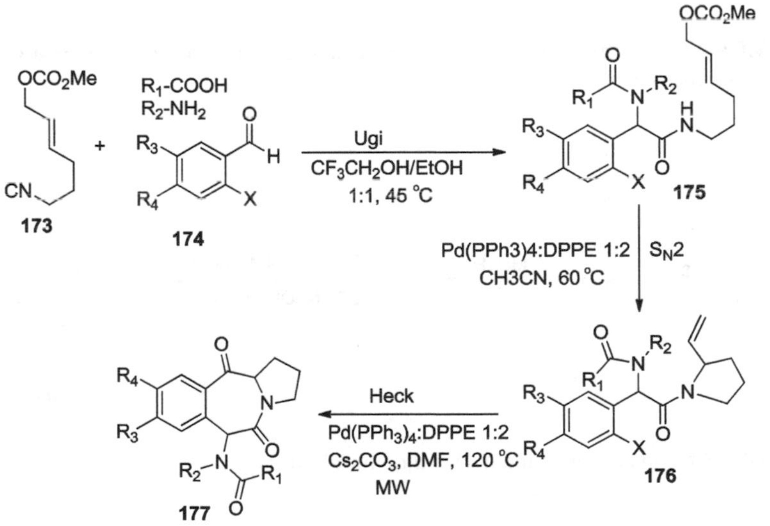

OCO$_2$Me CN 173 + R_1-COOH R_2-NH$_2$ R_3 R_4 H O X 174

Ugi, CF$_3$CH$_2$OH/EtOH 1:1, 45 °C → 175

Pd(PPh3)4:DPPE 1:2, CH3CN, 60 °C; S$_N$2 → 176

Heck, Pd(PPh$_3$)$_4$:DPPE 1:2, Cs$_2$CO$_3$, DMF, 120 °C, MW → 177

SCHEME 9.42 Synthesis of tricyclic *N*-heterocycles via Ugi reaction with a tandem SN^2–Heck double cyclization.

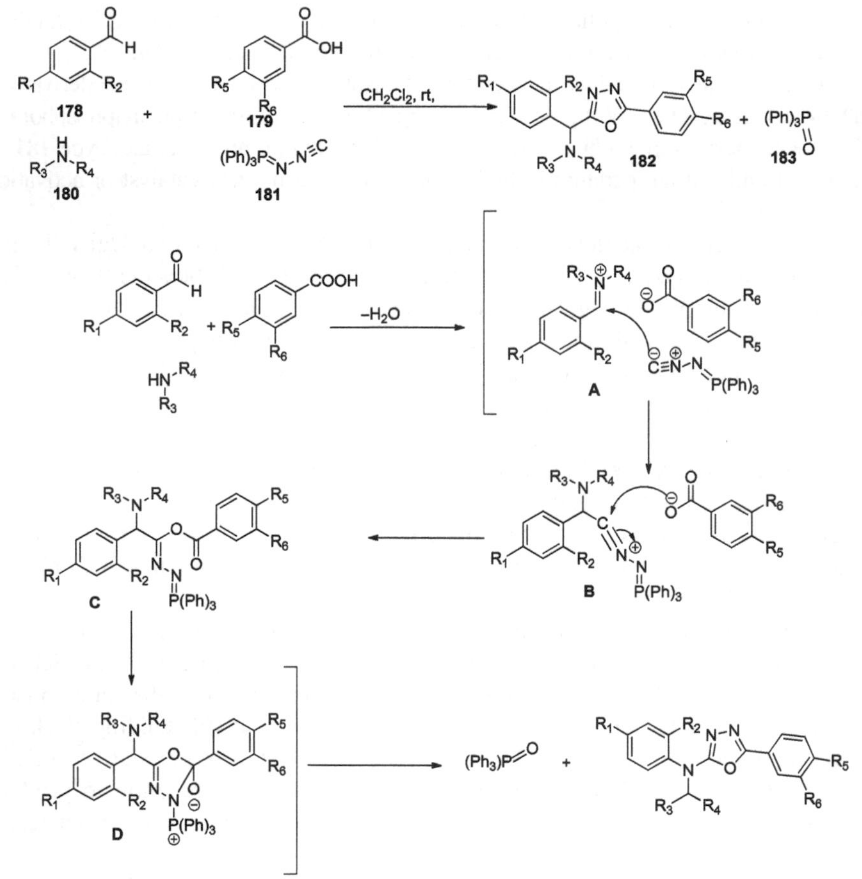

SCHEME 9.43 Synthesis of 2,5-disubstituted 1,3,4-oxadiazolederivatives by a Ugi-4CR/aza-Wittig sequence.

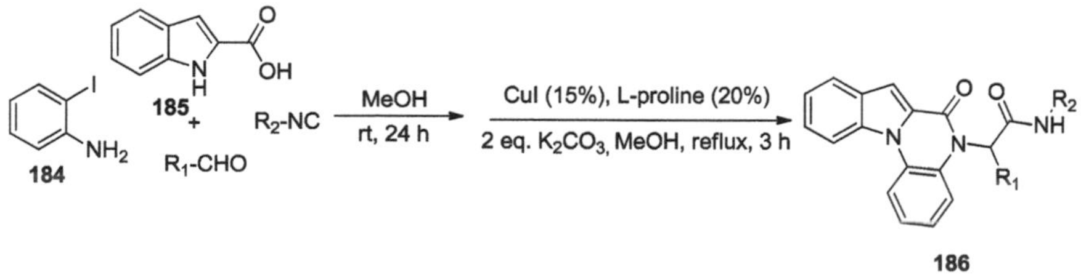

SCHEME 9.44 Synthesis of indolo[1,2-*a*]quinoxalinones.

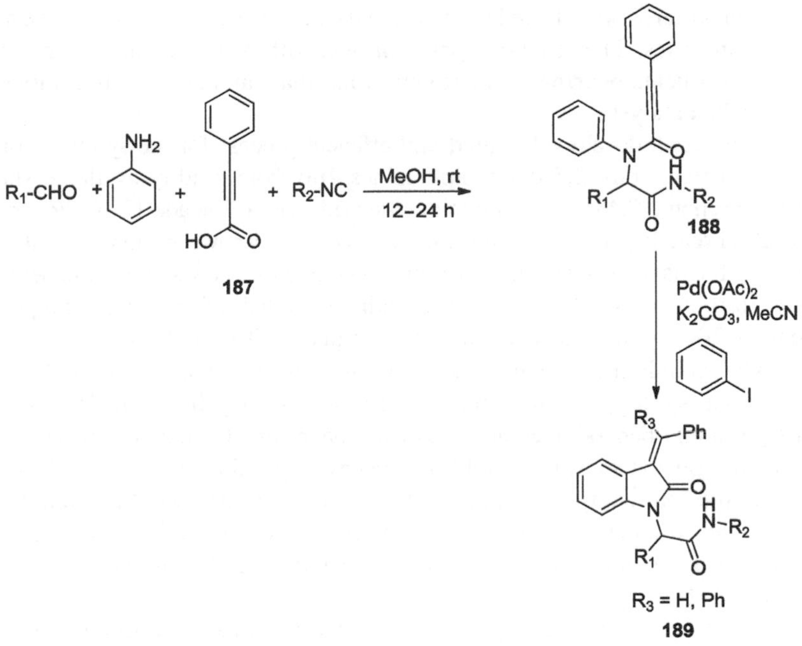

SCHEME 9.45 Stereoselective synthesis of 3-(diarylmethylene)-2-oxindoles and 3-(arylmethylene)-2-oxindoles.

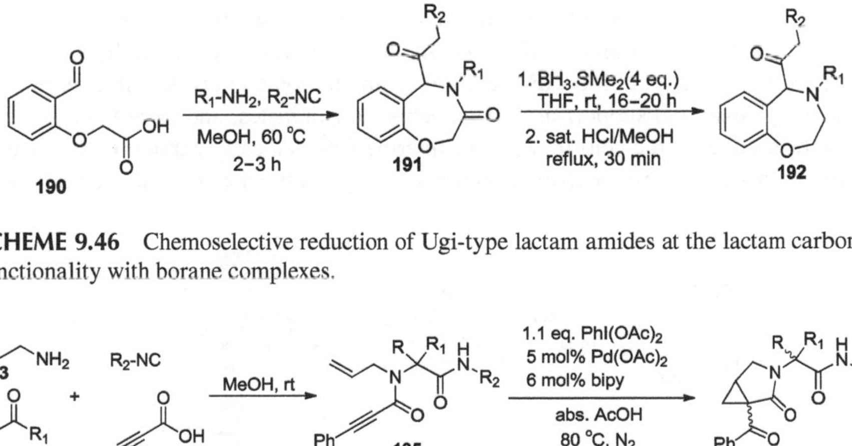

SCHEME 9.46 Chemoselective reduction of Ugi-type lactam amides at the lactam carbonyl functionality with borane complexes.

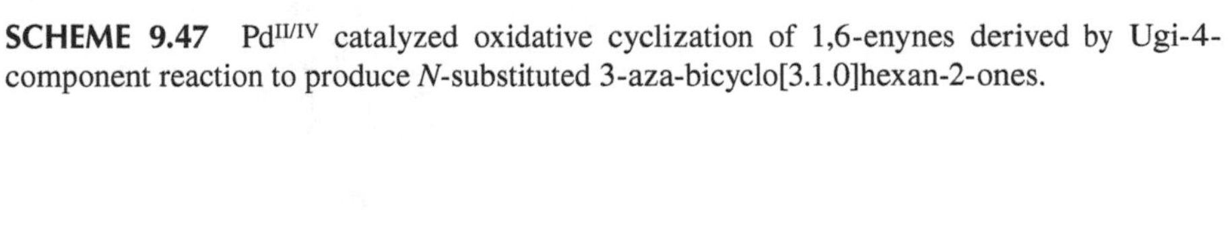

SCHEME 9.47 $Pd^{II/IV}$ catalyzed oxidative cyclization of 1,6-enynes derived by Ugi-4-component reaction to produce *N*-substituted 3-aza-bicyclo[3.1.0]hexan-2-ones.

substitution patterns were tested to examine the scope and limitations of the amide tethered substrates. The reactions perform best with benzylic amide substituents or with substituents bearing a functional group that can act as an intramolecular ligand for the catalyst.

Andreana et al. developed a rapid and efficient process for the synthesis of biologically relevant spiro-2,5-diketopiperazines **200** from readily available starting materials (Scheme 9.48).[67] The reaction proceeds via a cascade U-4CC/6-exo-trig aza-Michael reaction in water under the influence of microwaves and generates four contiguous bonds, a tertiary spiro carbon center, and one new stereogenic center.

The Torroba group described the synthesis of helix-forming pseudopeptidic hydantoins **206** by an Ugi 4CC/cyclization/reduction/Ugi 4CC sequence of reactions, giving mainly the L-adduct when benzoic acids were used (Scheme 9.49).[68]

A two-step strategy for the synthesis of arrays of tricyclic tetrazolo fused benzodiazepines and benzodiazepinones **211** has been investigated. The protocol uses *ortho-N*-Boc phenylisocyanides **207** and phenyl glyoxaldehydes or ethyl glyoxylate **208** in the four-component Ugi–azide reaction to afford MCR-derived adducts equipped with the desired diversity inputs. A subsequent acidic treatment (TFA/DCE) allows a simultaneous deprotection–cyclization, leading to the final products (Scheme 9.50).[69]

Seganish et al. reported the synthesis of dihydroimidazo[5,1-*a*]isoquinolines **214** using an Ugi–Bischler–Napieralski reaction sequence (Scheme 9.51).[70] A variety of modifications were tolerated on the phenyl ring. Electron-withdrawing and electron-donating groups were well tolerated in the reaction. The coupling–cyclization sequence also tolerated a variety of aromatic and heteroaromatic aldehydes.

This group also reported the concise synthesis of a series of 2,4,5-trisubstituted oxazoles **220** via a tandem Ugi/Robinson–Gabriel sequence (Scheme 9.52).[71] 2,4-Dimethoxybenzylamine **215** was used as an ammonia equivalent in combination with arylglyoxal and supporting Ugi reagents, an isonitriles, and carboxylic acid. As such the product of the acid-treated Ugi intermediate is ideally configured to undergo a Robinson–Gabriel cyclodehydration reaction to yield the desired oxazole scaffold.

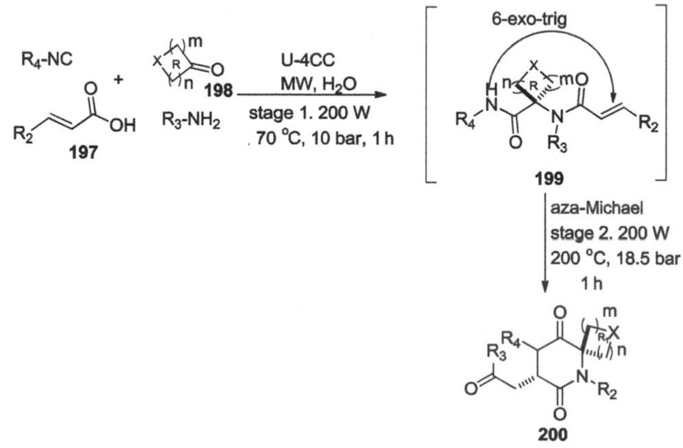

SCHEME 9.48 Synthesis of biologically relevant spiro-2,5-diketopiperazines.

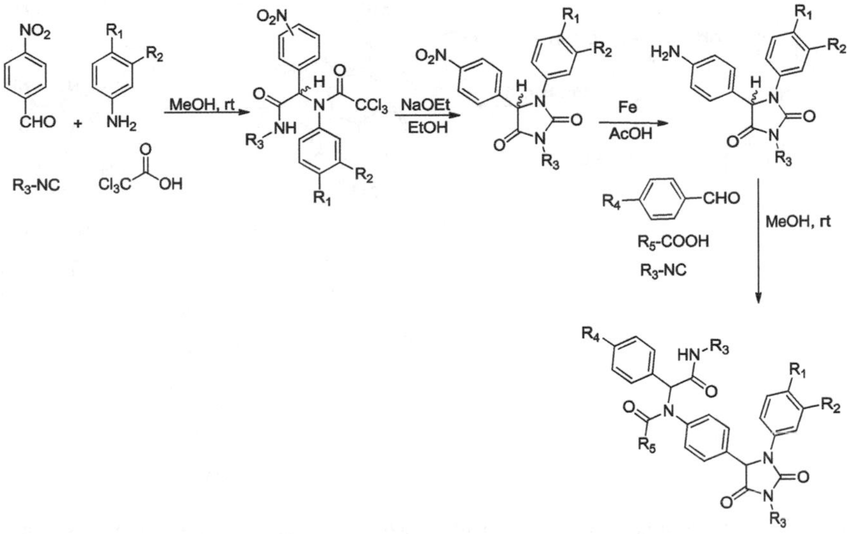

SCHEME 9.49 Synthesis of helix-forming pseudo-peptidic hydantoins by an Ugi 4CC/cyclization/reduction/Ugi 4CC sequence of reactions.

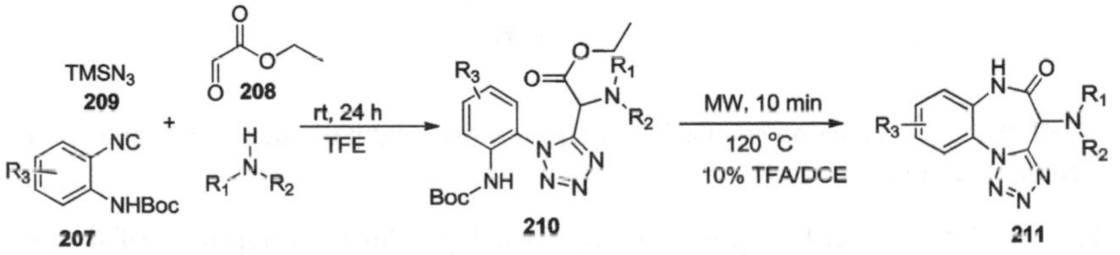

SCHEME 9.50 Synthetic route toward tetrazolo fused benzodiazepinones.

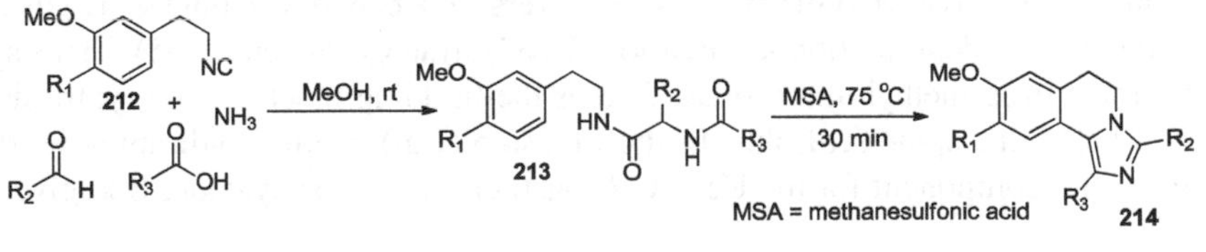

SCHEME 9.51 Synthesis of dihydroimidazo[5,1-*a*]isoquinolines using Ugi–Bischler–Napieralski reaction sequence.

Welsch et al. reported the synthesis substituted indazolones **224** via palladium- and copper catalyzed cyclizations of hydrazine-derived Ugi products (Scheme 9.53).[72] The cyclizing coupling can be applied to Boc protected, alkyl substituted, and free hydrazides **222**.

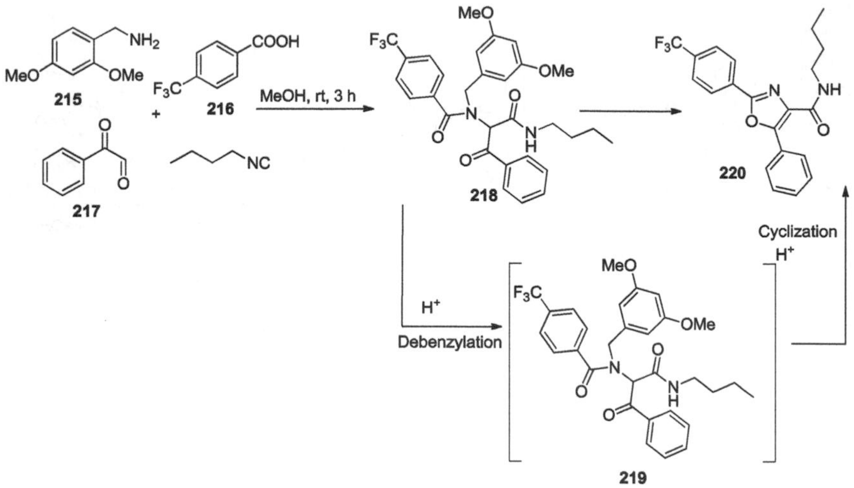

SCHEME 9.52 Synthesis of 2,4,5-trisubstituted oxazoles via a tandem Ugi/Robinson–Gabriel sequence.

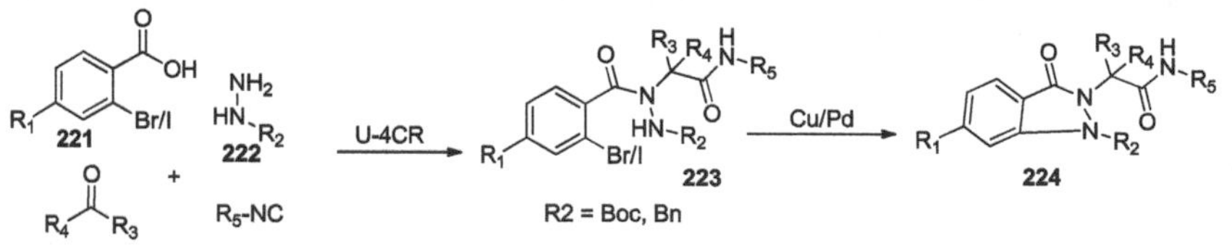

SCHEME 9.53 Indazolone synthesis via Ugi reaction and palladium- or copper catalyzed amidation/amination.

The synthesis of chloroquine analogs via Ugi–Smiles couplings of 4-hydroxy and mercapto pyridine derivatives **227** has been developed by Kaím et al. (Scheme 9.54).[73] Classical conditions were used, and methanol was selected as a solvent of choice to solubilize all the partners. The reduction could be performed using BH_3 $3Me_2S$ in THF or with Raney nickel in ethanol.

Domling et al. reported the synthesis of diverse 1,4-benzodiazepine scaffolds **231** utilizing the Ugi deprotection–cyclization (UDC) strategy (Scheme 9.55).[74] This synthetic route used methyl anthranilate **228** as the building block for the synthesis of the 1,4-benzodiazepine scaffolds. In the first step (Ugi), methyl anthranilate serves as an amine component for the Ugi-4CR together with an isocyanide, Boc-glycinal

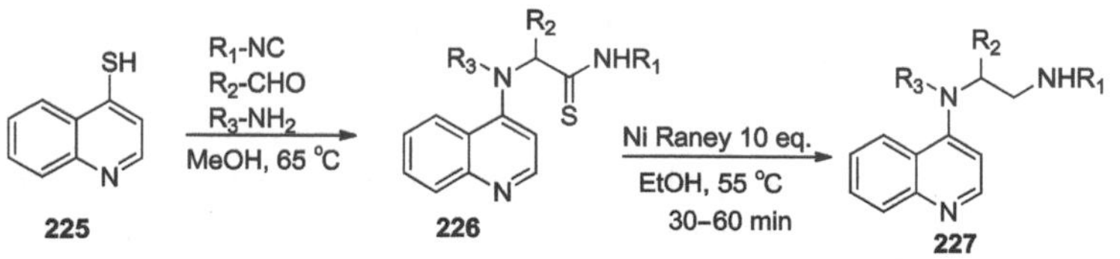

SCHEME 9.54 Synthesis of chloroquine analogs via Ugi–Smiles couplings.

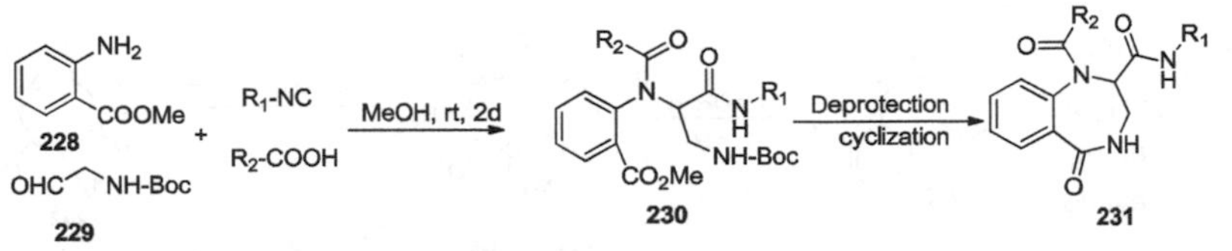

SCHEME 9.55 Synthesis of diverse 1,4-benzodiazepine scaffolds.

229, and a carboxylic acid. The Boc protection group is cleaved in the second step (deprotection), and then the free amine group is condensed with the orthogonal ester group to form the 1,4-diazepine ring in the third step.

Banfi et al. reported a tandem Ugi/Mitsunobu protocol, starting from *o*-aminophenols, α-hydroxy acids, amines, and aldehydes that gives benzo[*b*][1,4]oxazin-3-ones in two high-yielding steps (Scheme 9.56),[75] with the introduction of up to four diversity inputs. The mildness of the methodology allows the stereospecific synthesis of enantiomerically pure products as well as the introduction of additional functional groups. The overall procedure can also be carried out in a one-pot manner.

A new class of α,β-unsaturated pyran-2-carboxamides **238** has been synthesized by a multicomponent reaction, followed by a ring-closing metathesis (RCM) using a ruthenium catalyst (Scheme 9.57).[76] In the first step, α-acyloxy carboxamides **237** with two terminal double bonds were formed from terminal unsaturated carboxylic acids **235**, allyl ketones **236**, and isocyanides (Passerini reaction, P-3CR).

De Borggraeve et al. developed a strategy for synthesizing 3,4-dihydro-1*H*-pyrido[2,3-*e*][1,4]diazepine-2,5-dione compounds **243** starting from 2-hydroxynicotinic acid and by using an Ugi reaction as a key step in the synthesis (Scheme 9.58).[77] They used 2-isocyanophenyl benzoate instead of Armstrong's convertible isocyanide in this multicomponent reaction.

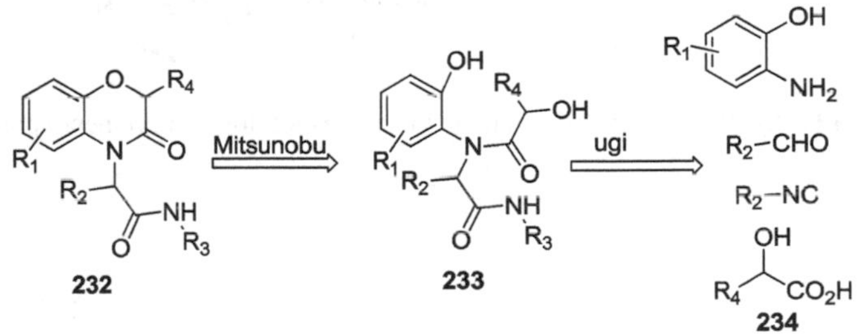

SCHEME 9.56 Two-step synthesis of the benzoxazinones.

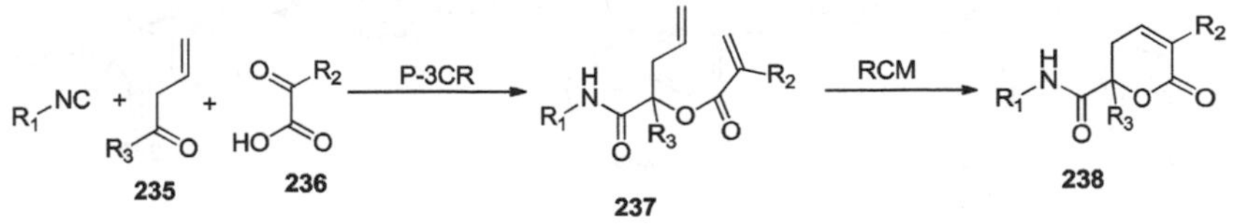

SCHEME 9.57 Synthesis of α,β-unsaturated lactones.

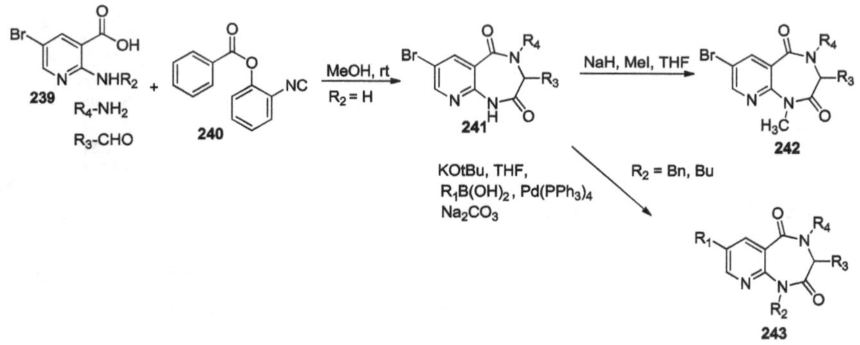

SCHEME 9.58 Synthesis of 3,4-dihydro-1*H*-pyrido[2,3-*e*][1,4]diazepine-2,5-dione.

A phosphite-mediated synthesis of benzimidazole **245** via a one-pot, four-component approach from nitrophenols has been described (Scheme 9.59).[78]

Eycken et al. reported a diversity-oriented approach to spiro indolines **250** via post-Ugi gold catalyzed diastereoselective domino cyclization (Scheme 9.60).[79]

An efficient multicomponent reaction toward synthesis of the schistosomiasis drug praziquantel **260** has been developed using the Ugi four-component reaction followed by the Pictet–Spengler reaction (Scheme 9.61).[80]

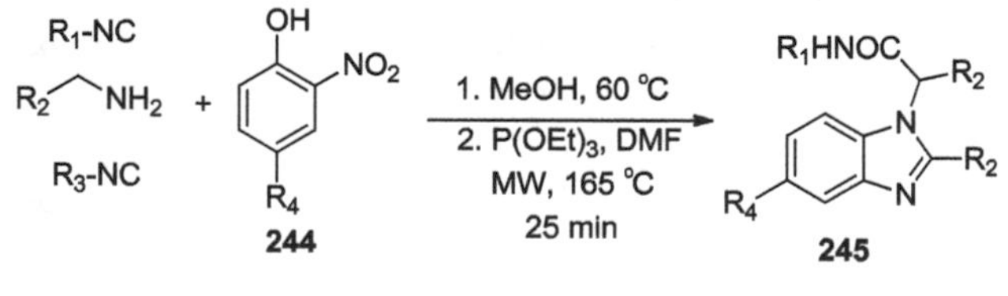

SCHEME 9.59 Synthesis of benzimidazole via a one-pot, four-component approach.

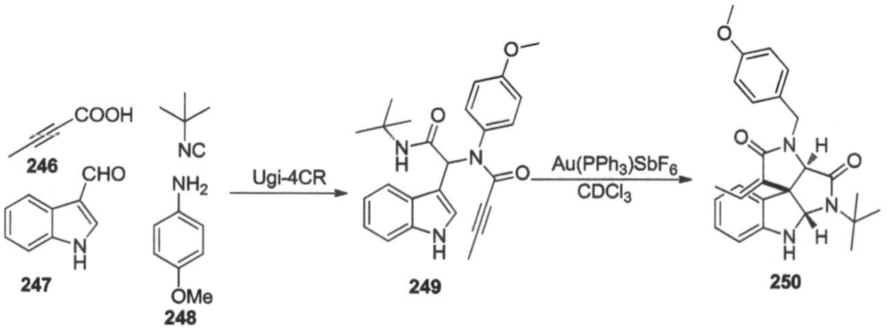

SCHEME 9.60 A diversity-oriented approach to spiro indolines.

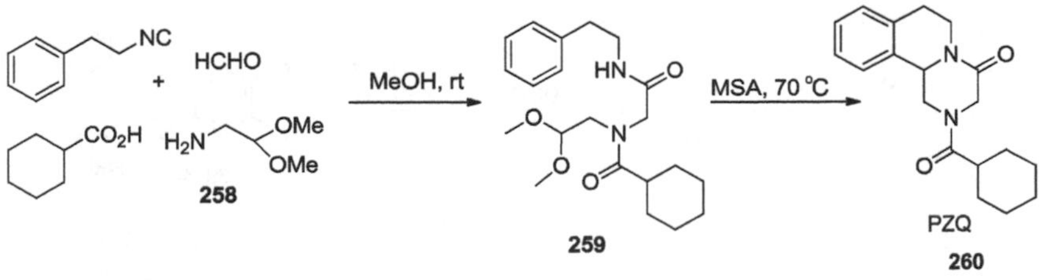

SCHEME 9.61 Stepwise Ugi four-component reaction and Pictet–Spengler reaction to yield the schistosomiasis drug praziquantel.

9.2.3 IMCRs Other Than Post-Ugi and -Paserini

Recently, various isocyanide based multicomponent reactions other than Ugi and Paserini have been reported concerning the synthesis of a diverse *N*-heterocycle. In this context, Soleimani et al. developed two new isocyanide based multicomponent reactions for the synthesis of a wide range of alkyl-2-(1-(alkylcarbamoyl)-2,2-dicyanoethyl)benzoate derivatives **262** and isochromeno[3,4-*b*]pyrrole derivates **263** from 2-formylbenzoic acid **261**, malononitrile, and isocyanides in dichloromethane and alcohol, respectively (Scheme 9.62).[81] This high atom economy reaction led to the construction of two carbon–carbon bonds, one amide, and one ester group in a single synthetic step.

This group also developed a synthetic strategy to prepare substituted imidazo[1,2-*a*]quinoxalines **267**. The conceptually new protocol includes two coupled IMCR processes. The simplicity of the reaction design and the possibility to synthesize novel heterocycles with four diversity elements in only four chemical operations (including two chromatographic purifications) makes the described methodology a tool of choice to construct these medicinally relevant compounds. In the first step, aromatic 1,2-diamines **264**, aldehydes, and isocyanide reactants led to redox-unstable 1,4-dihydroquinoxalines. These were quickly oxidized with DDQ into stable quinoxalines **266** containing three elements of diversity resulting from the three reactants. If a convertible isocyanide, such as *tert*-butylisocyanide, were used to construct quinoxalines using the same two-step sequence, then the R_{conv} group could be removed (still leaving behind two diversity elements from the first IMCR) and the resulting 2-aminoquinoxalines could serve as substrates for Groebke–Blackburn MCR (i.e., the second IMCR in a row) (Scheme 9.63).[82]

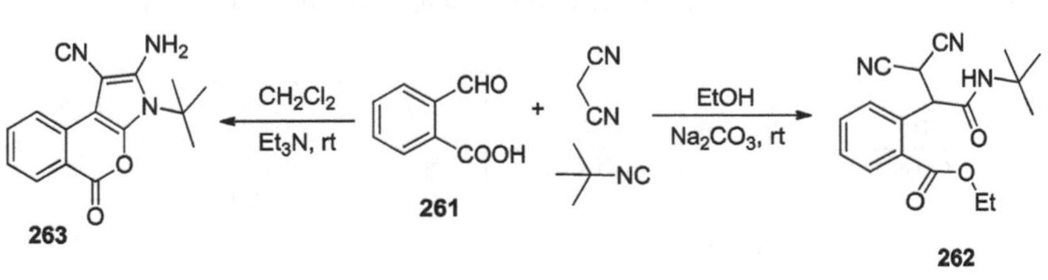

SCHEME 9.62 Synthesis of alkyl-2-(1-(alkylcarbamoyl)-2,2-dicyanoethyl)benzoate derivatives and isochromeno[3,4-*b*]pyrrole derivates.

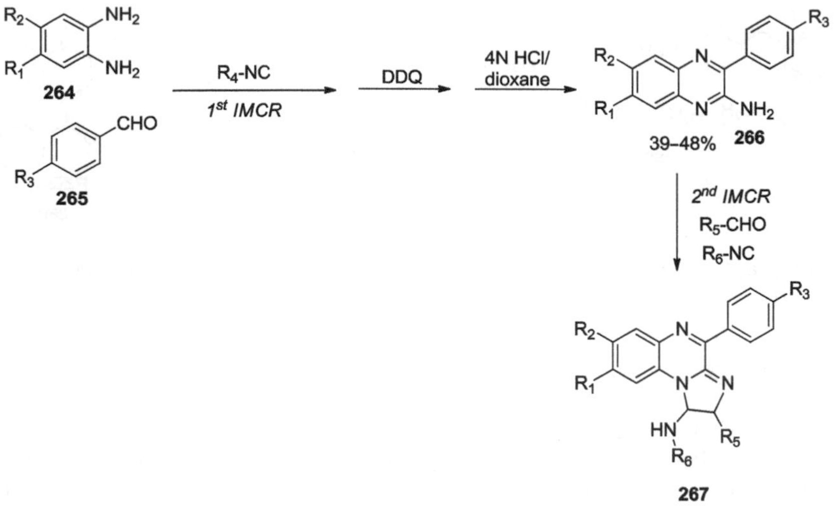

SCHEME 9.63 Synthesis of imidazo[1,2-*a*]quinoxalines accessed via two sequential isocyanide based multicomponent reactions.

The synthesis of an imino-pyrrolidine-thione **270** scaffold via the coupling of isocyanides, heterocyclic thiols **269**, and gem-dicyano olefins **268** has been developed (Scheme 9.64).[83] Smiles rearrangement followed by intramolecular cyclization leads directly to formation of the core structure. A water-acceleration effect is observed, promoting most of the reactions to go to completion within a short reaction time.

A three-component strategy starting from isocyanides allows a straightforward synthesis of a five-membered ring heterocycles **273** (Scheme 9.65).[84] This cascades reaction involves the addition of a nitrogenated nucleophile, an azide or a tetrazole,

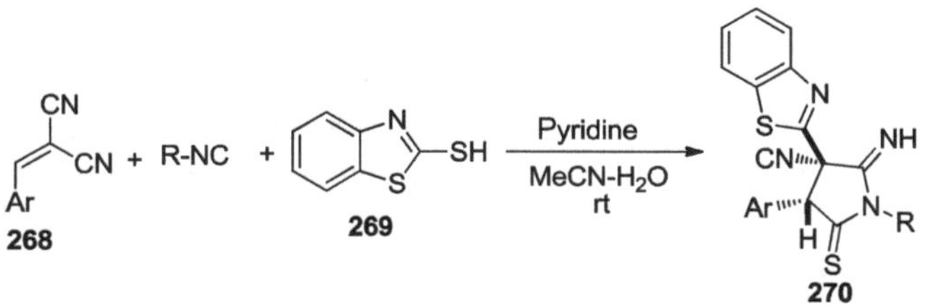

SCHEME 9.64 Synthesis of imino-pyrrolidine-thione scaffold via the Ugi reaction.

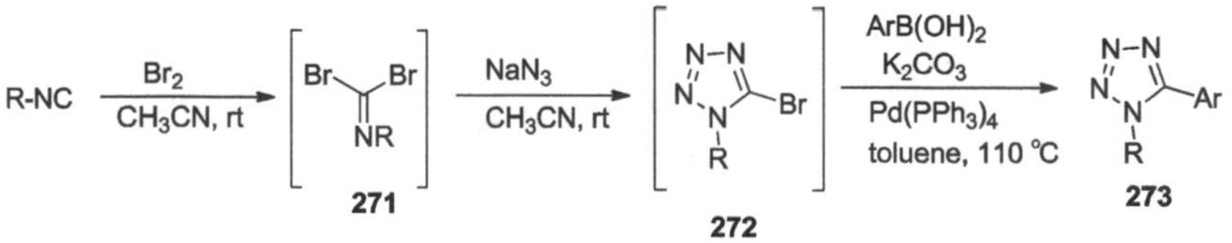

SCHEME 9.65 Three-component strategy starting from isocyanides allows a straightforward synthesis of five-membered ring heterocycles.

an isocyanide dibromides, an electrocyclization, and a Suzuki coupling, which afford new accesses to tetrazole and triazole scaffolds.

A three-component reaction toward the easy synthesis of structurally diverse 1-carboxamido-isoindoles **275** has been described (Scheme 9.66).[85]

Ukaji et al. reported a catalyst-free [5+1] cycloaddition reaction between isocyanides and C,N-cyclic N^1-acyl azomethine imines as the "isocyanophile" leading to novel heterocycles **277** (Scheme 9.67).[86] These reactions proceeded quickly and cleanly to afford the corresponding imine-1,3,4-oxadiazin-6-one derivatives in high to excellent yields. A wide range of C,N-cyclic N^1-acyl azomethine imines **276** and isocyanides were applicable to this reaction.

The synthesis of phenylimidazoquinoxalines **281** via a one-pot, two-step MCR process has been reported (Scheme 9.68).[87] In this reaction four new chemical bonds and two aromatic rings are formed in a one-pot fashion.

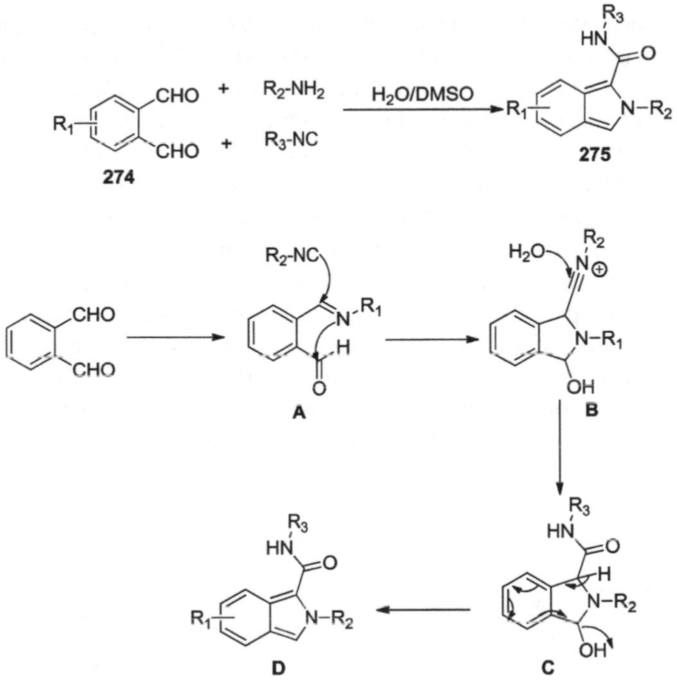

SCHEME 9.66 Synthesis of structurally diverse 1-carboxamido-isoindoles.

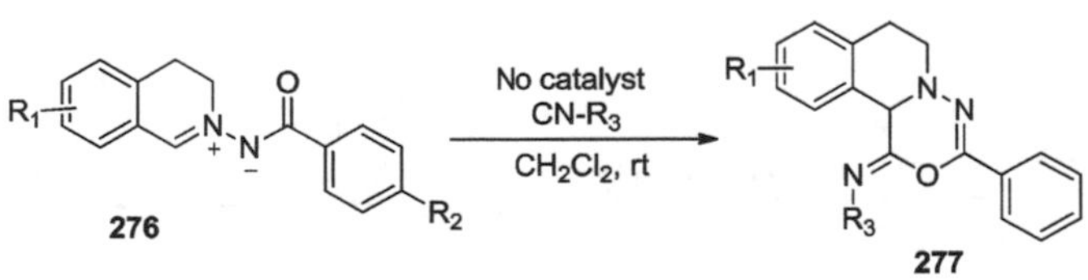

SCHEME 9.67 A catalyst-free [5+1] cycloaddition reaction between isocyanides and C,N-cyclic N^1-acyl azomethine imines.

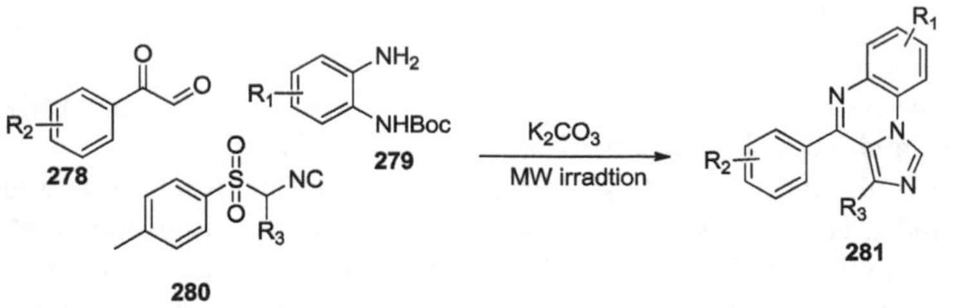

SCHEME 9.68 Synthesis of phenylimidazoquinoxalines via the Van Leusen 3-CR deprotection-cyclization in one-pot route.

A one-pot, three-component coupling was accomplished via the nucleophilic addition of an alkylsamarium(III) species to isocyanides and the subsequent addition of the resultant imidoyl samarium(III) species to isocyanates under mild conditions for the formation of α-iminocarboxamides **284**. The developed sequential C–C bond-forming procedure enabled the rapid synthesis of the α-iminocarboxamides in good to excellent yields from readily available starting materials (Scheme 9.69).[88]

A variety of functionalized imidazo[1,2-*a*]pyridines **288** have been synthesized via Namboothiri et al. through a one-pot, room-temperature, and reagent-free reaction between Morita–Baylis–Hillman (MBH) acetates of nitroalkenes **285** and 2-aminopyridines **286** (Scheme 9.70).[89] The reaction involves a cascade of inter/intramolecular double aza-Michael addition of 2-aminopyridines to MBH acetates. The methodology is marked by excellent yield, regioselectivity, and adaptability to synthesize imidazopyridine-based drug molecules such as alpidem and zolpidem.

Miranda et al. reported a practical two-step synthesis of 2,3-dihydropyrroles **290** from Ugi 4-CR/propargyl adducts (Scheme 9.71).[90] The protocol includes a base-mediated formation of an allenamide functional group and an *in situ* metal-free formal 5-endo cycloisomerization that occurs in a highly regioselective manner at the allenamide C–γ.

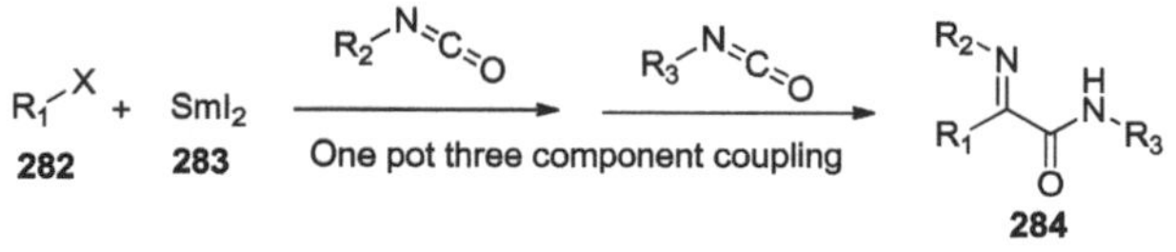

SCHEME 9.69 Synthesis of α-iminocarboxamides.

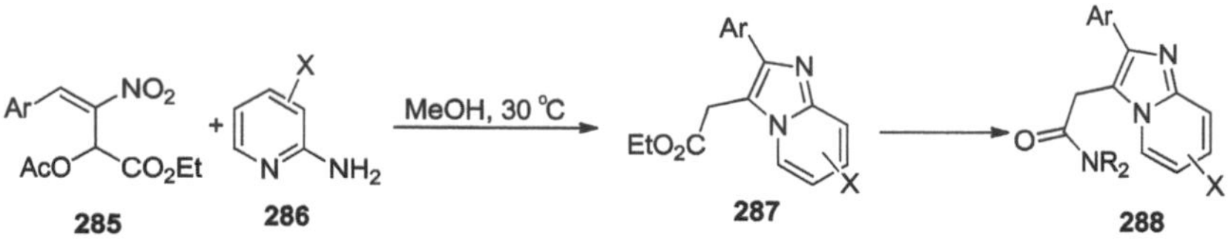

SCHEME 9.70 Synthesis of functionalized imidazo[1,2-*a*]pyridines.

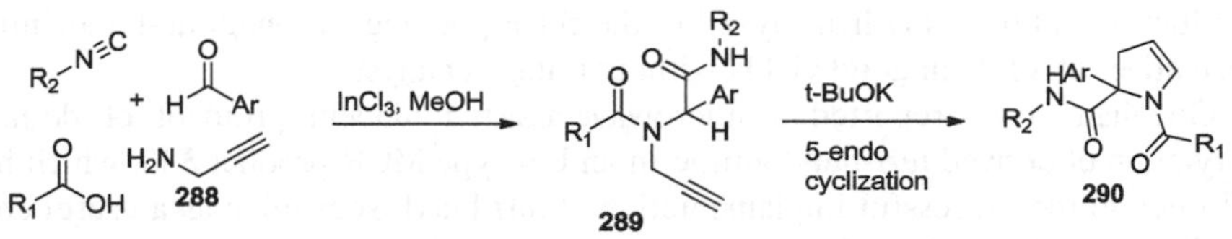

SCHEME 9.71 Two-step synthesis of 2,3-dihydropyrroles.

An unconventional oxazole formation **292** from isocyanides has been reported. The coupling of an acyl chloride with an isocyanide affords 2,5-disubstituted oxazoles under mild basic conditions instead of 4,5-disubstituted derivatives when using Schöllkopf conditions (butyllithium); this reaction constitutes a remarkable example of a base-induced chemoselective process in isocyanide chemistry (Scheme 9.72).[91]

A three-component reaction of the zwitterion was generated from dialkyl acetylenedicarboxylate **296** and an alkyl or aryl isocyanide with an alkyl or aryl sulfonamide to the synthesis of bifunctional sulfonamide-amide compounds (Scheme 9.73).[92] The reaction afforded the corresponding special type of ketenimine sulfonamide derivatives **295** in water without using any activation and modification. The

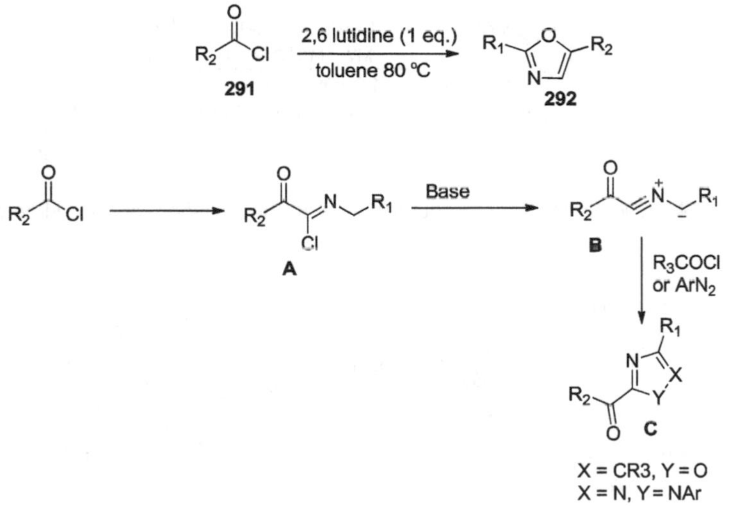

SCHEME 9.72 Oxazoles from acyl chlorides and isocyanides.

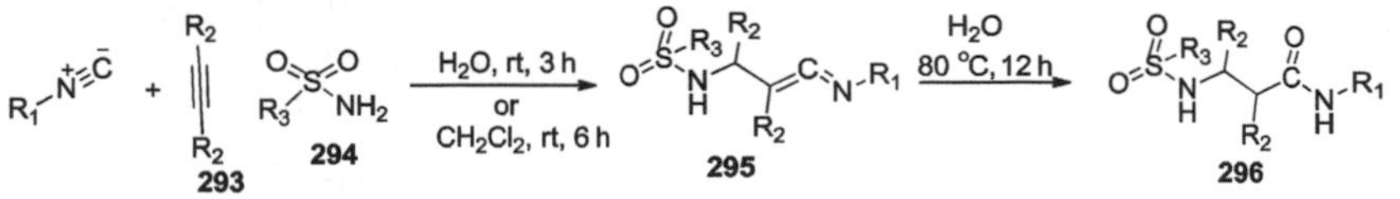

SCHEME 9.73 Synthesis of ketenimine sulfonamide derivatives.

products could be easily hydrolyzed to the corresponding sulfonamide-butanamide derivatives at 80 °C in good yields without using a catalyst.

Guchhait et al. reported a microwave-assisted tandem protocol of de-*tert*-butylation of derived *tert*-butyl amine in an Ugi-type MCR product **300**, which has afforded to the successful implementation of *tert*-butyl isocyanide as a useful convertible isonitrile. The method provides access to a diverse array of *N*-fused heterocycles (Scheme 9.74).[93]

A rapid access to the quinolin-2-(1*H*)-one scaffold **305** by a sequential four-component Ugi–Knoevenagel condensation of an amino phenyl ketone, an aromatic aldehyde possessing electron-donating moieties, cyanoacetic acid, and an aliphatic isocyanide has been reported (Scheme 9.75).[94] Interestingly, when the reaction was performed using aromatic aldehydes bearing electron-withdrawing moieties or isocyanides containing aromatic or ester units, a mixture of a quinolin-2-(1*H*)-one and an α-amino amide (Ugi three-component adduct) afforded in varying ratios. Further, when the reaction was performed utilizing a combination of an isocyanide-containing aromatic or carbonyl unit and an aldehyde possessing an electron-withdrawing functionality, the Ugi three-component adduct was exclusively afforded.

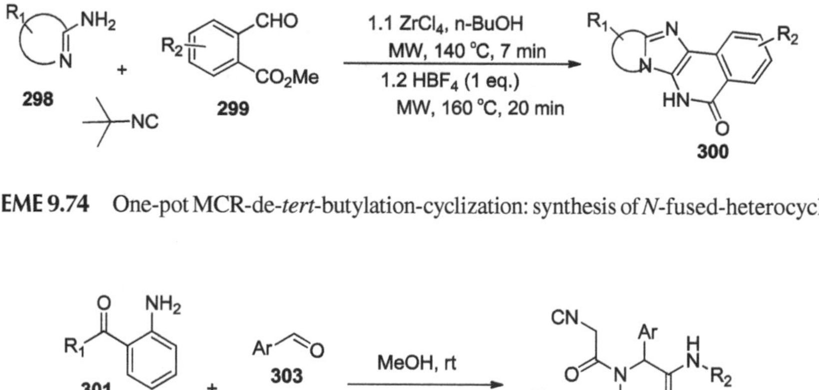

SCHEME 9.74 One-pot MCR-de-*tert*-butylation-cyclization: synthesis of *N*-fused-heterocycles.

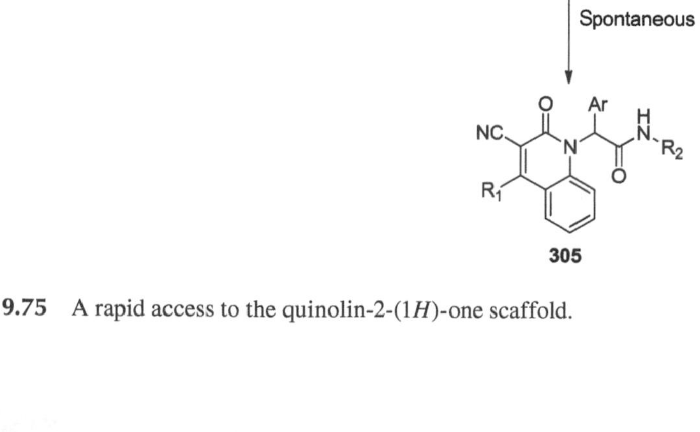

SCHEME 9.75 A rapid access to the quinolin-2-(1*H*)-one scaffold.

Ogawa et al. reported photochemical intramolecular cyclization of *o*-alkynylaryl isocyanides **306** with organic dichalcogenides leading to 2,4-bischalcogenated quinolines **307** (Scheme 9.76).[95] When a mixture of *o*-alkynylaryl isocyanides and organic dichalcogenides such as diselenides or ditellurides was irradiated with light of wavelength over 300 or 400 nm, the intramolecular cyclization of the isocyanides took place to afford the corresponding 2,4-bischalcogenated quinolines **308** selectively. The photochemical cyclization of 2 (phenylethynyl)phenyl isocyanide could also proceed in the presence of hydrogen transfer reagents such as tris(trimethylsilyl) silane, tributylgermyl hydride, alkanethiols, and benzeneselenol, providing the corresponding 3-phenylquinoline as the result of 2,4 dihydrogenation.

A strategy for the rapid construction of functionalized reduced indoles **311** starting from activated methylene isocyanides **310** and 1,9,5-dielectrophilic 5-oxohepta-2,6-dienoates **309** (and their equivalents) through a [5+1] annulation–isocyanide cyclization cascade under basic conditions has been developed (Scheme 9.77).[96] This strategy allows the synthesis of polysubstituted dihydroindolones and tetrahydroindolones in high to excellent yields under extremely mild conditions in a single step.

Pirali et al. reported a convenient synthesis of 3,8-diaminoimidazo[1,2-*a*]pyrazines **314** and **315** exploiting the isocyanide based multicomponent Blackburn reaction, followed by a nucleophilic aromatic substitution with ammonia or primary and secondary amines (Scheme 9.78).[97]

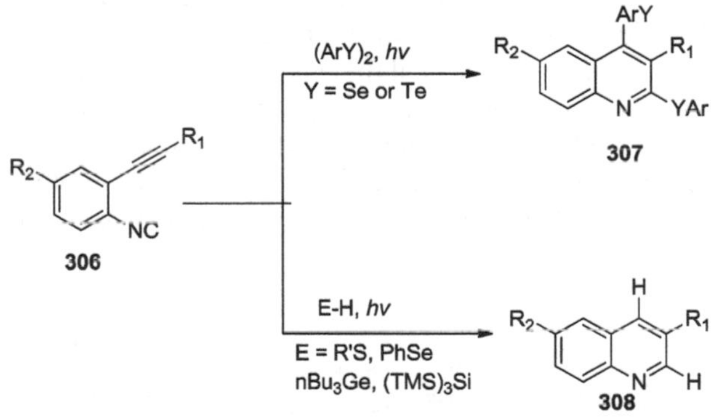

SCHEME 9.76 Photochemical cyclization of isocyanide.

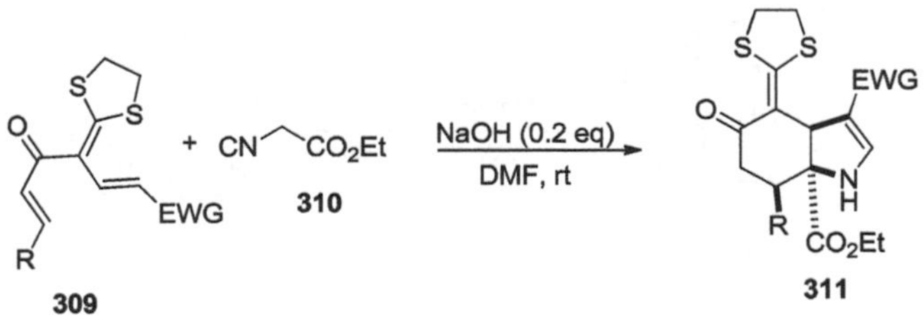

SCHEME 9.77 Synthesis of tetrahydroindolones.

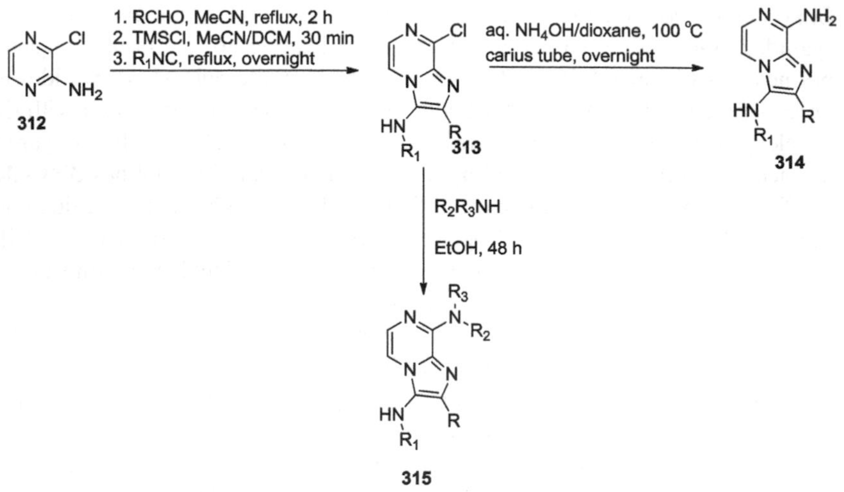

1. RCHO, MeCN, reflux, 2 h
2. TMSCl, MeCN/DCM, 30 min
3. R_1NC, reflux, overnight

aq. NH_4OH/dioxane, 100 °C
carius tube, overnight

R_2R_3NH
EtOH, 48 h

312 313 314 315

SCHEME 9.78 Synthesis of 3,8-diaminoimidazo[1,2-*a*]pyrazines.

Sheppard et al. reported that the synthesis of highly functionalized α-amino amides **307** and medium ring lactones **309** using four-component reactions between amino alcohols, aldehydes, isocyanides, and thiols proceeds rapidly under microwave or conventional heating at 60 °C in methanol (Scheme 9.79).[98] The reaction is successful with a wide range of components and gives access to potentially drug-like products containing amine, amide, and thioether functionality in moderate to excellent yield. The reaction conditions are also applicable to the synthesis of a range of 8- to 10-membered medium ring lactones via three-component reactions of amino alcohols, isocyanides, and acid-aldehydes. Incorporation of L-prolinol as the amino alcohol component in each case gives access to multicomponent products with moderate to high diastereoselectivity.

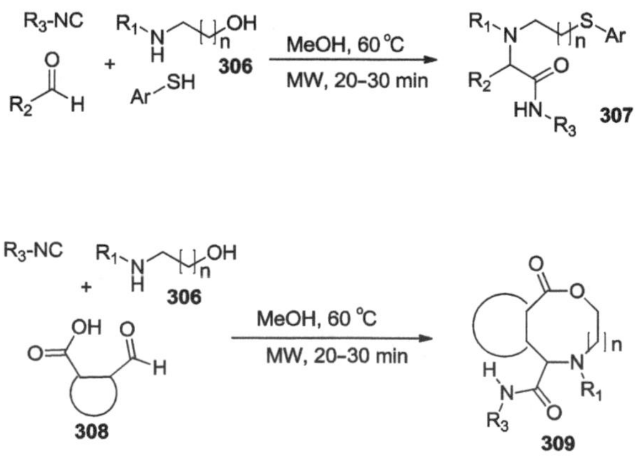

MeOH, 60 °C
MW, 20–30 min

306 307

MeOH, 60 °C
MW, 20–30 min

306 308 309

SCHEME 9.79 Synthesis of highly functionalized α-amino amides and medium ring lactones using four-component reactions.

That palladium catalyzed and CsF-promoted annulation reaction of bromoalkynes and isocyanides regioselectively affords a diverse set of 5-iminopyrrolone derivatives **311** has been reported by Orru et al. (Scheme 9.80).[99] This chemistry presumably proceeds through the bromoacrylamide intermediates, which can be readily prepared from the nucleophilic addition reaction of isocyanides to bromoalkynes in the presence of CsF.

Jiang and co-workers reported a robust route to 4-amine-benzo[*b*][1,4]oxazepines **314** relying upon a palladium catalyzed tandem reaction of *o*-aminophenols **312**, bromoalkynes **313**, and isocyanides (Scheme 9.81).[100] This chemistry presumably proceeds through the migratory insertion of isocyanides into the vinyl-palladium intermediate as a key step.

The intermolecular [2+2+1] multicomponent cycloadditions from readily available isocyanides, activated alkynes **315**, and isatins **316** has been reported by Jia et al. (Scheme 9.82).[101] This reaction proceeds by way of a Michael addition–nucleophilic

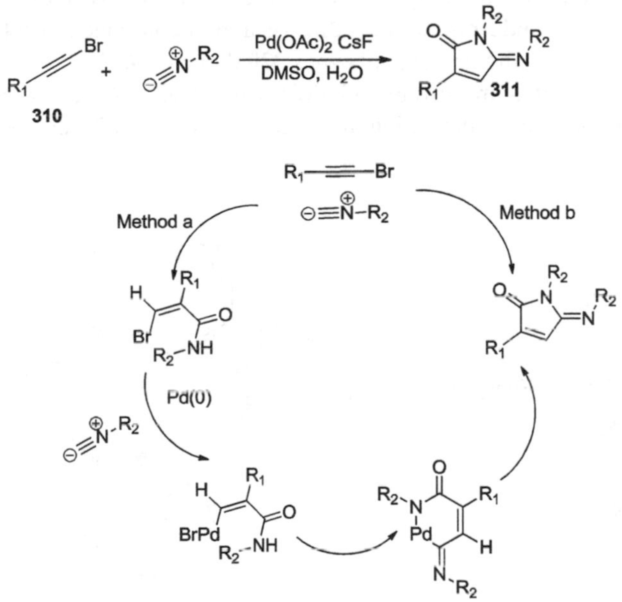

SCHEME 9.80 Synthesis of 5-iminopyrrolinone from bromoalkyne and isocyanide.

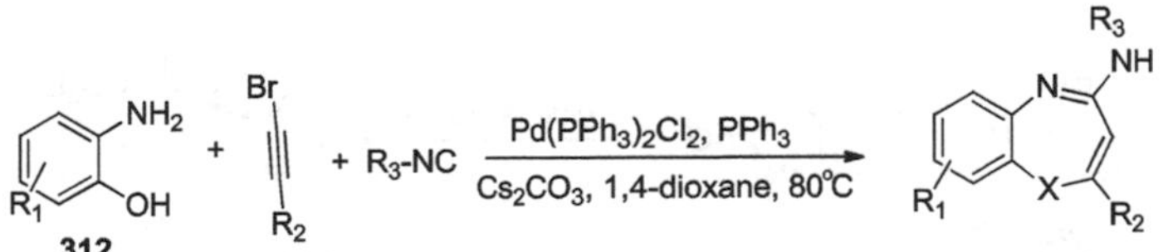

SCHEME 9.81 A synthetic route to 4-amine-benzo[*b*][1,4]oxazepines.

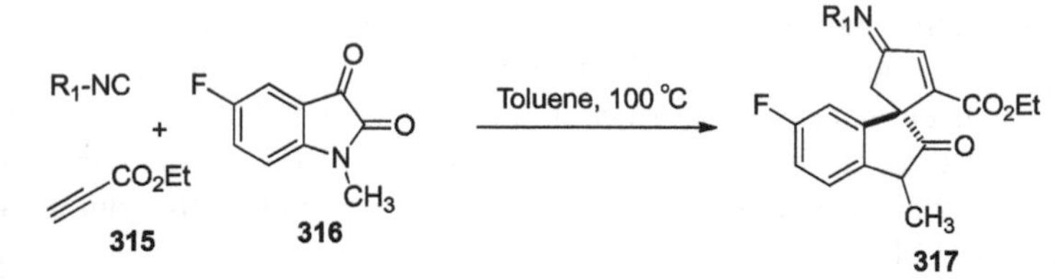

R1-NC + CO2Et 315 + 316 → Toluene, 100 °C → 317

SCHEME 9.82 Three-component [2+2+1] cycloaddition of isocyanide, ethyl propiolate, and isatin.

addition–intramolecular cyclization sequence, thus providing new access to spiro cyclic oxindole-butenolide **317** with exclusive stereoselectivity in an efficient and atom-economical manner. A broad range of isatins and isocyanides, including sterically demanding ones, are also found to be compatible with the present protocol, which offers an opportunity for the construction of a new compound library.

Huang et al. reported a highly chemo- and regioselective synthesis of polysubstituted pyridines **323** and isoquinolines **322** via a multicomponent reaction of arynes **318**, isocyanides **319**, and terminal alkynes **320** (Scheme 9.83).[102]

The syntheses of spiro cyclic oxindole-butenolides **326** by three-component cycloadditions of isocyanides, allenoates **324**, and isatins **325** has been described (Scheme 9.84).[103]

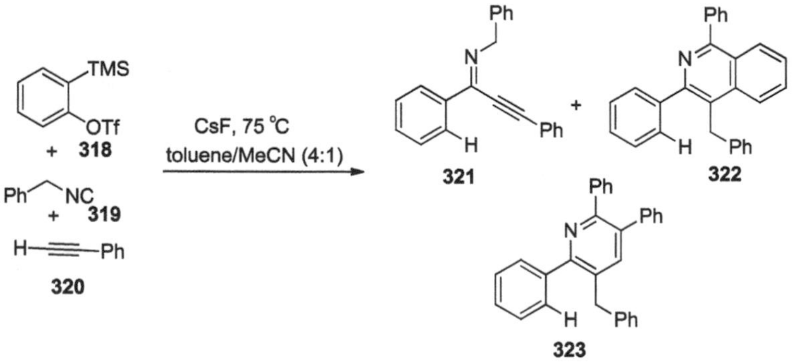

318 + 319 + 320 → CsF, 75 °C, toluene/MeCN (4:1) → 321 + 322 + 323

SCHEME 9.83 Synthesis of polysubstituted pyridines and isoquinolines.

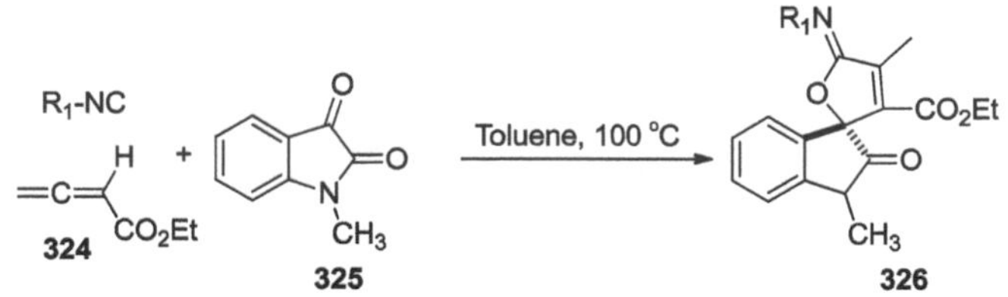

R1-NC + 324 + 325 → Toluene, 100 °C → 326

SCHEME 9.84 Synthesis of spiro cyclic oxindole-butenolides.

9.3 TRANSITION METAL CATALYZED ISOCYANIDE INSERTION REACTION

Isocyanides (RNCs) feature prominently in transition metal chemistry by virtue of their ability to serve as combined s-donor and p-acceptor ligands.[104] While this dual feature of the bonding is common to CO, an important distinction between RNC and CO is that the steric and electronic properties of isocyanides may be effectively tuned by modifying the substituent on nitrogen. Indeed, this aspect has been of benefit in the application of transition metal isocyanide complexes as catalysts for organic transformations.[105] Despite their widespread use, however, the majority of transition metal isocyanide complexes feature rather simple substituents on nitrogen, a reflection of the paucity of readily available commercial isocyanide compounds. In this context, Ji et al. reported a palladium catalyzed synthesis of isocoumarins **329** and phthalides **332** via *tert*-butyl isocyanide insertion (Scheme 9.85).[106] This process, providing one of the simplest methods for the synthesis of this class of valuable lactones, involves two steps including cyclization reaction and simple acid hydrolysis. The methodology is tolerant of a wide range of substrates and applicable to library synthesis.

The addition reaction of isocyanides to 3,4-dihydroisoquinoline *N*-oxides **333** in the presence of TMSCl, gave the corresponding 1,2,3,4-tetrahydroisoquinoline-1-carboxylamides **334** in moderate to high yields (Scheme 9.86).[107] A wide range

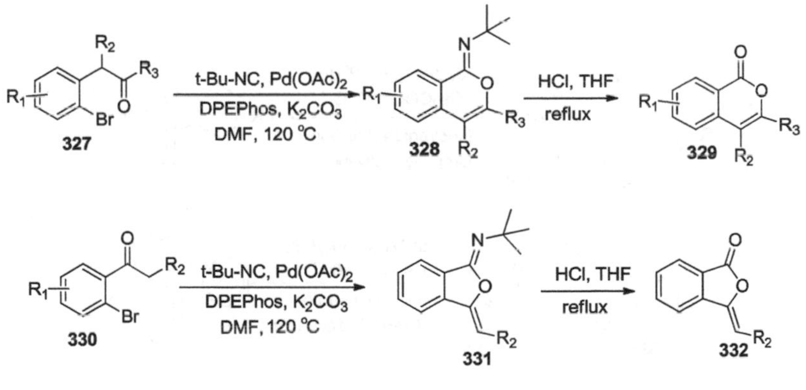

SCHEME 9.85 Palladium catalyzed synthesis of isocoumarins and phthalides via *tert*-butyl isocyanide insertion.

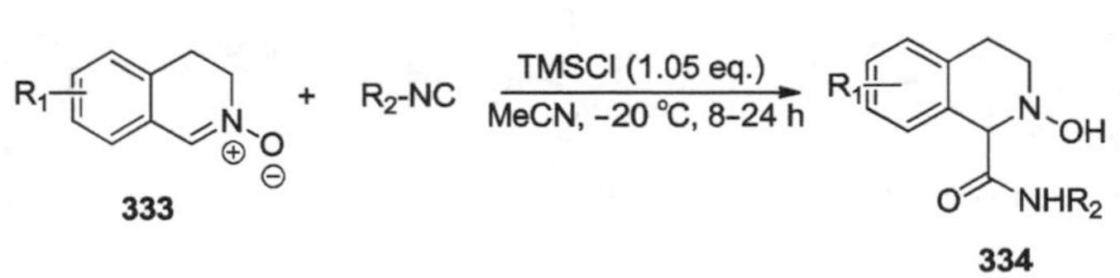

SCHEME 9.86 Addition reaction of isocyanides to 3,4-dihydroisoquinoline *N*-oxides.

of 3,4-dihydroisoquinoline *N*-oxides and isocyanides were applicable to this reaction.

Takemoto et al. reported the synthesis of the indole skeleton **336** using a Pd catalyzed cascade process consisting of isocyanide insertion and benzylic C(sp3)-H activation (Scheme 9.87).[108] It was found that slow addition of isocyanide is effective for reducing the amount of catalyst needed and Ad_2P^nBu is a good ligand for C(sp3)-H activation. The construction of the tetracyclic carbazole skeleton **340** was also achieved by a Pd catalyzed domino reaction incorporating alkyne insertion.

A palladium catalyzed regioselective C–H cyanation of heteroarenes **341** using *tert*-butyl isocyanide as "CN" source provided a new and unique strategy for the preparation of (hetero) aryl nitriles **342** and **343** (Scheme 9.88).[109] Indoles, pyrroles, and aromatic rings could be efficiently cyanated through C–H bond activation with high regioselectivity.

Zhu et al. also reported a palladium catalyzed cyanation of aromatic C–H bonds by using tertiary amine-derived isocyanide as a novel cyano source **346** (Scheme 9.89).[110] $Cu(TFA)_2$ was used as a requisite stoichiometric oxidant. Mechanistic studies suggest that a tertiary carbon cation-based intermediate is involved following the C–N bond breakage.

Wu et al. reported an efficient route for the preparation of quinazolin-4(3*H*)-imines **348** via a palladium catalyzed, three-component reaction of carbodiimide **347**, isocyanide, and nucleophile (Scheme 9.90).[111] The palladium catalyzed isocyanide insertion is believed to be the key step during the reaction process.

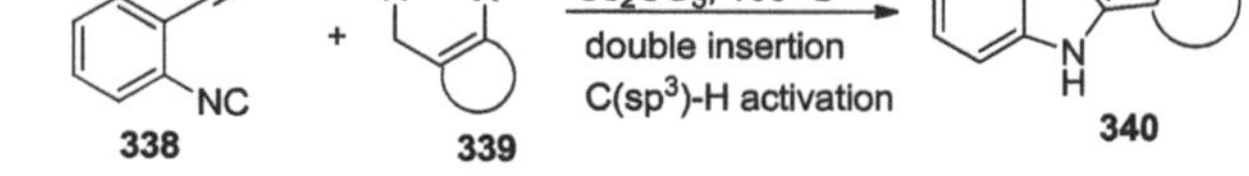

SCHEME 9.87 Synthesis of the indole and carbazol skeleton.

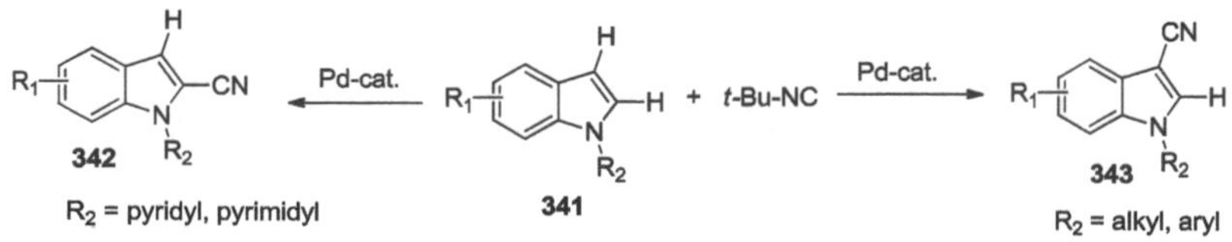

SCHEME 9.88 Palladium catalyzed regioselective C–H cyanation of heteroarenes using *tert*-butyl isocyanide as the "CN" source.

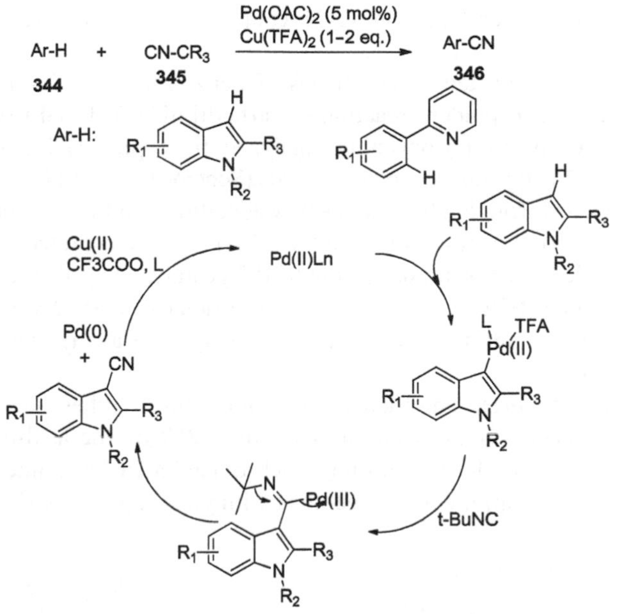

SCHEME 9.89 Palladium catalyzed cyanation of aromatic C–H bonds.

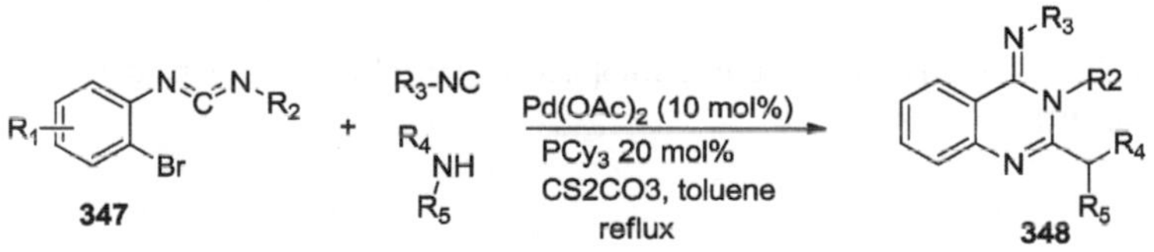

SCHEME 9.90 Synthesis of quinazolin-4(3*H*)-imines.

A selective C3 carboxamidation of indoles **349** including free (N–H) ones by palladium catalyzed sequential C–H activation–isocyanide insertion has been developed (Scheme 9.91).[112]

A palladium catalyzed, three-component reaction of bis-(2-iodoaryl)-carbodiimide **351**, isocyanide, and amine gave quinazolino-[3,2-*a*]quinazolines **352** and related compounds in good yields. Multibonds are formed in one pot through nucleophilic

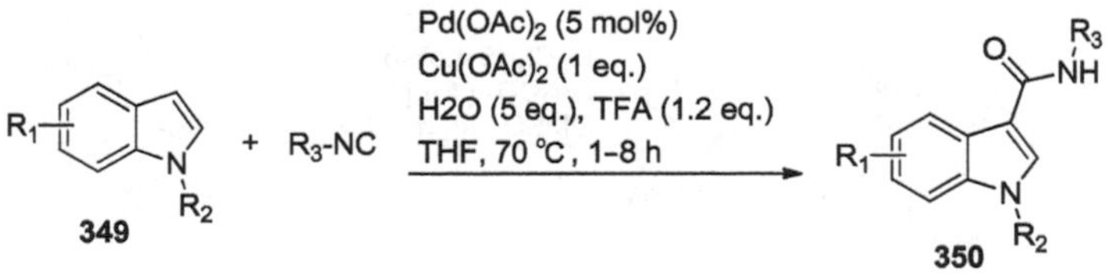

SCHEME 9.91 A selective C3 carboxamidation of indoles.

attack, isocyanide insertion, and C–N coupling during the reaction process (Scheme 9.92).[113]

This group also reported the synthesis of benzoimidazo[1,5-*a*]imidazoles **355** via a copper catalyzed tandem reaction of carbodiimide **353** and isocyanoacetate. Carbodiimide reacts with isocyanide catalyzed by copper(I) iodide, leading to benzoimidazo[1,5-*a*]imidazoles in good yields (Scheme 9.93).[114] This reaction is efficient and proceeds through a formal [3+2] cycloaddition and C–N coupling.

Zhu et al. reported a base-controlled synthesis of 2-substituted secondary and tertiary 1*H*-indole-3-carboxamides through $PdCl_2$ catalyzed cyclization of *o*-alkynyl trifluoro acetanilides followed by isocyanide insertion (Scheme 9.94).[115] The reaction proceeds smoothly at ambient temperature using O_2 in air as the sole oxidant of the palladium catalyst.

Kanizsai et al. reported an efficient, one-pot domino synthesis of new 2-amino-3-cyano-4*H*-chromene-4-carboxamide derivatives **359** by the acid-induced conjugate addition of isocyanides to 2-imino-2*H*-chromene-3-carboxamides, followed by an intramolecular O trapping rearrangement, with yields up to 92% (Scheme 9.95).[116]

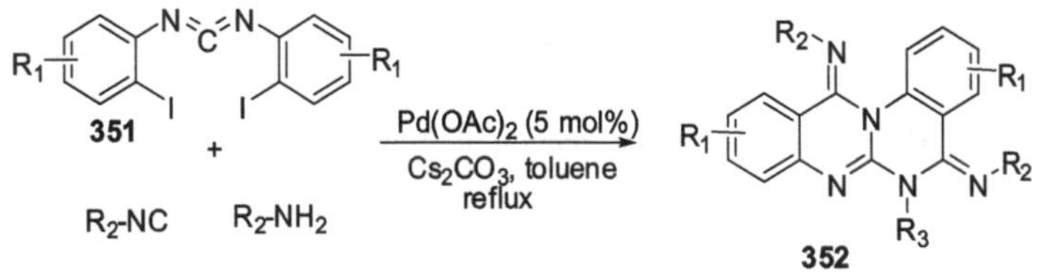

SCHEME 9.92 Synthesis of quinazolino[3,2-*a*]quinazolines via a palladium catalyzed three-component reaction.

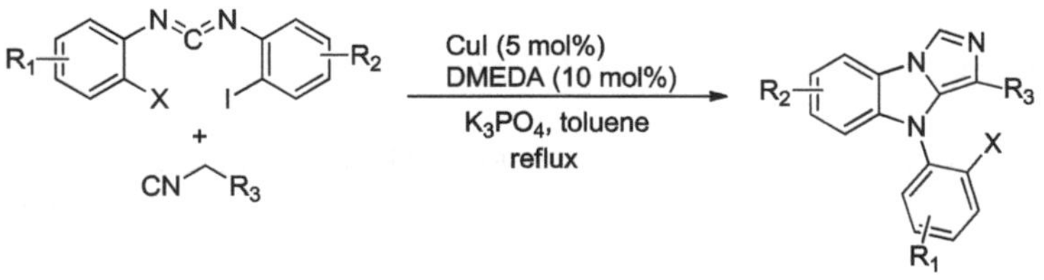

SCHEME 9.93 Synthesis of benzoimidazo[1,5-*a*]imidazoles.

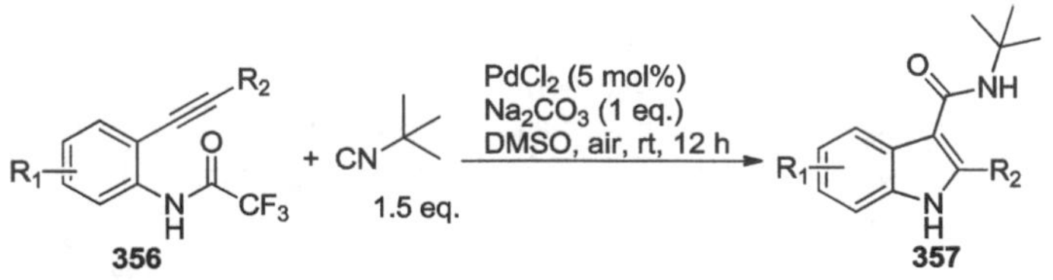

SCHEME 9.94 Synthesis of 2-substituted secondary and tertiary 1*H*-indole-3-carboxamides.

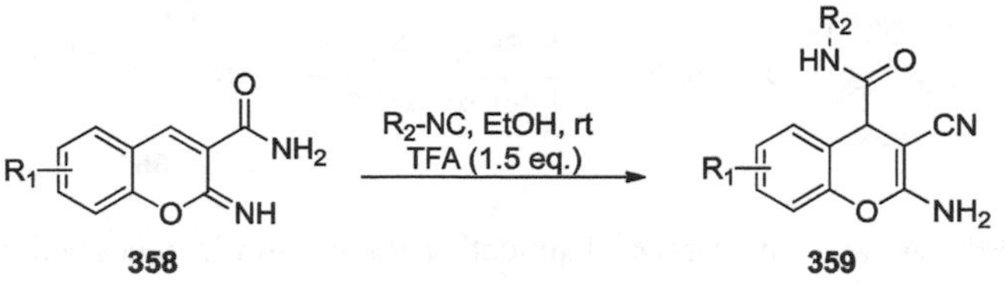

SCHEME 9.95 Synthesis of 2-amino-3-cyano-4*H*-chromene-4-carboxamide derivatives.

Multicomponent reactions of primary 1,2- and 1,3-diamines with carbonyl compounds and isocyanides resulted in the formation of diverse 2-amino-1,4-diazaheterocycles **361** and **362** (Scheme 9.96).[117] Lewis acids (LAs) promote the reactions effectively, and chlorotrimethylsilane (TMSCl) has been found to be a promoter of choice. The scope and limitations of the reactions with regard to each of the components are evaluated and discussed. Post-IMCR modifications of the synthesized heterocycles have been elaborated.

The synthesis of 1-substituted benzimidazoles **364** from *o*-bromophenyl isocyanide **363** and amines has been described (Scheme 9.97).[118] *o*-Bromophenyl isocyanide reacts with various primary amines under CuI catalysis to afford 1-substituted benzimidazoles in moderate to good yields. Analogously, 2-bromo-3-isocyanothiophene furnishes 3-substituted 3*H*-thieno[2,3-*d*]imidazoles.

Cai et al. reported a Pd catalyzed approach for the aminocarbonylation of *N*-tosylhydrazones **365** with isocyanides via ketenimine intermediates that avoids the use of stoichiometric organometallic reagents (Scheme 9.98).[119] IKt represents a general one-carbon extension transformation of carbonyl compounds into amides through the Pd catalyzed aminocarbonylation of *in situ* generated Pd-carbenes with isocyanides.

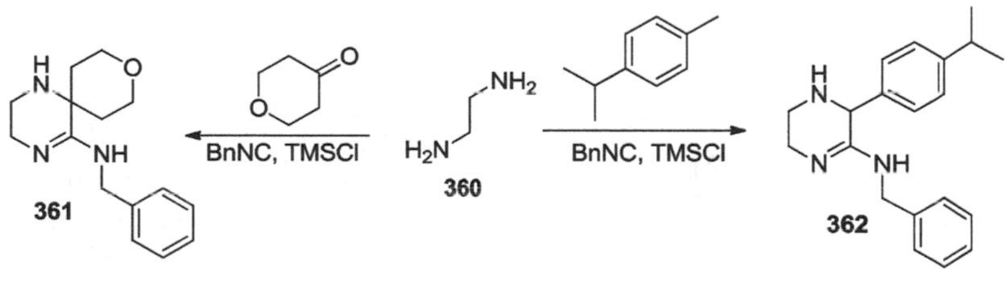

SCHEME 9.96 Formation of diverse 2-amino-1,4-diazaheterocycles.

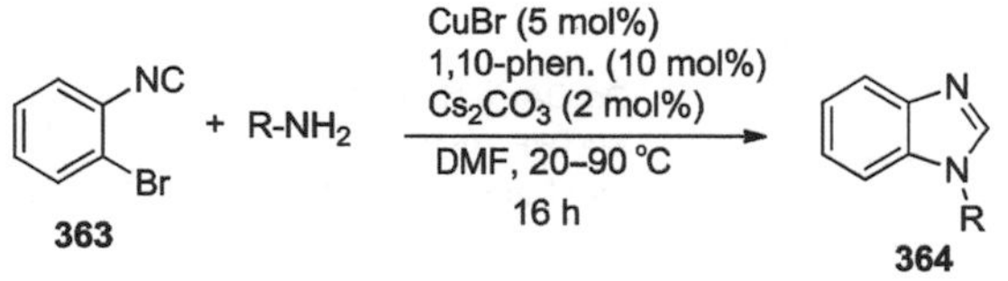

SCHEME 9.97 Synthesis of 1-substituted benzimidazoles.

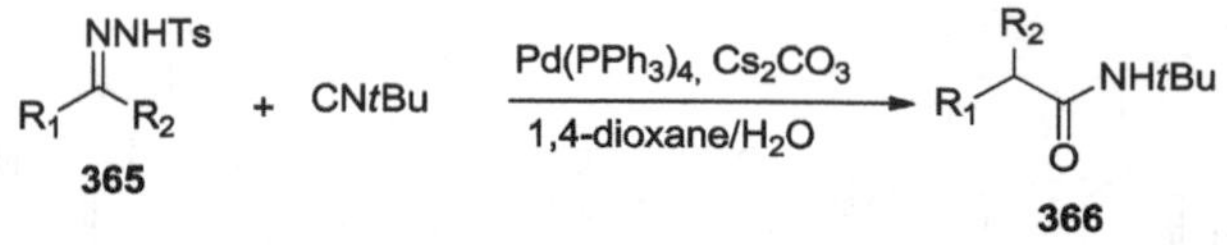

SCHEME 9.98 A palladium catalyzed amidation of *N*-tosylhydrazones with isocyanides.

Falck and co-workers reported the unprecedented rhodium catalyzed annulations of *N*-benzoylsulfonamide **367** with isocyanide through C–H activation (Scheme 9.99).[120] The transformation successfully suppresses the competitive reaction and is broadly compatible with *N*-benzoylsulfonamides as well as isocyanides with different electronic properties.

Orru et al. reported the intramolecular imidoylative cross coupling of *N*-(2-bromoaryl)amidines **370**, leading to 4-aminoquinazolines **371** (Scheme 9.100).[121] Various substituents are tolerated on the amidine and the isocyanide, providing efficient access to a broad range of diversely substituted 4-aminoquinazolines of significant pharmaceutical interest.

This group also reported a sustainable synthesis of diverse privileged heterocycles **373** by palladium catalyzed aerobic oxidative isocyanide insertion (Scheme 9.101).[122]

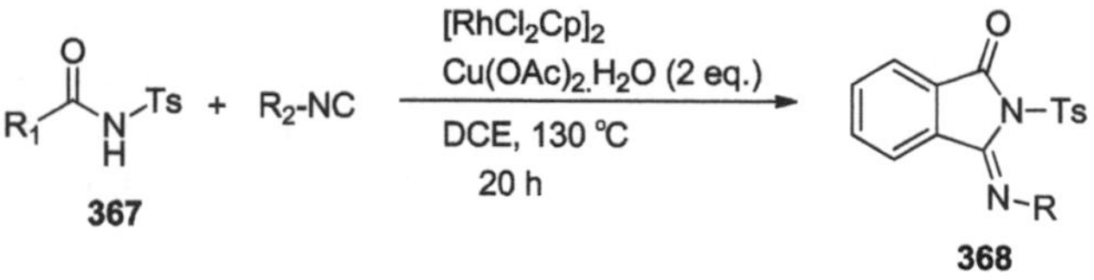

SCHEME 9.99 Annulation of *N*-benzoylsulfonamide with isocyanide.

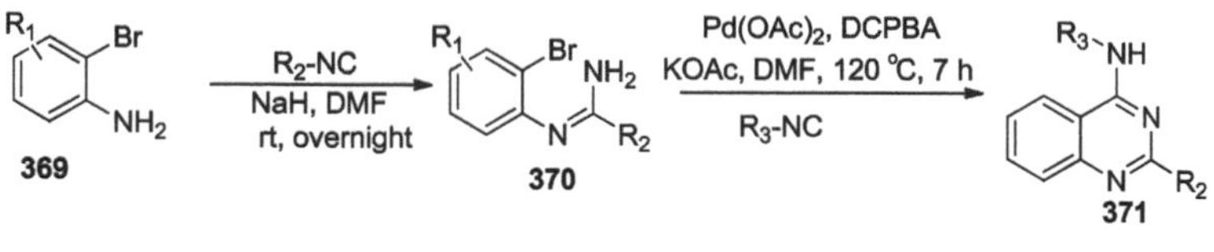

SCHEME 9.100 Synthesis of 4-aminoquinazolines.

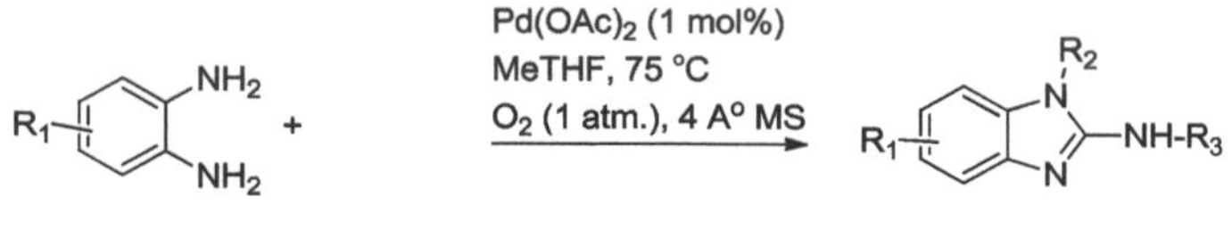

SCHEME 9.101 Synthesis of diverse privileged heterocycles.

9.4 SUMMARY

MCRs have offered increasingly important opportunities for synthesis of both chemically and medicinally useful compounds because of their environmentally friendly characteristics of atom economy and green chemistry. These reactions enable multistep processes to be achieved in a single step of operation to produce a diverse variety of products that include useful heterocycles with structural complexity. As compared with traditional multioperational synthesis, MCRs can dramatically reduce the generation of chemical waste and costs of starting materials. They can often shorten reaction times and substantially save the use of energy and manpower to give high overall chemical yields in a multistep processes. As illustrated in this review, a wide variety of heterocycles of different sizes and ring systems can be readily synthesized through MCR strategies that often result in a broad scope of applications. More novel and efficient multicomponent domino reactions need to be developed for the synthesis of leading structures, particularly of those complex products existing in nature.

REFERENCES

1. Suginome, M.; Ito, Y. *Stuttgart* **2004**, *19*, 445.
2. Lieke, W. *Annalen der Chemie and Pharmacie* **1859**, *112*, 316.
3. (a) Gautier, A. *Ann. Chem. Pharm.* **1868**, *146*, 119; (b) Hofmann, A. W. *Ann. Chem. Pharm.* **1867**, *144*, 114.
4. Dömling, A.; Ugi, I. *Angew. Chem. Int. Ed.* **2000**, *39*, 3168.
5. Lygin, A. V.; de Meijere, A. *Angew. Chem. Int. Ed.* **2010**, *49*, 9094.
6. DeSimone, R. W.; Currie, K. S.; Mitchell, S. A.; Darrow, J. W.; Pippin, D. A. *Comb. Chem.* **2004**, *7*, 473.
7. Leeson, P. D.; Springthorpe, B. *Nat. Rev. Drug Discovery* **2007**, *6*, 881.
8. Xu, W.; Fu, H. *J. Org. Chem.* **2011**, *76*, 3846.
9. Dömling, A.; Ugi, I. *Angew. Chem. Int. Ed.* **2000**, *39*, 3168.
10. See http://www.organic-chemistry.org/Highlights/mcr.shtm
11. Nishiyama, Y.; Katahira, C.; Sonoda, N. *Tetrahedron Lett.* **2004**, *45*, 8539.
12. Portlock, D. E.; Naskar, D.; West, L.; Ostaszewski, R.; Chen, J. J. *Tetrahedron Lett.* **2003**, *44*, 5121.
13. Tietze, L. F. *Chem. Rev.* **1996**, *96*, 115.
14. Kolb, J.; Beck, B.; Dömling, A. *Tetrahedron Lett.* **2002**, *43*, 6897.
15. Fayol, A.; Zhu, J. *Org. Lett.* **2005**, *7*, 239.
16. Passerini, M.; Simone, L. *Gazz. Chim. Ital.* **1921**, *51*, 126.
17. Ugi, I.; Meyr, R.; Fetzer, U.; Steinbrückner, C. *Angew. Chem.* **1959**, *71*, 386.
18. Ruijter, E.; Scheffelaar, R.; Orru, R. A. V. *Angew. Chem. Int. Ed.* **2011**, *50*, 6234.
19. Sperger, C. A.; Mayer, P.; Wanner, K. T. *Tetrahedron* **2009**, 65, 10463.
20. Plant, A.; Thompson, P.; Williams, M. *J. Org. Chem.* **2009**, *74*, 4870.
21. Mossetti, R.; Pirali, T.; Tron, G. C. *J. Org. Chem.* **2009**, *74*, 4890.
22. Krasavin, M.; Parchinsky, V.; Shumsky, A.; Konstantinov, I.; Vantskul, A. *Tetrahedron Lett.* **2010**, *51*, 1367.
23. Jida, M.; Malaquin, S.; Deprez-Poulain, R.; Laconde, G.; Deprez, B. *Tetrahedron Lett.* **2010**, *51*, 5109.
24. Brauch, S.; Gabriela, L.; Westermann, B. *Chem. Commun.* **2010**, *46*, 338.
25. Borisov, R. S.; Polyakov, A. I.; Medvedeva, L. A.; Khrustalev, V. N.; Guranova, N. I.; Voskressensky, L. G. *Org. Lett.* **2010**, *12*, 3894.
26. Marcaccini, S.; Menchi, G.; Trabocchi, A. *Tetrahedron Lett.* **2011**, *52*, 2673.

27. Riguet, E. *J. Org. Chem.* **2011**, *76*, 8143.
28. Aknin, K.; Gauriot, M.; Totobenazara, J.; Deguine, N.; Deprez-Poulain, R.; Deprez, B.; Charton, J. *Tetrahedron Lett.* **2012**, *53*, 458.
29. Shaabani, A.; Keshipour, S.; Shaabani, S.; Mahyari, M. *Tetrahedron Lett.* **2012**, *53*, 1641.
30. Ghandi, M.; Zarezadeh, N.; Taheri, A. *Tetrahedron Lett.* **2012**, *53*, 3353.
31. Soleymanifard, B.; Heravi, M. M.; Shiri, M.; Zolfigol, M. A.; Rafiee, M.; Kruger, H. G.; Naicker, T.; Rasekhmanesh, F. *Tetrahedron Lett.* **2012**, *53*, 3546.
32. Gunawan, S.; Petit, J.; Hulme, C. *ACS Comb. Sci.* **2012**, *14*, 160.
33. Zhu, D.; Xia, L.; Pan, L.; Li, S.; Chen, R.; Mou, Y.; Chen, X. *J. Org. Chem.* **2012**, *77*, 1386.
34. La Spisa, F.; Feo, A.; Mossetti, R.; Tron, G. C. *Org. Lett.* **2012**, *14*, 6044.
35. Znabet, A.; Blanken, S.; Janssen, E.; De Kanter, F. J. J.; Helliwell, M.; Turner, N. J.; Ruijter, E.; Orru, R. V. A. *Org. Biomol. Chem.* **2012**, *10*, 941.
36. Gulevich, A. V.; Shpilevaya, I. V.; Nenajdenko, V. G. *Eur. J. Org. Chem.* **2009**, 3801.
37. (a) Chen, W.; Li, J.; Fang, D.; Feng, C.; Zhang, C. *Org. Lett.* **2008**, *10*, 4565; (b) Simon, C.; Constantieux, T.; Rodriguez, J. *Eur. J. Org. Chem.* **2004**, *24*, 4957; (c) Balme, G.; Bossharth, E.; Monteiro, N. *Eur. J. Org. Chem.* **2003**, *21*, 4101; (d) Dömling, A.; Wang, W.; Wang, K. *Chem. Rev.* **2012**, *112*, 3083; (e) Candeias, N. R.; Montalbano, F.; Cal, P. M. S. D.; Gois, P. M. P. *Chem. Rev.* **2010**, *110*, 6169; (f) Rajarathinam, B.; Vasuki, G. *Org. Lett.* **2012**, *14*, 5204; (g) Zhu, S.-L.; Ji, S.-J.; Su, X.-M.; Sun, C.; Liu, Y. *Tetrahedron Lett.* **2008**, *49*, 1777; (h) Jin, H.; Zhou, B.; Wu, Z.; Shen, Y.; Wang, Y. *Tetrahedron* **2011**, *67*, 1178.
38. (a) Ma, Z.; Xiang, Z.; Luo, T.; Lu, K.; Xu, Z.; Chen, J.; Yang, Z. *J. Comb. Chem.* **2006**, *8*, 69; (b) Ilyin, A.; Kysil, V.; Krasavin, M.; Kurashvili, I.; Ivachtchenko, A. V. *J. Org. Chem.* **2006**, *71*, 9544; (c) Riva, R.; Banfi, L.; Basso A.; Cerulli, V.; Guanti, G.; Pani, M. *J. Org. Chem.* **2010**, *15*, 5134; (d) Guchhait, S. K.; Madaan, C. *Org. Biomol. Chem.* **2010**, *8*, 3631; (e) Erb, W.; Neuville, L.; Zhu, J. *J. Org. Chem.* **2009**, *74*, 3109; (f) El Kaïm, L.; Gizzi, M.; Grimaud, L. *Org. Lett.* **2008**, *10*, 3417; (g) Sunderhaus, J. D.; Dockendorff, C.; Martin, S. F. *Org. Lett.* **2007**, *9*, 4223; (h) El Kaïm, L.; Grimaud, L.; Wagschal, S. *J. Org. Chem.* **2010**, *75*, 5343; (i) Bonnaterre, F.; Bois-Choussy, M.; Zhu, J. *Org. Lett.* **2006**, *19*, 4351; (j) Hulme, C.; Ma, L.; Romano, J. J.; Morton, G.; Tang, S.-Y.; Cherrier, M.-P.; Choi, S.; Salvino, J.; Labaudiniere, R. *Tetrahedron Lett.* **2000**, *41*, 1889; (k) Gracias, V.; Moore, J. D.; Djuric, S. W. *Tetrahedron Lett.* **2004**, *45*, 417; (l) Zhu, J. *Eur. J. Org. Chem.* **2003**, *7*, 1133; (m) El Kaïm, L.; Grimaud, L.; Pravin, P. *Org. Lett.* **2012**, *14*, 477; (n) Zhu, D.; Xia, L.; Pan, L.; Li, S.; Chen, R.; Mou, Y.; Chen, X. *J. Org. Chem.* **2012**, *77*, 1386; (o) Kulsi, G.; Ghorai, A.; Chattopadhyay, P. *Tetrahedron Lett.* **2012**, *53*, 3619.
39. Wang, W.; Herdtweck, E.; Domling, A. *Chem. Commun.* **2010**, *46*, 770.
40. Wang, W.; Joyner, S.; Andrew, K.; Khoury, S.; Domling, A. *Org. Biomol. Chem.* **2010**, *8*, 529.
41. Liu, H.; Domling, A. *J. Org. Chem.* **2009**, *74*, 6895.
42. Znabet, A.; Zonneveld, J.; Janssen, E.; De Kanter, F. J. J.; Helliwell, M.; Turner, N. J.; Ruijter, E.; Orru, R. V. A. *Chem. Commun.* **2010**, *46*, 7706.
43. Coffinier, D.; El Kaim, L.; Grimaud, L. *Org. Lett.* **2009**, *11*, 995.
44. Malaquin, S.; Jida, M.; Gesquiere, J.-C.; Deprez-Poulain, R.; Deprez, B.; Laconde, G. *Tetrahedron Lett.* **2010**, *51*, 2983.
45. Bararjanian, M.; Balalaie, S.; Movassagh, B.; Bijanzadeh, H. R. *Tetrahedron Lett.* **2010**, *51*, 3277.
46. Lakontseva, E.; Krasavin, M. *Tetrahedron Lett.* **2010**, *51*, 4095.
47. Krasavin, M.; Parchinsky, V. *Tetrahedron Lett.* **2010**, *51*, 5657.
48. Islas-Jácome, A.; González-Zamora, E.; Gámez-Montaño, R. *Tetrahedron Lett.* **2011**, *52*, 5245.
49. Gordon, C. P.; Young, K. A.; Robertson, M. J.; Hill, T. A.; McCluskey, A. *Tetrahedron* **2011**, *67*, 554.

50. Shaw, A. Y.; McLaren, J. A.; Nichol, G. S.; Hulme, C. *Tetrahedron Lett.* **2012**, *53*, 2592.
51. Kulsi, G.; Ghorai, A.; Chattopadhyay, P. *Tetrahedron Lett.* **2012**, *53*, 3619.
52. Spatz, J. H.; Welsch, S. J.; Duhaut, D.-E.; Jäger, N.; Boursier, T.; Fredrich, M.; Allmendinger, L.; Ross, G.; Kolb, J.; Burdack, C.; Umkehrer, M. *Tetrahedron Lett.* **2009**, *50*, 1705.
53. Hulme, C.; Chappeta, S.; Griffith, C.; Lee, Y.-S.; Dietrich, J. *Tetrahedron Lett.* **2009**, *50*, 1939.
54. Tsirulnikov, S.; Nikulnikov, M.; Kysil, V.; Ivachtchenko, A.; Krasavin, M. *Tetrahedron Lett.* **2009**, *50*, 5529.
55. Erb, W.; Neuville, L.; Zhu, J. *J. Org. Chem.* **2009**, *74*, 3109.
56. Lecinska, P.; Corres, N.; Moreno, D.; García-Valverde, M.; Marcaccini, S.; Torroba, T. *Tetrahedron* **2010**, *66*, 6783.
57. Nayak, M.; Batra, S. *Tetrahedron Lett.* **2010**, *51*, 510.
58. Tyagi, V.; Khan, S.; Bajpai, V.; Gauniyal, H. M.; Kumar, B.; Chauhan, P. M. S. *J. Org. Chem.* **2012**, *77*, 1414.
59. Tyagi, V.; Khan, S.; Giri, A.; Gauniyal, H. M.; Sridhar, B.; Chauhan, P. M. S. *Org. Lett.* **2012**, *14*, 3126.
60. Pandey, S.; Khan, S.; Singh, A.; Gauniyal, H. M.; Kumar, B.; Chauhan, P. M. S. *J. Org. Chem.* **2012**, *77*, 10211.
61. Riva, R.; Banfi, L.; Basso, A.; Cerulli, V.; Guanti, G.; Pani, M. *J. Org. Chem.* **2010**, *75*, 5134.
62. Ramazani, A.; Rezaei, A. *Org. Lett.* **2010**, *12*, 2852.
63. Balalaie, S.; Bararjanian, M.; Hosseinzadeh, S.; Rominger, F.; Bijanzadeh, H. R.; Wolf, E. *Tetrahedron* **2011**, *67*, 7294.
64. Balalaie, S.; Motaghedi, H.; Bararjanian, M.; Tahmassebi, D.; Bijanzadeh, H. R. Tetrahedron **2011**, *67*, 9134.
65. Tsaloev, A.; Ilyin, A.; Tkachenko, S.; Ivachtchenko, A.; Kravchenko, D.; Krasavin, M. *Tetrahedron Lett.* **2011**, *52*, 800.
66. Welsch, S. J.; Umkehrer, M.; Ross, G.; Kolb, J.; Burdack, C.; Wessjohann, L. A. *Tetrahedron Lett.* **2011**, *52*, 6295.
67. Santra, S.; Andreana, P. R. *J. Org. Chem.* **2011**, *76*, 2261.
68. Sanudo, M.; García-Valverde, M.; Marcaccini, S.; Torroba, T. *Tetrahedron* **2012**, *68*, 2621.
69. Gunawan, S.; Ayaz, M.; De Moliner, F.; Frett, B.; Kaiser, C.; Patrick, N.; Xu, Z.; Hulme, C. *Tetrahedron* **2012**, *68*, 5606.
70. Seganish, W. M.; Bercovici, A.; Ho, G. D.; Loozen, H. J. J.; Timmers, C. M.; Tulshian, D. *Tetrahedron Lett.* **2012**, *53*, 903.
71. Shaw, A. Y.; Xu, Z.; Hulme, C. *Tetrahedron Lett.* **2012**, *53*, 1998.
72. Welsch, S. J.; Kalinski, C.; Umkehrer, M.; Ross, G.; Kolb, J.; Burdack, C.; Wessjohann, L. A. *Tetrahedron Lett.* **2012**, *53*, 2298.
73. El Kaim, L.; Grimaud, L.; Pravin, P. *Org. Lett.* **2012**, *14*, 476.
74. Huang, Y.; Khoury, K.; Chanas, T.; Domling, A. *Org. Lett.* **2012**, *14*, 5916.
75. Banfi, L.; Basso, A.; Giardini, L.; Riva, R.; Rocca, V.; Guanti, G. *Eur. J. Org. Chem.* **2011**, *1*, 100.
76. Schwäblein, A.; Martens, J. *Eur. J. Org. Chem.* **2011**, *23*, 4335.
77. Bogaert, M. V.; Nelissen, J.; Ovaere, M.; Meervelt, L. V.; Compernolle, F.; De Borggraeve, W. M. *Eur. J. Org. Chem.* **2010**, 5397.
78. El Kaïm, L.; Grimaud, L.; Purumandla, S. R. *Eur. J. Org. Chem.* **2011**, *31*, 6177.
79. Modha, S. G.; Kumar, A.; Vachhani, D. D.; Jacobs, J.; Sharma, S. K.; Parmar, V. S.; Meervelt, L. V.; Van der Eycken, E. V. *Angew. Chem. Int. Ed.* **2012**, *51*, 9572.
80. Cao, H.; Liu, H.; Dömling, A. *Chem. Eur. J.* **2010**, *16*, 12296.
81. Soleimani, E.; Zainali, M. *J. Org. Chem.* **2011**, *76*, 10306.
82. Krasavin, M.; Shkavrov, S.; Parchinsky, V.; Bukhryakov, K. *J. Org. Chem.* **2009**, *74*, 2627.
83. Zhu, X.; Xu, X.-P.; Sun, C.; Wang, H.-Y.; Zhao, K.; Ji, S.-J. *J. Comb. Chem.* **2010**, *12*, 822.

84. El Kaim, L.; Grimaud, L.; Patil, P. *Org. Lett.* **2011**, *13*, 1261.
85. Zhang, Y.; Lee, C. L.; Liu, H.; Li, X. *Org. Lett.* **2012**, *14*, 5146.
86. Soeta, T.; Tamura, K.; Ukaji, Y. *Org. Lett.* **2012**, *14*, 1226.
87. De Moliner, F.; Hulme, C. *Org. Lett.* **2012**, *14*, 1354.
88. Masui, H.; Fuse, S.; Takahashi, T. *Org. Lett.* **2012**, *14*, 4090.
89. Nair, D. K.; Mobin, S. M.; Namboothiri, I. N. N. *Org. Lett.* **2012**, *14*, 4580.
90. Polindara-García, L. A.; Miranda, L. D. *Org. Lett.* **2012**, *14*, 5408.
91. Santos, A.; El Kaím, L.; Grimaud, L.; Ronsseray, C. *Chem. Commun.* **2009**, 3907.
92. Shaabani, A.; Sarvary, A.; Ghasemi, S.; Rezayan, A. H.; Ghadaria, R.; Weng Ng, S. *Green Chem.* **2011**, *13*, 582.
93. Guchhait, S. K.; Madaan, C. *Org. Biomol. Chem.* **2010**, *8*, 3631.
94. Gordon, C. P.; Young, K. A.; Hizartzidis, L.; Deane, F. M.; McCluskey, A. *Org. Biomol. Chem.* **2011**, *9*, 1419.
95. Mitamura, T.; Iwata, K.; Nomoto, A.; Ogawa, A. *Org. Biomol. Chem.* **2011**, *9*, 3768.
96. Wang, H.; Zhao, Y.-L.; Ren, C.-Q.; Diallo, A.; Liu, Q. *Chem. Commun.* **2011**, *47*, 12316.
97. Guasconi, M.; Lu, X.; Massarotti, A.; Caldarelli, A.; Ciraolo, E.; Tron, G. C.; Hirsch, E.; Sorbaa, G.; Pirali, T. *Org. Biomol. Chem.* **2011**, *9*, 4144.
98. Bachman, M.; Mann, S. E.; Sheppard, T. D. *Org. Biomol. Chem.* **2012**, *10*, 162.
99. Znabet, A.; Blanken, S.; Janssen, E.; de Kanter, F. J. J.; Helliwell, M.; Turner, N. J.; Ruijter, E.; Orru, R. V. A. *Org. Biomol. Chem.* **2012**, *10*, 941.
100. Liu, B.; Li, Y.; Yin, M.; Wu, W.; Jiang, H. *Chem. Commun.* **2012**, *48*, 11446.
101. Li, J.; Liu, Y.; Li, C.; Jiea, H.; Jia, X. *Green Chem.* **2012**, *14*, 1314.
102. Sha, F.; Huang, X. *Angew. Chem. Int. Ed.* **2009**, *48*, 3458.
103. Li, J.; Liu, Y.; Li, C.; Jia, X. *Chem. Eur. J.* **2011**, *17*, 7409.
104. (a) Ma, B.; Wang, Y.; Peng, J.; Zhu, Q. *J. Org. Chem.* **2011**, *76*, 6362; (b) Wu, X.-F.; Anbarasan, P.; Neumann, H.; Beller, M. *Angew. Chem. Int. Ed.* **2010**, *49*, 7316.
105. (a) Tobisu, M.; Imoto, S.; Ito, S.; Chatani, N. *J. Org. Chem.* **2010**, *75*, 4835; (b) Park, S.; Shintani, R.; Hayashi, T. *Chem. Lett.* **2009**, *38*, 204–205; (c) Zhang, W.-X.; Nishiura, M.; Hou, Z. *Angew. Chem. Int. Ed.* **2008**, *47*, 9700.
106. Fei, X.-D.; Ge, Z.-Y.; Tang, T.; Zhu, Y.-M.; Ji, S.-J. *J. Org. Chem.* **2012**, *77*, 10321.
107. Soeta, T.; Fujinami, S.; Ukaji, Y. *J. Org. Chem.* **2012**, *77*, 9878.
108. Nanjo, T.; Tsukano, C.; Takemoto, Y. *Org. Lett.* **2012**, *14*, 4270.
109. Xu, S.; Huang, X.; Hong, X.; Xu, B. *Org. Lett.* **2012**, *14*, 4614.
110. Peng, J.; Zhao, J.; Hu, Z.; Liang, D.; Huang, J.; Zhu, Q. *Org. Lett.* **2012**, *14*, 4966.
111. Qiu, G.; Liu, G.; Pu, S.; Wu, J. *Chem. Commun.* **2012**, *48*, 2903.
112. Peng, J.; Liu, L.; Hu, Z.; Huang, J.; Zhu, Q. *Chem. Commun.* **2012**, *48*, 3772.
113. Qiu, G.; He, Y.; Wu, J. *Chem. Commun.* **2012**, *48*, 3836.
114. Qiua, G.; Wu, J. *Chem. Commun.* **2012**, *48*, 6046.
115. Hu, Z.; Liang, D.; Zhao, J.; Huang, J.; Zhu, Q. *Chem. Commun.* **2012**, *48*, 7371.
116. Gyuris, M.; Madácsi, R.; Puskás, R. G.; Tóth, G. K.; Wölfling, J.; Kanizsai, I. *Eur. J. Org. Chem.* **2011**, *5*, 848.
117. Kysil, V.; Khvat, A.; Tsirulnikov, S.; Tkachenko, S.; Williams, C.; Churakova, M.; Ivachtchenko, A. *Eur. J. Org. Chem.* **2010**, *8*, 1525.
118. Lygin, A. V.; de Meijere, A. *Eur. J. Org. Chem.* **2009**, *30*, 5138.
119. Zhou, F.; Ding, K.; Cai, Q. *Chem. Eur. J.* **2011**, *17*, 12268.
120. Zhu, C.; Xie, W.; Falck, J. R. *Chem. Eur. J.* **2011**, *17*, 12591.
121. Baelen, G. V.; Kuijer, S.; Rycêk, L.; Sergeyev, S.; Janssen, E.; De Kanter, F. J. J.; Maes, B. U. W.; Ruijter, E.; Orru, R. V. A. *Chem. Eur. J.* **2011**, *17*, 15039.
122. Vlaar, T.; Cioc, R. C.; Mampuys, P.; Maes, B. U. W.; Orru, R. V. A.; Ruijter, E. *Angew. Chem. Int. Ed.* **2012**, *51*, 1.

10 Recent Progress on One-Pot, Multicomponent Reaction for Pyridine Synthesis

Vijaykumar Paike, Padmakar Suryavanshi, Sandeep More, Sandeep B. Mane, Rajendra P. Pawar, and K. L. Ameta**

CONTENTS

10.1 INTRODUCTION

10.1.1 Multicomponent Reaction (MCR)

Synthesis of heterocyclic complex and highly diverse structures through one-pot, tandem, domino, or cascade reactions has become a significant area of research in organic chemistry. Successful heterocycles containing medicinally important products and natural product development rely on high productivity, atom economy, and step efficiency, and have high exploratory power with regard to chemical space. Thus, short and effective synthetic sequences for the generation of libraries of compounds are required (Figure 10.1).[1] Multicomponent reactions (MCRs) serve as a rapid and efficient tool for the synthesis of versatile heterocycles, particularly those containing structural diversity and complexity, by a one-pot operation.

Multicomponent reactions (MCRs) are defined as the reaction process of at least three components in one pot that involves forming a single product that assimilates essentially all the atoms of the starting materials (Figure 10.2).

* Corresponding authors: vv.paike@yahoo.co.in and klameta77@hotmail.com.

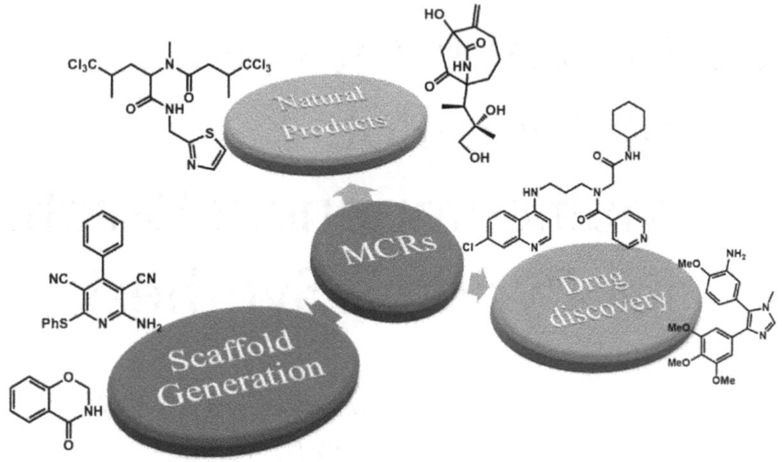

FIGURE 10.1 MCRs valuable for the synthesis of structural diversity and complexity in a one-pot operation.

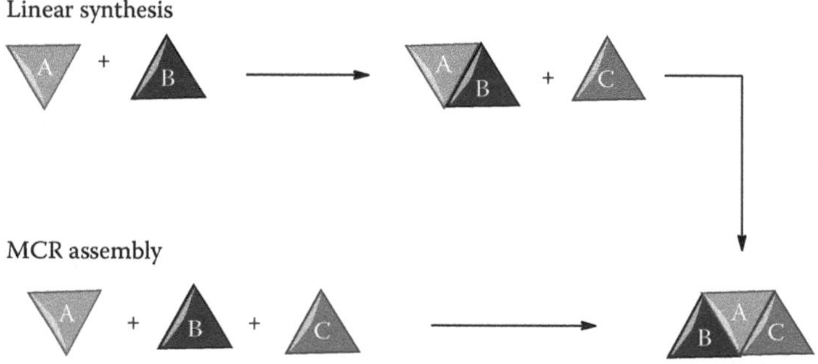

FIGURE 10.2 MCR and linear synthesis.

Multicomponent reactions that offer an intrinsically more attractive transformation for the synthesis of heterocycles through covalent bonds have steadily gained importance in synthetic organic chemistry. One key aspect of these MCRs is that they are a significant source of complex and highly diverse structures created in a one-pot fashion. For example, a three-component MCR will provide up to 1,000 products when 10 variants of every component are employed.[2]

MCRs can be subcategorized mainly into two common classes: isocyanide-based and non-isocyanide-based reactions. The most well-known general MCRs include those of Strecker (1850), Hantzsch's dihydropyridine synthesis (1882), Biginelli (1891), and Mannich (1912). The first isocyanide-based 3CRs was introduced by Passerini in 1921 and the isocyanide-based four-component Ugi reaction in 1959. MCRs are efficient and effective methods in the sustainable and diversity-oriented synthesis of heterocycles.

Heterocycles are ubiquitous in natural products, pharmaceuticals, organic materials, and numerous functional molecules. Six-membered *N*-heterocycles are of immense importance as intermediates as well as building blocks in medicinal and organic chemistry. They play a key role in catalyzing both biological and chemical systems. Among the heterocyclic systems, the pyridine nuclei are the most basic structural moiety; it has considerable significance owing to its presence in biologically active compounds, active pharmaceuticals, and broad range of natural products and functional materials.

10.1.2 Pyridine

In the pharmaceutical industry, pyridine forms the nucleus of over thousands existing drugs. In 1846, Anderson found the picoline, the first pyridine base from bone oil. Later, the correct structure of pyridine was proposed by Korner and Dewar, who independently formulated a monoaza analog of benzene. Furthermore, the demand for pyridine and its derivatives has been improved over the last few years by the discovery of many bioactive pyridine-containing compounds by several pharmaceutical industries. For example, Figure 10.3 shows a recently isolated natural product containing the pyridine core as well as highly substituted pyridine moiety. In addition to these, substituted pyridines are prominent building blocks in supramolecular chemistry due to their *p*-stacking and directed H-bond forming ability.[3–6] The ongoing interest for developing a new methodology for the syntheses of pyridine has always attracted huge attention from both synthetic and medicinal chemists.[7]

Literature studies revealed that these polysubstituted pyridincs show pharmaceutically significant properties, such as antibacterial,[8] anticancer,[9] potassium channel opener for treatment of urinary incontinence,[10] anti-hepatitis B virus (HBV) infection,[11] Parkinson's disease, hypoxia, asthma, cancer, kidney disease, etc.[12–14]

The pyridine molecule is a prominent structural moiety found in various synthetic compounds and natural products with vital biological activities. Considerable attention has been directed toward design and development of new approaches for the synthesis of pyridines by both synthetic and medicinal chemists. Particularly, multicomponent reactions are masterpieces of synthetic efficiency and reaction design.[15]

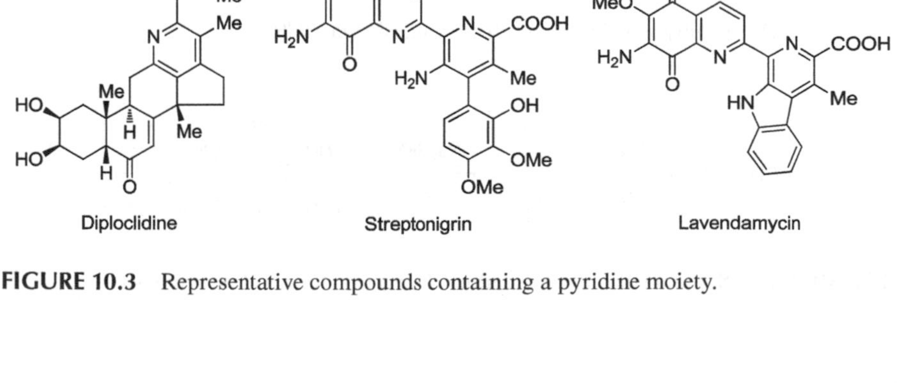

Nicotine   Anabasine   Nicotinic acid   Polyvinyl Pyridine

Diploclidine   Streptonigrin   Lavendamycin

FIGURE 10.3 Representative compounds containing a pyridine moiety.

In this chapter we will review a large variety of MCRs for the synthesis of polysubstituted pyridine. Nowadays, there have been a number of synthetic applications performed with acid, base, and metal heterogeneous catalysts.

10.2 PYRIDINE SYNTHESIS

10.2.1 DICARBONITRILE- AND THIOL-SUBSTITUTED PYRIDINE

Singh et al. have developed one-pot MCRs for highly substituted pyridines under microwave irradiation (Scheme 10.1).[16] KF/alumina was used as a heterogeneous green catalyst in order to give 3,5-dicarbonitrile substituted pyridine. The present work describes an efficient signal pot MCR for polysubstituted pyridine through a condensation of malononitrile, aldehyde, and thiols via conventional as well as microwave irradiation.

Heravi et al. developed a method for the synthesis of polysubstituted pyridine dicarbonitrile molecules by using sodium silicate catalyst at the ambient temperature (Scheme 10.2).[17] The procedure offers several advantages, including high yields, operational simplicity, cleaner reaction, minimal environmental impact, and low cost, which make it a useful and attractive process for the synthesis of these compounds This process has cleaner reaction operational simplicity, making it a feasible and attractive process for the synthesis of various dicarbonitrile substituted pyridine analogs.

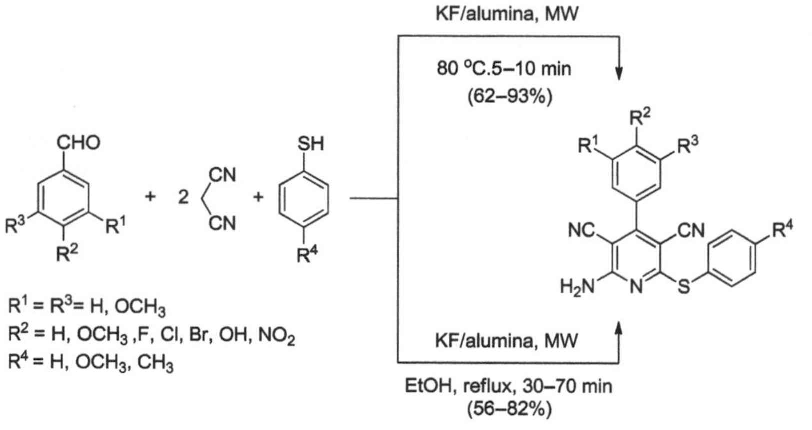

KF/alumina, MW
80 °C.5–10 min
(62–93%)

$R^1 = R^3 = H, OCH_3$
$R^2 = H, OCH_3, F, Cl, Br, OH, NO_2$
$R^4 = H, OCH_3, CH_3$

KF/alumina, MW
EtOH, reflux, 30–70 min
(56–82%)

SCHEME 10.1 Highly substituted pyridine synthesis via MCR.

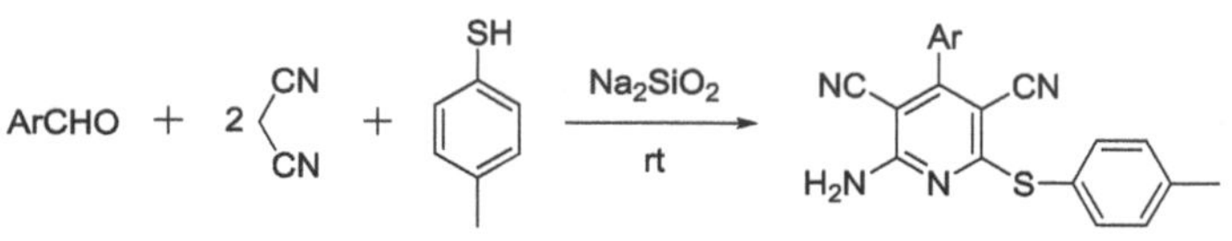

ArCHO + 2 malononitrile + 4-methylthiophenol —Na_2SiO_2, rt→

SCHEME 10.2 Sodium silicate catalyzed MCR reaction for pyridine synthesis.

The feasibility of tetrabutylammonium hydroxide (TBAH) was illustrated in the synthesis of pyridine-3,5-dicarbonitriles, via an MCR between malonitrile, thiophenol, and various aldehydes (Scheme 10.3).[18–20] In the case of ionic[18] and nitrogenated[19] bases, they afforded similar yields of pyridine via multicomponent reaction with elongated reaction time. The desired pyridine-3,5-dicarbonitriles were achieved after only 60 min in refluxing acetonitrile in the presence of TBAH (50 mol%); however, 24 h were essential in the presence of piperidine (30 mol%) in order to obtain similar yields.

Kantam and co-workers have reported one-pot, three-component synthesis of 2-amino-4-aryl-3,5-dicyano-6-sulfanylpyridines are from readily accessible starting materials (Scheme 10.4).[21] Simply heating an ethanolic solution in the presence of nanocrystalline magnesium oxide, various aldehydes with numerous thiols, and malononitrile provides the substituted analogous pyridine. The advantage of the current method is that the catalyst can be recovered and reused with consistent activity up to four cycles for the synthesis of pyridine derivatives.

Perumal et al. have reported a simple protocol for the effective synthesis of 2-(1*H*-Indol-3-yl)-6-methoxy-4-arylpyridine-3,5-dicarbonitrile through one-pot, multicomponent reaction under reflux conditions in the presence of basic conditions (Scheme 10.5).[22] Newly synthesized pyridine-3,5-dicarbonitrile showed a good

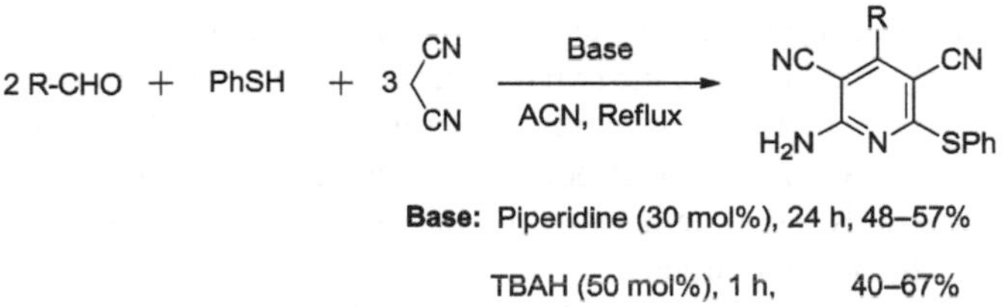

2 R-CHO + PhSH + 3 $CH_2(CN)_2$ → (Base; ACN, Reflux) 2-amino-4-R-6-(phenylthio)pyridine-3,5-dicarbonitrile

Base: Piperidine (30 mol%), 24 h, 48–57%

TBAH (50 mol%), 1 h, 40–67%

SCHEME 10.3 Basic ionic liquid catalyzed multicomponent synthesis of pyridine-3,5-dicarbanitriles.

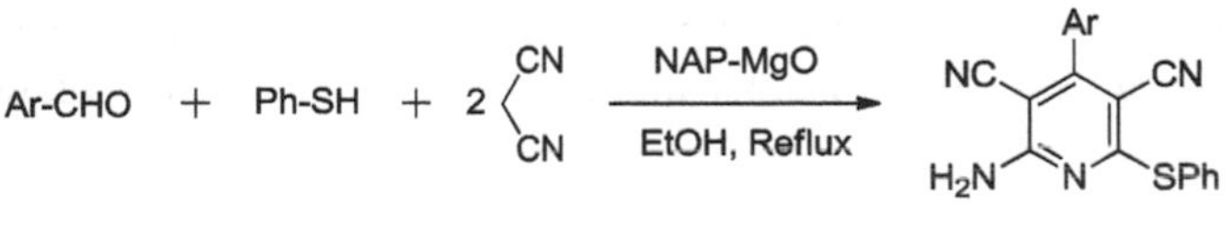

Ar-CHO + Ph-SH + 2 $CH_2(CN)_2$ → (NAP-MgO; EtOH, Reflux) 2-amino-4-Ar-6-(phenylthio)pyridine-3,5-dicarbonitrile

SCHEME 10.4 One-pot synthesis of substituted pyridines catalyzed by NAP-MgO.

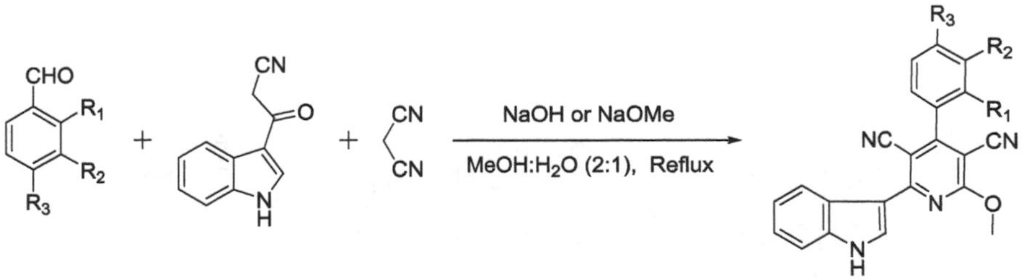

Substituted benzaldehyde (R_1, R_2, R_3) + 3-(cyanoacetyl)indole + $CH_2(CN)_2$ → (NaOH or NaOMe; MeOH:H_2O (2:1), Reflux) 2-(1*H*-indol-3-yl)-6-methoxy-4-aryl-pyridine-3,5-dicarbonitrile

SCHEME 10.5 Synthesis of indolyl-3-yl pyridine derivatives.

anti-inflammatory activity in comparison with the standard drug indomethacin. Also, a series of bis-Hantzsch dihydropyridine derivatives were synthesized and they exhibited analgesic activity when compared with aspirin.

The efficient synthesis of pyridine dicarbonitriles via one-pot, three-component condensation of aldehydes, malononitrile, and thiophenols in the presence of bases (e.g., triethyl amine, morpholine, DIPEA, DMAP, etc.) has been reported by Chen and co-workers (Scheme 10.6).[23] However, exploration of the reaction conditions did not lead to very high yields.

Ranu et al. have reported the basic ionic liquid (such as [bmim]OH) feasible for promoting a one-pot, three-component condensation reaction for the highly substituted pyridines in high yields at ambient temperature (Scheme 10.7).[24] The benefits of this process were the recovery and reuse of ionic liquid for the subsequent reaction without loss of activity. The method offers many advantages, such as high productivity, operational simplicity, minimal environmental impact, and low cost, making it a useful and attractive procedure for the preparation of these derivatives.

The effects of an ionic base such as tetrabutylammoniumhydroxide (TBAH), and an amine base such as piperidine, on the direct synthesis of pyridine-3,5-dicarbonitriles through a multicomponent reaction of malononitrile, aldehydes, and thiols was systematically investigated by Chen and co-workers (Scheme 10.8).[25] The amine base exhibited better effects when the MCR was accomplished in ethanol, whereas using the ionic base in acetonitrile resulted in comparable yields but in a much shorter reaction time. Mechanistically, the two different types of base (ionic base and amine base) as catalysts were found to exhibit different selectivity of oxidant in the final step of the MCR. A Knoevenagel adduct intermediate plays an important role in the amine base catalyzed system; however, in the presence of an ionic base, aerobic oxygen acts as the prime oxidant.

Recently, Reissig and co-workers have reported the practical approach to highly functionalized 4-hydroxypyridine analog with stereogenic side chains in the positions 2 and 6 (Scheme 10.9).[26] This process employs a multicomponent reaction of alkoxyallenes, nitriles, and carboxylic acids to provide β-methoxy-β ketoenamides, which are converted into 4-hydroxypyridines followed by cyclocondensation. The reagents were successfully consumed and converted into the corresponding pyridine

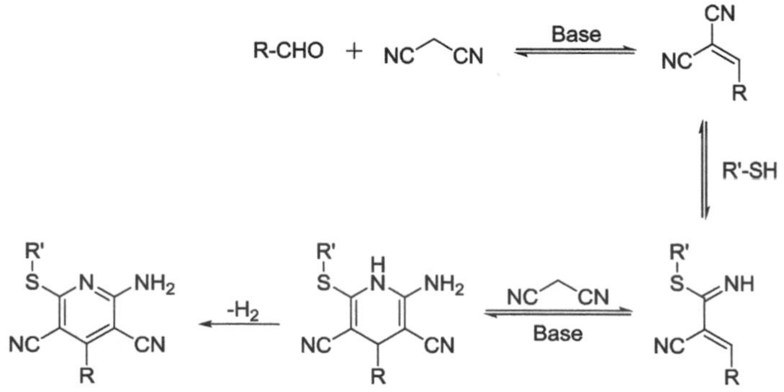

R-CHO + NC CN Base NC CN R R'-SH R' S NH NC R NC CN Base R' H S N NH2 NC CN R $-H_2$ R' S N NH2 NC CN R

SCHEME 10.6 Reported mechanism for pyridine dicarbanitrile synthesis.

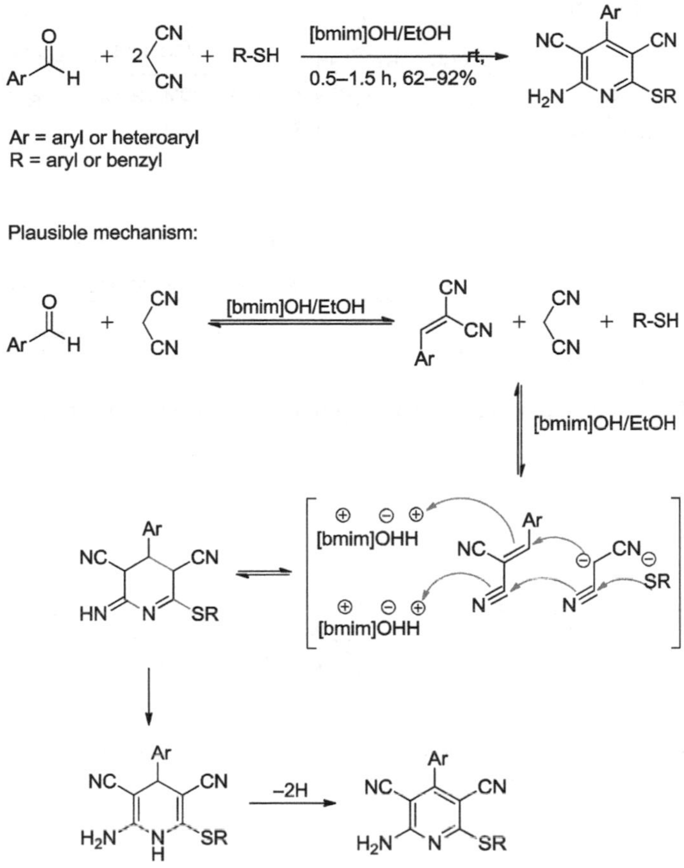

SCHEME 10.7 Synthesis of highly substituted pyridines and plausible mechanistic pathways.

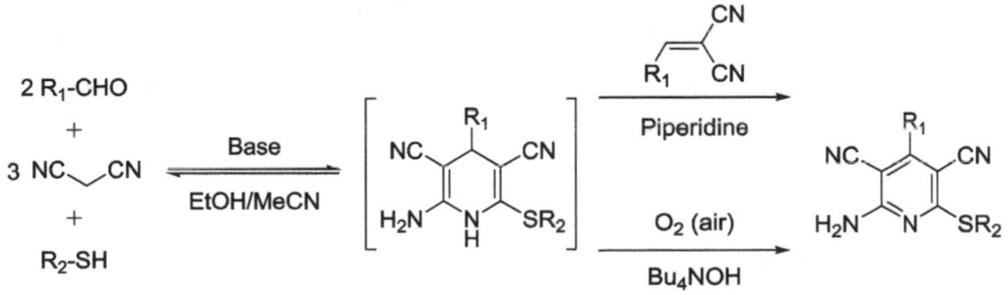

SCHEME 10.8 Synthesis of pyridine-3,5-dicarbonitriles via MCRs.

molecules without loss of enantiopurity to produce enantiopure 4-hydroxypyridine derivatives with stereogenic side chains at C-2 and C-6. The advantage of this technique is that the 4-hydroxy group allows further variations (Scheme 10.10).

Rodriguez and co-workers have developed a simple metal-free and regioselective three-component reaction for the one-pot synthesis of polysubstituted pyridines (Scheme 10.11).[27] This synthetic process worked well via Michael addition

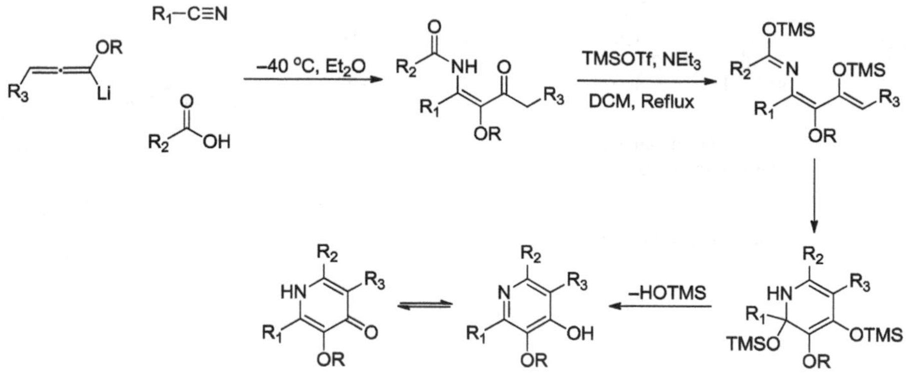

SCHEME 10.9 One-pot, two-step MCR to prepare 4-hydroxypyridines and its functionalization.

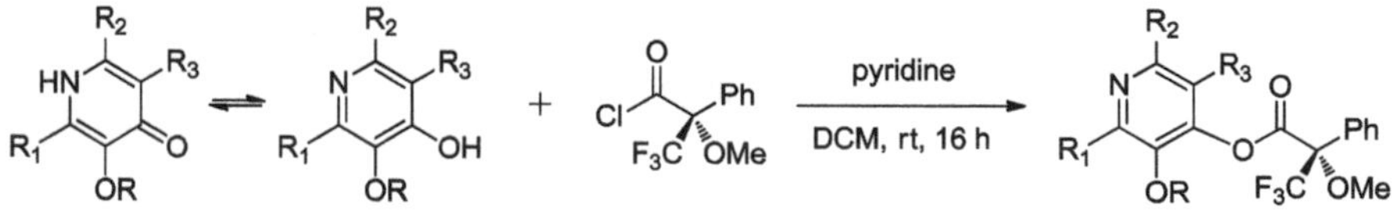

SCHEME 10.10 Esterification of different pyridinol derivatives with the 3,3,3-trifluoro-2-methoxy-2-phenylpropanoic acid.

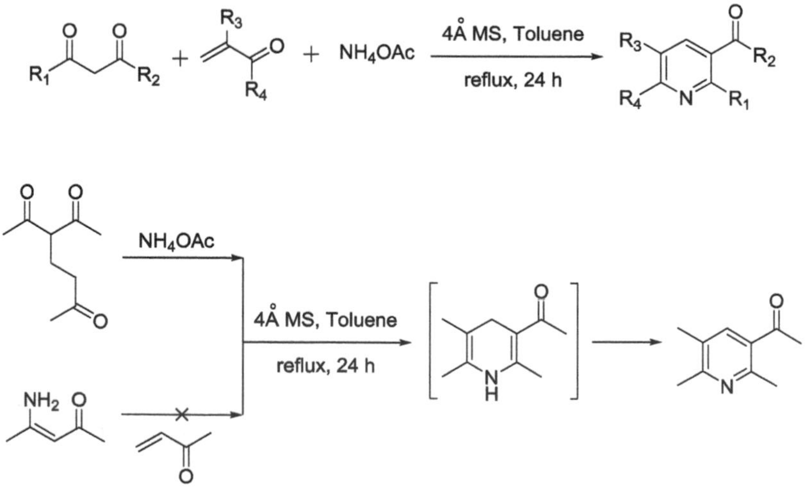

SCHEME 10.11 MCR synthesis of polysubstituted pyridines and mechanistic investigations.

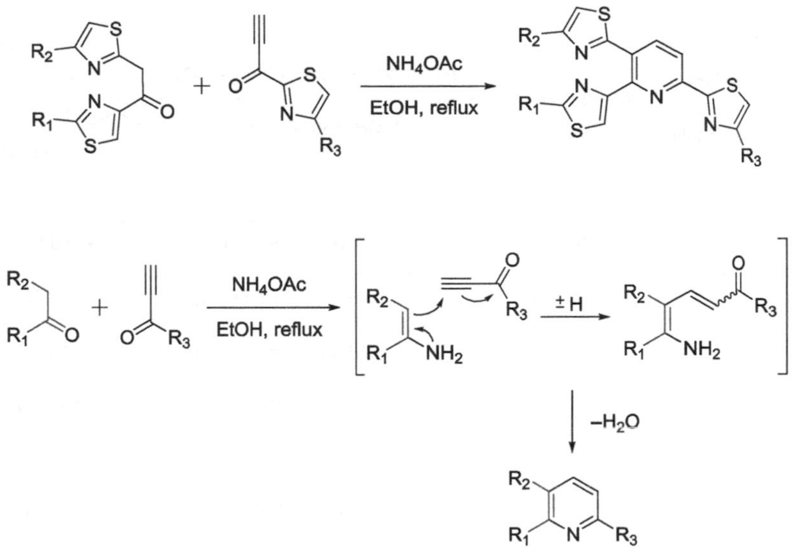

SCHEME 10.12 Synthesis of pyridine–thiazole through the Bagley variant of the Bohlmann–Rahtz reaction.

reaction multicomponent oxidative cyclodehydration to allow invention of functionalized heterocycles of both synthetic and biological interest. A molecular sieve (4Å) stimulated-Michael addition initiated a domino three-component MCR reaction between a Michael acceptor, a 1,3-dicarbonyl, and ammonia as a nitrogen source. This pyridine method is like a biomimetic that does not require any harmful reagent or metal-based catalyst.

Ciufolini et al. have reported new route for the synthesis of pyridine–thiazole through the Bagley variant of the Bohlmann–Rahtz reaction by using simple starting materials—enolizable ketone, ynone, and NH_4OAc—in refluxing ethanol (Scheme 10.12).[28]

Jun et al. have developed a new protocol for the construction of regiocontrolled pyridines that employs sequential Rh(I) catalyzed chelation-assisted hydroacylation of alkynes with aldehydes and consequent Rh(III) promoted annulation of the resulting α,β-enone with another alkyne and ammonia (Scheme 10.13).[29]

10.3 FUSED PYRIDINE SYNTHESIS

The coupling isomerization reaction of electron-poor (hetero)-aryl halide and terminal propargyl *N*-tosylamine furnishes enamine, which undergoes [4+2] cycloaddition with added N,S ketene acetal with inverse electron demand and gives rise to cycloadducts such as pyrrolo[2,3-*b*]pyridines, [1,8]naphthyridines, and pyrido[2,3-*b*]-azepines (Scheme 10.14).[30] The obtained cycloadduct undergoes twofold elimination of tolylsulfiinate and methylmecaptane and finally aromatizes to give rise to the expected fused pyridine.

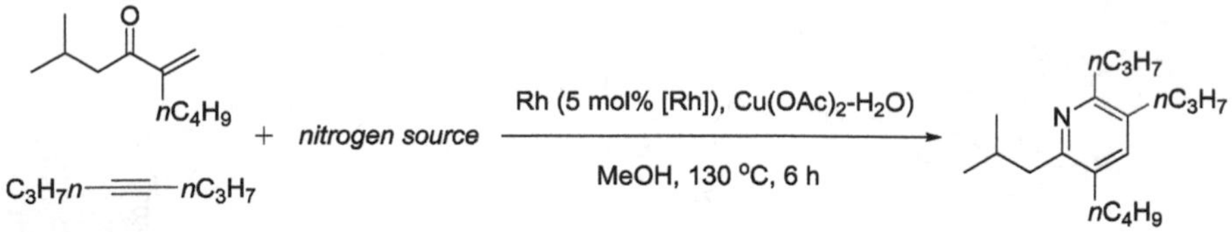

Entry	Rh	Nitrogen Source	Yield (%)[a]
1	$[Cp^*RhCl_2]_2$	7N NH_3 in MeOH (1 eq)	87 (91)
2		7N NH_3 in MeOH (4 eq)	95 (99)
3		NH_4OAc	94 (97)
4		$NH_2OH.HCl$	42 (66)
5		$PhNH\text{-}NH_2$	32 (53)
6		$NH_2\text{-}NH_2.xH_2O$	18 (27)
7		NH_4NO_3	32 (58)
8		NH_4Cl	11 (24)
9	$Cp^*Rh(MeCN)_3(SbF_6)_2$	7N NH_3 in MeOH	75 (82)
10	$(Ph_3P)_3RhCl$		(0)

[a] Isolated yields. GC yields are given in parentheses.

SCHEME 10.13 Rh(I) catalyzed pyridine synthesis.

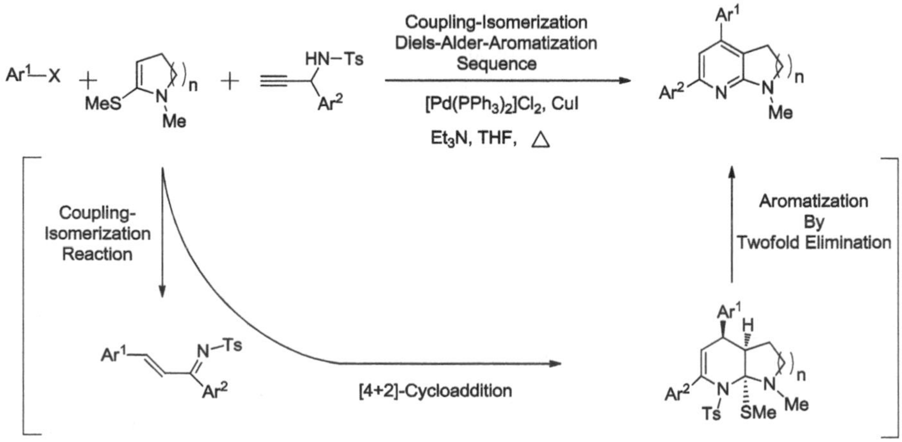

SCHEME 10.14 Synthesis of pyrrolo[2,3-*b*]pyridines.

If we consider indozolines as fused or annelated pyridines, many of the natural or synthetic indozolines have shown biological activities that can be useful for pharmaceutical products[31,32] for the synthesis of some interesting indozolines. Yan and Liu have developed a multicomponent approach catalyzed by gold (Scheme 10.15).[33] In their report they have explained the multicomponent coupling/cycloisomerization reaction of heteroaryl aldehydes, amines, and alkynes catalyzed by gold under solvent-free conditions or in water. No loss of enantiomeric purity was also an important achievement of this methodology when enantiomerially enriched amino acids were used.

The proposed mechanism for this reaction shows that the indozolines were formed via NR^1–NR^2 propargylic pyridine intermediate, which was formed by Mannich–Grignard reaction of pyridine-2-carboxaldehyde, amine, and alkyne. During this process, increased electrophilicity of alkyne due to the coordinated gold catalyst favors the attack of a nitrogen lone pair and forms cation 5. Finally, the deprotonation followed by demetallation affords respective indozoline.

The indenopyridine nucleus is present in the 4-azafluorenone alkloids and also plays an important role in the treatment of neurogenerative disorders and inflammation-related diseases (Scheme 10.16).[34,35] Due to such characteristics, the one-pot, multicomponent reaction approach for the synthesis of indenopyridines has attracted the attention of the synthetic organic community. In a current approach, indeno[1,2-*b*] pridines were synthesized at room temperature by a one-pot, multicomponent

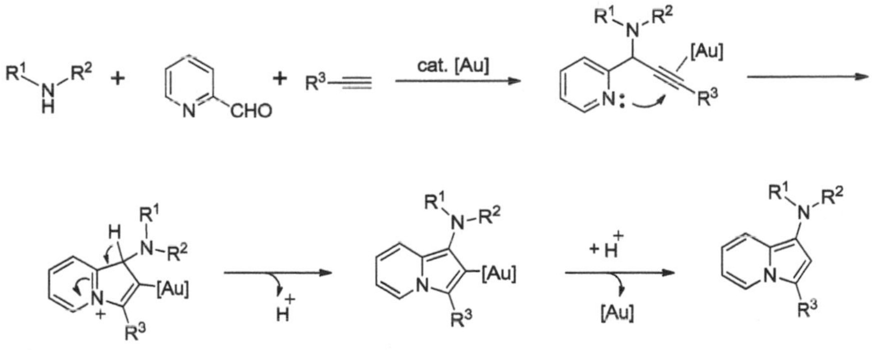

SCHEME 10.15 Gold catalyzed MCRs for indozolines as fused or annelated pyridines.

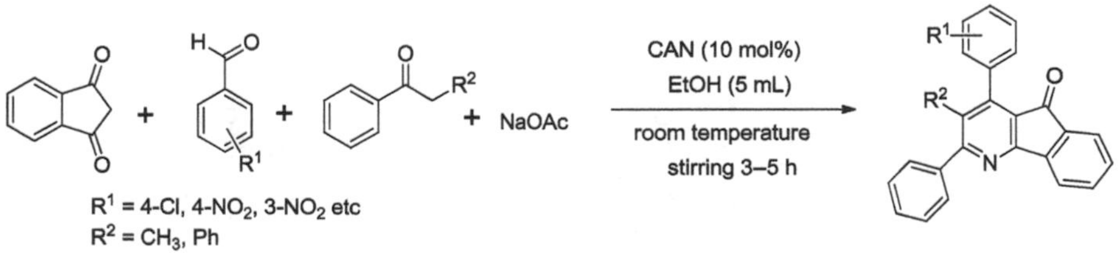

SCHEME 10.16 Synthesis of indenopyridine molecules.

reaction of 1,3-indandione, aldehydes, propiophenone, or 2-phenylacetophenone and ammonium acetate using cerric ammonium nitrate (CAN) as a catalyst.[36]

With an efficient and environmentally benign process, series of indeno[2′,1′:5,6] pyrido[2,3-*d*]pyrimidine derivatives were synthesized (Scheme 10.17) via the three-component reaction of an aldehyde, 6-aminopyrimidine-2,4-dione and 5,5-dimethyl-1,3-cyclohexanedione in ionic liquid 1-*n*-butyl-3-methylimidazolium bromide ([bmim]Br) (Scheme 10.18).[37]

In a similar way, a series of indeno[2′,1′:5,6]-pyrido[2,3-*d*]pyrimidines was obtained by using indanedione instead of 5,5-dimethyl-1,3-cyclohexanedione.

With the proposed mechanism it has been shown that the reactions were preceded by condensation, addition, cyclization, and dehydration and aromatization reaction in a sequence (Scheme 10.19).

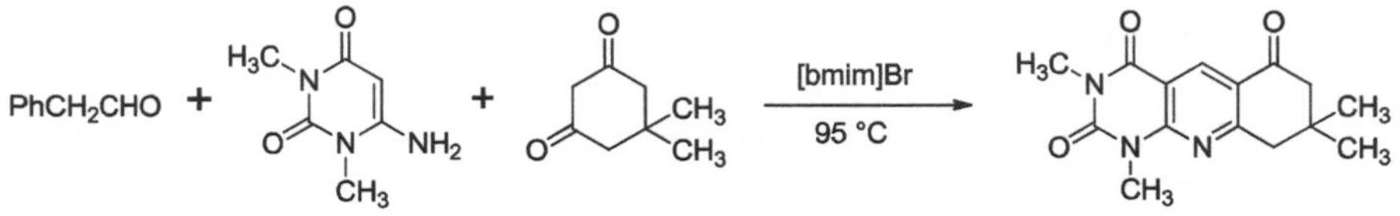

SCHEME 10.17 Synthesis of indeno[2′,1′:5,6]pyrido[2,3-*d*]pyrimidine.

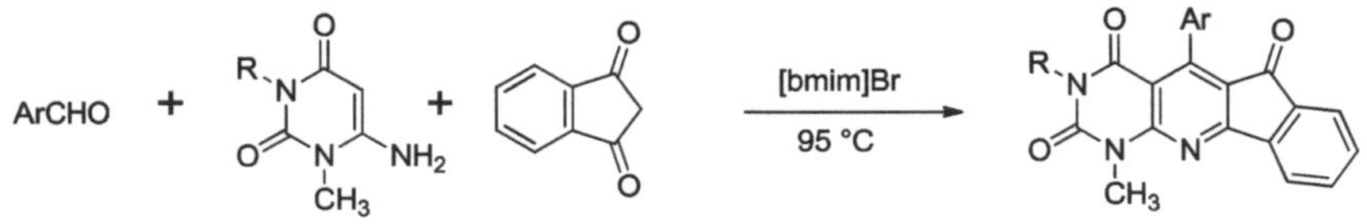

SCHEME 10.18 Synthesis of indeno[2′,1′:5,6]-pyrido[2,3-*d*]pyrimidines.

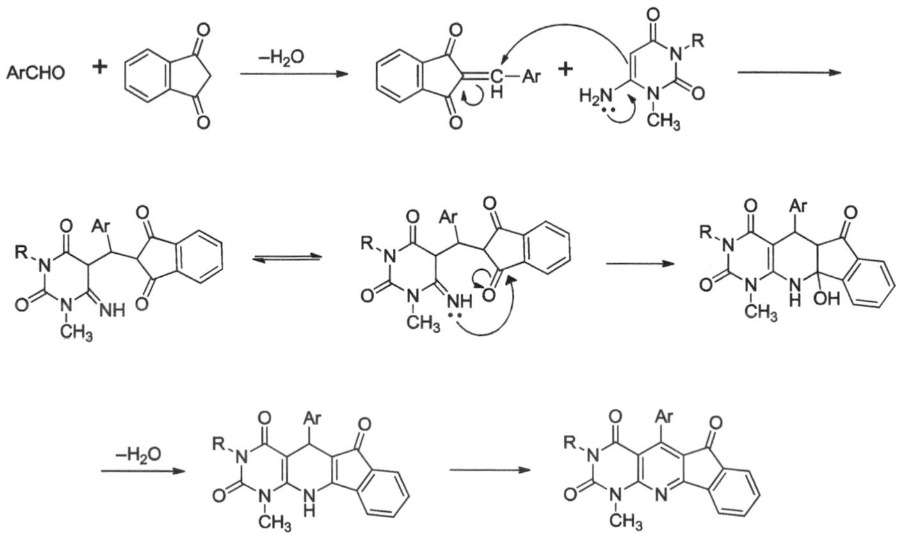

SCHEME 10.19 Plausible mechanism indeno[2′,1′:5,6]-pyrido[2,3-*d*]pyrimidines.

Initially, the aldehyde and used indanedione form an adduct that undergoes additional reaction with 6-aminopyrimidine-2,4-dione to form the intermediate. This formed intermediate then undergoes intramolecular cyclization and dehydration. Finally, aromatization of obtained species results in indeno[2′,1′:5,6]-pyrido[2,3-*d*]pyrimidines.

The ionic liquid 1-*n*-butyl-3-methylimidazolium bromide ([bmim]Br) has been also used in a similar way to synthesize fused pyridine derivatives such as pyrazolo[3,4-*b*]pyridine and pyrido[2,3-*d*]pyrimidine (Scheme 10.20).[38] For this, the three components used were aldehyde, acyl acetonitrile, and electron-rich amino heterocycles including aminopyrazoles and aminouracils.

The reaction proceeds via Knovenagel condensation (Scheme 10.21) of aldehyde and acyl acetonitrile to give the intermediate **A**; then the Michael addition of electron-rich amino heterocycles to this intermediate takes place to form **B**. This intermediate **B** undergoes dehydration and subsequent aromatization to give the final product (Scheme 10.22).

The biological properties shown by pyrido[2,3-*b*]indoles, also known as α-carbinols, have attracted attention of medicinal and natural product chemistry worlds. Apart

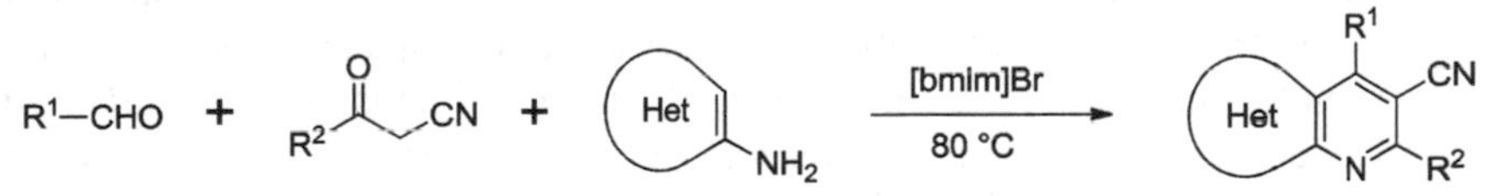

SCHEME 10.20 Ionic liquid catalyzed MCR for synthesis pyrazolo[3,4-*b*]pyridine.

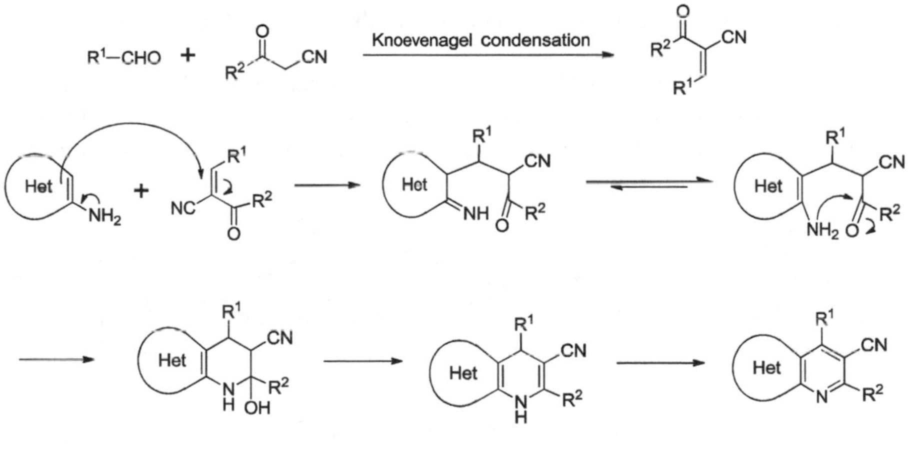

SCHEME 10.21 Proposed mechanistic for fused pyridine.

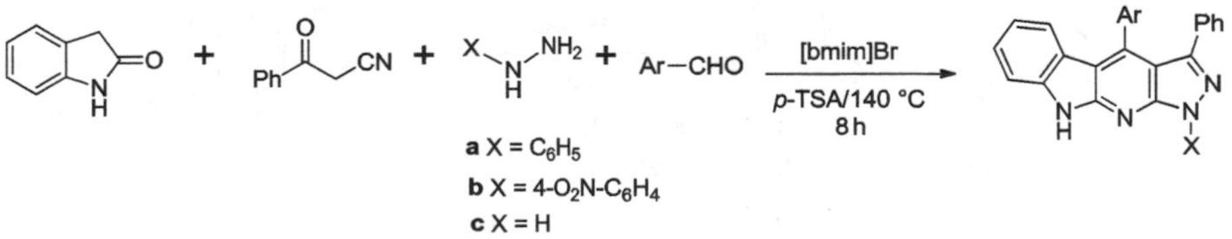

SCHEME 10.22 Synthesis of pyrido[2,3-*b*]indoles via MCRs.

from conventional linear-type synthesis, a multicomponent synthetic route for α-carbinol derivatives under green conditions has been developed.[39] In this multicomponent approach, reaction of indolin-2-one, 3-oxo-3-phenylpropanenitrile, phenylhydrazine, and benzaldehyde under different conditions was carried out. The best result was obtained in presence of *p*-toluenesulphonic acid at 140 °C in [bmim]Br.

Similarly, functionalized thiochromones and their fused analogs have their own importance in drug chemistry due to their interesting biological activities.[40–43] There are very few methods for synthesis of thiochromeno[2,3-*b*]pyridines derivatives and MCR reported by Wen et al. would be a more efficient way for it due to the access to a wide range of substituents.[44] In this protocol, β-(2-chloroaroyl) thioacetanilide, a polyfunctional moiety with four active reaction sites, has been reacted with activated 4-arylidene-2-phenyloxazol-5(4*H*)-ones or aromatic aldehydes and ethyl 2-cyanoacetate under microwave irradiation, which undergoes [3+3] annulation and S_NAr, respectively, to yield corresponding fused tricyclic thiochromeno[2,3-*b*]pyridines (Scheme 10.23).

Among the fused pyridine derivatives, thiopyranopyridines have their own importance due to their biological activities, which makes them a valuable component of drug. Hence, the synthesis of a library comprising thiopyranopyridine scaffold by one-pot, multicomponent reaction got special attention.[45] This method starts with, aromatic aldehyde, ammonium acetate, 3-methyl-1-phenyl-1*H*-pyrazol-5-amine, and 1,3-dicarbonyl compounds, including ethyl-3-oxobutanoate, methyl-3-oxobutanoate, pentane-2,4-dione, 5,5-dimethylcy-clohexane-1,3-dione, and 2*H*-thiopyran-3,5(4*H*,6*H*)-dione (Scheme 10.24). After several screens for suitable solvent and temperature, a glacial acetic condition at room temperature was the most yielding condition for this protocol.

Using these conditions, a library of thiopyranopyridines has been synthesized with moderate to excellent yields. These compounds were synthesized from different types of aliphatic as well as aromatic aldehydes along with a variety of substituents.

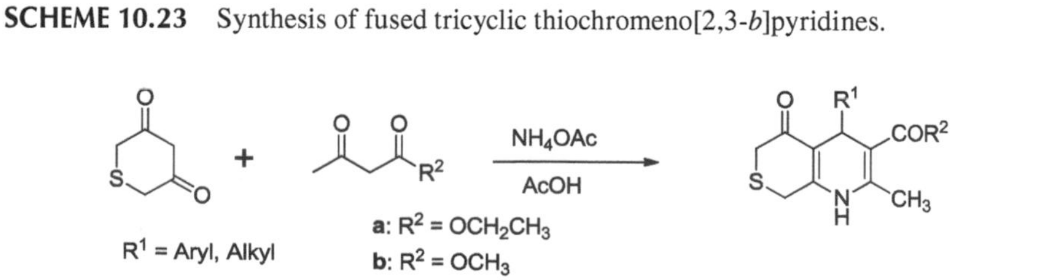

SCHEME 10.23 Synthesis of fused tricyclic thiochromeno[2,3-*b*]pyridines.

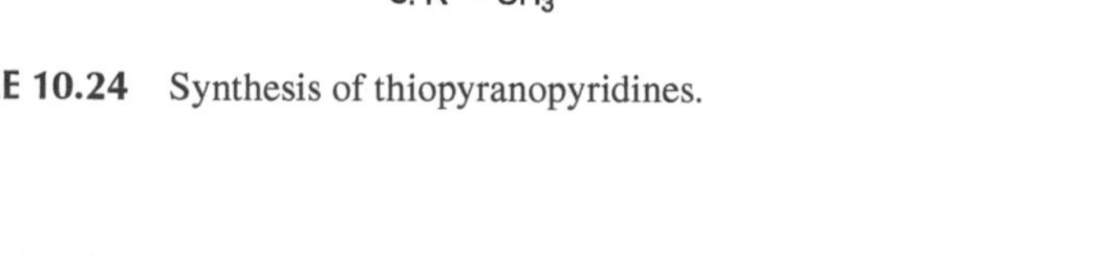

SCHEME 10.24 Synthesis of thiopyranopyridines.

In this way this one-pot, multicomponent reaction approach has been proved to be one of the reliable ways to synthesize thiopyranopyridines under mild reaction conditions with good yields.

One of the common and important features of a variety of natural products and pharmaceutical agents is represented by indole (Scheme 10.25). This property or characteristic of indole makes it an important heterocycle to be focused on by chemists worldwide. The enhancement of this indole unit can also be achieved by active participation by 3-carbon of indole in spiroindoline derivative formations.[46–49] A very well reported one-pot, pseudo-four-component methodology initiates with 1,3-indandione, aniline, isatin, and *p*-TSA as a catalyst.[50] The reaction has been monitored using various solvents and Lewis acid catalysts but acetonitrile and *p*-TSA provided the most favorable condition for the formation of spiro[diindeno[1,2-*b*:2′,1′-*e*]pyridine-11,3′-indoline]-2′,10,12-trione derivatives.

The mechanism proposed on the basis of spectroscopic data suggests that, in the presence of *p*-TSA, 1,3-indandione reacts with isatin to form the intermediate **A**, which finally reacts with aniline to form the desired product (Scheme 10.26). Unfortunately, this methodology was not very successful with aliphatic amines such as *n*-propylamine or ethylamine.

Coumarins are very well known naturally occurring compounds and they are famous for their various pharmacological activities.[51–53] Among various types of derivativcsm 3-aminocoumarin is important due to its participation as a fundamental skeleton in lots of naturally occurring alkaloids and antibiotics like novobiocin. There are many reports on the synthesis of pyrido[2,3-*c*] coumarins.[54–58]

The method reported by Khan and co-workers using molecular iodine is special, as the protocol does not need aqueous workup and chromatographic purification.[59]

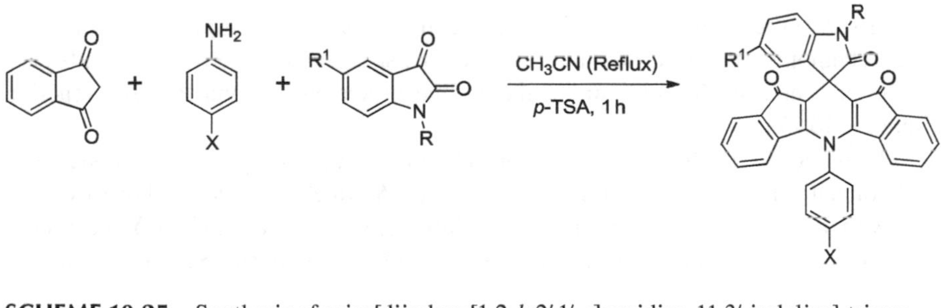

SCHEME 10.25 Synthesis of spiro[diindeno[1,2-*b*:2′,1′-*e*]pyridine-11,3′-indoline]-trione.

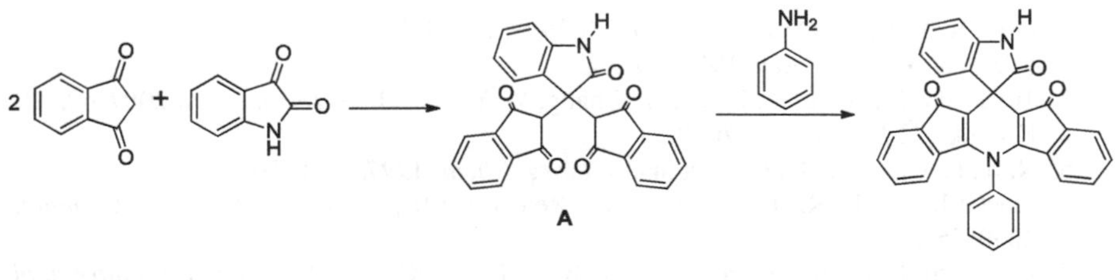

SCHEME 10.26 Mechanistic pathways for spiro[diindeno[1,2-*b*:2′,1′-*e*]pyridine-11,3′-indo line]-trione.

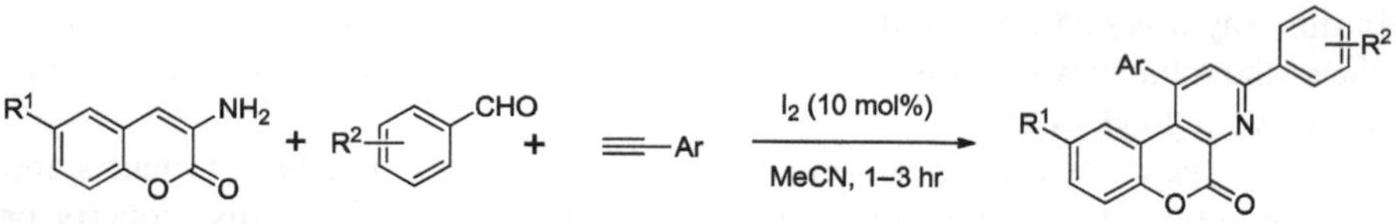

SCHEME 10.27 Iodine catalyzed pyrido[2,3-*c*]coumarins via Povarov reaction.

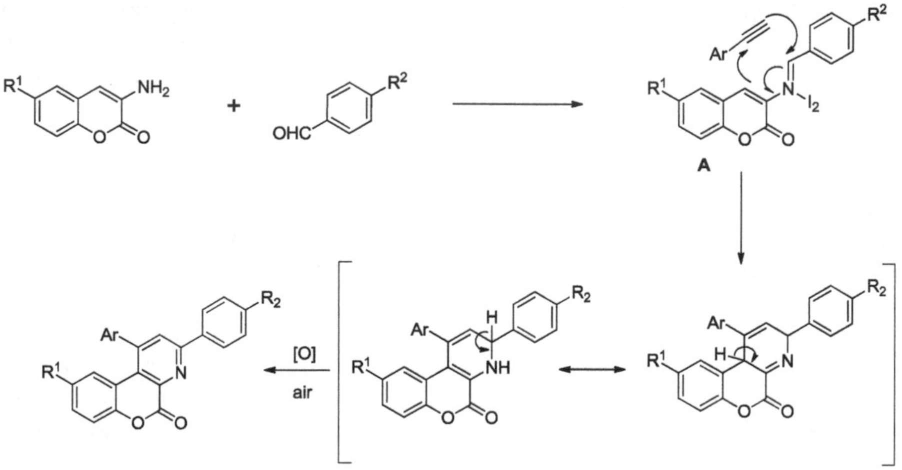

SCHEME 10.28 Mechanism for the synthesis of pyrido[2,3-*c*] coumarine derivatives.

In this report the synthesis of pyrido[2,3-*c*]coumarins was accomplished by one-pot, three component reaction of 3-amino-coumarins, aromatic aldehydes, and alkynes in the presence of 10 mol% molecular iodine in acetonitrile via Povarov reaction (Scheme 10.27).

The proposed mechanism for the formation of product suggests that the aldehyde and coumarin react to for the imine intermediate **A** which undergoes a Povarov reaction with the dienophilic alkyne to give the dihydropyridine intermediate **Y** that undergoes aerial oxidation and gives respective pyrido[2,3-*c*]coumarin (Scheme 10.28).

REFERENCES

1. Orru, R. V. A.; de Greef, M. *Synthesis* **2003**, *10*, 1471.
2. Eilbracht, P. *Chem. Rev.* **1999**, *99*, 3329.
3. Tu, S.; Jia, R.; Jiang, B.; Zhang, J.; Zhang, Y.; Yao, C.; Ji, S. *Tetrahedron* **2007**, *63*, 381.
4. Krohnke, F. *Synthesis* **1976**, *01*, 1.
5. Neve, F.; Crispini, A.; Campagna, S. *Inorg. Chem.* **1997**, *36*, 6150.
6. MacGillivray, L. R.; Diamente, P. R.; Reid J. L.; Ripmeester, J. A. *Chem. Commun.* **2000**, 359.
7. Fredholm, B. B.; Ijzerman, A. P.; Jacobson, K. A.; Klotz K. N.; Linden, J. *Pharmacol. Rev.* **2001**, *53*, 527.
8. Srivastava, S. K.; Tripathi R. P.; Ramachandran, R. *J. Biol. Chem.* **2005**, *280*, 30273.

9. Azuine, M. A.; Tokuda, H.; Takayasu, J.; Enjyo, F.; Mukainaka, T.; Konoshima, T.; Nishino, H.; Kapadia, G. *J. Pharmacol. Res.* **2004**, *49*, 161.
10. Harada, H.; Watanuki, S.; Takuwa, T.; Kawaguchi, K.; Okazaki, T.; Hirano, Y.; Saitoh, C. PCT Int. Appl., WO 2002006237 A1 20020124, 2002.
11. Chen, H.; Zhang, W.; Tam, R.; Raney, A. K. PCT Int. Appl., WO 2005058315 A1 20050630, 2005.
12. Quintela, J. M.; Peinador, C.; Veiga, M. C.; Botana, L. M.; Alfonso, A.; Riguera, R. *Eur. J. Med. Chem.* **1998**, *33*, 887.
13. Attia, A. M. E.; El-H, A.; Ismail, A. A. *Tetrahedron* **2003**, *59*, 1749.
14. Chang, L. C. W.; von Frijtag Drabbe Kunzel, J. K.; Mulder-Krieger, T.; Spanjersberg, R. F.; Roerink, S. F.; van den Hout, G.; Beukers, M. W.; Brussee, J.; Ijzerman, A. P. *J. Med. Chem.* **2005**,*48*, 2045.
15. Fredholm, B. B.; Ijzerman, A. P.; Jacobson, K. A.; Klotz, K. N.; Linden, J. *Pharmacol. Rev.* **2001**, *53*, 527.
16. Singh, K. N.; Singh S. K. *ARKIVOC* **2009**, *xiii*, 153.
17. Heravi, M. M.; Khorshidi, M.; Beheshtia, Y. S.; Baghernejad, B. *Bull. Korean Chem. Soc.* **2010**, *31*, 1343.
18. Ranu, B. C.; Jana, R.; Sowmiah, S. *J. Org. Chem.* **2007**, *72*, 3152.
19. Evdokimov, N. M.; Magedov, I. V.; Kireev A. S.; Kornienko, A. *Org. Lett.* **2006**, *8*, 899.
20. Guo, K.; Thompson, M. J.; Chen, B. *J. Org. Chem.* **2009**, *74*, 6999.
21. Kantam, M. L.; Mahendar, K.; Bhargava, S.; *J. Chem. Sci.* **2010**, *122*, 63.
22. Thirumurugan, P.; Mahalaxmi, S.; Perumal, P. T.; *J. Chem. Sci.* **2010**, *122*, 819.
23. Reddy, T. R. K.; Mutter, R.; Heal, W.; Guo, K.; Gillet, V. J.; Pratt, S.; Chen, B. *J. Med. Chem.* **2006**, *49*, 607.
24. Ranu, B. C.; Jana, R.; Sowmiah, S. *J. Org. Chem.* **2007**, *72*, 3152.
25. Guo, K.; Thompson, M. J.; Chen, B. *J. Org. Chem.* **2009**, *74*, 6999.
26. Eidamshaus, C.; Kumar, R.; Bera M. K.; Reissig, H. U. *Beilstein J. Org. Chem.* **2011**, *7*, 962.
27. Lieby-Muller, F.; Allais, C.; Constantieux, T.; Rodriguez, J. *Chem. Commun.* **2008**, *35*, 4207.
28. Aulakh V. S.; Ciufolini, M. A. *J. Org. Chem.* **2009**, *74*, 5750.
29. Sim, Y. K.; Lee, H.; Park, J. W.; Kim, D. S.; Jun, C. H. *Chem. Commun.* **2012**, *48*, 11787.
30. Nee Dediu, O. G. S.; Oeser, T.; Muller, T. J. J. *J. Org. Chem.* **2006**, *71*, 3494.
31. Micheal, J. P. *Alkaloids* **2001**, *55*, 91.
32. Micheal, J. P. *Nat. Prod. Rep.* **2002**, *19*, 742.
33. Yan, B.; Liu, Y. *Org. Lett.* **2007**, *9*, 4323.
34. Heintzelman, G. R.; Averill, K. M.; Dodd, J. H.; Demarest, K. T.; Tang, Y.; Jackson, P. F. PCT Int. Appl. WO 2003088963, 2003.
35. Heintzelman, G. R.; Averill, K. M.; Dodd, J. H.; Demarest, K. T.; Tang, Y.; Jackson, P. F. U.S. Pat. Appl. Publ. US 2004082578, 2004.
36. Tapaswi, P. K.; Mukhopadhyay, C, *ARKIVOC* **2011**, *x*, 287.
37. Shi, D.-Q.; Ni, S.-N.; Yang, F.; Shi, J.-W.; Dou, G.-L.; Li, X.-Y.; Wang X.-S.; Ji, S.-J. *J. Heterocycl. Chem.* **2008**, *45*, 693.
38. Huang, Z.; Hu, Yu.; Zhou, Y.; Shi, D. *ACS Comb. Sci.* **2011**, *13*, 45.
39. Ghahremanzadeh, R.; Ahadi, S.; Bazgi, A. *Tetrahedron Lett.* **2009**, *50*, 7379.
40. Geissler, J. F.; Roesel, J. L.; Meyer, T.; Trinks, U. P.; Traxler, P.; Lydon, N. B. *Cancer Res.* **1992**, *52*, 4492.
41. Nakazumi, H.; Ueyama, T.; Kitao, T. *J. Heterocycl. Chem.* **1984**, *21*, 193.
42. Wang, H.; Bastow, K. F.; Cosentino, L. M.; Lee, K. *J. Med. Chem.* **1996**, *39*, 1975.
43. Holshouser, M. H.; Loeffler, L. J.; Hall, I. H. *J. Med. Chem.* **1981**, *24*, 853.
44. Wen, L. R.; Sun, J. H,; Li, M.; Sun, E. T.; Zhang, S. S. *J. Org. Chem.* **2008**, *73*, 1852.

45. Yao, C-S.; Wang, C. H.; Jiang, B.; Tu, S. J. *J. Comb. Chem.* **2010,** *12*, 472.
46. Joshi, K. C.; Chand, P. *Pharmazie* **1982**, *37*, 1.
47. Da- Silva, J. F. M.; Garden, S. J.; Pinto, A. C. *J. Braz. Chem. Soc.* **2001**, *12*, 273.
48. Abdel-Rahman, A. H.; Keshk, E. M.; Hanna, M. A.; El-Bady, Sh. M. *Bioorg. Med. Chem.* **2004**, *12*, 2483.
49. Zhu, S.-L.; Ji, S.-J.; Yong, Z. *Tetrahedron* **2007**, *63*, 9365.
50. Ghahremanzadeh, R.; Shakibaei, G. I.; Ahadi, S.; Bazgir, A. *J. Comb. Chem.* **2010**, *12*, 191.
51. O'Kennedy, R.; Thornes, R. D. *Coumarins: Biology, applications and mode of action.* J. Wiley & Sons: Chichester, England, 1997.
52. Fylaktakidou, K. C.; Hadjipavlou-Litina, D. J.; Litinas, K. E.; Nicolaides, D. N. *Curr. Pharm. Des.* **2004**, *10*, 3813.
53. Zhang, W.; Pugh, G. *Tetrahedron Lett.* **2001**, *42*, 5613.
54. Khan, M. A.; Gremal, A. L. *J. Heterocycl. Chem.* **1977**, *14,* 1009
55. Pave, G.; Chalard, P.; Viaud-Massuard, M.-C.; Troin, Y.; Guillaumet, G. *Synlett* **2003**, *07*, 987.
56. Majumdar, K. C.; Chattopadhyay, B.; Taher, A. *Synthesis* **2007**, *23*, 3647.
57. Kudale, A. A.; Kendall, J.; Miller, D. O.; Collins, J. L.; Bodwell, G. L. *J. Org. Chem.* **2008**, *73*, 8437.
58. Kudale, A. A.; Miller, D. O.; Dawe, L. N.; Bodwell, G. *J. Org. Biomol. Chem.* **2011**, *9*, 7196.
59. Khan, A. T.; Das, D. K.; Islam, K.; Das, P. *Tetrahedron Lett.* **2012**, *53*, 6418.

11 Synthesis of Functionalized Piperidine Derivatives Based on Multicomponent Reaction

Padmakar Suryavanshi, Vijaykumar Paike, Sandeep More, Rajendra P. Pawar,* Sandeep B. Mane, and K. L. Ameta*

CONTENTS

11.1 INTRODUCTION

Nitrogen-containing heterocycles are one of the extensively explored heterocyclic compounds in various fields like natural products, pharmaceuticals, and agrochemicals.[1] Piperidines and their derivatives have become increasingly popular building blocks in vast array of synthetic protocols.[2] Several review reports in the literature have described various strategies for the piperidine synthesis, but none have highlighted the synthesis of piperidines through multicomponent reactions (MCRs).[3–5] In recent years, MCRs have become a powerful methodology in organic chemistry.[6,7] Plenty of heterocyclic molecules are synthesized by this methodology and, day by day, demands for new developments in this area are steadily increasing. MCRs can be defined as convergent processes whereby more than two reagents are combined in a sequential manner to give the final product in a one-pot chemical operation that retains significant portions of all starting material (Figure 11.1). Therefore, they have to be proved to be very effective and efficient bond-forming tools in synthetic organic chemistry.[8]

* Corresponding authors: padmakarchem@gmail.com and rppawar@yahoo.com.

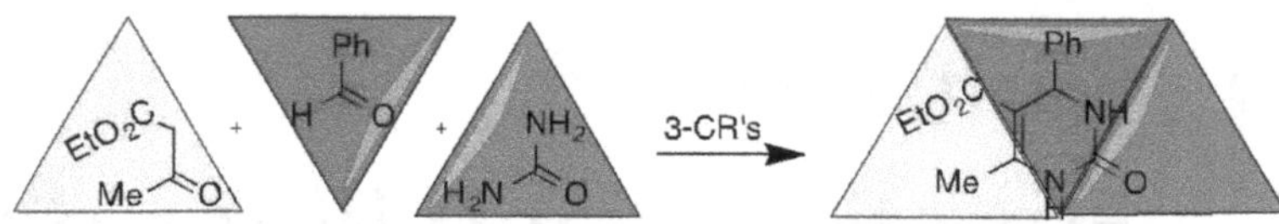

FIGURE 11.1 Multicomponent reactions (MCRs).

The multicomponent reaction approach is based on a simple reaction procedure, shorter reaction time, simple and readily available starting materials, experimental simplicity, and favorable economic and environmental factors such as energy saving and use of environmentally benign solvents. In a single synthetic operation it can generate high diversity and complexity in molecules and retains high chemo-, regio-, and steteroselectivity. Many interesting name reactions were reported based on MCR's pathway (Figure 11.2). The first multicomponent reaction reported by Strecker in 1850[9] and it is well known as Strecker synthesis for the α-aminonitriles, which is a capable intermediate for the synthesis of amino acid. While in an MCR a large part of the work developed on isonitrile (also called isocynide)[10] substrate because of its interesting functional group "–CN," it undergoes both electrophilic and nucleophilic reaction and also occurs in many natural products as a functional group.[11]

In MCRs isonitrile is used as one of the starting materials and products leading to peptide-like structures. There are two important multicomponent reactions based on isonitriles: the three-component Passerine reaction[12,13] reported in 1921 to generate α-acyloxy carboxamides, and the four-component Ugi reaction,[14] reported in 1959 to produce α-acylamino carboxamides.

Piperidine, a six-membered heterocyclic amine, is found in many important natural products and is a significant pharmacophore present in numerous interesting biologically active synthetic compounds.[15] In fact, more than half the alkaloids known today contain piperidine substructures. Many research groups have gained attention for the synthesis of piperidine derivatives through several routs. Indeed, over the last two decades various functionalized piperidine derivatives have been successfully entered into preclinical and clinical trials.[16] Some representative examples of piperidine alkaloids are summarized in Figure 11.3.

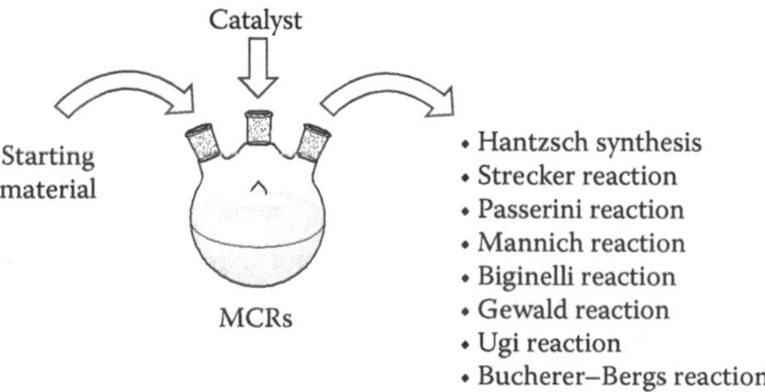

FIGURE 11.2 Name reactions based on MCR pathway.

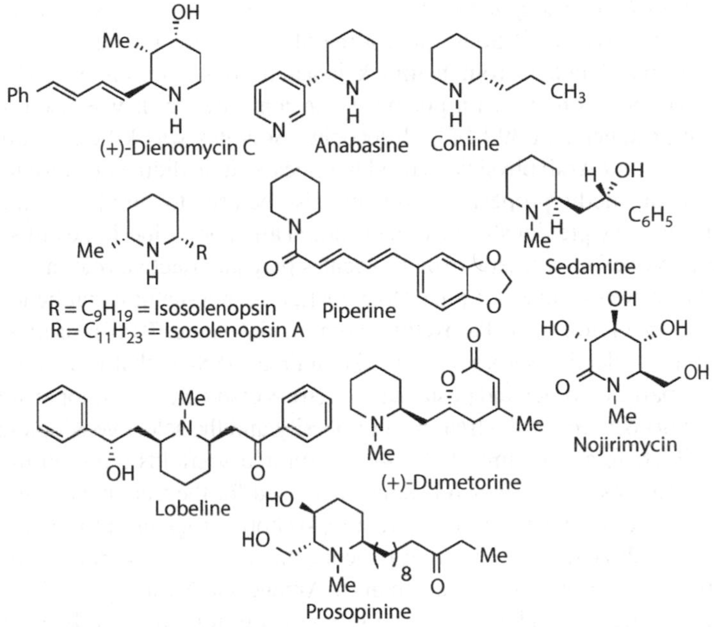

FIGURE 11.3 Some bioactive piperidine natural alkaloids.

Dienomycin C is an alkaloid with three chiral centers that has been extracted from *Streptomyces* strain (MC.67-Cl) and shows antibiotic activity against some strains of mycobacteria.[17] Anabasine is found in the tree tobacco (*Nicotiana glauca*) plant and shows activity as a nicotinic acetylcholine receptor agonist; initially it was used as insecticide.[18] Sedamine and solenopsin-A alkaloids are found not only in plants but also in the animal kingdom, typically in insects. These alkaloids inhibit angiogenesis via the phosphoinositol-3-kinase (PI3-K) pathway and it exhibits cytotoxic, insecticidal, antibacterial, and antifungal as well as anti-HIV properties.[19] The alkaloids isosolenopsin and isosolenopsin-A are extracted from fire ant venom and have demonstrated a variety of activities such as antibiotic, necrotoxic, hemolytic, antifungal and anti-HIV activities.[20] The *Lobelia* genus contains more than 50 species of alkaloids in which *Lobelia inflata* itself has high concentrations of more than 20 piperidine alkaloids. In early years it was used to produce vomiting and relieve conditions such as colic, rheumatism, and fever. Currently, it is used to treat drug abuse and neurological disorders and for smoking cesstion.[21] The Dumetorine (+) alkaloid has been isolated from *Discorea dumetorum* and is often used in African folk medicine.[22] The alkaloid piperidine is extracted from black pepper and extensively used in traditional medicine.[23] Numerous polysubstituted alkaloids have exhibited a high range of biological activities. As a polysubstituted natural glucose mimic produced by *Streptomyces roseochromogenes* R-468 and *S. lavendulae* SF-425, nojirimycin was first described as an antibiotic; recently, it has been found to be a potent inhibitor of α- and β-glucosidase. Currently, different analogs of nojirimycin exhibited a diverse range of biological activities.[24] Prosopinine and prosophylline alkaloids are extracted from *Prosopis africana* Taub leaves and these alkaloids exhibit antibiotic and anesthetic activities.[25] Some piperidine alkaloids are potent poison. Coniine, a neuromuscular blocker from hemlock isolated from *Conium maculatum*, is a plant that has been known as a poison; in ancient Greece, it was used to execute condemned prisoners. In 399 BC, philosopher Socrates was killed by this poison. Ancient Greek and Arab physicians used hemlock as a medicine because it has sedative and antispasmodic properties and could also be used to cure joint pain.[26]

In addition to the previously mentioned natural alkaloids, piperidine units present in a number of bioactive synthetic drug molecules play an effective role in curing many diseases. Selective examples of piperidine containing non-natural synthetic drugs are summarized in Figure 11.4. Paroxetine (commercial name, Paxil), used as an antidepressant, is a selective serotonin reuptake inhibitor (SSRI) that is also used for the treatment of panic disorder and posttraumatic stress disorder.[27] Alvimopan (other trade name, Entereg) is used for the treatment of obesity and the selective K opioid receptor antagonist.[28] There are a number of 1,4-disubstituted scaffolds reported for different biological activities, one of which is donepezil, used for the treatment of Alzheimer's disease.[29] Sertindole and risperidone are antipsychotic drugs used to manage schizophrenia.[30] For schizoaffective disorder patients, these drugs are very effective.

Naratriptan is a triptan drug (trade names, Amege and Naramig) used for the treatment of migraine headaches.[16] Several potent μ-opinoid agonist drugs have been

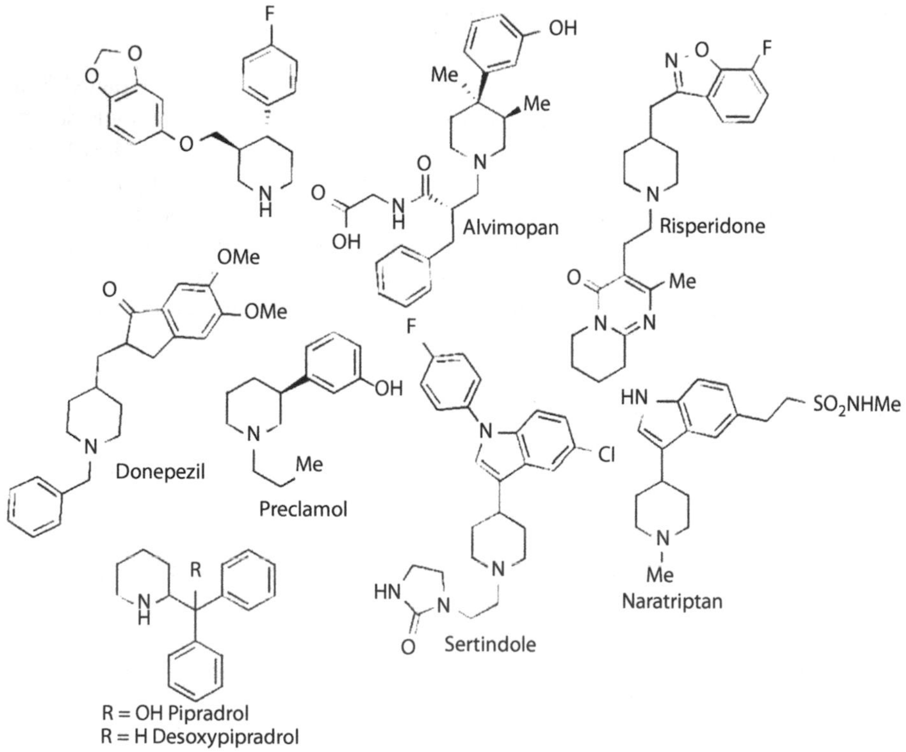

FIGURE 11.4 Some drug molecules comprising piperidine scaffold.

reported, such as Remifentanil and Carfentanil, which are used as short-acting ultra-potent analgesic drugs and in veterinary medicine as potent opioids,[31] respectively. Pipradrol and desoxypipradrol are used as central nervous system stimulants, particularly norepinephrine–dopamine reuptake inhibitors.[32] Methylphenidate is structured similarly to the compounds pipradrol and desoxypipradrol; it acts as a central nervous system stimulant and is used for the treatment of postural orthostatic tachycardia syndrome, narcolepsy, and hyperactivity disorder (ADHD).[33]

11.2 SYNTHESIS

11.2.1 Mono-, Di-, and Trisubstituted Piperidine Derivatives by MCR

A four-component transformation between anilines, diols, aldehydes, and formic acid has provided an efficient entry into mono-, di-, and trisubstituted piperidines as shown by Bruneau and co-workers. This transformation was carried out in toluene in the presence of catalyst **B** as an iridium complex having a phosphanesulfonate chelate and camphorsulfonic acid (CSA) as a Brønsted acid; it afforded the expected substituted piperidine with excellent regio- and promising diastereoselectivities. Functionalization used variety of aromatic aldehydes, different electron-rich anilines, and substituted diols. In this transformation the authors successfully applied diethyl carbonate, which is an environmentally benign substrate (Scheme 11.1).[34]

The Hui group developed a three-component tandem aza-Michael–aldol reaction, starting between α,β-unsaturated enone and 4-methylbenzesulfonamide or trifluoromethanesulfonamide in acetonitrile as a solvent; DBU was used as base for this

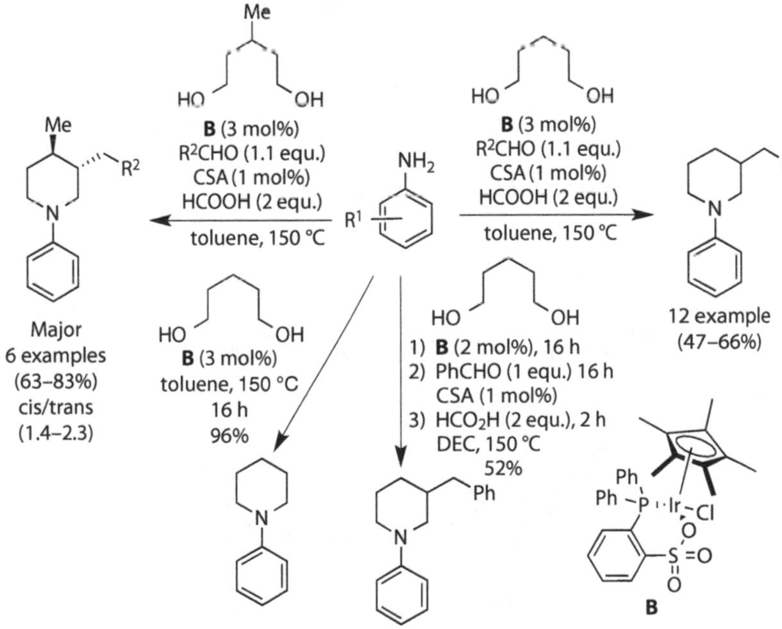

SCHEME 11.1 Synthesis of mono-, di-, and trisubstituted piperidines through MCR.

transformation. The reaction involved α,β-unsaturated enones and substituted sulfonamides; a variety of α,β-unsaturated enones with aromatic, heteroaryl and alicyclic groups were applied for the reaction. 4-Methylbenzesulfonamide or trifluromethanesulfonamide was used as a nucleophilic reagent for the reaction (Scheme 11.2).[35]

A mechanistic proposal involved first the aza-Michael addition reaction of α,β-unsaturated enones I with TsNH2 to give the enolate II, which further attacked another molecule of α,β-unsaturated enones I to generate enolate III. Then intermediate III moved next to an intramolecular aldol reaction to form piperidine derivatives (Scheme 11.3).

A three-substrate multicomponent coupling was between pipecolic acid, aldehydes, and isonitriles via a metal-free decarboxylative pathway to give regioselective synthesis of *N*-substituted piperidine-2-carboxamides by the Batra group. This was a one-pot, multicomponent entry via a cascade process. The reaction proceeded as sequential imine formation, decarboxylation, then isonitrile insertion, and, finally, through hydrolysis to afford the piperidine derivatives (Scheme 11.4).[36] The reaction became very impressive because it proceeded without any metal catalyst or any additive. From this transformation, 2-substituted piperidine derivative was synthesized; toluene played a solvent role for this reaction.

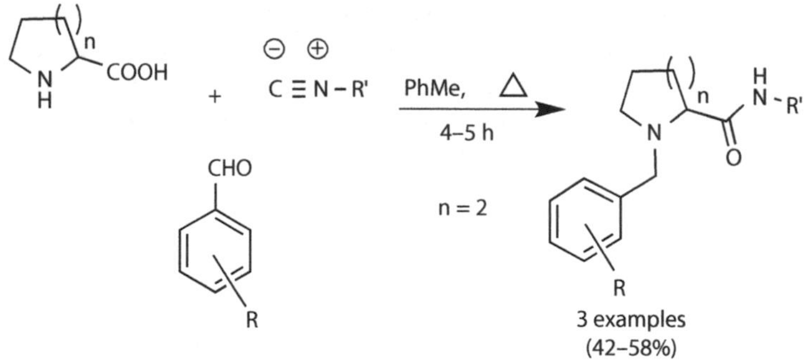

SCHEME 11.2 MCR-based aza-Michael-aldol reaction.

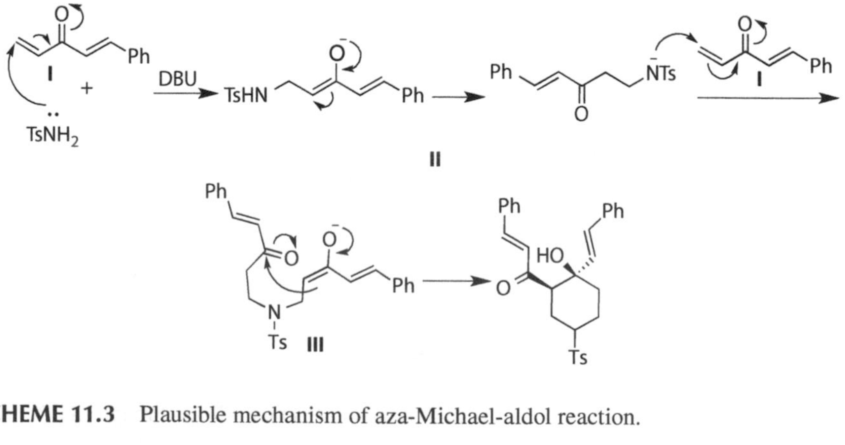

SCHEME 11.3 Plausible mechanism of aza-Michael-aldol reaction.

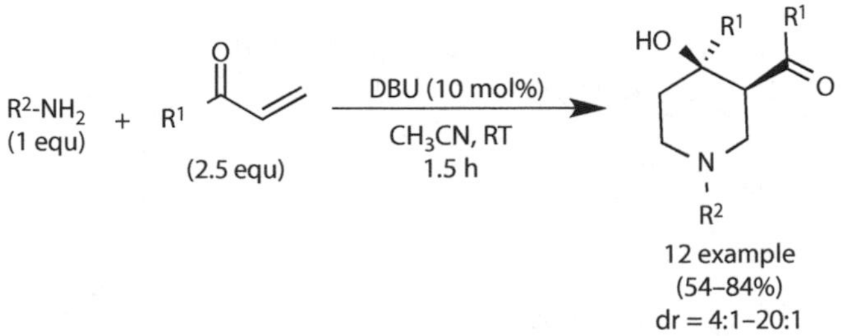

R2-NH2 (1 equ) + R1 (2.5 equ) DBU (10 mol%) CH3CN, RT 1.5 h
HO R1 R1 O N R2
12 example (54–84%) dr = 4:1–20:1

SCHEME 11.4 One-pot, multicomponent synthesis of piperidine-2-carboxamides.

Ramapanicker and co-workers reported a one-pot, three-component Mannich reaction between 5-bromopentanal, acetone, and *p*-anisidine in presence of proline catalyst to offered 2-substituted piperidine derivatives (Scheme 11.5).[37]

The reaction proceeded at room temperature in methanol to give quantitative conversion with solvent-free conditions in the presence of one equivalent Et3N offered 75% yield. The significance of this methodology is the short synthesis of natural alkaloids pelletrine, sedridine, allosedridine, and coniine. These proline catalyzed reactions proceeded as initially acetone and proline reacted to form enamine, which then underwent a Michael reaction to generate iminium ion intermediate A. Then cyclization of this intermediate via hydrolysis gave a piperidine compound (Scheme 11.6).

The Ugi reaction played a very important role for the construction of different heterocycles via multicomponent pathways. Laconde and co-workers successfully used the Ugi four-component reaction started from propionic acid, aniline, 4-phenylethylpiperidone, and 1-cyclohexenyl isocyanide in methanol at room temperature for the synthesis of substituted piperidine derivatives. This multicomponent strategy has led to synthesis of two drugs—carfentanil and remifentanil—in a one-step sequence with good yields (Scheme 11.7).[38]

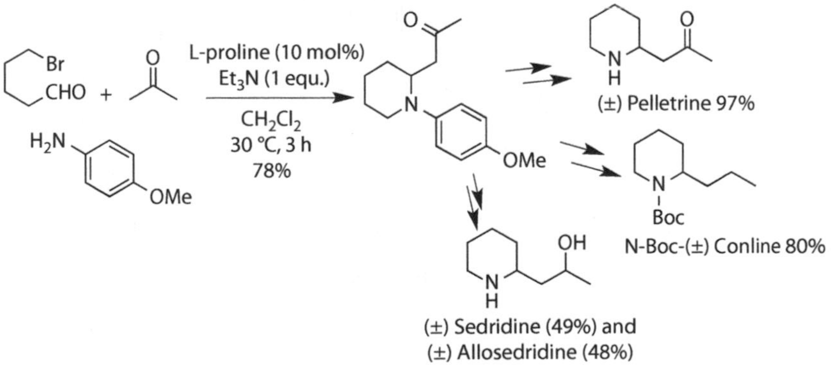

Br CHO + O H2N OMe
L-proline (10 mol%) Et3N (1 equ.) CH2Cl2 30 °C, 3 h 78%
O N OMe
N H O (±) Pelletrine 97%
N Boc N-Boc-(±) Conline 80%
N H OH (±) Sedridine (49%) and (±) Allosedridine (48%)

SCHEME 11.5 One-pot, three-component Mannich reaction.

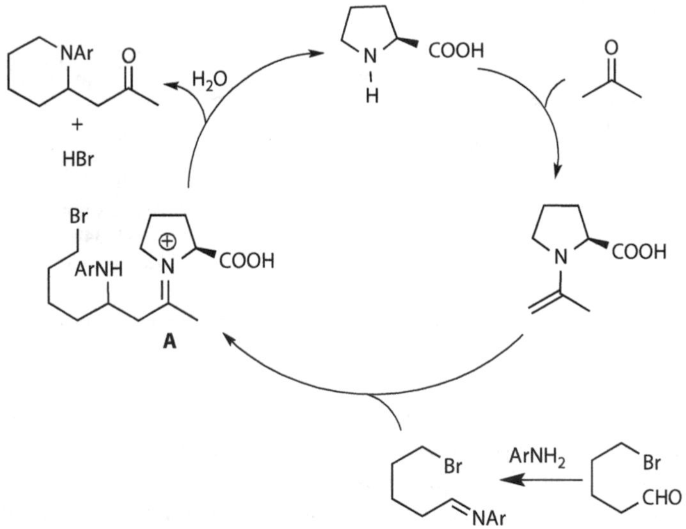

SCHEME 11.6 Plausible mechanism of Mannich reaction.

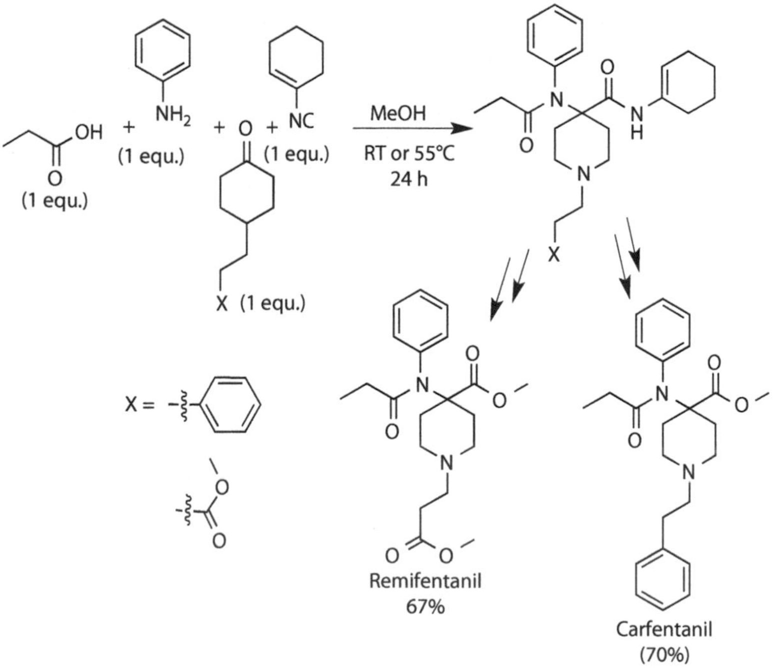

SCHEME 11.7 Multicomponent synthesis of drug molecules.

11.2.2 Polysubstituted Piperidine Derivatives by MCR

A variety of functionalized piperidine nuclei are present in many natural alkaloids and drug molecules. Various synthetic methodologies are present for synthesis of the polysubstituted piperidine derivatives. Among them, one of the most important is the multicomponent reaction. Clarke and co-workers reported a five-component condensation between methyl acetoacetate, aldehyde, and aniline in the presence of InCl3 as Lewis acid to give highly substituted piperidine derivatives (Scheme 11.8).[39] For this reaction, acetonitrile performed as a solvent. This transformation is one of the best examples for pot, atom, and step economic (PASE) synthesis.

Recently, many groups have investigated bromodimethylsulfonium bromide (BDMS) in organic synthesis as a useful reagent. It worked as brominating agent and as a catalyst for organic transformations. Khan et al. recently used BDMS as a catalyst for the multicomponent reaction. They successfully developed a three-component reaction starting between benzaldehyde, aniline, and 1,3-dicarbonyl compound in acetonitrile in the presence of BDMS as a catalytic at room temperature. They used different 1,3-dicarbonyl compounds and substituted aniline for this conversion (Scheme 11.9).[40]

The same group described a one-pot, multicomponent reaction by using tetrabutylammonium tribromide (TBATB) as a catalyst for the synthesis of highly

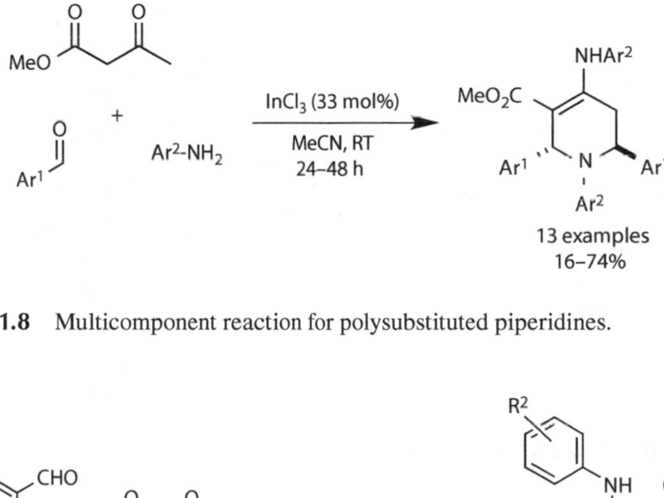

SCHEME 11.8 Multicomponent reaction for polysubstituted piperidines.

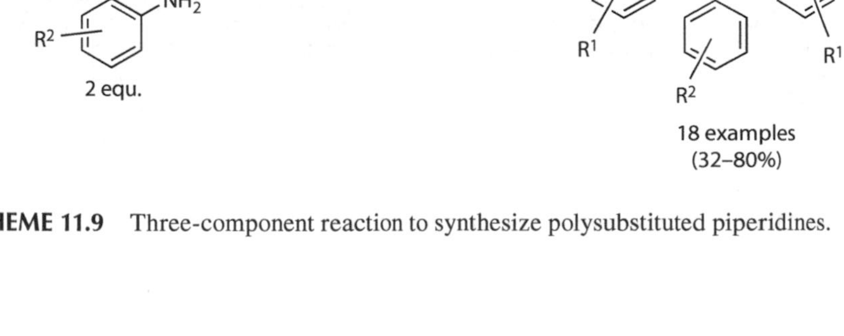

SCHEME 11.9 Three-component reaction to synthesize polysubstituted piperidines.

functionalized piperidine derivatives. Many research groups have used TBATB as a significant brominating reagent and catalyst for a variety of organic transformations. The reaction started from three components between 1,3-dicarbonyl compounds, aromatic aldehydes, and amines as a starting material at room temperature.

After numerous solvent optimization studies, ethanol has been chosen as solvent for the reaction. For this transformation, substituted 1,3-dicarbonyl compounds, substituted aniline, and substituted aldehydes were used to explore the reaction scope. This method became effective due to its mild reaction conditions, good yields, and use of cheap, readily available starting materials (Scheme 11.10).[41]

An iodine catalyzed one-pot, multicomponent reaction has been reported by Khan and co-workers for the synthesis of highly functionalized piperidine. In the literature, molecular iodine was investigated as an environmentally benign catalyst, which is nontoxic, nonmetallic, and readily available for variety of organic transformations. The reaction started from five components of 1,3-dicarbony compounds, amines, and aromatic aldehydes using 10 mol% iodine as a catalyst in methanol at room temperature and offered polysubstituted piperidine derivatives. The structure was confirmed by x-ray crystallographic analysis. To explore the reaction scope, the authors used a different substituent on β-position of 1,3-dicarbonyl compound and used various substituted aromatic aldehyde and aniline having electron-donating and electron-withdrawing groups (Scheme 11.11).[42]

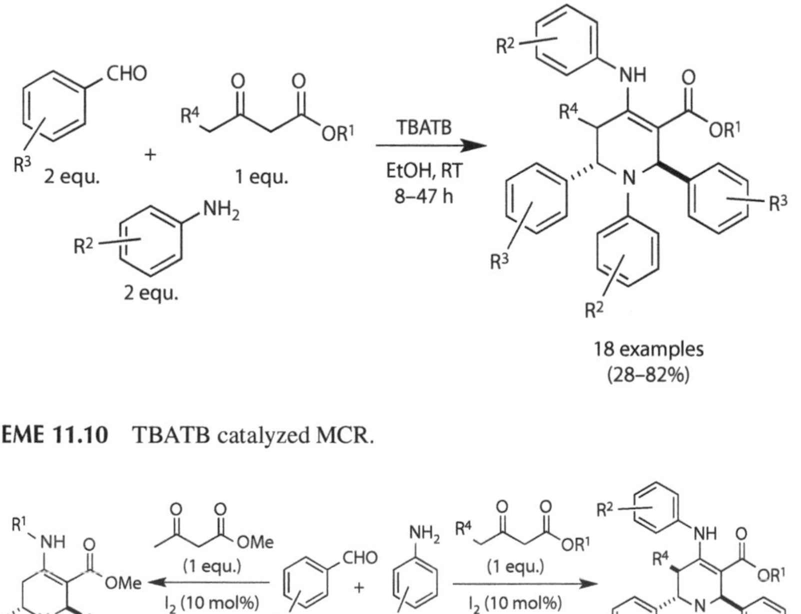

SCHEME 11.10 TBATB catalyzed MCR.

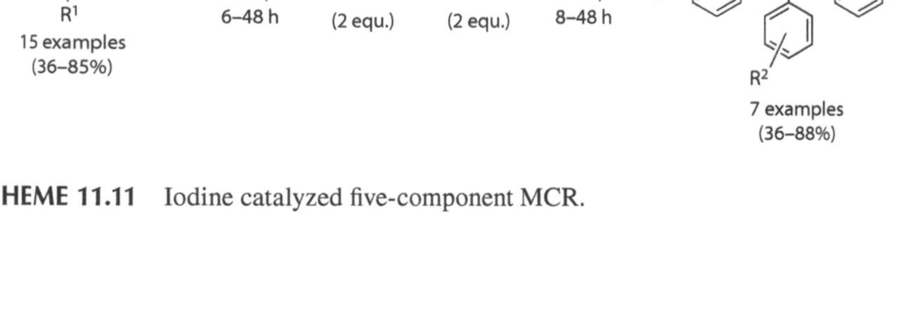

SCHEME 11.11 Iodine catalyzed five-component MCR.

The Ghosh group described a one-pot, three-component transformation between aromatic aldehydes, amines, and acetoacetic esters in ethanol at reflux conditions for the synthesis of functionalized piperidine scaffolds based on an aqua-compatible $ZrOCl_2.8H_2O$ catalyst. The product and stereochemistry were confirmed by x-ray crystallography. The catalyst was recovered and recycled after reaction. For this transformation, the authors screened various catalysts and solvents; they found that $ZrOCl_2.8H_2O$ as a catalyst and ethanol as a solvent gave the best result. Recently, many research groups have reported $ZrOCl_2.8H_2O$ as a catalyst for the organic transformation (Scheme 11.12).[43]

A wet picric acid catalyzed reaction has been developed by the Mukhopadhyay group for the synthesis of *syn*-diastereoisomer in the diastereoselective synthesis of highly functionalized piperidine derivatives. A reaction started from aromatic aldehydes 1,3-dicarbonyl compounds and aromatic amines via a one-pot condensation. Reaction conditions were established according to pure *anti* and pure *syn* isomers. They confirmed that anti-isomers are more stable than *syn* on the basis of DFT calculation and ORTEP diagrams of a single crystal of each isomer. For reaction, they used MeOH–water (60:40) as a medium after a variety of solvents were optimized (Scheme 11.13).[44]

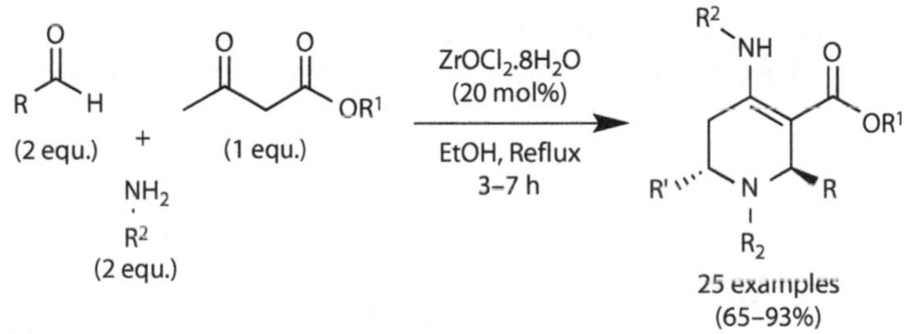

SCHEME 11.12 $ZrOCl_2{\cdot}8H_2O$ catalyzed three-component MCR.

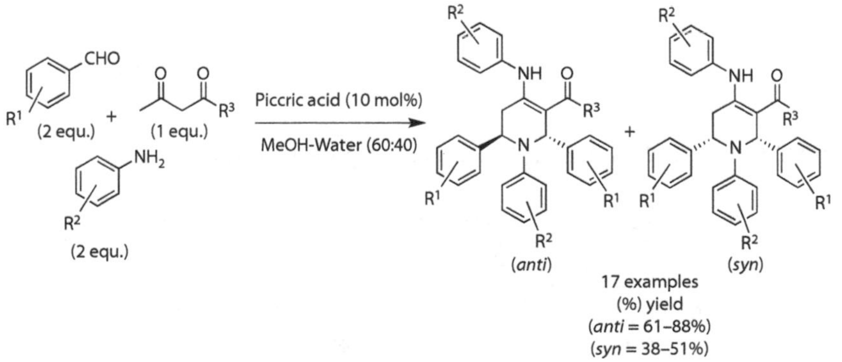

SCHEME 11.13 Wet picric acid catalyzed three-component MCR.

In 2011, Sain and co-workers described the PEG-embedded potassium tribromide [(K^+PEG)Br_3^-] catalyzed three-component reaction combining aniline, aldehyde, and β-keto ester for the synthesis of highly substituted piperidines in a one-pot operation at room temperature in ethanol. After reaction, [(K^+PEG)Br_3^-] was readily recycled, regenerated, and used for fresh reaction. For this transformation, the authors successfully installed different substituents on aniline, aldehyde, and β-keto ester. The product of this MCR was confirmed on x-ray crystallography (Scheme 11.14).[45]

PEG-embedded potassium tribromide [(K^+PEG)Br_3^-] has been synthesized by host–guest complexation with an alkali metal cation concept. The preparation started from equimolar amounts of PEG400 and KBr to form [(K^+PEG)Br_3^-]; then, subsequent reaction with Br2 offered a dark orange-red colored viscous liquid that dried under vacuum and was used without any purification for synthesis. It was assumed that the structure of [(K^+PEG)Br_3^-] was similar to [18-crown-6]KBr_3 because of poly(ethylene)glycols have similar structures of crown ethers (Scheme 11.15).[46]

The Brahmachari group developed a simple, straightforward, one-pot synthesis of pharmaceutically interesting functionalized piperidine derivatives from 1,3-dicarbonyl compounds, aromatic aldehydes, and various amines in presence of $Bi(NO_3)2.5H_2O$ in ethanol at room temperature. They screened different catalysts for this transformation and found $Bi(NO_3)2.5H_2O$ to give the best results (Scheme 11.16).[47]

Organic catalysts play a very important role in synthetic chemistry. A number of catalysts have been developed for the different transformations including

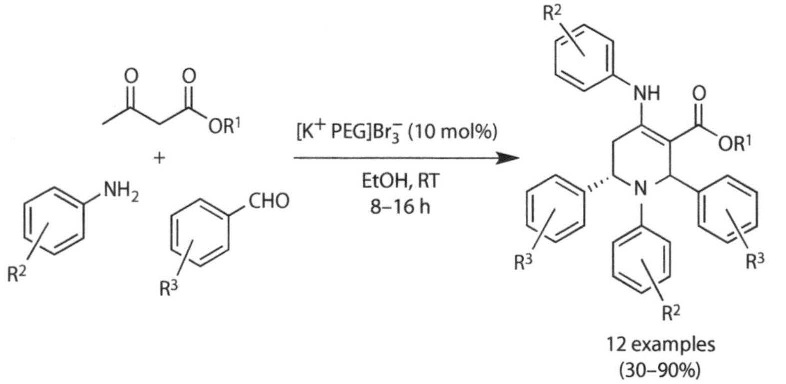

SCHEME 11.14 (K^+PEG)Br_3^- catalyzed three-component MCR.

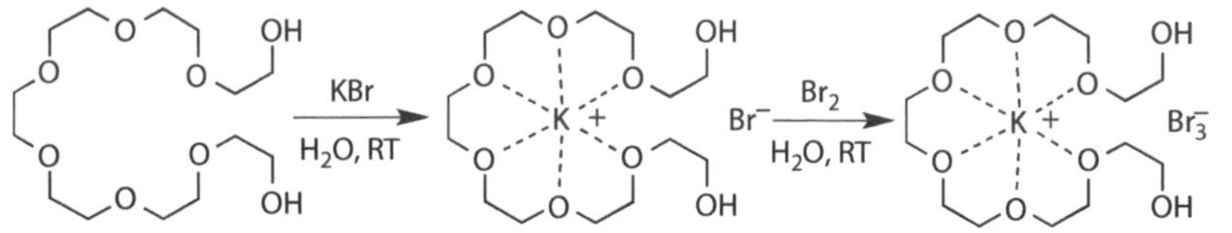

SCHEME 11.15 (K^+PEG)Br_3^- catalyst for MCR.

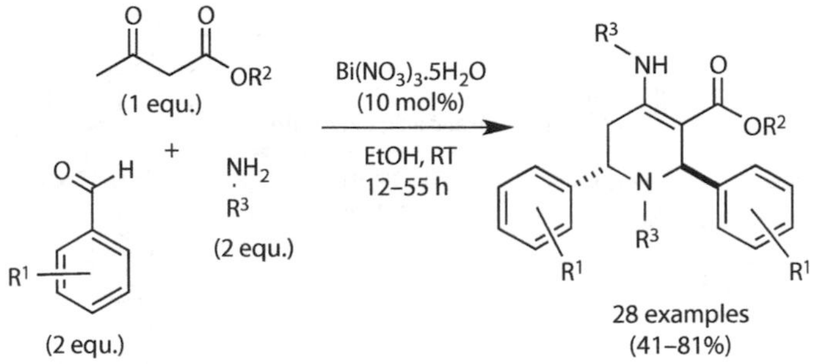

SCHEME 11.16 $Bi(NO_3)2.5H_2O$ catalyzed three-component MCR.

multicomponent synthesis. As shown in Scheme 11.17,[48] *p*-toluenesulfonic acid monohydrate (*p*-$TsOH.H_2O$) is used for the synthesis of highly functionalized piperidines via one-pot, multicomponent reaction from aldehydes, amines, and β-keto esters in EtOH at ambient temperature.

Five-component, one-pot synthesis of highly functionalized piperidine has been reported by the Maghsoodlou group starting from aromatic aldehydes, anilines, and β-keto esters in the presence of oxalic acid dihydrates as a catalyst in ethanol as a solvent at ambient temperature. For this transformation, they used different substituents on aldehydes, on a benzene ring having OMe, Me, F, Cl, and Br and methyl and/or ethyl acetoacetate with the same reaction condition; good yields were found. The relative stereochemistry was confirmed by single x-ray crystallographic analysis (Scheme 11.18).[49]

Recently, several groups efficiently developed and applied many lanthanide series catalysts, including lanthanum, cerium, samarium, and ytterbium for a number of synthetic transformations. Lu et al. used a $LaCl_3.7H_2O$ catalyst for the one-pot synthesis of 3,4-dihydropyrimidinones. Sathiyanarayana and co-worker also reported a $LaCl_3.7H_2O$ for multicomponent synthesis of highly functionalized piperidine derivatives. Reaction starting from 1,3-dicarbonyl compounds condensed with aromatic

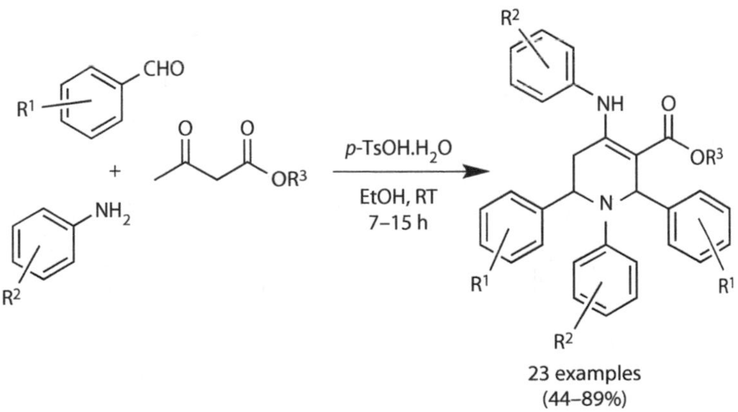

SCHEME 11.17 PTSA catalyzed three-component MCR.

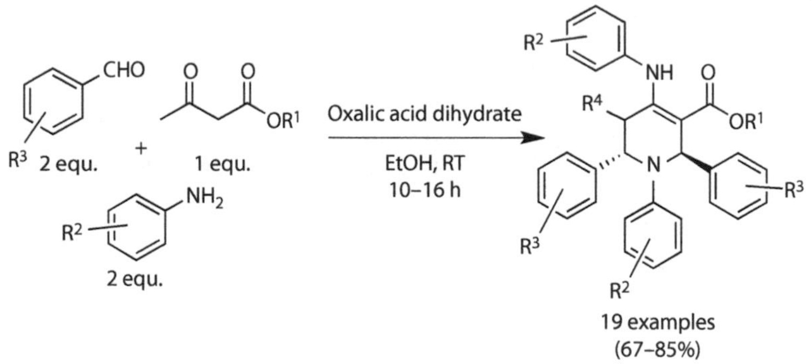

SCHEME 11.18 Oxalic acid dehydrate catalyzed five-component MCR.

aldehydes and aniline in the presence of $LaCl_3.7H_2O$ catalyst in methanol at room temperature to give highly substituted piperidine derivatives. They optimized several reaction conditions for this transformation and, on the basis of best result, picked $LaCl_3.7H_2O$ as a catalyst and methanol as a solvent for this reaction. To explore the reaction scope, they chose different substituents on aldehyde and aniline. Noticeably, aliphatic aldehydes failed to undergo this reaction because of the electron-releasing nature of alkyl groups. Product structure and relative stereochemistry are confirmed by single x-ray crystallographic analysis (Scheme 11.19).[50]

Ghashang and co-workers have used zinc hydrogen sulfate for the synthesis of highly substituted piperidine derivatives via three-component cyclocondensation starting from aldehydes, aromatic amines, and acetoacetic esters in acetonitrile at room temperature (Scheme 11.20).[51]

Several solvents and catalysts were tested for this transformation. They implemented several substituted functional groups on aldehyde and on amine to offered good yield. The product and relative stereochemistry of the 2,6-position as *trans* was confirmed by ^{1}H NMR study compared with reported authentic spectra.

Amberlite IRA400-Cl resin/I2/KI catalyzed one-pot synthesis has been described by Harichandran and co-workers for the functionalized piperidine derivatives. The

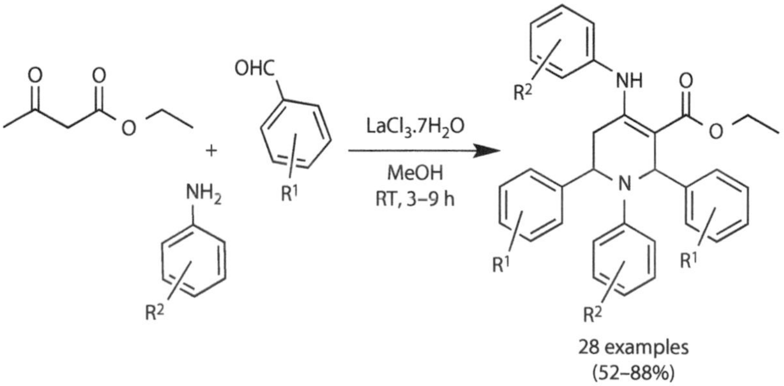

SCHEME 11.19 LaCl3.7H_2O catalyzed three-component MCR.

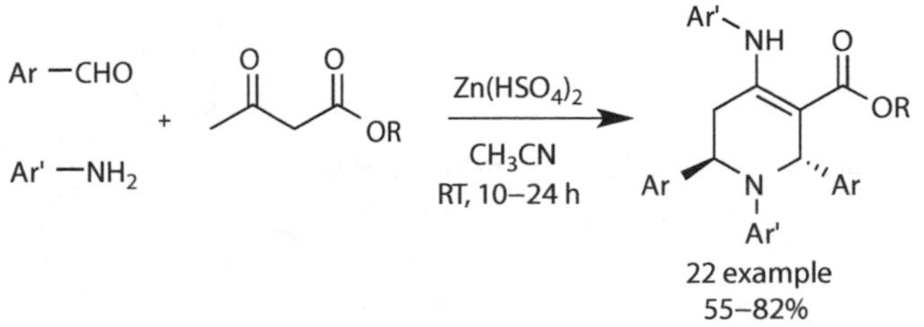

SCHEME 11.20 Zinc hydrogen sulfate catalyzed three-component MCR.

multicomponent reaction was carried out between various substituted aldehydes, aromatic amines and 1,3-dicarbonyl compounds in methanol at room temperature. The authors introduced Amberlite resin as a solid supported catalyst for this transformation, which is very effective and ecofriendly for the synthesis of functionalized piperidine derivatives (Scheme 11.21).[52]

Brahmachari and co-workers reported an efficient one-pot, five-component synthesis of β-keto esters, aromatic aldehydes, and anilines in ethanol at room temperature in the presence of $Ni(ClO_4)2{\cdot}6H_2O$ catalyst to give functionalized piperidine scaffolds. They tested several catalysts and solvents for this reaction. To explore the reaction scope, they used different substituted functional groups on aromatic aldehydes and on anilines, including electron-donating and electron-withdrawing groups, which formed good yields except for the nitro substituent, which transformed moderate yields (Scheme 11.22).[53]

A magnesium hydrogen sulfate [$Mg(HSO_4)_2$] catalyzed four-component condensation has been outlined by the Ghashang group from the aldehydes, aromatic amines, and acetoacetic esters in acetonitrile at room temperature for the synthesis of highly functionalized piperidines via a cyclocondensation reaction.

A variety of catalysts were screened and it was found that $Mg(HSO_4)_2$ was an effective and efficient catalyst for this transformation. The relative stereochemistry of the 2,6-position was confirmed as a *trans* by spectroscopic data compared

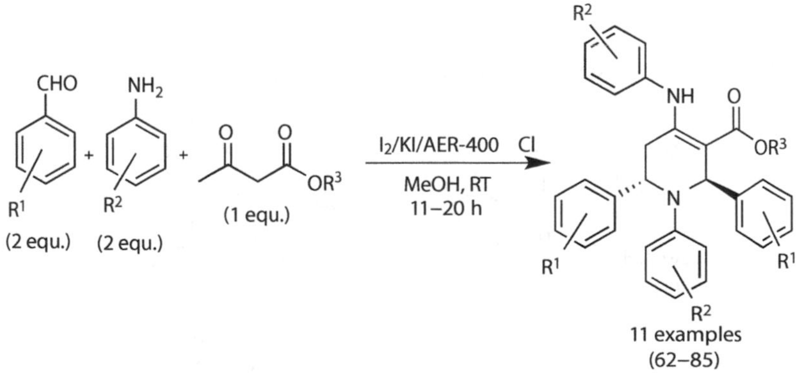

SCHEME 11.21 Amberlite catalyzed three-component MCR.

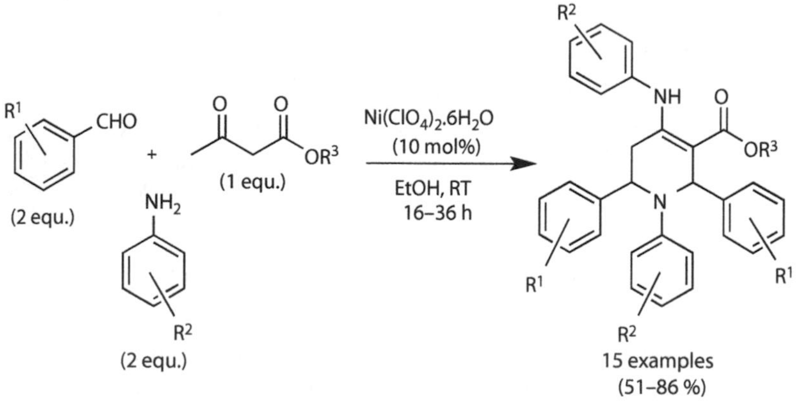

SCHEME 11.22 $Ni(ClO_4)_2{\cdot}6H_2O$ catalyzed five-component MCR.

with authentic sample. They used different substituents on aldehydes and aromatic amines for exploring the reaction scope (Scheme 11.23).[54]

Many Lewis and Bronstead acid catalysts have been screened for the synthesis of functionalized piperidine. Maghsoodlou et al. reported one-pot, five-component reactions between β-keto esters, aromatic aldehydes, and various amines in the presence of acetic acid, which is used for multitasks like solvents, catalysts, and additives in this reaction for the synthesis of functionalized piperidine. The relative stereochemistry of the obtained product has been confirmed by single-crystal x-ray crystallography. The reaction proceeded successfully using a variety of aromatic aldehydes and anilines (Scheme 11.24).[55]

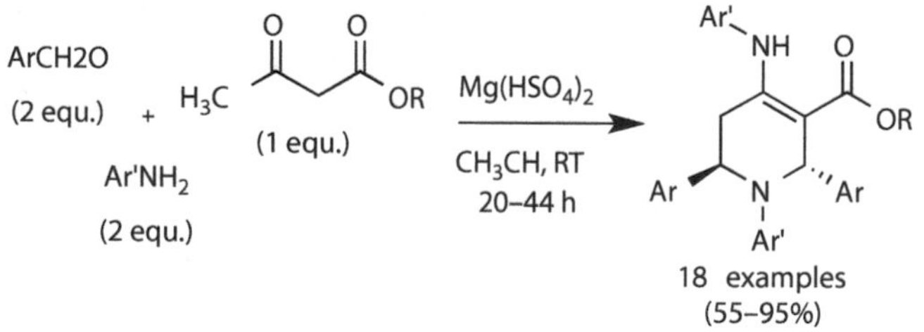

SCHEME 11.23 $Mg(HSO_4)_2$ catalyzed five-component MCR.

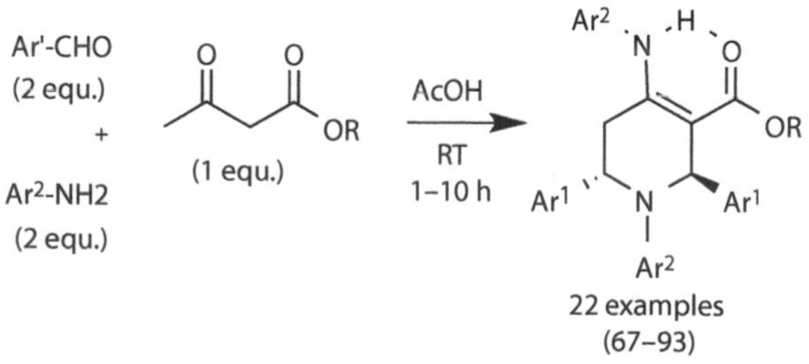

SCHEME 11.24 Acetic acid catalyzed five-component MCR.

The Ghorbanian group successfully introduced nanocrystalline acid catalysts for the synthesis of functionalized piperidines. A one-pot, multicomponent reaction was performed between 1,3-dicarbonyls, amines, and aromatic aldehydes in ethanol at room temperature. They used nanosulfated zirconia, nanostructured ZnO, nano-γ-alumina, and nano-ZSM-5 zeolites as nanostructured crystalline solid acid catalysts for this transformation. After the reaction, the catalyst was recovered and reused for the next reaction without any loss of efficiency. Recently, these catalysts received considerable attention in synthetic chemistry owing to their recyclability. The authors also compared the efficiency of nanocrystalline sulfated zirconia (SZ) with other well-known reported catalysts and found that SZ showed better efficiency than other catalysts except $ZrOCl_2.8H_2O$, which showed slightly better results. They tested several solvents for this transformation and found ethanol was the best solvent for reaction (see Scheme 11.25[56] and Table 11.1).

Many rare earth metal triflates have been reported for organic reactions because of their low toxicity, moisture stability, recyclability and easy commercial availability. $Bi(OTf)_3$ catalyzed synthesis of functionalized piperidine was reported by Debache et al. The one-pot condensation starting between ethyl acetoacetate, benzaldehydes and anilines in ethanol at reflux condition to give high diastereoselectivity of highly functionalized piperidine derivatives. To explore the reaction scope, the group successfully utilized different substituted aldehydes, anilines, and 1,3-dicarbonyl compounds for this transformation to give moderate to good yields (Scheme 11.26).[57]

Recently, a solvent-free synthesis of a substituted heterocyclic system based on the silica promoted one-pot, multicomponent condensation has been developed by Hajra and co-workers for the synthesis of highly substituted piperidine derivatives. The reaction was carried out between aldehydes, amines, and β-keto ester at room temperature. The reaction took place smoothly and without any chromatographic purification; analytically pure products were obtained (Scheme 11.27).[58]

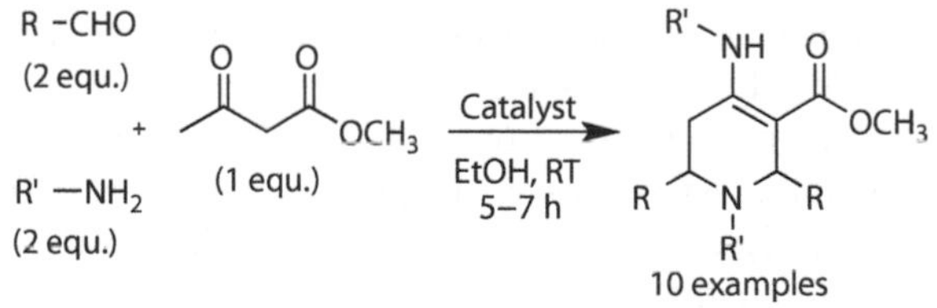

R -CHO
(2 equ.)
+
R' —NH_2
(2 equ.)
OCH_3
(1 equ.)
Catalyst
EtOH, RT
5–7 h
R'
NH
O
OCH_3
R
N
R
R'
10 examples

SCHEME 11.25 Nanocrystalline acid catalyzed five-component MCR.

TABLE 11.1
Catalyst and Yield of the MCR

Catalyst	Yield (%)
Nano-ZnO	60–80
Nano-γ-alumina	64–80
Nano-ZMS-5	63–83
Nanocrystalline SZ	65–87

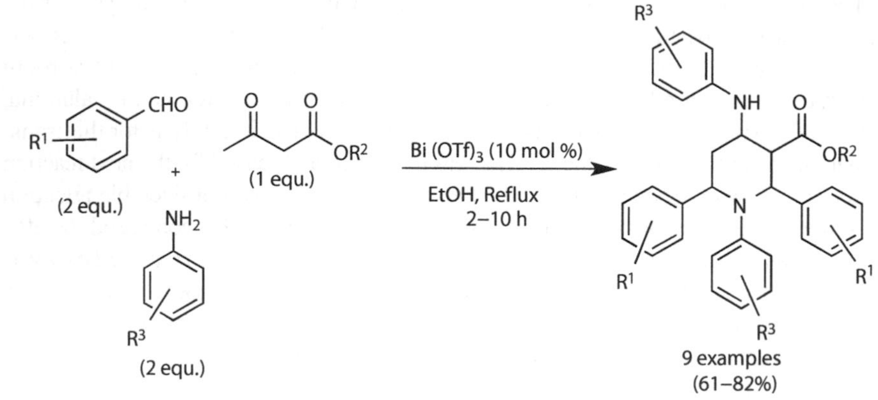

SCHEME 11.26 Bi$(OTf)_3$ catalyzed five-component MCR.

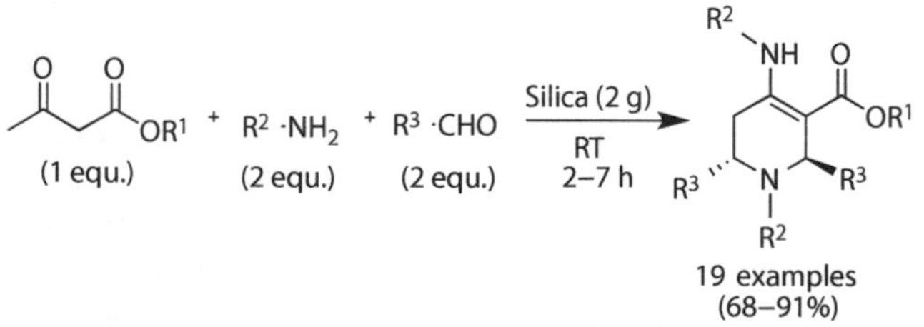

SCHEME 11.27 Silica catalyzed solvent-free five-component MCR.

One of the advantages of this method is the reusability of silica gel for the next reaction, and it can also be significantly applied in the large-scale synthesis of piperidine derivatives. For reaction scope, this method was applied successfully for a variety of substituted aldehydes and amines.

Abbas and co-workers pointed out one-pot, five-component synthesis of highly functionalized piperidine derivatives in the presence of silica sulfuric acid (SSA) as a heterogeneous catalyst. Aldehydes, amines, and β-keto esters were treated in the presence of SSA in methanol to yield the functionalized piperidines. The synthesized compounds were screened for the biological activities *in vitro* and showed significant antibacterial activity. They tested several catalysts and solvents for this transformation and found SSA and methanol as the best catalyst and solvent combination for this reaction in terms of yield and time. The catalyst was recovered and reused at least four times for fresh reactions under the same reaction conditions to achieve the desired products without any loss in catalyst activity (Scheme 11.28).[59]

The possible mechanism of this five-component transformation proceeds through the initial attack of SSA on β-keto ester, amine and aldehyde to generate β-enaminone **I** and imine **II**, respectively. Then, addition of β-enaminone as an intermolecular Mannich reaction to the imine **II** took place to give intermediate **III**. Next, activated aldehyde acted with the intermediate **III** to generate intermediate **IV**; by tautamerizaion it gives intermediate **V**, which undergoes an intramolecular Mannich-type

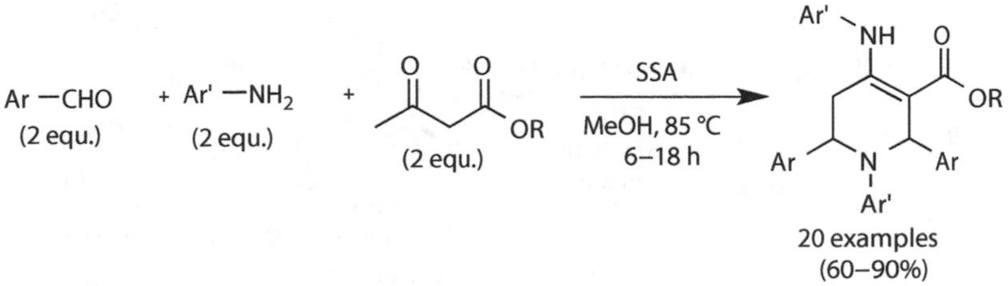

SCHEME 11.28 Silica sulfuric acid catalyzed five-component MCR.

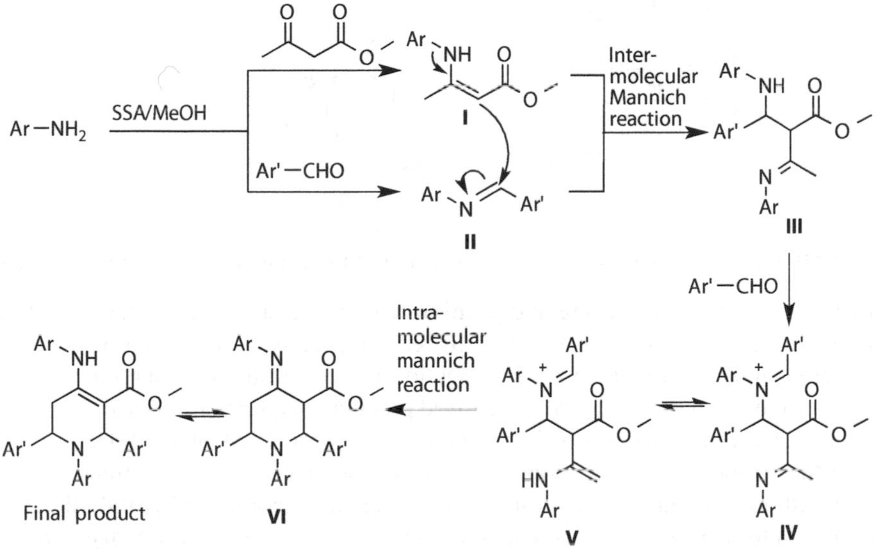

SCHEME 11.29 Proposed mechanism of silica sulfuric acid catalyzed five-component MCR.

reaction, offering intermediate **VI**. Then, intermediate **VI** tautomerizes to form the expected substituted piperidine derivatives as depicted in Scheme 11.29.

11.2.3 Synthesis of Chiral Substituted Piperidine Derivatives by MCR

Several natural alkaloids consist of different numbers of chiral centers in the backbone. It is a challenging job to install the chirality in a molecule. Many approaches have been reported, such as chiral pools, chiral reagents, chiral auxiliaries, use of chiral building blocks, etc. Moreover, only a handful of reports mention the generation of chirality in substituted piperidine derivatives via one-pot, multicomponent strategies.

Terada and co-workers (Scheme 11.30)[60] developed a one-pot diastereo- and enantioselective tandem aza-ene type reaction/cyclization cascade to form enantioenriched substituted piperidine derivatives. The transformation was catalyzed by biphenol-derived phosphoric acid A or B between *N*-Boc aldimine with *N*-Cbz

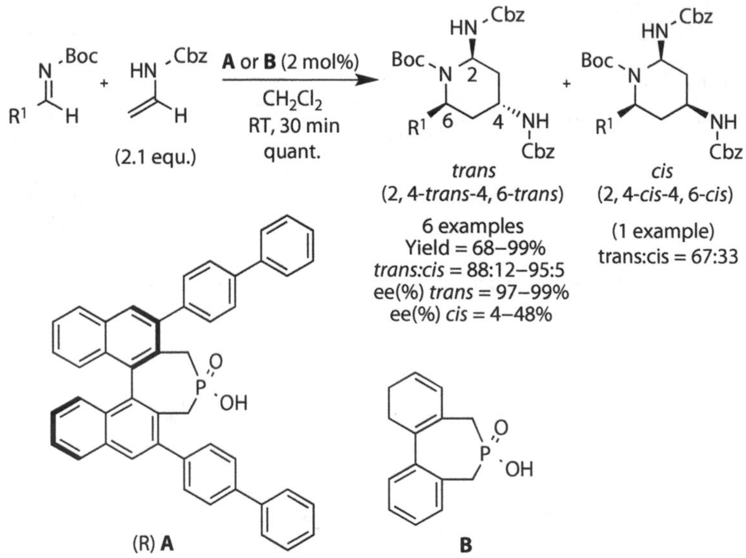

SCHEME 11.30 Synthesis of enantio-enriched substituted piperidine derivatives by MCR.

enecarbamate to give diastereomeric mixture of the desired piperidine derivatives having three stereogenic centers. For this reaction, catalysts A and B provided good diastereo- and enantioselectivity. The reaction proceeded smoothly at room temperature in dichloromethane to give the desired products in quantitative yield.

The Hayashi group described the method for generation of enantioselective polysubstituted piperidine derivatives. The reaction sequence started by treating aldehyde with nitroalkene in the presence of diphenylprolinol trimethylsilyl ether as a catalyst in toluene and then sequential addition of imine and triethylsilane to afford the chiral piperidine derivatives.

For this one-pot, four-component transformation, the authors used diphenylprolinol silyl ether as a proline based catalyst. The reaction proceeded as a Michael/aza-Henry/hemiaminalization/allylation or cyanation pathway to yield the chiral piperidine derivatives. The scope of this reaction was examined by utilizing different substituents on aldehyde, nitroalkene, and on imine (Scheme 11.31).[61]

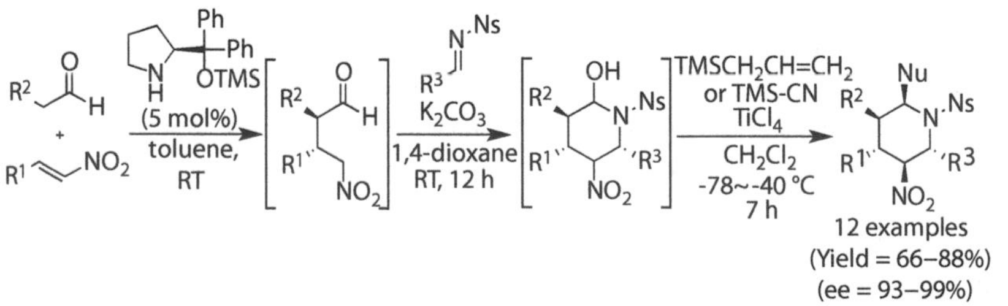

SCHEME 11.31 Synthesis of enantioselective polysubstituted piperidines by MCR.

The proposed reaction sequence proceeded initially as a diphenylprolinol silyl ether mediated Michael reaction of aldehyde and nitroalkene to generate γ-nitroaldehyde followed by an aza-Henry reaction. Then, a subsequent hemiaminalization reaction offered 2-hydroxy piperidine derivatives, which reacted further with a nucleophile to give highly substituted piperidine derivatives with high enantioselectivity. The relative configuration of the product was confirmed by coupling constant and nuclear Overhauser enhancement spectroscopy (NOESY) experiments (Scheme 11.32).

Isocyanoacetates are very interesting synthons in organic chemistry due to presence of the cyano with intriguing reactivity. It can undergo both electrophilic and nucleophilic reactions. Sello and co-workers published the chiral isocyanoacetates in multicomponent reactions as a Joullie–Ugi three-component condensation. The reaction proceeded from the starting material isocyanide and carboxylic acid reacted with cyclic imine in methanol at room temperature with the stated time to form the substituted piperidine product with the isocyanoacetate epimerization between 17% and 28% (Scheme 11.33).[62]

Brase et al. investigated asymmetric organocatalytic three-component enantioselective synthesis, which starts from 3-vinylindoles and imino esters in the presence of a chiral phosphoric acid catalyst in dichloromethane as a solvent at room

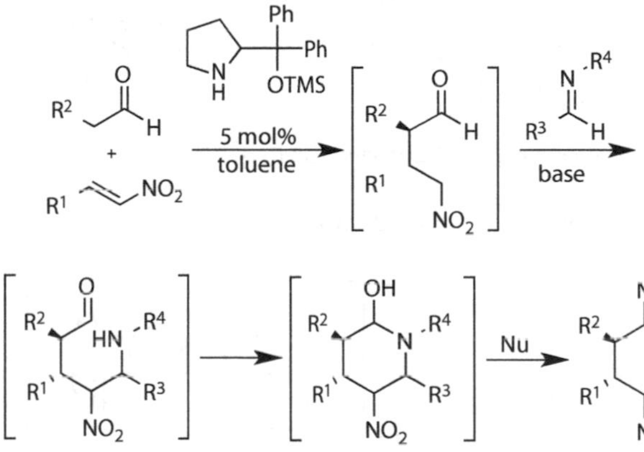

SCHEME 11.32 Proposed mechanism for synthesis of enantioselective polysubstituted piperidines.

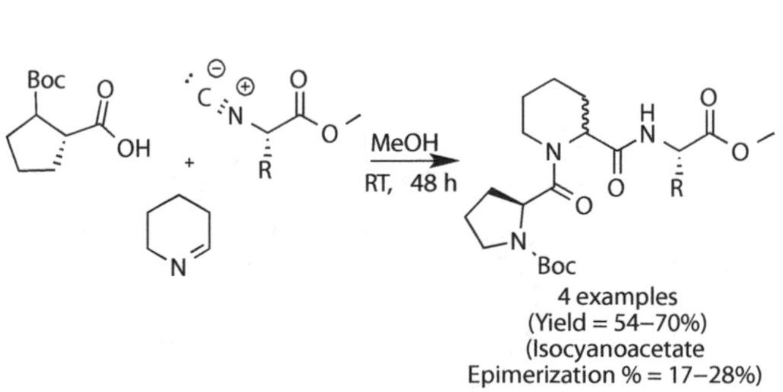

SCHEME 11.33 Joullie–Ugi three-component MCR.

temperature to yield bisindole piperidine derivatives with high enantioselectivity. They utilized different substituents on 3-vinylindoles and imino esters, including electron-donating as well as electron-withdrawing groups. They achieved good yields and enantiomeric excess (Scheme 11.34).[63]

The possible mechanism for this transformation proceeded initially as addition of 3-vinyl-indole on imino ester to generate intermediate **I**; then a second molecule of 3-vinyl-indole attacked stabilized intermediate **I** to form intermediate **II**. This step moved considerably faster than nucleophilic attack of the aromatic phenyl ring at position 4, and

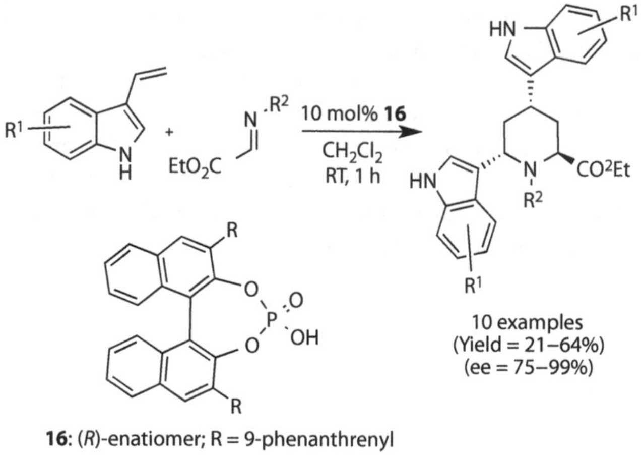

SCHEME 11.34 Phosphoric acid catalyzed MCR.

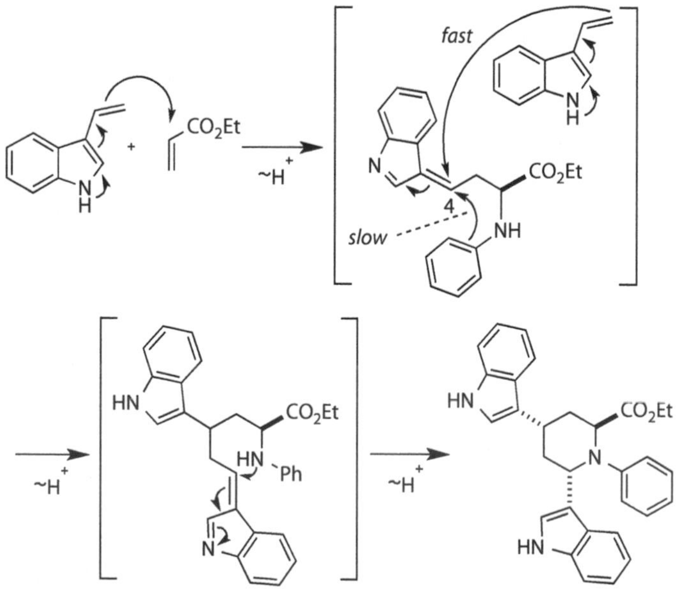

SCHEME 11.35 Proposed mechanism for phosphoric acid catalyzed MCR.

traces of the related Povarov reaction product were obtained. Then the intermediate **II** intramolecular ring closure formed the substituted piperidine derivatives (Scheme 11.35).

11.2.4 Synthesis of Spirosubstituted Piperidine Derivatives by MCR

Many spiropiperidine derivatives exhibit important activity as pharmacophores in several biologically active compounds. Several multicomponent approaches have been reported for the synthesis of these spiropiperidine derivatives, which have been isolated from various plant alkaloids and animal toxins. They have also been synthesized chemically utilizing MCR.

Kadutskii et al. have developed a three-component reaction to synthesize 3,5-dispirosubstituted piperidine derivatives. The substituted anilines were reacted with dimedone and formaldehyde in ethanol at reflux conditions. The reaction proceeded effectively without any catalyst. The scope of this reaction was verified by using a variety of functional groups on anilines for this transformation (Scheme 11.36).[64]

Iodine catalyzed synthesis of spiropiperidine derivatives, developed recently by the Song group, was based on a four-component reaction between ethyl trifluoroacetoacetate with 1,3-indanedione, ammonium acetate, and aromatic aldehyde under mild reaction conditions. It yielded spiropiperidine derivatives as a major product along with minor compound II. Both structures, were confirmed by x-ray diffraction analysis of a single crystal. The spiropiperidine derivatives contain a variety of a fluorine functional group. In the literature many fluorine compounds have showed different biological activities (Scheme 11.37).[65]

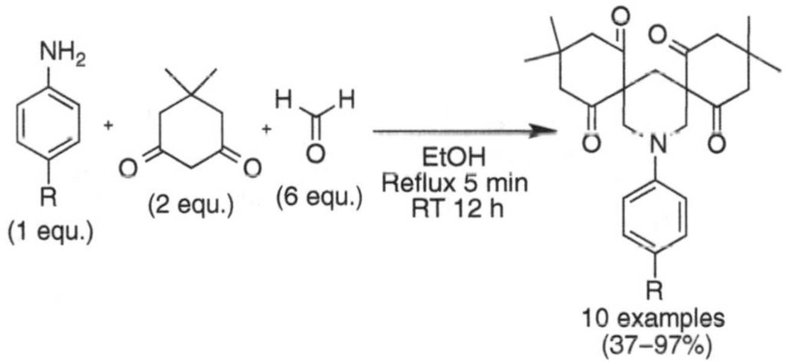

NH_2 (1 equ.) + (2 equ.) + H(C=O)H (6 equ.) → EtOH, Reflux 5 min, RT 12 h; 10 examples (37–97%)

SCHEME 11.36 Synthesis of 3,5-dispirosubstituted piperidine derivatives by MCR.

F_3C–CO–CH$_2$–CO–OEt + $ArCH_2O$ + 1,3-indanedione + NH_4OAc → I_2 (10 mol%), EtOH, RT, 4–10 h → I (Major) + II (Minor); 8 examples (Major = 53–75%) (Minor = 8–11%)

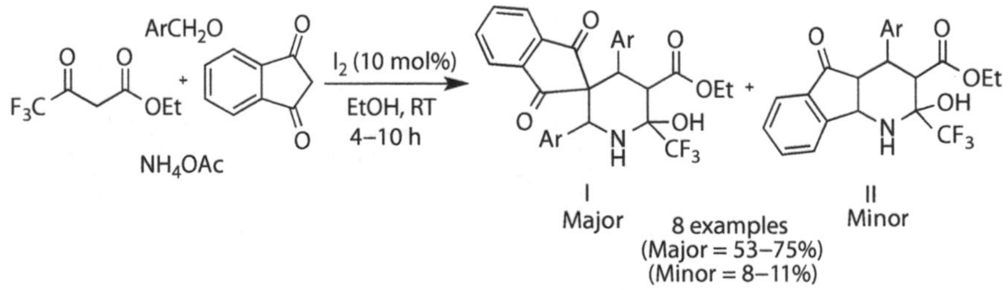

SCHEME 11.37 Iodine catalyzed synthesis of spiropiperidines by MCR.

A three-component, one-pot cyclocondensation reaction starting from aromatic amines, formaldehyde, and dimedone has been investigated by the Jeong group using silica supported tungstic acid (STA) as a heterogeneous catalyst for the synthesis of 3,5-dispirosubstituted piperidines. The catalyst was recovered and reused up to six times without any loss of its catalytic activity. The group used a different substituent on aniline to explore the reaction scope (Scheme 11.38).[66]

This three-component reaction proceeded as a domino sequence of Knoevenagel, Michael, and double Mannich reactions. Initially, Knoevenagel condensation of dimedone with formaldehyde gives intermediate I and then subsequent Michael addition of another dimedone molecule offers intermediate II. Next, the Mannich reaction of aniline and formaldehyde with intermediate II forms intermediate III, on which a second Mannich reaction of formaldehyde takes place to offer the spiropiperidine derivatives (Scheme 11.39).

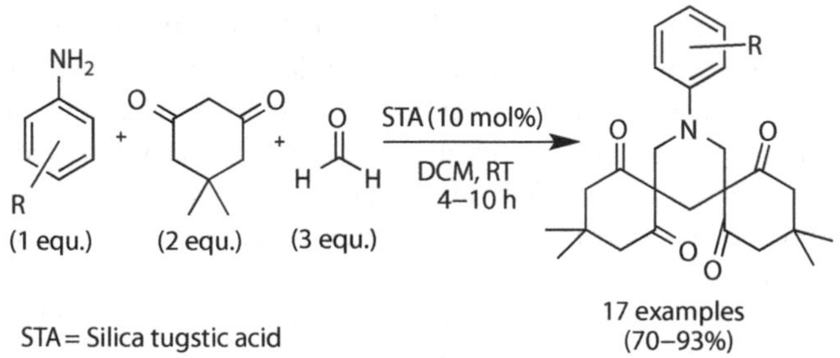

SCHEME 11.38 STA catalyzed synthesis of spiropiperidines by MCR.

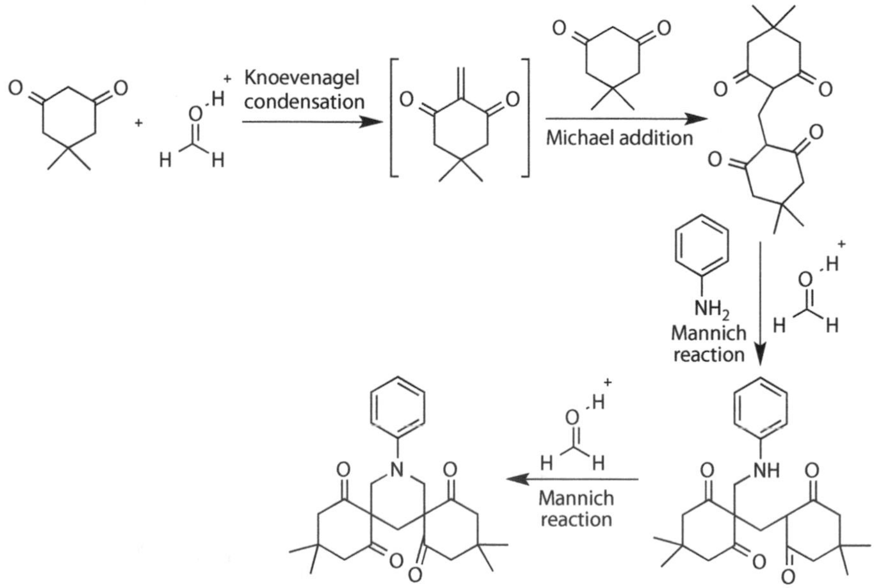

SCHEME 11.39 Proposed mechanism for STA catalyzed synthesis of spiropiperidines.

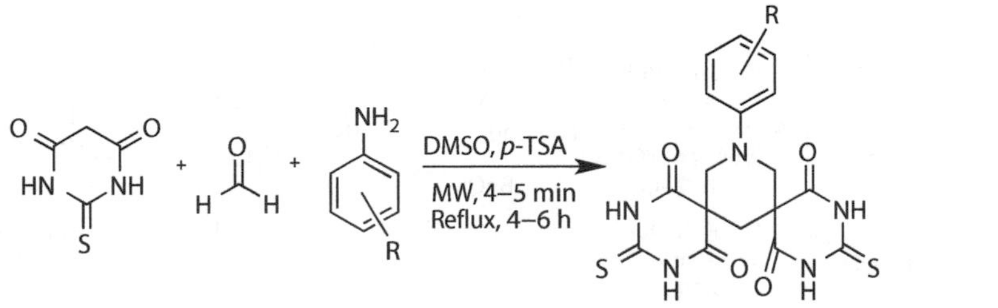

Thermal method		MW method	
Time (h)	Yield (%)	Time (Min)	Yield (%)
4–6	68–74	4–5	91–96

SCHEME 11.40 Microwave-assisted synthesis of spiropiperidines by MCR.

The Meshram group investigated a one-pot, three-component, microwave-assisted reaction for the synthesis of sulfur-containing 3,5-dispirosubstituted piperidine derivatives. 2-Thiobarbituric acid, formaldehyde, and aromatic amines were treated in the presence of catalytic amounts of *p*-toluenesulfonic acid in dimethylsulfoxide. For comparison, they also checked this transformation under thermal reaction conditions. Under thermal heating, the reaction took a long time to complete and achieve high yields compared to microwave condition. To further check the reaction scope, they applied different substituents on aniline including electron-withdrawing and electron-donating groups (Scheme 11.40).[67]

11.2.5 Synthesis of Piperidone Derivatives by MCR

Numerous piperidone derivatives play an important role as intermediates in a variety of bioactive alkaloids and pharmaceutically active compounds. There are few reports for construction of piperidone via a multicomponent reaction. Landais and co-workers described a one-pot radical cascade of three-component transformation starting from α-iodoester, olefins, and oximes in the presence of Et_3B/O_2 in dichloromethane at room temperature. This formed piperidinones in good yields and sequential four-component reactions starting between xanthate, olefin, and oximes to provided an oxime intermediate, which reacted *in situ* with an alkyl iodide and a Lewis acid (BF_3-OEt_2) to give highly substituted piperidinones with *trans*-diastereomer as a major isomer (Scheme 11.41).[68]

Wilson and co-workers developed a four-component reaction starting between diketene, tosyl imine, methanol, and aldehyde in the presence of $TiCl_4$ in dichloromethane at –78 °C to generate 2,6-disubstituted nonsymmetrical piperid-4-ones in the form of *cis-/trans*-diastereomers in good yields. Conversion can take place through epimerization with K_2CO_3 to a single 2,6-*cis*-diasteromer. The stereochemistry of 2,6-disubstituted piperid-4-ones was confirmed by single crystal x-ray analysis. They used different substituents on imines and on aldehydes to explore the reaction scope (Scheme 11.42).[69]

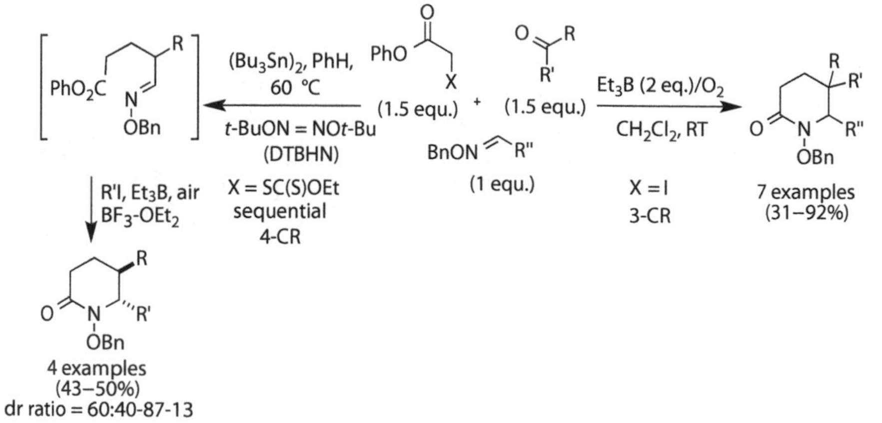

SCHEME 11.41 One-pot MCR for synthesis of piperidones.

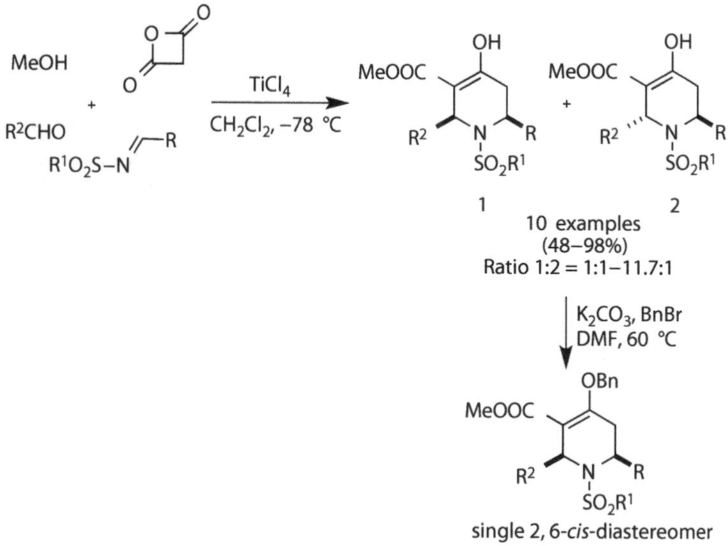

SCHEME 11.42 Four-component MCR for synthesis of substituted piperidones.

The Barker group reported a double Mannich reaction of β-keto esters and bisaminol ethers in the presence of trichloromethylsilane as the Lewis acid in acetonitrile at room temperature to give the polysubstituted 4-piperidone derivatives. They confirmed the product stereochemistry by a spectral method with coupling constant and NOESY study. They used a variety of substituents on β-keto esters and bisaminol ethers for this transformation (Scheme 11.43).[70]

A $Yb(OTf)_3$ and AgOTf cocatalyzed synthesis of polysubstituted piperidone derivatives was reported by the Zhang group starting from dimethyl malonate and formaldehyde O-benzyl oxime in dichloromethane. They confirmed the product by

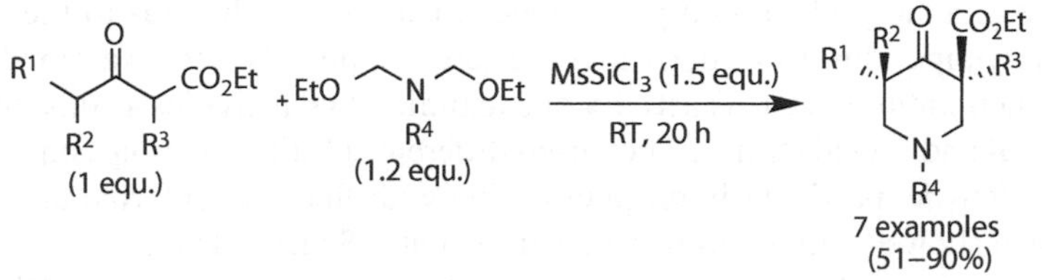

SCHEME 11.43 Synthesis of polysubstituted piperidones.

spectroscopy analysis including correlation spectroscopy (COSY) and heteronuclear multiple quantum coherence (HMQC) experiments (Scheme 11.44).[71]

A very efficient and effective route for the synthesis of complex piperidone scaffolds has been developed by the Ghosez group, based on a four-component process between aldehyde, ammonia, acyl chloride, and dienophile under the basic reaction conditions in toluene as the reaction medium. They applied different functionality on acyl chloride, aldehyde, and dienophile to explore the reaction scope yielding mono- and polycyclic piperidone derivatives. By use of different cyclic or acyclic dienophile components, different functionalization and stereochemical diversity was achieved. This MCR has been extended further to access the synthesis of an intermediate for aspidosperma alkaloid analogs. The stereochemical assignments of polycyclic piperidone scaffolds were confirmed by x-ray diffraction analysis (Scheme 11.45).[72]

Kabilan and co-workers described one-pot, four-component condensation between ketone, two molecules of benzaldehyde, and ammonium acetate in ethanol as

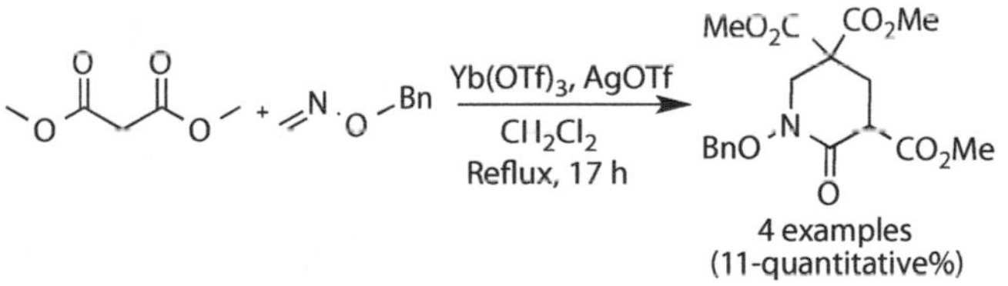

SCHEME 11.44 Synthesis of polysubstituted piperidones catalyzed by metal triflates.

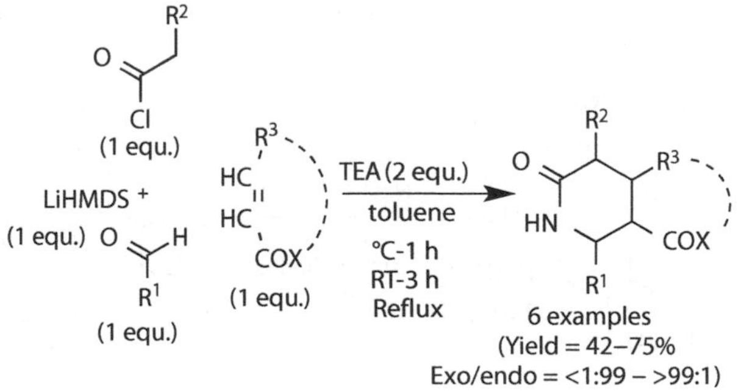

SCHEME 11.45 Four-component MCR for synthesis of mono- or polysubstituted piperidones.

solvent to give polysubstituted piperidone scaffolds, which leads to the synthesis of 2,6-diarylpiperidin-4-one derivatives. These compounds were examined successfully for their antituberculosis activity and antimicrobial activities against five familiar bacterial and fungal strains. They used different substituted groups on ketone and on benzaldehyde, particularly on phenyl ring with fluorine; chlorine and methoxy substitution showed better antituberculosis activity (Scheme 11.46).[73]

A base catalyzed, four-component reaction has been developed by the Wang group starting from nitrostyrenes, aromatic aldehydes, dimethyl malonate, and formamide in methanol as a solvent for the synthesis of trans-4,6-diaryl-5-nitropiperidin-2-ones. Initially, for optimization they used a variety of bases, solvents, and temperature for this transformation and the best results were obtained with sodium hydroxide and methanol as a base and solvent for the reaction, respectively. They utilized various functional groups on nitrostyrenes and aromatic aldehydes for this transformation to extend the reaction scope. They determined the structure and *trans*-relative configuration of the piperidone compound by spectral data and x-ray diffraction study (Scheme 11.47).[74]

Wang's group has published an article on a base catalyzed, one-pot, pseudo-five-component reaction starting from two equivalents of aromatic aldehyde,

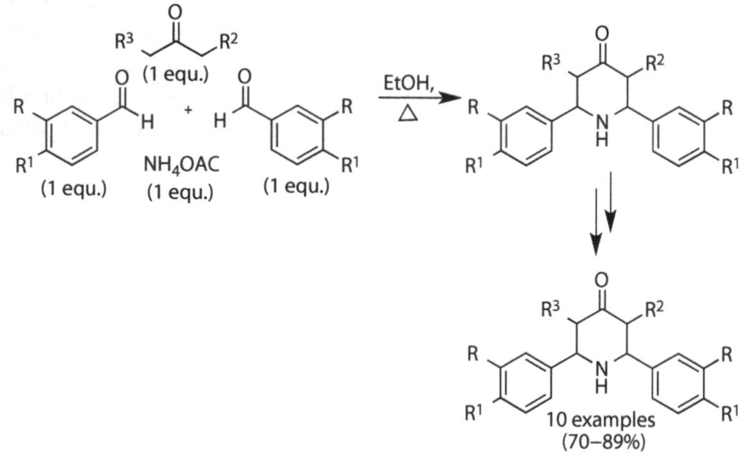

R³ R² O (1 equ.)

R R¹ (1 equ.) + NH₄OAC (1 equ.) (1 equ.)

EtOH, Δ

10 examples (70–89%)

SCHEME 11.46 One-pot, four-component MCR for synthesis of polysubstituted piperidones.

NO₂ R¹ (1 equ.) CHO R² (1 equ.) + (1 equ.) HCONH₂ (1.5 equ.)

NaOH (4 equ.) MeOH 0 °C to Reflux 40 h

23 examples (51–73%)

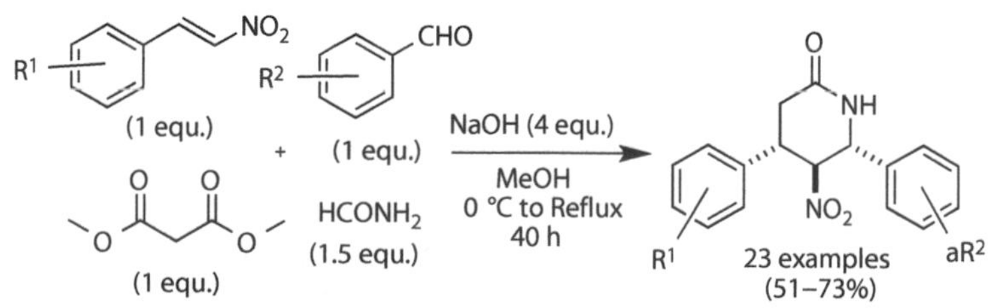

SCHEME 11.47 Base catalyzed four-component MCR for synthesis of polysubstituted piperidones.

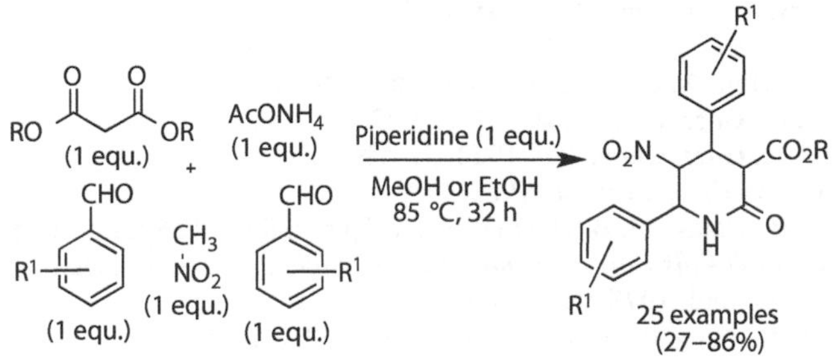

SCHEME 11.48 One-pot, pseudo-five-component MCR for synthesis of polysubstituted piperidones.

nitromethane, ammonium acetate, and dialkyl malonates in methanol or ethanol as a solvent for the synthesis polysubstituted 2-piperidinone derivatives. For this transformation, they optimized various bases and solvents and chose piperidine as a base and methanol as a solvent based on the results (Scheme 11.48).[75]

11.3 CONCLUSION

The purpose of this chapter is to highlight the significance of multicomponent reactions for the development of the piperidine skeleton. As we have mentioned before, piperidine is a highly important heterocycle in organic chemistry and present in many natural alkaloids. It has also occurred in a variety of drug molecules that are very effective in curing several diseases. Piperidine synthesis via multicomponent fashion continues to be a very active field and will continue to be an important endeavor in the future. We hope that this chapter will serve as a guideline for researchers to develop new and creative multicomponent routes to the piperidine heterocyclic frame. The multicomponent reaction approach provides a high-speed path for the rapid synthesis of piperidine libraries with a high degree of structural diversity. There are great expectations for this approach, as it has the potential to transform the future of drug discovery.

REFERENCES

1. Lue, M.; Sibi, M. P. *Tetrahedron* **2002**, *58*, 7991.
2. Comins, D. L.; Joseph, S. P. In *Advances in nitrogen heterocycles*, vol. 2 (ed.: C. J. Moody), JAI Press, Greenwich, CT. **1996**, 251.
3. Bailey, P. D.; Millwood, P. A.; Smith, P. D. *Chem. Commun.* **1998**, *6*, 633.
4. Weintraub, P. M.; Sabol, J. S.; Kane, J. M.; Borcherding, D. R. *Tetrahedron* **2003**, *59*, 2953.
5. Buffat, M. G. P. *Tetrahedron* **2004**, *60*, 1701.
6. For a monograph on multicomponent reactions, see Zhu, J.; Bienayme, H. (eds.), *Multicomponent reactions*. Wiley-VCH, Weinheim, Germany, **2005**.
7. For a special issue on this topic, see Menendez, J. C. (ed.), *Curr. Org. Chem.* **2013**, *17*, issue 18.
8. For a review of multiband forming reaction, see Coquerel, Y.; Boddaert, T.; Presset, M.; Mailhol, D.; Rodriguez, J.; Ideas in chemistry and molecular sciences, in *Advances in synthetic chemistry* (ed.: B. Pignataro), Wiley-VCH, Weinheim, Germany, vol. 1, chap. 9, **2010**.

9. Strecker, A. *Ann. Chem. Pharm.* **1850**, *75*, 27.
10. Domling, A.; Ugi, I. *Angew. Chem., Int. Ed. Engl.* **2000**, *39*, 3168.
11. Hulme, C.; Gore, V. *Curr. Med. Chem.* **2003**, *10*, 51.
12. Passerini, M. *Gazz. Chim. Ital.* **1921**, *51*, 126.
13. Passerini, M. *Gazz. Chim. Ital.* **1924**, *54*, 529.
14. Ugi, I. *Angew. Chem., Int. Ed.* **1962**, *1*, 8.
15. Fodoe, G. B.; Colasanti, B. The pyridine and piperidine alkaloids: Chemistry and pharmacology, in *The alkaloids: Chemical and biological perspectives* (ed.: S. W. Pelletier), Pergamon, Oxford, **1975**, vol. 3, p. 1.
16. Watson, P. S.; Jiang, B.; Scott, B. *Org. Lett.* **2000**, *2*, 3679.
17. Ripoche, I.; Canet, J. L.; Gelas, J.; Troin, Y. *Eur. J. Org. Chem.* **1999**, *9*, 1517.
18. Beckett, A. H.; Sheikh, A. H. *J. Pharm. Pharmacol.* **1973**, *25*, 171.
19. Leclercq, S.; Daloze, D.; Braekman, J. C. *Org. Prep. Proc. Int.* **1996**, *28*, 499.
20. Gouault, N.; Rock, L. R.; Pinto, G. C.; David, M. *Org. Biomol. Chem.* **2012**, *10*, 5541.
21. Felpin, F.-X.; Lebreton, J. *Tetrahedron* **2004**, *60*, 10127.
22. Riva, E.; Rencurosi, A.; Gagliardi, S.; Passarella, D.; Martinelli, M. *Chem. Eur. J.* **2011**, *17*, 6221.
23. Okwute, S. K.; Egharevba, H. O. *Int. J. Chem.* **2013**, *5*, 3.
24. Asano, N.; Nash, R. J.; Molyneux, R. J.; Fleet, W. J. *Tetrahedron: Asymmetry* **2000**, *11*, 1645.
25. Ojima, I.; Vidal, E. *J. Org. Chem.* **1998**, *63*, 7999.
26. Voituriez, A.; Ferreira, F.; Chemla, F. *J. Org. Chem.* **2007**, *72*, 5358.
27. Serretti, A.; Zanardi, R.; Cusin, C.; Rossini, D.; Lorenzi, C.; Smeraldi, E. *Eur. Neuropsychopharmacol* **2001**, *11*, 5, 375.
28. Husbands, S. M.; Furkert, D. P. *Org. Lett.* **2007**, *9*, 19, 3769.
29. Das, S.; Brahmachari, G. *J. Org. Biomol. Chem.* **2013**, *1*, 33.
30. Hert, M. D.; Mittoux, A.; He, Y.; Peuskens, J. *Eur. Arch. Psychiatry Clin. Neurosci.* **2011**, *261*, 231.
31. Malaquin, S.; Jida, M.; Gesquiere, J.-C.; Deprez-Poulain, R.; Deprez, B.; Laconde, G. *Tetrahedron Lett.* **2010**, *51*, 2983.
32. Davidson, C.; Ramsey, J. *J. Psychopharmacol.* **2012**, *26*, 1036.
33. Leffingwell, J. C. *Leffingwell Rep.* **2003**, *3*, 1.
34. Yuan, K.; Jiang, F.; Sahli, Z.; Achard, M.; Roisnel, T.; Bruneau, C. *Angew. Chem. Int. Ed.* **2012**, *51*, 8876.
35. Huang, F.-P.; Dong, Z.-W.; Zhang, H.-R.; Hui, X.-P. *J. Hetercycle Chem.* **2014**, *51*, 532.
36. Dighe, S. U.; Anil Kumar, K. S.; Srivastava, S.; Shukla, P.; Singh, S.; Dikshit, M.; Batra, S. *J. Org. Chem.* **2015**, *80*, 99.
37. Chacko, S.; Ramapanicker, R. *Tetrahedron Lett.* **2015**, *56*, 2023.
38. Malaquin, S.; Jida, M.; Gesquiere, J.-C.; Deprez-Poulain, R.; Deprez, B.; Laconde, G. *Tetrahedron Lett.* **2010**, *51*, 2983.
39. Clarke, P. A.; Zaytzev, A. V.; Whitwood, A. C. *Tetrahedron Lett.* **2007**, *48*, 5209.
40. Khan, A. T.; Parvin, T.; Choudhary, L. H. *J. Org. Chem.* **2008**, *73*, 8398.
41. Khan, A. T.; Lal, M.; Khan, M. *Tetrahedron Lett.* **2010**, *51*, 4419.
42. Khan, A. T.; Khan, Md. M.; Bannuru, K. K. R. *Tetrahedron* **2010**, *66*, 7762.
43. Mishra, S.; Ghosh, R. *Tetrahedron Lett.* **2011**, *52*, 2857.
44. Mukhopadhyay, C.; Rana, S.; Butcher, R. J.; Schmiedekamp, A. M. *Tetrahedron Lett.* **2011**, *52*, 5835.
45. Verma, S.; Jain, S. L.; Sain, B. *Beilstein J. Org. Chem.* **2011**, *7*, 1334.
46. Zolfigo, M. A.; Chehardoli, G.; Salehzadeh, S.; Adams, H.; Ward, M. D. *Tetrahedron Lett.* **2007**, *48*, 7969.
47. Brahmachari, G.; Das, S. *Tetrahedron Lett.* **2012**, *53*, 1479.

48. Sajadikhah, S. S.; Maghsoodlou, M. T.; Hazeri, N.; Habibi-Khorassani, S. M.; Shams-Najafi, S. J. *Monatsh Chem.* **2012**, *143*, 939.
49. Sajadikhah, S. S.; Maghsoodlou, M. T.; Hazeri, N.; Habibi-Khorassani, S. M.; Willis, A. C. *Chin. Chem. Lett.* **2012**, *23*, 569.
50. Umamahesh, B.; Sathesh, V.; Ramachandran, G.; Sathishkumar, M.; Sathiyanarayanan, K. *Catal. Lett.* **2012**, *142*, 895.
51. Ghashang, M. *Lett. Org. Chem.* **2012**, *9*, 497.
52. Harichandran, G.; Amalraj, S. D.; Shanmugam, P. *J. Heterocyclic Chem.* **2013**, *50*, 539.
53. Das, S.; Brahmachari, G. *J. Org. Biomol. Chem.* **2013**, *1*, 33.
54. Shafiee, M. R. M.; Najafabadi, B. H.; Ghashang, M. *Res. Chem. Intermed.* **2013**, *39*, 3753.
55. Lashkari, M.; Maghsoodlou, M. T.; Hazeri, N.; Habibi-Khorassani, S. M.; Sajadikhah, S. S.; Doostmohamadi, R. *Synthetic Commun.* **2013**, *43*, 635.
56. Teimouri, A.; Chermahini, A. N.; Ghorbanian, L. *Bulgarian Chem. Commun.* **2014**, *46*, 3, 523.
57. Chouguiat, L.; Boulcina, R.; Debache, A. *J. Chem. Pharm. Res.* **2014**, *6*, 6, 79.
58. Das, S.; Bagdi, A. K. Santra, S.; Majee, A.; Hajra, A. *Res. Chem. Intermed.* DOI: 10.1007/s1164-014-1774-7.
59. Basyouni, W. M.; El-Bayouki, K. A. M.; Tohamy, W. M.; Abbas, S. Y. *Synthetic Commun.* **2015**, *45*, 1073.
60. Terada, M.; Machioka, K.; Sorimachi, K. *J. Am. Chem. Soc.* **2007**, *129*, 10336.
61. Urushima, T.; Sakamoto, D.; Ishikawa, H.; Hayashi, Y. *Org, Lett.* **2010**, *12*, 20, 4588.
62. Carney, D. W.; Truong, J. V.; Sello, J. K. *J. Org. Chem.* **2011**, *76*, 10279.
63. Zhong, S.; Nieger, M.; Bihlmeier, A.; Shi, M.; Brasc, S. *Org. Biomol. Chem.* **2014**, *12*, 3265.
64. Kozlov, N. G.; Kadutskii, A. P. *Tetrahedron Let.* **2008**, *49*, 4560.
65. Dai, B.; Duan, Y.; Liu, X.; Song, L.; Zhang, M.; Cao, W.; Zhu, S.; Deng, H.; Shao, M. *J. Fluorine Chem.* **2012**, *133*, 127.
66. Atar, A. B.; Jeong, Y. T. *Tetrahedron Lett.* **2013**, *54*, 1302.
67. Chidurala, P.; Jetti, V.; Meshram, J. S. *J. Heterocyclic Chem.* **2015**, *53*, 389.
68. Godineau, E.; Landalis, Y. *J. Am. Chem. Soc.* **2007**, *129*, 12662.
69. Clarke, P. A.; Zaytsev, A. V.; Morgan, T. W.; Whitwood, A. C.; Wilson, C. *Org. Lett.* **2008**, *10*, 13, 2877.
70. Chan, Y.; Guthmann, H.; Brimble, M. A.; Barker, D. *Synlett* **2008**, *17*, 2601.
71. Lian, G.-Y.; Lin, F.; Yu, J.-D.; Zhang, D.-W. *Synthetic Commun.* **2008**, *38*, 4321.
72. Zhu, W.; Mena, M.; Jnoff, E.; Sun, N.; Pasau, P.; Ghosez, L. *Angew. Chem. Int. Ed.* **2009**, *48*, 5880.
73. Rani, M.; Parthiban, P.; Ramachandran, R.; Kabilan, S. *Med. Chem. Res.* **2012**, *21*, 653.
74. Zhou, Z.; Liu, H.; Sun, Q.; Li, Y.; Yang, J.; Liu, J.; Yan, P.; Wang, C. *Synlett* **2012**, *23*, 2255.
75. Liu, H.; Sun, Q.; Zhou, Z,; Liu, J.; Yang, J.; Wang, C. *Monatsh Chem.* **2013**, *144*, 1031.

12 Malononitrile: A Key Reagent for the Synthesis of Medicinally Promising Fused and Spiro Pyridine Derivatives in Multicomponent Reactions

*Ruby Singh and K. L. Ameta**

CONTENTS

* Corresponding author: klameta77@hotmail.com.

12.1 INTRODUCTION

Recent research in organic synthesis has focused on atom economy.[1] The development of rapid and selective synthetic strategies toward focused libraries of functionalized heterocyclic structural units is of great importance to both medicinal as well as organic chemists, and still constitutes a challenge from academic and industrial points of view. In modern organic chemistry, owing to the increasing economic and ecological pressure, investigations are now concentrated on the discovery of methods that largely take into account the criterion of sustainable chemistry.[2]

In this point of view, multicomponent reactions (MCRs),[3] which involve domino processes[4] combining at least three or more substrates in a one-pot operation, have emerged as an important tool that pairs substrate-directed synthetic alternatives to other known methods.[5–8] Domino reactions have emerged as powerful tools to allow the rapid increase of molecular complexity. These processes avoid the excessive handling and isolation of intermediates generated, produce less waste, and thus contribute toward "green chemistry."[9] Domino reactions (also known as cascade or tandem reactions) have been reported broadly in the literature and have become state of the art in synthetic organic chemistry.[10] Moreover, these transformations amalgamate various classical concerns such as efficiency, selectivity, molecular complexity, and diversity.[11,12]

Pyridine derivatives are an important class of aza-heterocycles found in many natural products, active pharmaceuticals, and functional materials.[13] Atazanavir[14] (Reyataz) and imatinib mesylate[15] are pyridine-containing drugs used for the treatment for human immunodeficiency virus (HIV) and chronic myelogenous leukemia, respectively. A pyridine ring is also found in several potent inhibitors of phosphodiesterase (PDE4) (e.g., piclamilast, roflumilast, etc.) and has been investigated for its applications to the treatment of conditions such as chronic obstructive pulmonary disease (COPD), bronchopulmonary dysplasia, and asthma.[16] Roflumilast (trade names Daxas, Daliresp) is a drug that acts as a selective, long-acting inhibitor of the enzyme PDE-4. In addition to these, various substituted pyridines are prominent building blocks in supramolecular chemistry due to their π-stacking and directed H-bond forming ability.[17]

In recent years, the synthesis of fused heterocyclic pyridine derivatives has received significant attention, owing to the derivatives' therapeutic and pharmacological properties (Figure 12.1). Diploclidine[18] is an example of recently isolated and structurally diverse natural products containing the pyridine core. Similarly, streptonigrin and lavendamycin are antibiotic natural products having a highly substituted fused pyridine structural motif.[19]

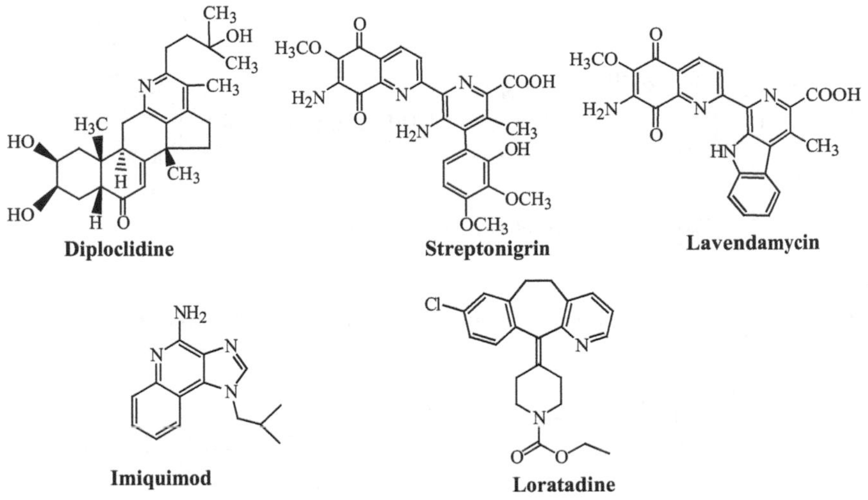

Diploclidine Streptonigrin Lavendamycin

Imiquimod Loratadine

FIGURE 12.1 Some therapeutic and pharmacologically important fused pyridine derivatives.

Imiquimod is a patient-applied cream used to treat certain diseases of the skin, including skin cancers (basal cell carcinoma, Bowen's disease.[20] Loratadine (INN) is a second-generation[21] H_1 histamine antagonist drug used to treat allergies.

In the last decades, organic cyano compounds have found extensive utilization in the synthesis of heterocycles. Malononitrile and malononitrile derivatives are versatile reagents and their chemistry has been studied in the past[22] and still attracts considerable interest. Malononitrile, also known as propanedinitrile, has an active methylene group with the formula $CH_2(CN)_2$ and is a smart synthetic building block for the construction of diverse types of heterocyclic compounds. Malononitrile is relatively acidic, with a pK_a of 11.2 in water and exhibits extraordinary reactivity because the strongly electron-withdrawing cyano groups make the methylene group active.[23] The polar nitrile group is found to be suitable for nucleophilic addition reactions and also acts as a good leaving group for substitution reactions.

In this chapter, we wish to present a complementary compilation of the synthetic routes of the fused and spiro pyridine derivatives via multicomponent reaction using malononitrile as one of the key reagents. We believe that this combination will give a comprehensive overview that will be valuable for students and academic or medicinal chemists interested in applied chemical synthesis.

12.2 PYRIDINE FUSED WITH SEVEN-MEMBERED RINGS

12.2.1 BENZOTHIEPINO[5,4-*B*]PYRIDINE-3-CARBONITRILES

The multicomponent reaction of 4-arylmethylene-3,4-dihydro-[1]-benzothiepin-5(2*H*)-ones, malononitrile, and alcohol in the presence of sodium to afford the 2-alkoxy-4-aryl-5,6-dihydro-[1]-benzothiepino[5,4-*b*]pyridine-3-carbonitriles **1** was studied by Girgis et al.,[24] in a high regioselective manner (Scheme 12.1). Anti-inflammatory screening of the prepared compounds was also done using

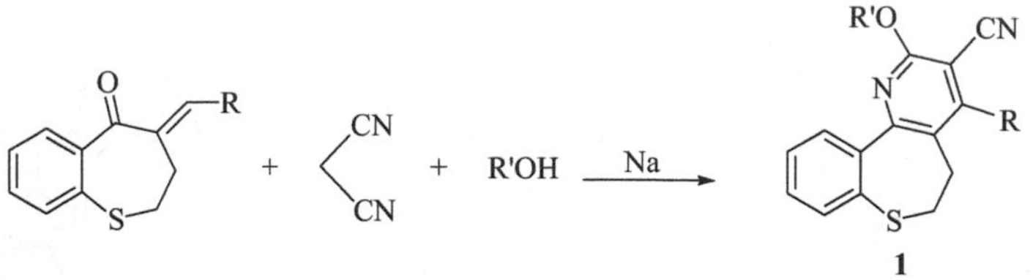

R = 4-Cl(C_6H_4); 4-Br(C_6H_4); 4-CH_3(C_6H_4); 2-thienyl; 2-furanyl
R' = CH_3; C_2H_5

SCHEME 12.1 Synthesis of benzothiepino[5,4-*b*]pyridine-3-carbonitriles **1**.

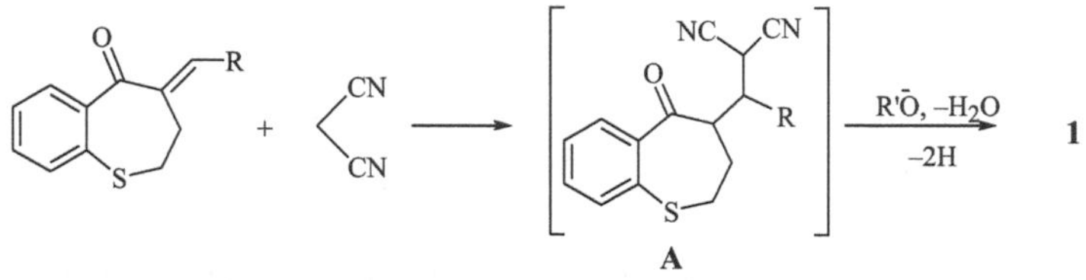

SCHEME 12.2 Proposed mechanism for the synthesis of **1**.

in vivo acute carrageenan-induced paw edema in rats and results showed that all tested compounds possessed significant activity. In addition, few synthesized derivatives revealed notable anti-inflammatory properties comparable with indomethacin, which was used as a reference during the pharmacological screening studies.

The proposed reaction take place via Michael addition of active methylene malononitrile to the β-carbon of an unsaturated system of 4-arylmethylene-3,4-dihydro-[1]-benzothiepin-5-(2*H*)-ones to afford Michael adduct **A**. In the next step, alkoxide nucleophilic attack at Michael adduct **A**, followed by dehydration and subsequent dehydrogenation, afforded finally the corresponding [1]-benzothiepino[5,4-*b*] pyridine-3-carbonitriles **1** (Scheme 12.2).

12.2.2 Cycloocta[*b*]Pyridines

The cycloocta[*b*]pyridines form the center of the antipsychotic drug blonanserin, which has been approved for the treatment of schizophrenia in Japan and Korea.[25] Kumar et al.[26] have synthesized novel alkoxy substituted cycloocta[*b*]pyridines **2a** by the four-component reaction of cyclooctanone, malononitrile, aromatic aldehydes, and MeOH or EtOH using as base LiOEt (Scheme 12.3).

The proposed mechanism for the synthesis of 4-arylcycloocta[*b*]pyridines **2a** involves first Knoevenagel condensation to give arylidene malononitrile **A**, which in turn undergoes Michael addition with cyclooctanone to afford the intermediate **B**, which further rearranged to intermediate ketene imine **C**. Then, after nucleophilic attack of the alkoxide anion generated from the solvent methanol or ethanol over the intermediate **C**, a new intermediate **D** is given. Further, the intramolecular *N*-cyclization, dehydration, and subsequent oxidative aromatization yields the desired product alkoxy substituted cycloocta[*b*]pyridines **2a** (Scheme 12.4).

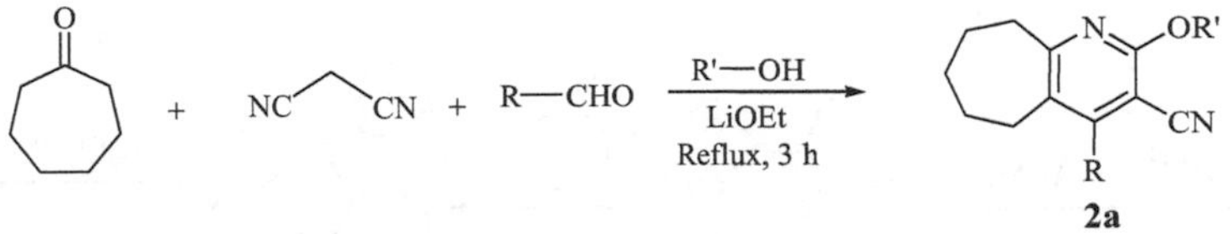

$R = C_6H_5$; 4-Cl(C_6H_4); 4-Me(C_6H_4); 4-Br(C_6H_4);2-Cl(C_6H_4); 3-F(C_6H_4); 4-OMe(C_6H_4); 4-ʼPrʼ(C_6H_4); Naphthyl; Propionaldehyde; Butyraldehyde; Pivalaldehyde

$R' = CH_3, C_2H_5$

SCHEME 12.3 Synthesis of cycloocta[*b*]pyridines **2a**.

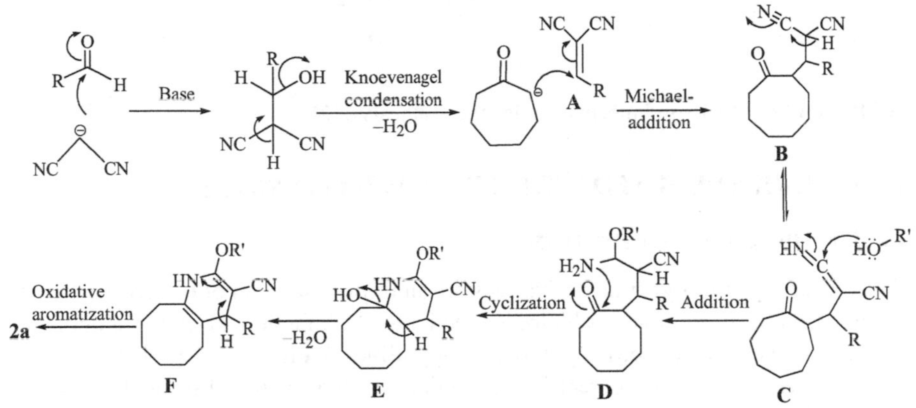

SCHEME 12.4 Proposed mechanism for the synthesis of **2a**.

Authors Jonnalagadda et al.[27] have reported the synthesis of amino substituted cycloocta[*b*]pyridines **2b** with four-component reactions of benzaldehyde, malononitrile, cycloheptanone, and ammonium acetate using a novel recoverable heterogeneous catalyst Au/MgO (Scheme 12.5).

The formation of amino substituted cycloocta[*b*]pyridines **2b** can be rationalized by the initial formation of arylidenemalononitrile (**A**) as an intermediate via the Knoevenagel condensation. The Michael type addition of ketone to the activated double bond of **A** gives **B**, which is by intramolecular cyclization, and subsequent oxidation yielded target products **2b** (Scheme 12.6).

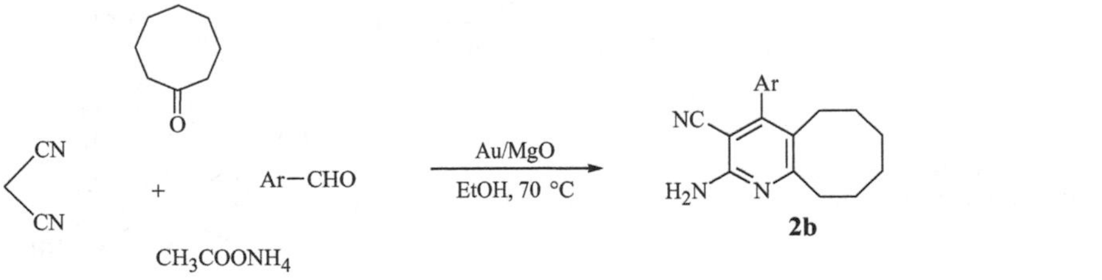

$Ar = C_6H_5$; 4-OCH_3(C_6H_4); 4-Br(C_6H_4); 4-Cl(C_6H_4); 2-Br(C_6H_4); 2-OCH_3(C_6H_4); 4-$N(CH_3)_2$(C_6H_4)

SCHEME 12.5 Synthesis of cycloocta[*b*]pyridines **2b**.

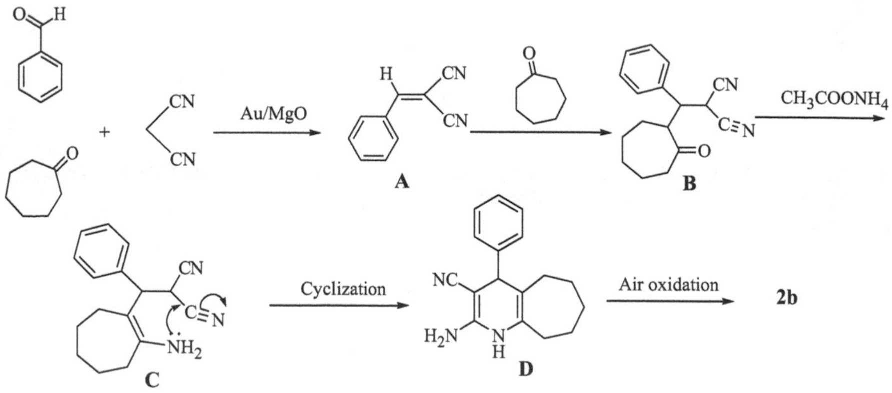

SCHEME 12.6 Proposed mechanism for the synthesis of **2b**.

12.3 PYRIDINE FUSED WITH SIX-MEMBERED RINGS

12.3.1 Benzopyrano-Pyridines

Fused benzopyrano-pyridine derivatives are an important class of heterocyclic compounds having miscellaneous biological activities.[28] A facile, one-pot, multicomponent synthesis of novel benzopyrano[2,3-*b*]pyridines **3/4** has been studied by Olyaei et al.[29] under catalyst and solvent-free conditions using various substituted naphthols, salicylaldehydes, and malononitrile at 110 °C temperature. The mild reaction conditions, high yields of products, generality, short reaction times, and operational simplicity are the advantages of this procedure (Scheme 12.7).

The proposed mechanism involves the Knoevenagel condensation of salicylaldehyde and malononitrile. Intermediate 2-arylidene malononitrile **A** was formed,

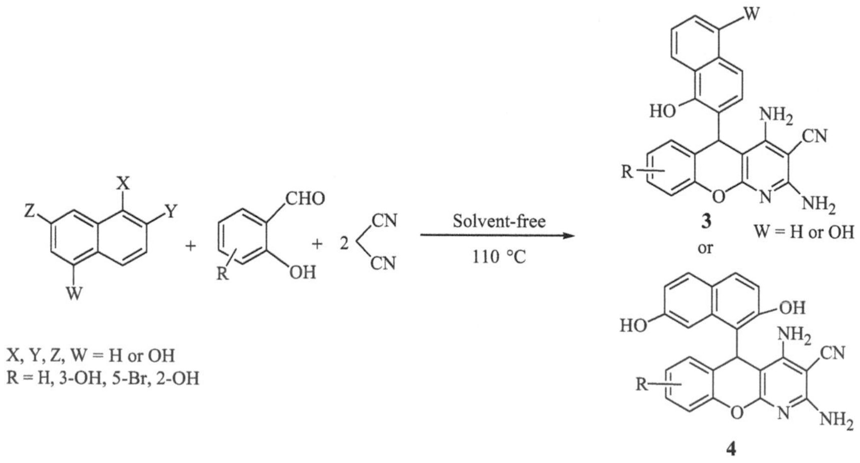

SCHEME 12.7 Synthesis of benzopyrano[2,3-*b*]pyridines **3/4**.

which subsequently underwent a Pinner reaction to give an intermediate iminocoumarin **B**. Next, iminocoumarin **B** attacked by the naphthol via Michael addition to produce intermediate **C**. Finally, the cyano group of intermediate **C** was attacked by another molecule of malononitrile, followed by cyclization to afford the product **3/4** (Scheme 12.8).

The present work also extended with 2-hydroxynaphthalene carbaldehyde instead of salicylaldehyde under the same reaction conditions.

Ghosh et al.[30] described an efficient K_2CO_3 (20 mol%) catalyzed one-pot synthesis of chromeno[2,3-*b*]pyridines **5** from the reaction of substituted salicylaldehyde or 2-hydroxynaphthylaldehyde with a variety of thiophenols and malononitrile in refluxing aqueous ethanol. Later, the present multicomponent reaction was studied by Guo et al.[31] using pyridine as base catalyst (Scheme 12.9).

The proposed mechanism involves the formation of chromene intermediate **A** generated from salicylaldehyde and malononitrile in the first step. Then intermediate **B** was formed by 1,4-addition of **A** with thiol and followed by reaction with manolonitrile, leading to the formation of desired product **5** (Scheme 12.10).

A novel one-pot, four-component coupling of resorcinol, malononitrile, aromatic aldehydes, and cyclohexanone to prepare pyrano[2,3-*b*]pyridine **6** derivatives was reported by Damavandi et al. using 4-nitro-2,6-diacetylpyridinebis(2,4,6-trimethylanil)

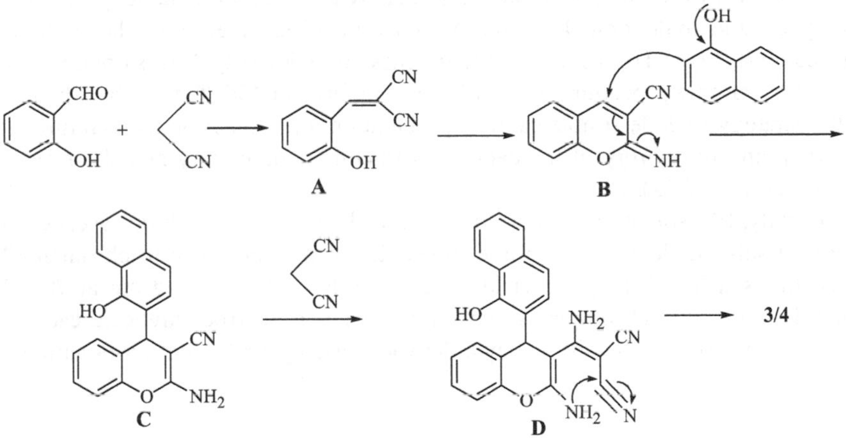

SCHEME 12.8 Proposed mechanism for the synthesis of **3/4**.

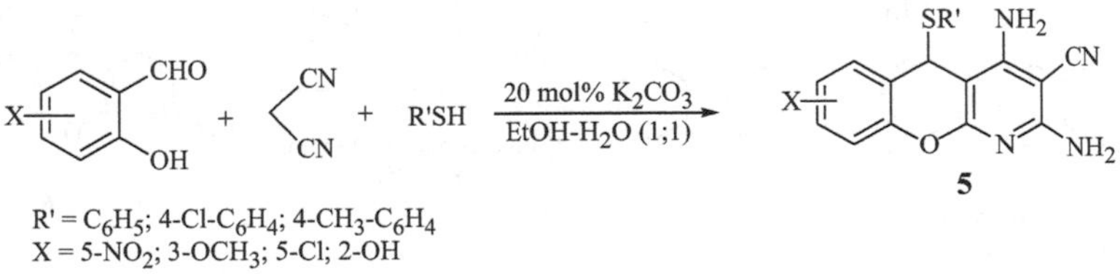

R' = C_6H_5; 4-Cl-C_6H_4; 4-CH_3-C_6H_4
X = 5-NO_2; 3-OCH_3; 5-Cl; 2-OH

SCHEME 12.9 Synthesis of chromeno[2,3-*b*]pyridines **5**.

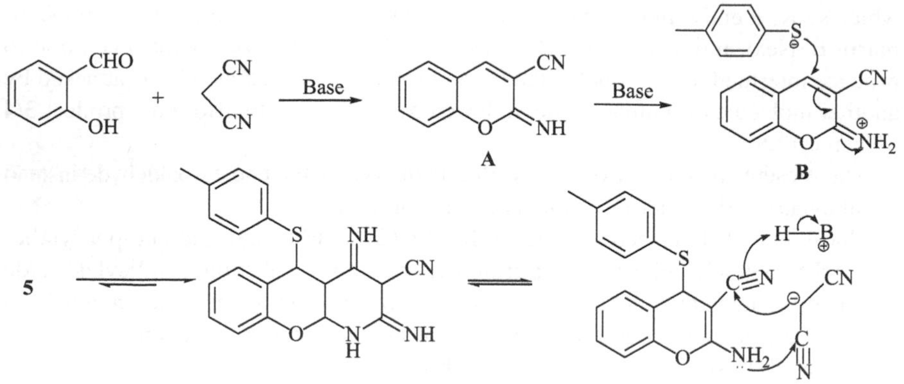

SCHEME 12.10 Proposed mechanism for the synthesis of **5**.

$FeCl_2$ as a catalyst[32] in solvent-free conditions under sonication. The catalyst exhibited high activity and the reactions were clean and highly selective, affording exclusively pyrano[2,3-*b*]pyridines **6** in moderate to excellent yields (Scheme 12.11).

Chromeno[3,4-*c*]pyridines **7/8** were synthesized by Mohammadzadeh and Sheibani[33] from the one-pot, multicomponent reaction of salicylaldehyde, ketones/aldehydes, and malononitrile in the presence of a highly effective heterogeneous base catalyst, MgO. High surface area magnesium oxide (MgO) was obtained using a novel but simple procedure by rehydrated $Mg(OH)_2$ at 450 °C for 2 h. Advantages of the strategy include simple catalyst preparation, mild reaction temperature, easy recovery, and reusability of the catalyst with consistent activity and short reaction times (Scheme 12.12).

Recently, Elinson et al.[34] reported a chemical and electrocatalytic cascade cyclization of salicylaldehyde with three molecules of malononitrile to chromeno[2,3-*b*]pyridine scaffold **9** in good yield. The electrolysis was carried out at 78 °C in ethanol using sodium bromide as electrolyte. The new electrocatalytic cascade reaction smoothly proceeds with salicylaldehydes bearing both electron-donating and

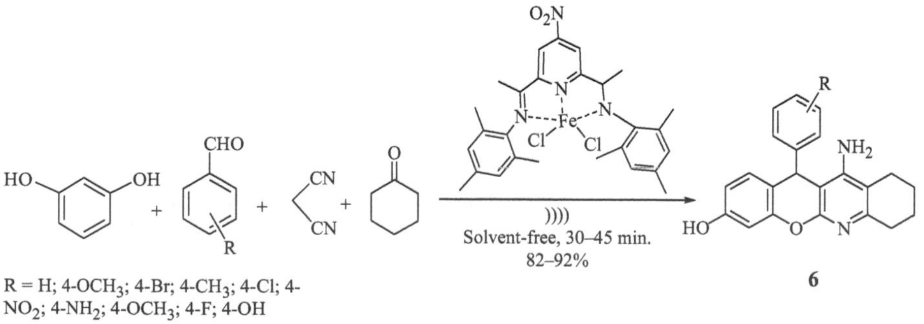

$R = H; 4\text{-}OCH_3; 4\text{-}Br; 4\text{-}CH_3; 4\text{-}Cl; 4\text{-}NO_2; 4\text{-}NH_2; 4\text{-}OCH_3; 4\text{-}F; 4\text{-}OH$

Solvent-free, 30–45 min. 82–92%

SCHEME 12.11 Synthesis of pyrano[2,3-*b*]pyridines **6**.

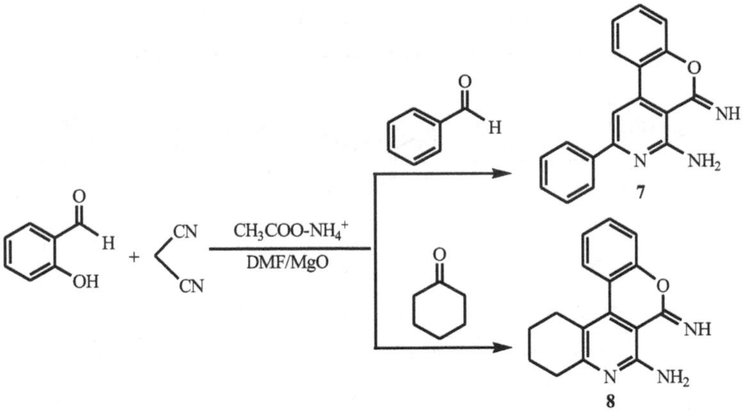

SCHEME 12.12 Synthesis of chromeno[3,4-*c*]pyridines **7/8**.

electron-withdrawing groups, and allows for the selective and combined one-step introduction of a wide range of medicinally promising functionalities into a privileged chromeno[2,3-*b*]pyridine **9** framework (Scheme 12.13).

The proposed mechanism for the synthesis of chromeno[2,3-*b*]pyridine **9** involves the first deprotonation of an ethanol as the cathode leads to the formation of ethoxide anion. The subsequent reaction in solution between ethoxide anion and malononitrile gives rise to a malononitrile anion. Then Knoevenagel condensation of the malononitrile anion and salicylaldehyde leads to the formation of Knoevenagel adduct **A** with the elimination of the hydroxide anion. The subsequent intramolecular cyclization of Knoevenagel adduct **A** followed by the addition of a second malononitrile anion leads to (4*H*-chromen-4-yl)malononitrile **B** with regeneration of the malononitrile anion, which attacks the nitrile group in the pyrane ring of **B** with the formation of anion **C**. Further, cyclization of anion **C** leads to chromeno [2,3-*b*]pyridine **9** with the rebirth of the malononitrile anion (Scheme 12.14).

The one-pot, three-component reaction of 4-chloro-3-formylcoumarins, sodium azide, and alkyl/aryl acetonitriles such as malononitrile, cyanoacetamide, etc. was studied by Bhuyan et al.[35] in the synthesis of tetrazole fused pyrido[2,3-*c*]coumarin derivative **10** (Scheme 12.15).

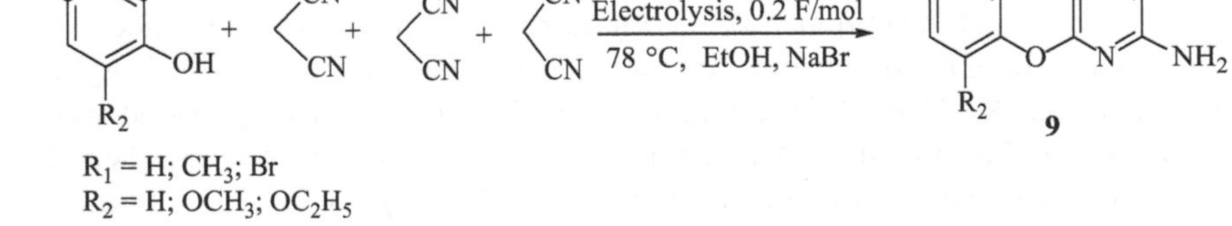

R_1 = H; CH_3; Br
R_2 = H; OCH_3; OC_2H_5

SCHEME 12.13 Synthesis of chromeno[2,3-*b*]pyridine **9** by electrocatalytic method.

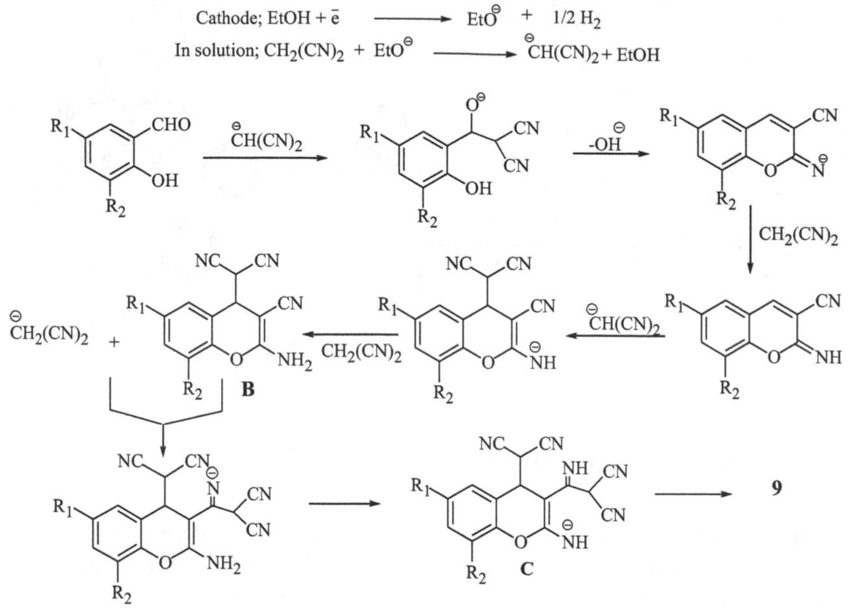

SCHEME 12.14 Proposed mechanism for the synthesis of **9** using NaBr as electrolyte.

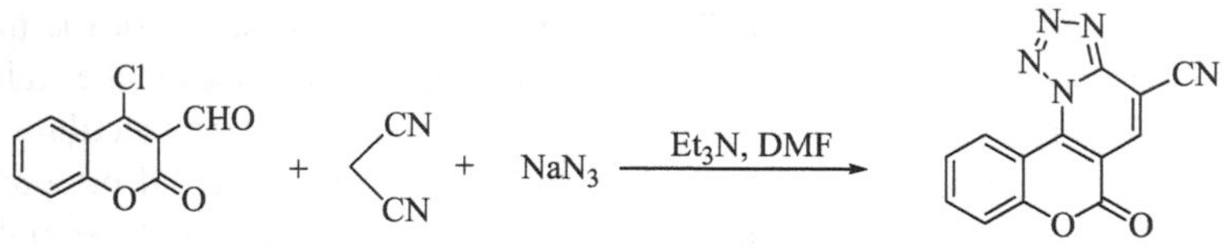

SCHEME 12.15 Synthesis of pyrido[2,3-*c*]coumarins **10**.

12.3.2 THIOCHROMENO[2,3-*B*]PYRIDINES

The interest in the *N,S*-containing heterocycle thiochromones has been significantly increased since a wide range of biological activities have been associated with them.[36] Li et al.[37] synthesized thiochromeno[2,3-*b*]pyridines **11** by a simple MCR of *ortho*-chloro-β–aroyl thioamides, aldehydes, and malononitrile using catalyst KF/neutral Al_2O_3 cooperated with PEG 6000 under microwave irradiations (Scheme 12.16).

Construction of thiochromeno[2,3-*b*]pyridine **11** proceeds by initial Knoevenagel condensation of aldehydes with malononitrile, leading to the formation of 2-arylidene malononitrile **A**. Then, Michael addition with simultaneous cyclocondensation between **A** and ortho-chloro-β-aroylthioamides give intermediate **B** and then **C**, in which the formation of a pyridine ring is involved. Finally, the formation of the final product proceeds by a nucleophilic aryl substitution (SNAr) reaction of the mercapto group attacking a carbon atom linked to *ortho*-chloro on the phenyl ring of the intermediate **D** (Scheme 12.17).

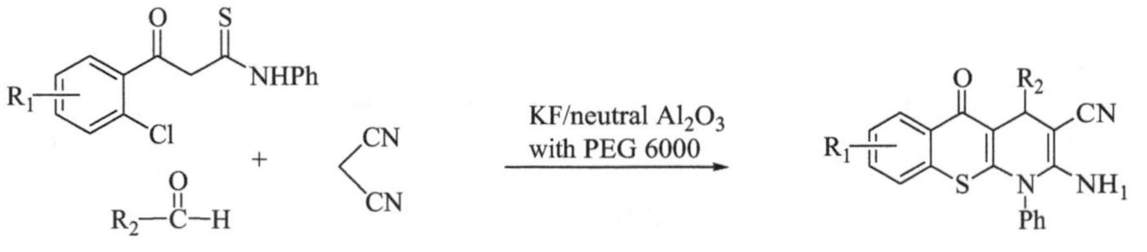

R_1 = H; 4-Cl; 5-Cl
R_2 = CH_3; C_6H_5; 3-NO_2.C_6H_4; 4-Cl.C_6H_4; 3,4-(OCH_2O)C_6H_3; 3,4-$(OCH_3)_2C_6H_3$; 2,4-$Cl_2C_6H_3$; 2-Cl.C_6H_4; furfural

SCHEME 12.16 Synthesis of thiochromeno[2,3-*b*]pyridines **11**.

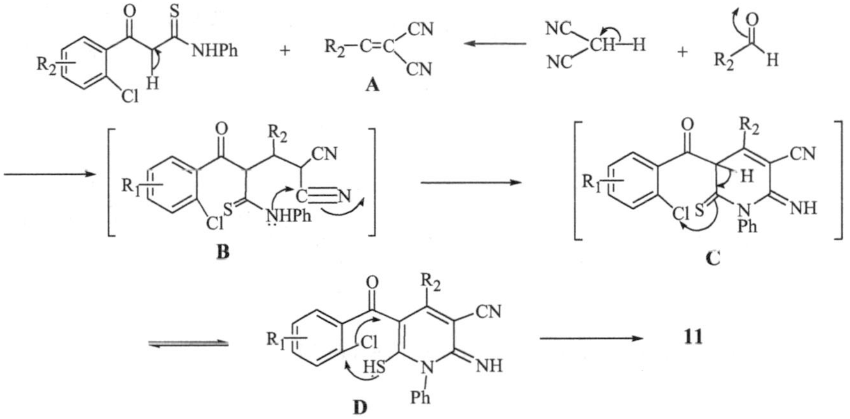

SCHEME 12.17 Proposed mechanism for synthesis of **11**.

In the present reaction PEG 6000 adsorbed on the surface of Al_2O_3 may be chelated with K^+, forming the chelate cation; F^- acts as a balance ion. The three basic species generated through a cooperative action of KF on the surface Al_2O_3 and PEG 6000 would take part in nucleophilic reaction and display high catalytic activity.

12.3.3 Hydroquinoline-3-Carbonitriles

Quinoline moiety is found in a large variety of naturally occurring compounds and also chemically useful synthons bearing diverse bioactivities,[38] including EGFR inhibitory activity.[39] Wu et al.[40] reported the synthesis of N2-methyl- or aryl substituted 2-amino-4-aryl-5,6,7,8-tetrahydroquinoline-3-carbonitriles **12** via a four-component, one-pot reaction of aromatic aldehyde, cyclohexanone, malononitrile, and amines in basic ionic liquid [bmim]OH with good to excellent yields (Scheme 12.18).

The possible mechanism for the formation of **12** is shown in Scheme 12.19. Ionic liquid [bmim]OH plays a crucial role in suppressing the formation of other by-products.

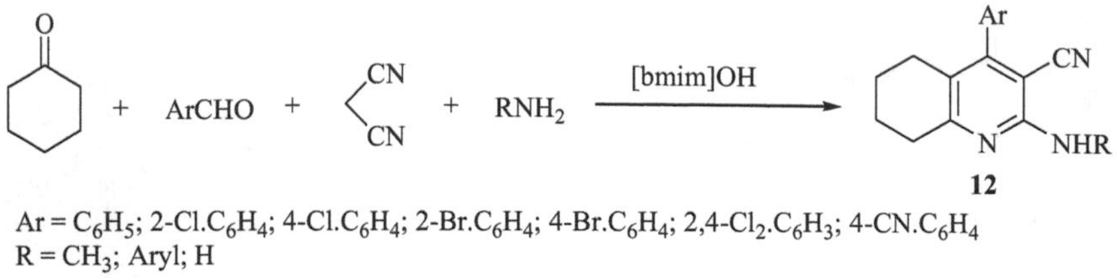
[bmim]OH
ArCHO
RNH2
CN
NHR
Ar
12
Ar = C6H5; 2-Cl.C6H4; 4-Cl.C6H4; 2-Br.C6H4; 4-Br.C6H4; 2,4-Cl2.C6H3; 4-CN.C6H4
R = CH3; Aryl; H

SCHEME 12.18 Synthesis of hydroquinoline-3-carbonitriles **12**.

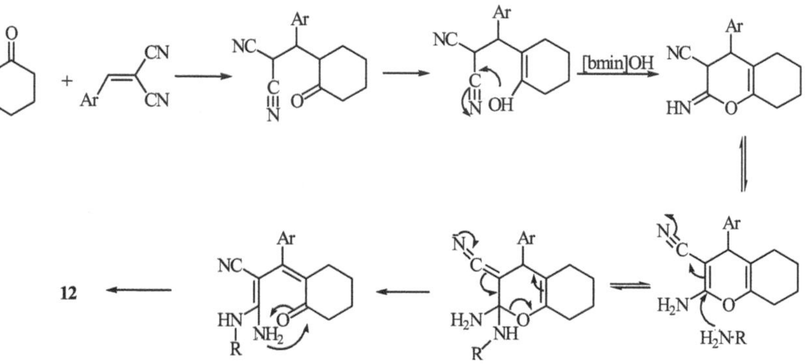
[bmin]OH
12

SCHEME 12.19 Proposed mechanism for synthesis of **12**.

Synthesis of biquinoline–pyridine hybrids **13** has been reported by Zhu et al.[41] from multicomponent reaction of tetrazolo[1,5-*a*]quinoline-4-carbaldehyde, malononitrile/methylcyanoacetate/ethylcyano acetate and β-pyridinyl enaminone using piperidine as a base catalyst (Scheme 12.20). Synthesized compounds were tested for *in vitro* anticancer activities against two cancer cell lines, A549 (adenocarcinomic human alveolar basal epithelial) and Hep G2 (liver cancer). Enzyme inhibitory activities of all compounds were carried out against EGFR and HER-2 kinase.

12.3.4 Polyhydroisoquinoline

The hydroisoquinoline skeleton was found in a wide range of natural alkaloids, such as reserpine[42] and yohimbine.[43] Isoquinolines have also been used as ligands in a number of catalytic asymmetric transformations.[44]

Recently, Shi et al. reported[45] the synthesis of polyhydroisoquinoline derivatives **14** via the three-component domino reaction of glutaraldehyde and malononitrile with a series of β-keto amides under microwave irradiations in the presence of a catalytic amounts of Et_3N (10 mol%) via a C–N bond cleavage reaction without the need for a multistep reaction process. Various solvents, including EtOH, THF, DMF, acetonitrile, toluene, and water were used in this multicomponent reaction, but EtOH was found to be the best solvent.

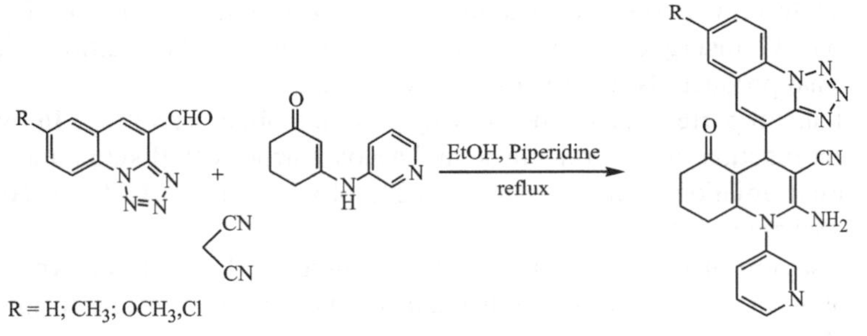

SCHEME 12.20 Synthesis of biquinoline–pyridine derivatives **13**.

Several catalysts, including Na_2CO_3, piperidine, Et_3N, and *p*-TsOH, were also evaluated for their impact on the yield. The results of these screening experiments revealed that Et_3N (10 mol%) provided superior catalytic efficiency compared with all of the other catalysts tested.

The use of 3-oxo-3-aryl-*N*-alkylpropanamide in place of 3-oxo-3-aryl-*N*-arylpropanamide under the optimized reaction conditions leads to the formation of isomerization products of **14** (i.e., octahydroisoquinoline derivatives) and **15** in good yields (Scheme 12.21).

The mechanism involved in this transformation is Knoevenagel condensation of glutaraldehyde with malononitrile to give intermediate **A**. The consequent Michael addition of β-keto amide to intermediate **A** gives intermediate **B**, which then undergoes an intramolecular nucleophilic addition reaction to form intermediate **C**. Intramolecular cyclization of intermediate **C** gives intermediate **D**, which undergoes

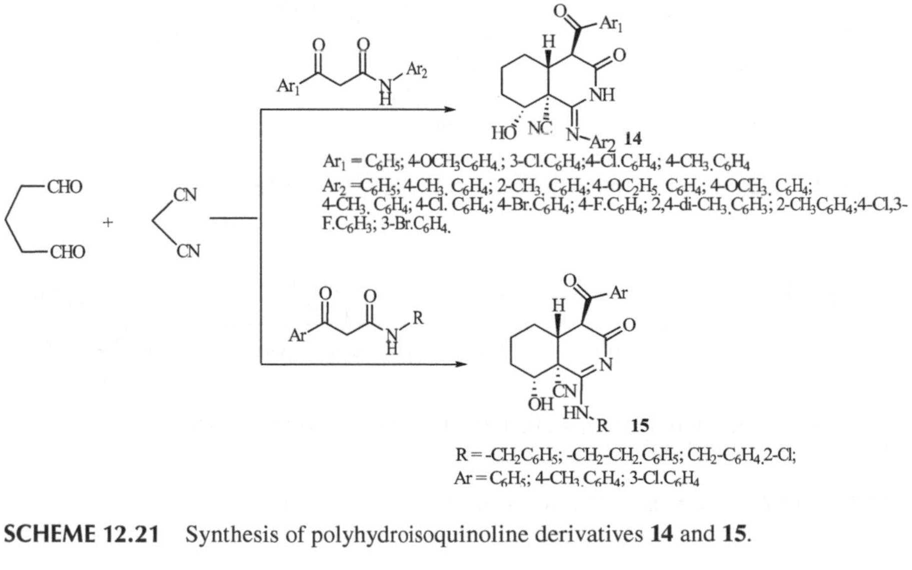

SCHEME 12.21 Synthesis of polyhydroisoquinoline derivatives **14** and **15**.

a Et_3N catalyzed ring-opening reaction to give the ketene intermediate **G**. Lastly, intermediate **G** undergoes a second intramolecular nucleophilic addition reaction to give the final products **14** and then **15** (Scheme 12.22).

Shi et al.[46] reported the synthesis of hydroisoquinoline derivatives **16** by multicomponent reaction of glutaraldehyde and malononitrile with β-keto amides under microwave irradiations in the presence of a 20 mol% NaOH via a C–C bond cleavage reaction (Scheme 12.23).

The first step of this reaction proceeds through NaOH catalyzed Knoevenagel condensation of glutaraldehyde with malononitrile or β-keto amide to give intermediate **A** or **A′**. Subsequent Michael addition of β-keto amide or malononitrile to intermediate **A** or **A′** gives intermediate **B**, which via intramolecular nucleophilic addition reaction converts to intermediate **C**. After cyclization and subsequent isomerization, the intermediate **C** converts to another one, **E**, which undergoes a NaOH catalyzed C–C bond cleavage reaction to give desired product **16** via the loss of its R_1COO^- and hydroxy groups (Scheme 12.24).

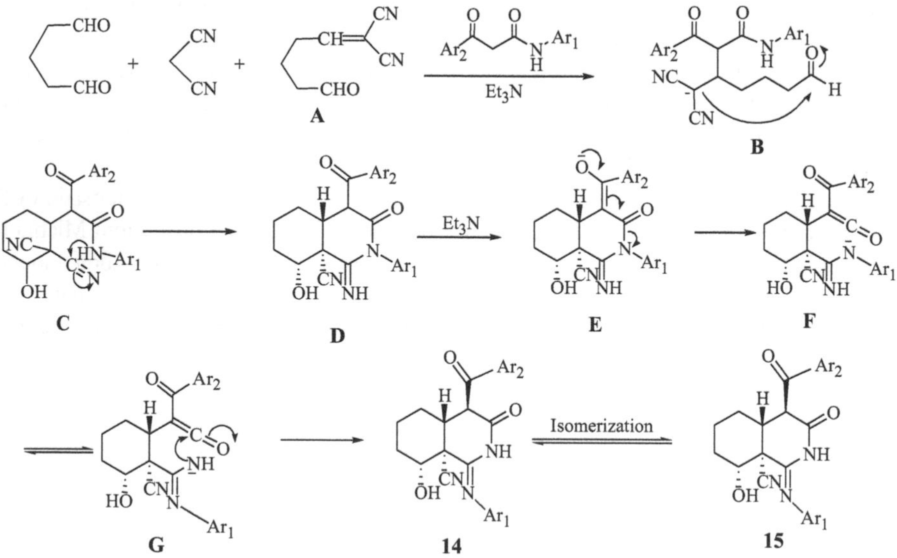

SCHEME 12.22 Proposed mechanism for synthesis of **14** and **15**.

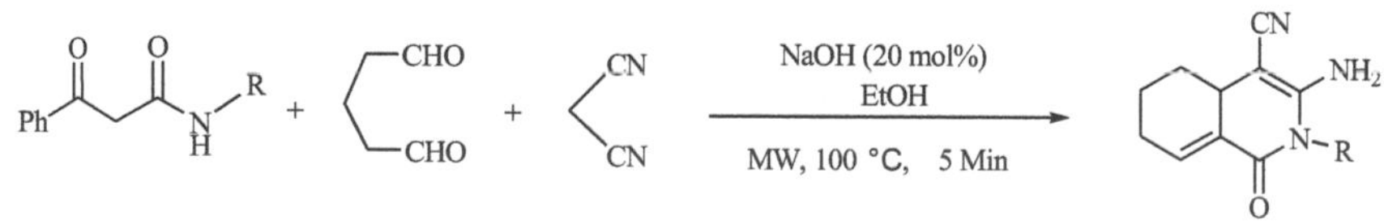

R = 4-$(CH_3)_3C.C_6H_4$; 4-$(CH_3)_2CH.C_6H_4$; 4-$CH_3CH_2OC_6H_4$; 4-$CH_3O.C_6H_4$; 4-$CH_3.C_6H_4$; 3,5-$(CH_3)_2.C_6H_3$; 4-$F.C_6H_4$; 4-$Cl.C_6H_4$

SCHEME 12.23 Synthesis of hydroisoquinoline derivatives **16**.

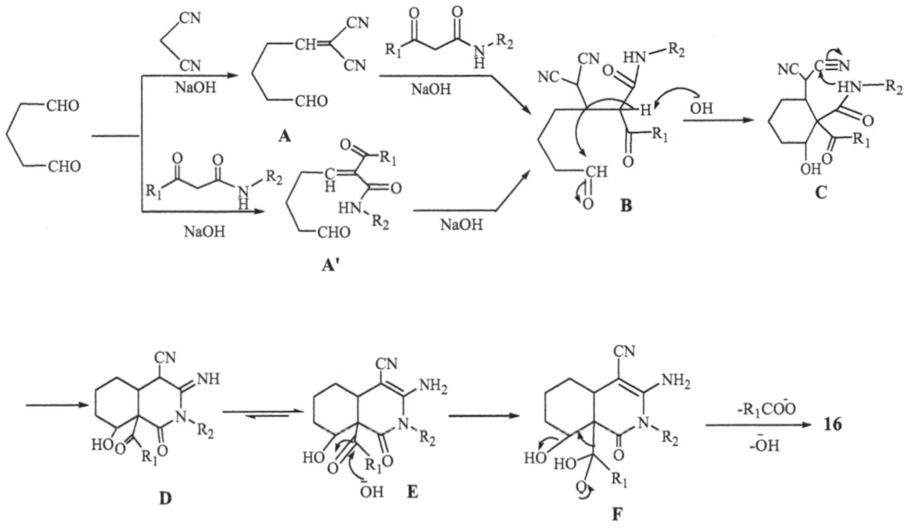

SCHEME 12.24 Proposed mechanism for synthesis of **16**.

12.3.5 Pyrido[2,3-a]Carbazoles

The pyrido[2,3-*a*]carbazole motif is a privileged pharmacophore scaffold found in many naturally occurring products.[47] Pyrido[2,3-*a*]carbazoles[48] **17** were prepared in good yields by the multicomponent reaction of 6-methyl-2,3,4,9-tetrahydro-1*H*-carbazol-1-one, benzaldehyde, malononitrile, and ammonium acetate using L-proline as organocatalyst in dry ethanol (Scheme 12.25).

A plausible mechanism for the synthesis of pyrido[2,3-*a*]carbazoles **17** involves the Knoevenagel condensation followed by Michael addition and prototropic shift, and final aerial oxidation leads to pyrido[2,3-*a*]carbazoles **17** (Scheme 12.26). Few synthesized compounds showed considerable activity toward both Gram-positive and Gram-negative bacterial strains. The cytotoxicity was evaluated by sulforhodamine B assay against A-549, B16F10, HCT-15, SKMel2, and SKOV3 cell lines and compared with the standard drug cisplatin.

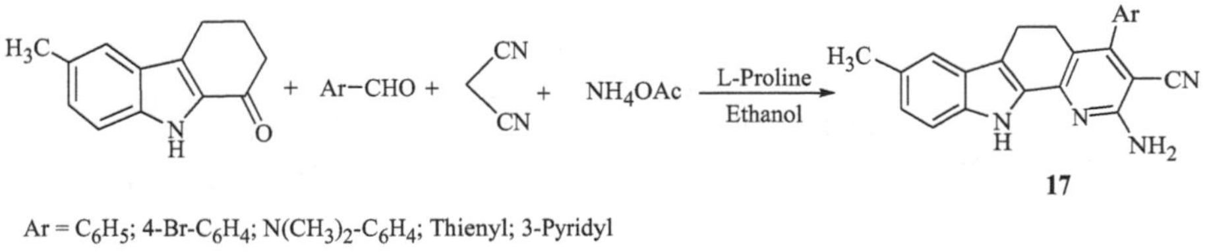

Ar = C_6H_5; 4-Br-C_6H_4; $N(CH_3)_2$-C_6H_4; Thienyl; 3-Pyridyl

SCHEME 12.25 Synthesis of pyrido[2,3-*a*]carbazoles **17**.

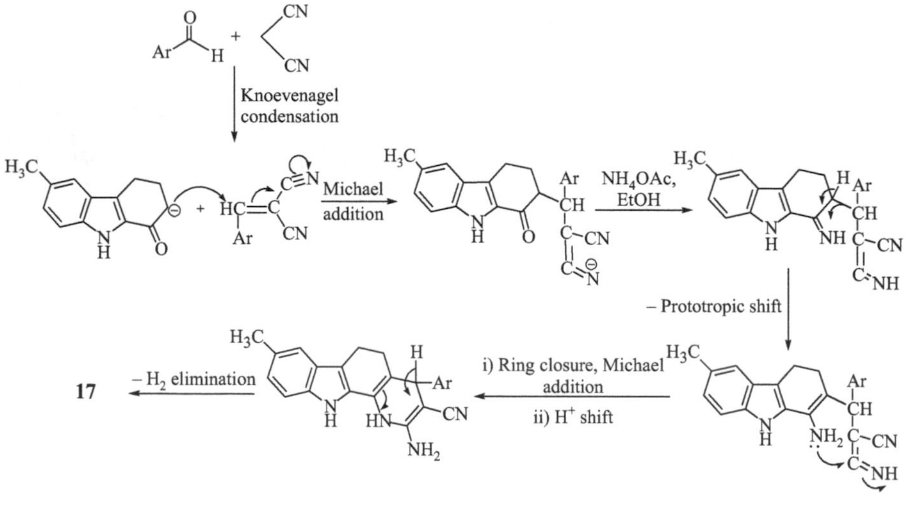

SCHEME 12.26 Proposed mechanism for synthesis of **17**.

12.3.6 Pyrido[2,3-*d*]Pyrimidine

Siddiqui et al.[49] described efficient and straightforward synthesis of 1,2,3,4-tetrahydropyrido[2,3-*d*]pyrimidine-6-carbonitrile **18** using 6-amino uracil, various aromatic aldehydes, and malononitrile under aqueous medium at 90 °C in the presence of thiamine hydrochloride (VB1) as a reusable, green catalyst (Scheme 12.27).

The mechanism involved in this reaction involves, first, Knoevenagel condensation of aldehyde and malononitrile and then subsequent Michael addition of 6-amino-1-methyluracil on intermediate **A**, leading to the formation of another intermediate, **C**, through an unstable intermediate **B**. Intermediate **C** undergoes intramolecular cyclization and converts into the final product via air oxidation. The product was isolated by simple filtration because VB1 is soluble in water and the product is insoluble. The filtrate containing VB1 was again used for further reaction (Scheme 12.28).

Afterward Ziarani et al.[50] described the preparation of pyrido[2,3-*d*]pyrimidine-6-carbonitrile derivatives **18** using 6-amino uracil, various aromatic aldehydes, and malononitrile in the presence of sulfonic acid functionalized SBA-15.

Das et al.[51] have developed a highly convergent and efficient protocol for the facile synthesis of a library of pyridopyrimidine **18** by applying a Lewis base

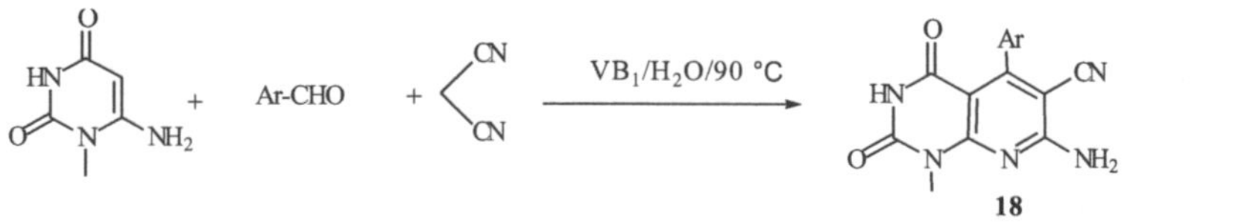

Ar = C_6H_5; 4-$NO_2.C_6H_4$; 2-Cl.C_6H_4; 4-$OCH_3.C_6H_4$; 3-OH.C_6H_4; 3-$NO_2.C_6H_4$; 4-$CH_3.C_6H_4$; 2,4-$Cl_2.C_6H_3$

SCHEME 12.27 Synthesis of pyrido[2,3-*d*]pyrimidine-6-carbonitriles **18**.

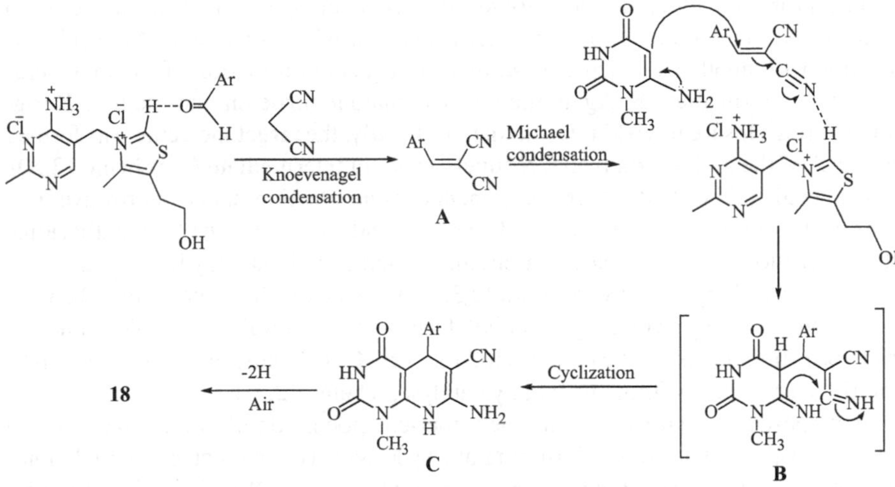

SCHEME 12.28 Proposed mechanism for synthesis of **18**.

surfactant-combined catalyst (LBSC) triethanolamine (TEOA). In this multicomponent reaction, TEOA is a new Lewis base surfactant combined catalyst (LBSC) that acts both as a catalyst to activate the substrate molecules and as a surfactant to form colloidal particles.

12.3.7 Naphthyridines

Functionalized naphthyridines and their heterofused analogs are present in numerous products of marine origins[52] and possess remarkable biological properties such as antiproliferative activity, and act as HIV-1 integrase inhibitors, allosteric inhibitors of Akt1 and Akt2, and selective antagonists of 5-HT4 receptors.[53]

Mukhopadhyay et al.[54] reported a new one-pot, catalyst-free, pseudo-five-component synthesis of 1,2-dihydro[1,6]naphthyridines **19** from methyl ketones, amines, and malononitrile using water as an ecofriendly solvent (Scheme 12.29).

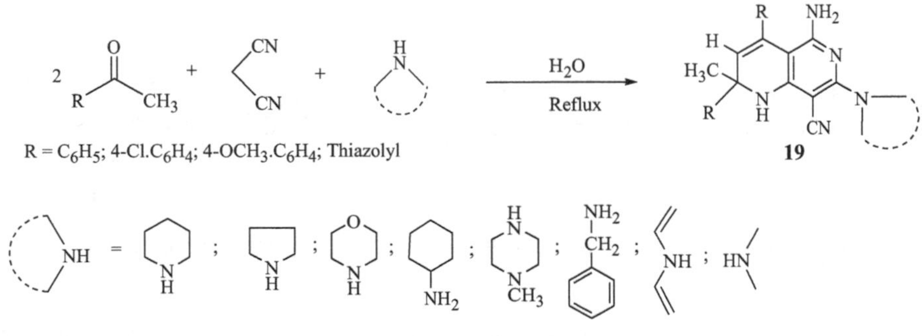

R = C_6H_5; 4-Cl.C_6H_4; 4-OCH_3.C_6H_4; Thiazolyl

SCHEME 12.29 Synthesis of 1,2-dihydro[1,6]naphthyridines **19**.

The proposed mechanism involves the Knoevenagel condensation between ketones and malononitrile to form intermediate **A**, which undergoes Michael-type reaction with another molecule of **A**, and subsequent elimination of the malononitrile gives intermediate **B**. Again the attack of malononitrile on intermediate **B** triggers the ring closure to yield intermediate **C**. Finally, the target molecule was formed by attack of the amino group on CN functionality in intermediate **C** (Scheme 12.30).

Cao et al.[55] described a four-component domino reaction under microwave irradiations for synthesis of new octahydrobenzo[*b*]indeno[1,2,3-de][1,8] naphthyridine **20** via reaction of enaminones, malononitrile, and *o*-phthalaldehyde using DMF as a solvent. Similarly, decahydropyrido[2,3,4-gh]phenanthridine derivatives **21** were also synthesized by replacing *o*-phthalaldehyde by glutaraldehyde. This one-pot transformation, involved the constructed four new C–C bonds, two new C–N bonds, and three new rings without using any catalyst (Scheme 12.31).

A proposed mechanism for this new four-component domino reaction involved Knoevenagel condensation of *o*-phthalaldehyde with two molecules of malononitrile to gives the intermediate **A**. The second intermediate **B** is formed by Michael addition of enaminone to intermediate **A** and cyclization. Intermediate **A** to **B** is a reversible process. Therefore, from **A** to **B**, there may be two isomers formed. One is *syn*-**B**, the other is *anti*-**B**. The *syn*-**B** is more stable than *anti*-**B**. Moreover, *syn*-**B** is favored over *anti*-**B** to cyclize the third ring, so *anti*-**B** will go back to **A**. Finally all **A** is transformed to **B**. The *syn*-**B** undergoes imine-enamine tautomerization to give

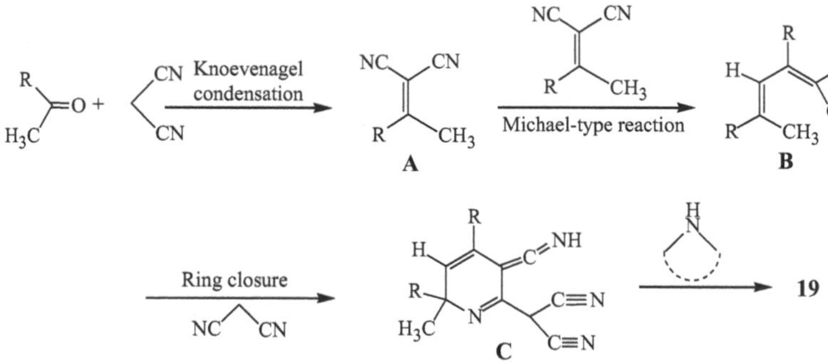

SCHEME 12.30 Proposed mechanism for synthesis of **19**.

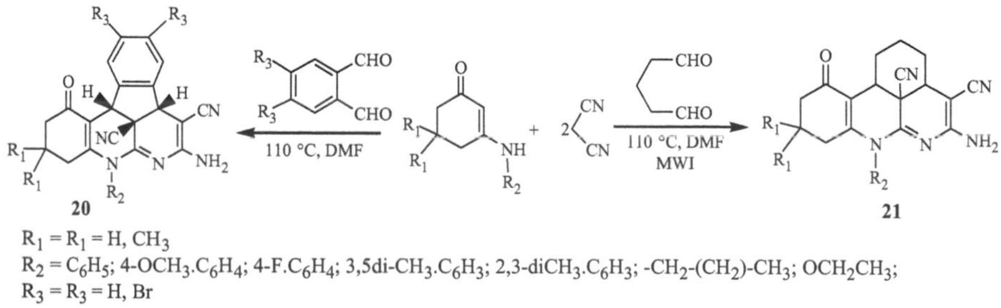

$R_1 = R_1 = H, CH_3$
$R_2 = C_6H_5$; 4-$OCH_3.C_6H_4$; 4-F.C_6H_4; 3,5di-$CH_3.C_6H_3$; 2,3-di$CH_3.C_6H_3$; -CH_2-(CH_2)-CH_3; OCH_2CH_3;
$R_3 = R_3 = H$, Br

SCHEME 12.31 Synthesis of benzo[*b*]indeno[1,2,3-de][1,8]naphthyridine **20** and decahydropyrido[2,3,4-gh]phenanthridine derivatives **21**.

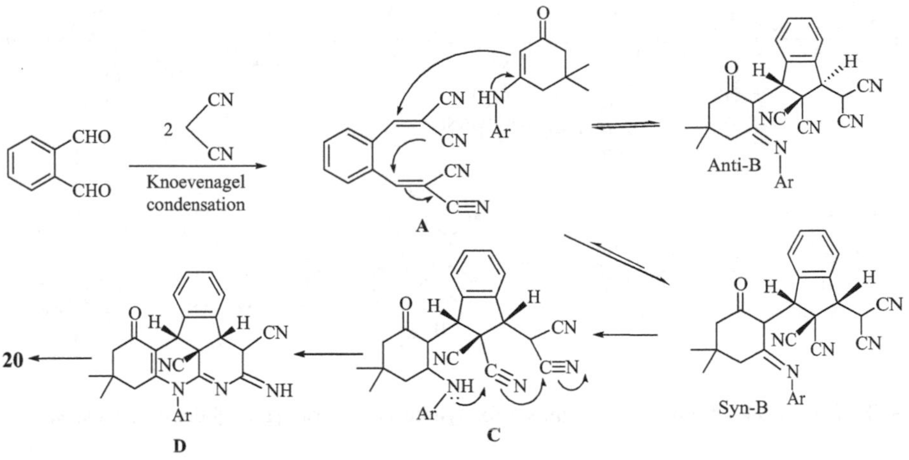

SCHEME 12.32 Proposed mechanism for synthesis of **20**.

intermediate **C**, which subsequently reacts in an intramolecular cyclization to give intermediate **D**. In the last step, intermediate **D** undergoes imine-enamine tautomerization to give the product **20** (Scheme 12.32).

12.4 PYRIDINE FUSED WITH FIVE-MEMBERED RINGS

12.4.1 Pyrido[1,2-a]Benzimidazole

Pyrido[1,2-*a*]benzimidazole motifs constitute various natural products, biologically active compounds, and synthetic intermediates.[56] Yan et al.[57] reported a very interesting procedure for the synthesis of polysubstituted pyrido[1,2-*a*]benzimidazole derivatives **22** by four-multicomponent reaction of pyridine or 3-picoline, chloroacetonitrile, malononitrile, and aromatic aldehyde (Scheme 12.33).

When a mixture of chloroacetonitrile, malononitrile, benzaldehyde, and an excess of pyridine was heated in acetonitrile for several hours, the unexpected nitrogen-containing heterocyclic compound pyrido[1,2-*a*]benzimidazole **22** was isolated in 35% yield. Similarly, various aromatic aldehydes were screened under the same conditions, and the corresponding pyrido[1,2-*a*]benzimidazole **22** were produced in

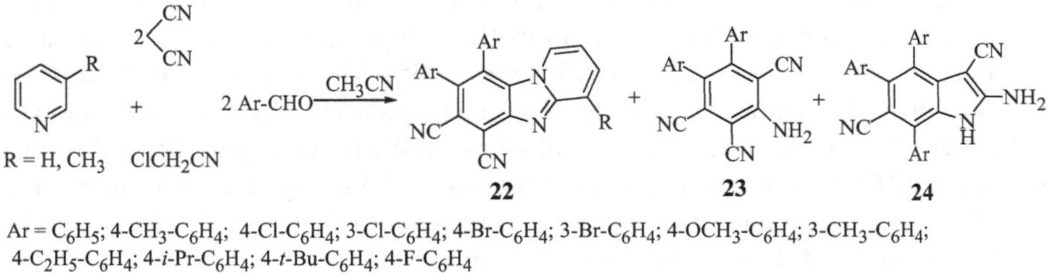

Ar = C_6H_5; 4-CH_3-C_6H_4; 4-Cl-C_6H_4; 3-Cl-C_6H_4; 4-Br-C_6H_4; 3-Br-C_6H_4; 4-OCH_3-C_6H_4; 3-CH_3-C_6H_4; 4-C_2H_5-C_6H_4; 4-*i*-Pr-C_6H_4; 4-*t*-Bu-C_6H_4; 4-F-C_6H_4

SCHEME 12.33 Synthesis of pyrido[1,2-*a*]benzimidazole **22** along with other products.

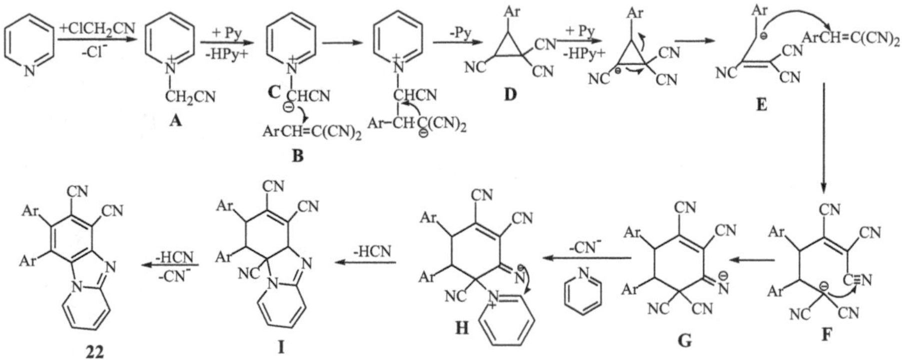

SCHEME 12.34 Proposed mechanism for synthesis of pyrido[1,2-*a*]benzimidazole **22**.

moderate yields. In addition to affording pyrido[1,2-*a*]benzimidazole derivatives as main products, the reactions of *p*-chloro- and *p*-isopropylbenzaldehyde also yield polysubstituted indole products **22** (17% and 14%, respectively) in lower yield. From the reaction mixture of *p*-methoxybenzaldehyde and *p*-fluorobenzaldehyde, another by-product polysubstituted benzene derivative **24** could be separated in 10% and 14% yields, respectively. The reaction of *p*-methylbenzaldehyde gave all three products: **22** (20%), **23**(17%), and **24** (15%). The results of the reaction are very interesting and show that three different kinds of products can be formed in the reaction. The structures of all pyrido[1,2-*a*]benzimidazoles **22**, polysubstituted benzenes **23**, and polysubstituted indoles **24** were fully characterized by various spectral analyses and by single-crystal x-ray diffraction studies.

The mechanism of this novel reaction involved the formation of polysubstituted benzene with subsequent substitution and annulation reaction of pyridine (Scheme 12.34).

12.4.2 Thiazolo[3,2-a]Pyridine

Thiazolo[3,2-*a*]pyridines contain two fused heterocyclic moieties in one molecule. They have many important bioactivities, such as inhibiting β-amyloid production[58] and acting as a potent CDK2-cyclin A inhibitor[59] and an α-glucosidase inhibitor.[60]

Zheng et al. reported[61] a green chemoselective synthesis of thiazolo[3,2-*a*]pyridines **25** and arylidene thiazolo[3,2-*a*]pyridines **26** via microwave-assisted three-component reaction of malononitrile, aromatic aldehydes, and 2-mercaptoacetic acid with different molar ratios in water (Scheme 12.35). Three-component reaction of malononitrile, 4-bromobenzaldehyde, and 2-mercaptoacetic acid with molar ratio of 2:1:1.5 at 90 °C under microwave irradiations (MW) in water gave **25** exclusively. Compound **26** was formed when molar ratio was 2:2.2:1 and the reaction temperature 100 °C. These compounds were subject to screen antioxidant activity and cytotoxicity to carcinoma HCT-116 cells and mice lymphocytes. Nearly all of the tested compounds possessed potent capacities for scavenging free radicals. In addition, most of these compounds showed cytotoxicity to HCT-116 cells and mice lymphocytes with

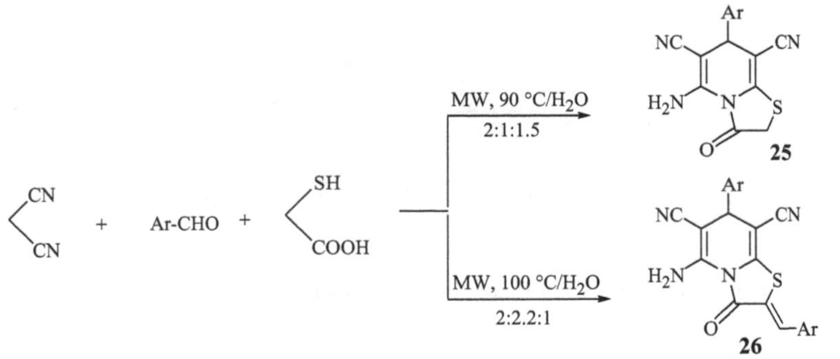

Ar = C_6H_5; 4-F.C_6H_4; 4-Br.C_6H_4; 2-Cl.C_6H_4; 2,4-Cl_2.C_6H_3; 4-CH_3.C_6H_4; 4-OH,3-NO_2-C_6H_3; 2-Thienyl

SCHEME 12.35 Synthesis of thiazolo[3,2-*a*]pyridines **25** and arylidene thiazolo[3,2-*a*] pyridines **26**.

no selectivity. Of these, only thiazolo[3,2-*a*]pyridine **25** derivative suggested selective cytotoxicity to tumor cell line HCT-116 cells.

A mechanism to account for the formation of **25** and **26** is demonstrated in Scheme 12.36.

Altug et al. have synthesized a series of thiazolo[3,2-*a*]pyridines **27** using a multicomponent reaction between aromatic aldehydes, 2-nitromethylenethiazolidine, and malononitrile, in the presence of base Et_3N under mild reaction conditions with high yields of product.[62] One of the compounds shows promising anticancer activity across a range of cancer cell lines (Scheme 12.37).

This reaction most probably proceeds by an initial Knoevenagel condensation of the aldehyde and malononitrile followed by conjugate addition of the enamine, 2-nitromethylenethiazolidine, and then cyclization (Scheme 12.38).

Later, Altug et al.[63] described the synthesis of 5-amino-7-aryl-3-oxo-8-(phenylsulfonyl)-thiazolo[3,2-*a*]pyridine-6-carbonitriles **28** via multicomponent reactions of enamine 2-phenylsulfonylmethylenethiazolidin-4-one with various aromatic aldehydes and malonitrile (Scheme 12.39).

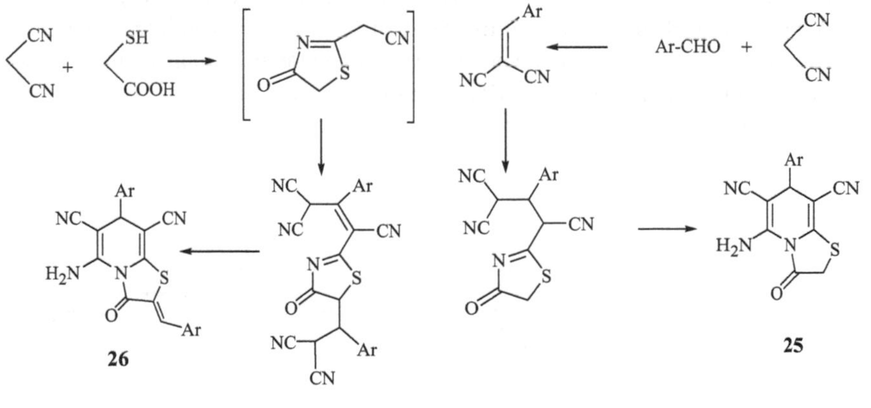

SCHEME 12.36 Proposed mechanism for synthesis of **25** and **26**.

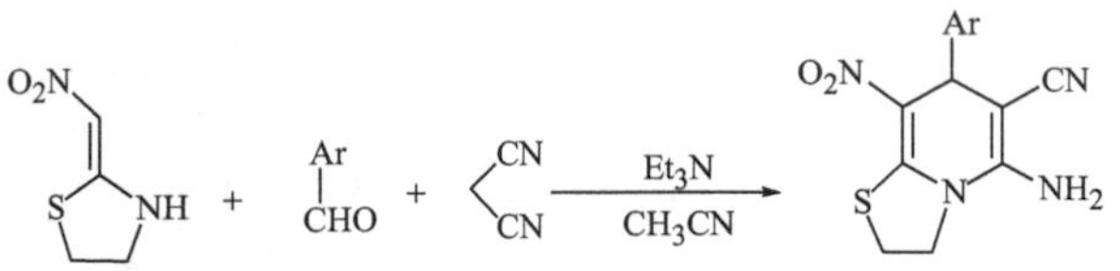

Ar = C_6H_5; 4-F.C_6H_4; 4-Cl.C_6H_4; 4-Br.C_6H_4; 2-F.C_6H_4; 2,4-Cl_2.C_6H_3; 2,6-$Cl_2C_6H_3$; 2-Br-C_6H_4; 2-HO,5-Br.C_6H_3; 4-NO_2.C_6H_4; 4 -$(CH_3)_2N$.C_6H_4

SCHEME 12.37 Synthesis of of thiazolo[3,2-*a*]pyridines **27**.

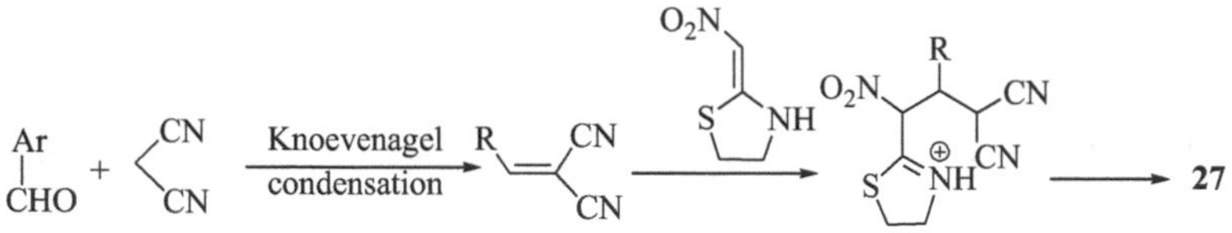

SCHEME 12.38 Proposed mechanism for synthesis of **27**.

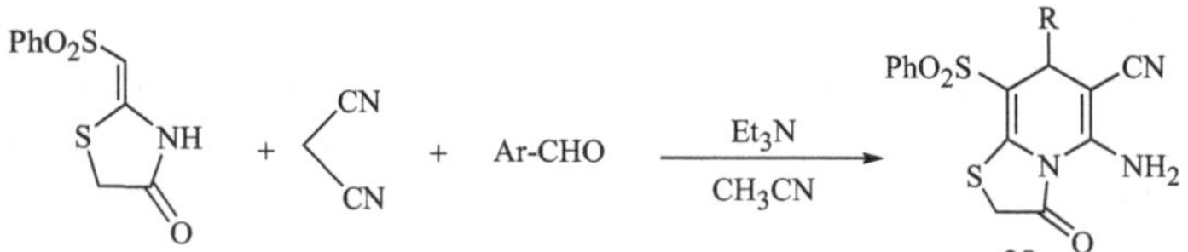

Ar = C_6H_5; 4-NC.C_6H_4; 4-Cl.C_6H_4; 2,6-diCl.C_6H_3; 2,4-diCl.C_6H_3; 4-O_2N.C_6H_4

SCHEME 12.39 Synthesis of thiazolo[3,2-*a*]pyridines **28**.

The mechanistic path of the present reaction proceeded through the Knoevenagel condensation of the aldehyde with malononitrile, and benzylidenemalononitrile **A** was formed, which, after addition with enamine, gave intermediate **B**. Thiazolo[3,2-*a*] pyridine-6-carbonitriles **28** was formed by attack of the secondary amine on one of nitrile groups in intermediate **B** with a ring cyclization (Scheme 12.40).

Jonnalagadda et al.[64] have synthesized thiazolo[4,5-*b*]pyridines **29** by four-component reaction of substituted benzaldehydes, malononitrile, thiazolidine-2,4-dione, and ammonium acetate using (AAPTMS/m-ZrO_2), an acidic/basic nature catalyst using an aqueous medium (Scheme 12.41).

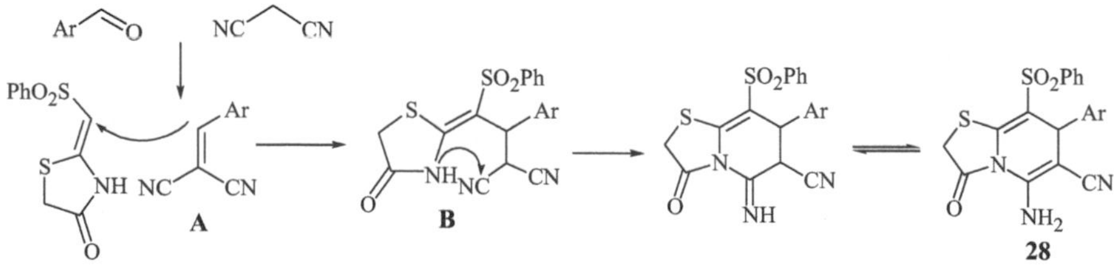

SCHEME 12.40 Proposed mechanism for synthesis of **28**.

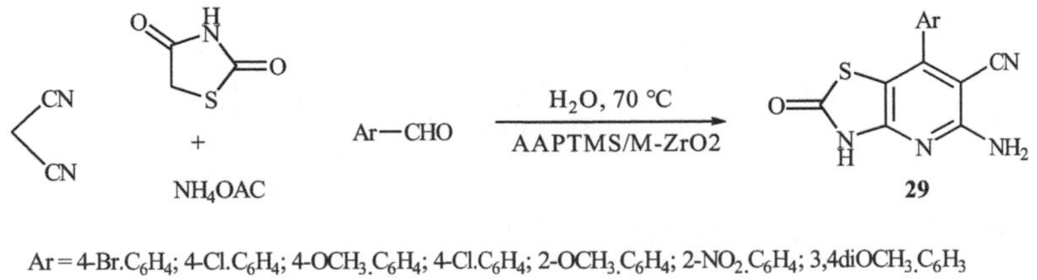

Ar = 4-Br.C_6H_4; 4-Cl.C_6H_4; 4-OCH_3.C_6H_4; 4-Cl.C_6H_4; 2-OCH_3.C_6H_4; 2-NO_2.C_6H_4; 3,4diOCH_3.C_6H_3

SCHEME 12.41 Synthesis of thiazolo[4,5-*b*]pyridines **29**.

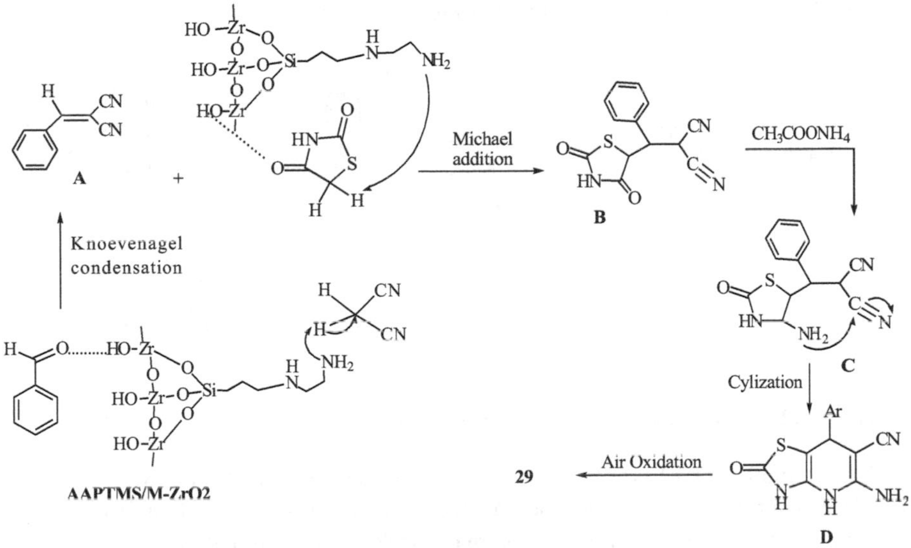

SCHEME 12.42 Proposed mechanism for synthesis of **29**.

The AAPTMS/m-ZrO_2 catalyzed present reaction most likely proceeds through a cyclic transition state with a catalyst. The reaction is possibly facilitated by the highly effective acid–base bifunctional surface character of a catalyst that is capable of mediating both the Knoevenagel condensation **A** and Michael addition **B** of carbonyl compounds (Scheme 12.42).

12.4.3 Pyrazolo[3,4-*b*]Pyridines

Pyrazolopyridines have been reported as a potent cyclin-dependent kinase (CDK1) inhibitor,[63] HIV reverse transcriptase inhibitors,[65] and protein kinase inhibitors.[66] Three-multicomponent synthesis of fused 6-amino-3-methyl-4-aryl-1*H*-pyrazolo[3,4-*b*]pyridine-5-carbonitrile **30** from 3-amino-5-methylpyrazole, malononitrile, and substituted aldehydes has been described by Zare et al.[67] in catalyst-free reaction conditions under ultrasound irradiation in short reaction times (8–10 min) with good yields (85%–98%) (Scheme 12.43).

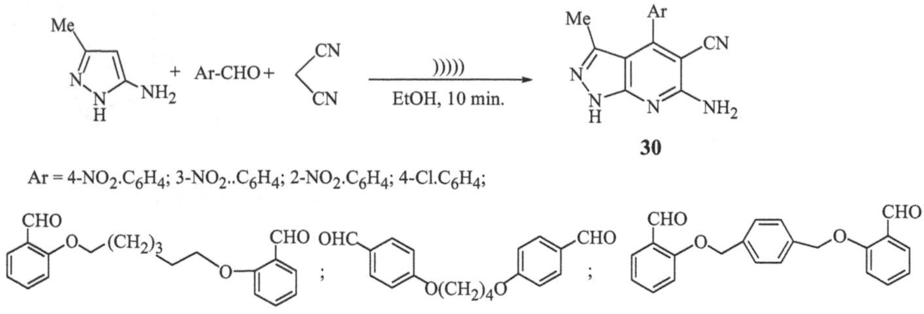

Ar = 4-NO2.C6H4; 3-NO2..C6H4; 2-NO2.C6H4; 4-Cl.C6H4;

SCHEME 12.43 Synthesis of pyrazolo[3,4-*b*]pyridines **30**.

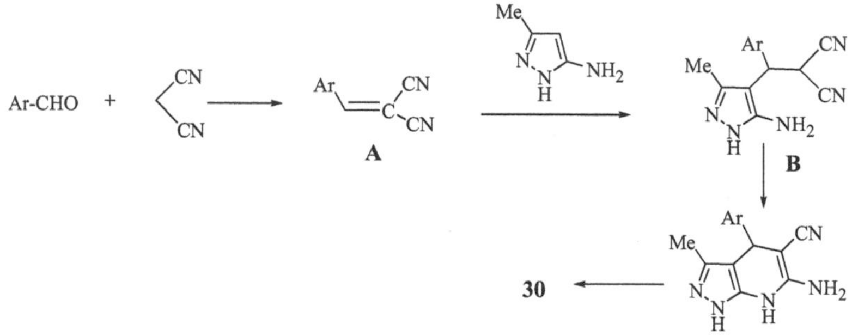

SCHEME 12.44 Proposed mechanism for synthesis of **30**.

The formation of **17** proceeds first by Knoevenagel condensation between malononitrile and substituted benzaldehyde (**A**) and then Michael-type addition (**B**) from the free ring carbon atom in 3-amino-5-methylpyrazole to the activated double bond of the Knoevenagel adduct, which subsequently cyclizes to give final product **30** (Scheme 12.44).

12.4.4 Polyhydroimidazo[1,2-a]Pyridines

Tetrahydroimidazo[1,2-*a*]pyridine motifs **31** are of general interest in medicinal chemistry with therapeutic properties, and a series of substituted variants of 2-(2,4-dichlorophenyl) methyleneimidazolidine have been reported as a basis for anticancer,[68] analgesics, and anti-inflammatory agents.[69] Li et al.[70] have reported the aza-ene type reaction by three-component reaction of the precursor 2-(2,4-dichlorophenyl)methyleneimidazolidine with substituted benzaldehyde and malononitrile leading to tetrahydroimidazo[1,2-*a*]pyridine **31** using Et_3N as base and CH_3CN as solvent (Scheme 12.45).

A possible mechanism for the formation of tetrahedron imidazo[1,2-*a*]pyridines **31** involves, first, Knoevenagel condensation to give intermediates **A**. Then, the heterocyclic ketene aminal, acting as hetero-ene components due to the strong nucleophilicity at the α-position of the ketene *N,N*-acetals, reacts with **A** to form the intermediates **B**, which undergo a rapid imine-enamine tautomerization to give intermediate **C**. Next, intramolecular cyclization of **C** leads to the formation of fused heterocyclic imidazo[1,2-*a*]pyridine **31** motifs (Scheme 12.46).

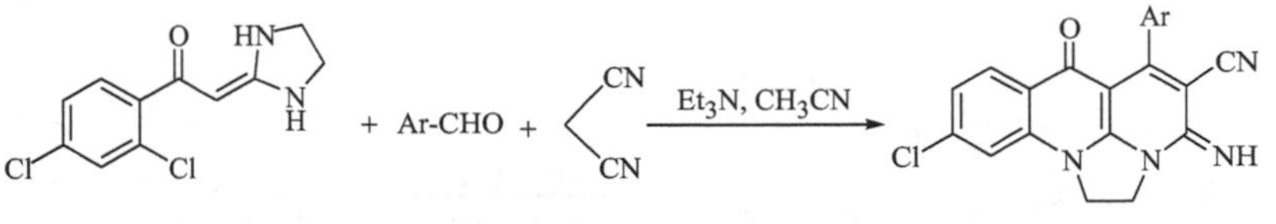

Ar =C_6H_5; 4-Cl.C_6H_4; 2,4-Cl_2.C_6H_3; 4-Br.C_6H_4; 2-Br.C_6H_4; 4-F.C_6H_4; 2-F.C_6H_4; 3-F.C_6H_4; 4-NO_2.C_6H_4; 3-NO_2.C_6H_4; 4-CH_3.C_6H_4; 4-OCH_3.C_6H_4; 3-CH_3.C_6H_4; 3-OCH_3.C_6H_4

SCHEME 12.45 Synthesis of tetrahydroimidazo[1,2-*a*]pyridines **31**.

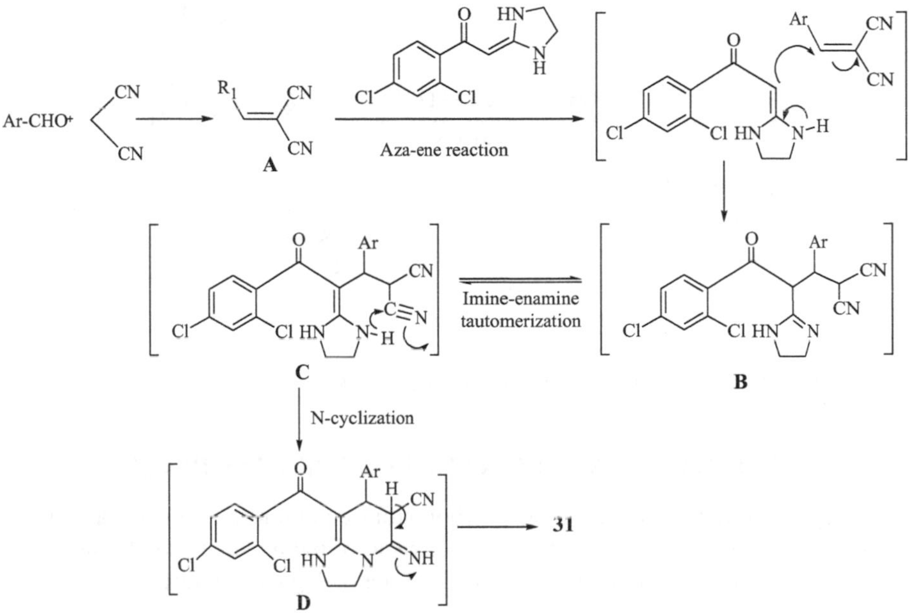

SCHEME 12.46 Proposed mechanism for synthesis of **31**.

Li et al.[71] reported the synthesis of hexahydroimidazo[1,2-*a*]pyridine **32** via convenient and practical three-component reaction of malononitrile, benzaldehyde, and 2-chloro-5-((2-(2-(furan-2-yl)-1-nitrovinyl)-4,5-dihydro-1*H*-imidazol-1-yl)methyl)-thiazole at room temperature (Scheme 12.47). The optimal catalyst and solvent reported for this reaction are piperidine and dichloromethane, respectively.

The present reaction was not suitable for benzyl, phenyl, or ethyl substituted starting material nitrovinyl-imidazolidines because they are unstable and difficult to isolate. Thus, an alternative approach was proposed to synthesized **32** based on an aza-hydro-allyl addition of compounds 1-benzyl/phenyl//ethyl-2-(nitromethylene) imidazolidine on the five-membered aromatic heterocyclic aldehydes, followed by aza-Diels–Alder cycloaddition with electron-deficient dienophiles synthesized *in situ* by the Knoevenagel condensation of malononitrile and benzaldehyde (Scheme 12.48).

Synthesized compounds were screened for their insecticidal activity against cowpea aphids. Most of the title compounds exhibited good activities.

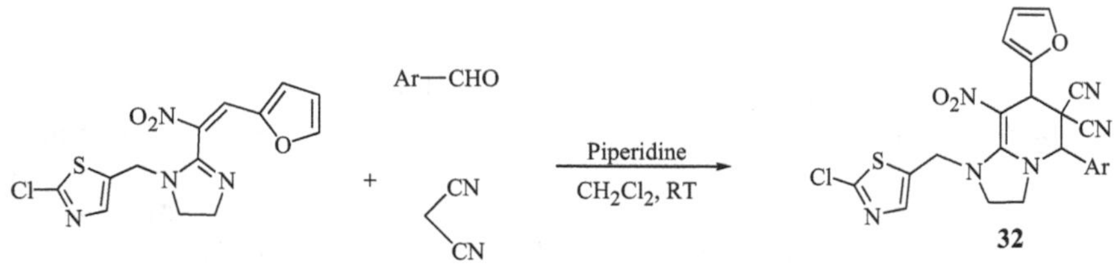

SCHEME 12.47 Synthesis of hexahydroimidazo[1,2-*a*]pyridines **32**.

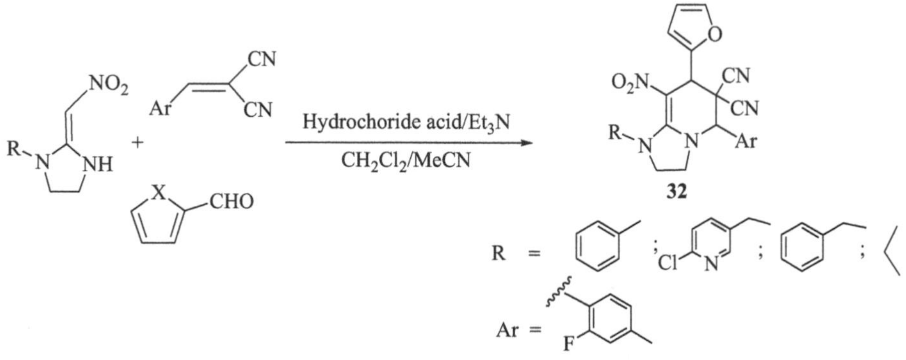

SCHEME 12.48 Alternative approach for synthesis of **32**.

12.5 SPIRO DERIVATIVES CONTAINING PYRIDINE RINGS

Spiro cyclic systems containing one carbon atom common to two rings are structurally interesting. The asymmetric characteristic of the molecule due to the chiral spiro carbon is one of the important criteria of the biological activities.[72]

12.5.1 Spiro[Indoline-3,4′-Quinoline] Derivatives

A simple and efficient synthesis of spiro[indoline-3,4′-quinoline] derivatives **33** was investigated by Ji et al.[73] from a one-pot reaction of isatin, malononitrile, and enaminone under microwave irradiations. The advantages of this protocol are short reaction time, high yields of products, broad substrate scope, and easy workup procedure. Among the different polar solvents screened, such as ethanol, acetic acid, glycol, dimethylformamide (DMF), and water, the best yield was found when ethanol was employed as solvent. The most suitable temperature was found to bc 80 °C (Scheme 12.49).

The process represents a typical cascade reaction in which the isatin first condenses with malononitrile to afford isatylidene malononitrile **A**. This step was regarded as a fast Knoevenagel condensation. Then, **A** is attacked via Michael addition with enaminone to give the intermediate **B**, followed by the intramolecular cyclization to form the desired product **33** (Scheme 12.50).

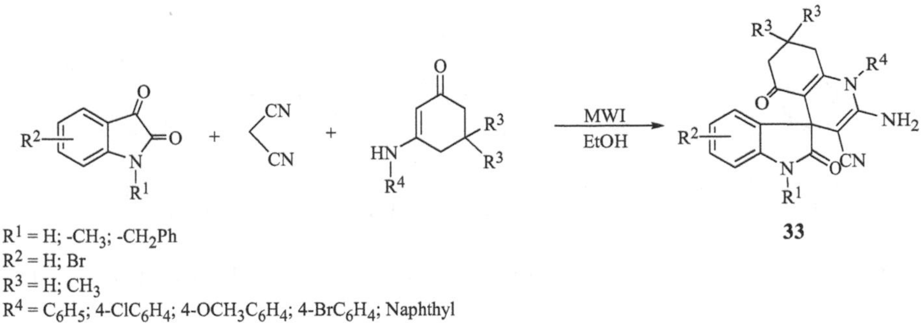

R^1 = H; -CH_3; -CH_2Ph
R^2 = H; Br
R^3 = H; CH_3
R^4 = C_6H_5; 4-ClC_6H_4; 4-O$CH_3C_6H_4$; 4-BrC_6H_4; Naphthyl

SCHEME 12.49 Synthesis of spiro[indoline-3,4′-quinoline] derivatives **33**.

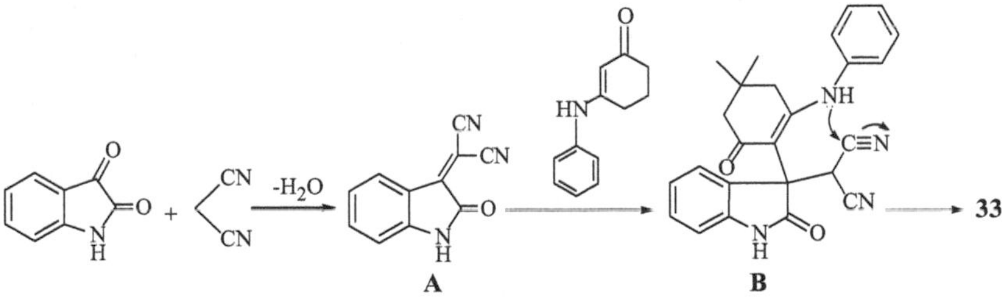

SCHEME 12.50 Proposed mechanism for synthesis of **33**.

12.5.2 Spiro[furo[3,4-*b*]pyridine-4,3′-indoline]-3-carbonitriles

Synthesis of 1*H*-spiro[furo[3,4-*b*]pyridine-4,3′-indoline]-3-carbonitriles **34** has been studied by Naeimi et al. via three component condensation reaction of isatins, malononitrile, and anilinolactones in the presence of a catalytic amount of Et_3N as basic catalyst in THF under ultrasound irradiations.[74]

To study the effect of catalyst amount, the reactions were carried out in the presence of different amounts of Et_3N ranging from 10 mol% to 20 mol%. It was found that 15 mol% Et_3N in THF is sufficient to drive this reaction forward. To optimize the reaction temperature, the reaction was studied at 25 °C, 40 °C, 45 °C, 50 °C, and 55°C under ultrasonic irradiation in THF; the most suitable reaction temperature was found to be 50°C for the synthesis of these spiro oxindole derivatives (Scheme 12.51).

According to the proposed mechanism a base catalyzed condensation of isatin with malononitrile produced the adduct **A**. Secondly, Michael addition of phenylaminofuran-2(3*H*)-one to adduct **A** afforded the intermediate **B**, which tautomerized to intermediate **C**, followed by nucleophilic attack of the amine tautomer to the cyanide group and cyclization to **D**, then to form **34** as target product (Scheme 12.52).

12.5.3 Spiro[cyclopenta[*b*]pyridine-4,2′-indenes]

Spiro[cyclopenta [*b*]pyridine-4,2′-indenes] **35** were successfully synthesized by Zhang et al. via one-pot domino reaction of ninhydrin, malononitrile with

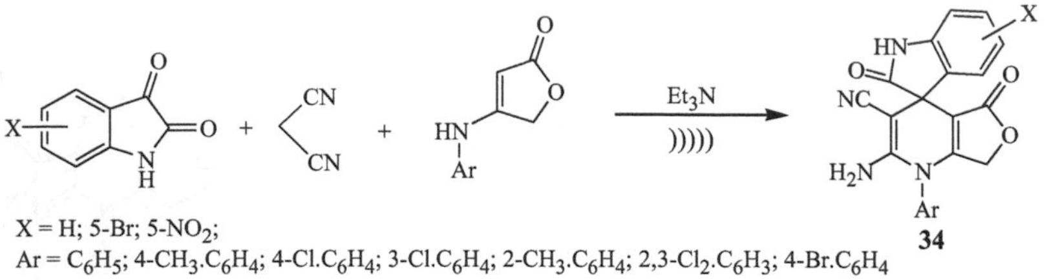

$X = H; 5\text{-}Br; 5\text{-}NO_2;$
$Ar = C_6H_5; 4\text{-}CH_3.C_6H_4; 4\text{-}Cl.C_6H_4; 3\text{-}Cl.C_6H_4; 2\text{-}CH_3.C_6H_4; 2,3\text{-}Cl_2.C_6H_3; 4\text{-}Br.C_6H_4$

SCHEME 12.51 Synthesis of 1*H*-spiro[furo[3,4-*b*]pyridine-4,3′-indoline] derivatives **34**.

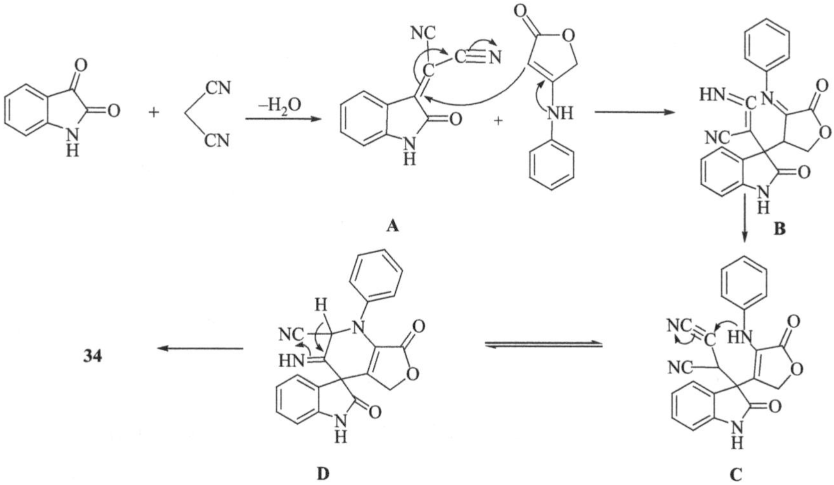

SCHEME 12.52 Proposed mechanism for synthesis of **34**.

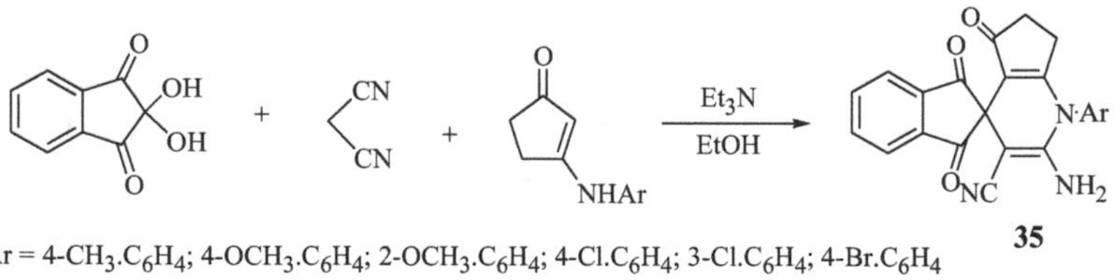

$Ar = 4\text{-}CH_3.C_6H_4; 4\text{-}OCH_3.C_6H_4; 2\text{-}OCH_3.C_6H_4; 4\text{-}Cl.C_6H_4; 3\text{-}Cl.C_6H_4; 4\text{-}Br.C_6H_4$

SCHEME 12.53 Synthesis of spiro[cyclopenta[*b*]pyridine-4,2'-indenes] **35**.

3-arylamino-2-cyclopentenones in the presence of triethylamine using ethanol at room temperature[75] (Scheme 12.53).

The proposed mechanism involves, first, Knoevenagel condensation; second, Michael addition; and then cyclization to afford the target compound (Scheme 12.54).

12.5.4 Spiro[Indoline-3,4′-Pyrazolo[3,4-*b*]Pyridine]-5′-Carbonitrile

Dandia et al.[76] described a direct and efficient approach for the preparation of medicinally promising pyrazolopyridinyl spiro oxindoles **36** through a one-pot, three-component

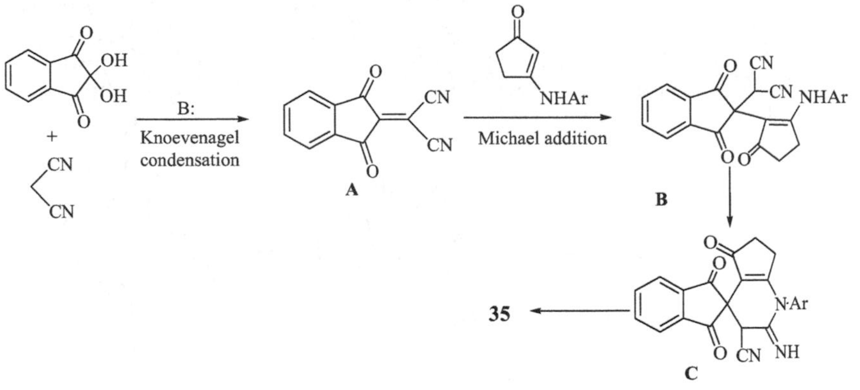

SCHEME 12.54 Proposed mechanism for synthesis of **35**.

reaction of isatin, malononitrile, and 5-amino-3-methylpyrazole catalyzed by sodium chloride in water. Desired products were obtained in high to excellent yields using a simple workup procedure. The present green synthesis shows attractive characteristics such as the use of water as the reaction medium, one-pot conditions, a short reaction period, easy workup/purification, and reduced waste production, without use of any acid or metal promoters (Scheme 12.55).

Due to their dislike for water, all reactants come close enough to react due to the hydrophobic effect. This effect is further increased by the presence of a significant concentration of NaCl due to the salting-out effect. It was found that when the amount of NaCl increased from 2.5 mol%, 5 mol%, and 10 mol%, the yields increased from 85% to 89% and 93%, respectively.

Knoevenagel condensation reaction of isatin and malononitrile is proposed to give the intermediate **A**. Michael addition of an electron-rich amino heterocycle to **A** should then occur to provide intermediates **B**, which undergo intramolecular cyclization resulting in spiro ring systems **C** and then **D**. Finally, compounds **D** undergo tautomerization to produce 3′methyl-2,6′-dioxo-2′,5′,6′,7′-tetrahydrospiro[indoline-3,4′-pyrazolo[3,4-*b*]pyridine]-5′-carbonitrile **36** (Scheme 12.56).

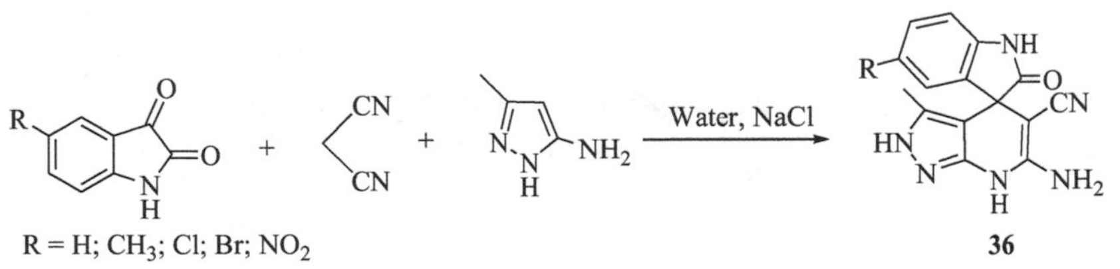

SCHEME 12.55 Synthesis of spiro[indoline-3,4′-pyrazolo[3,4-*b*]pyridine] derivatives **36**.

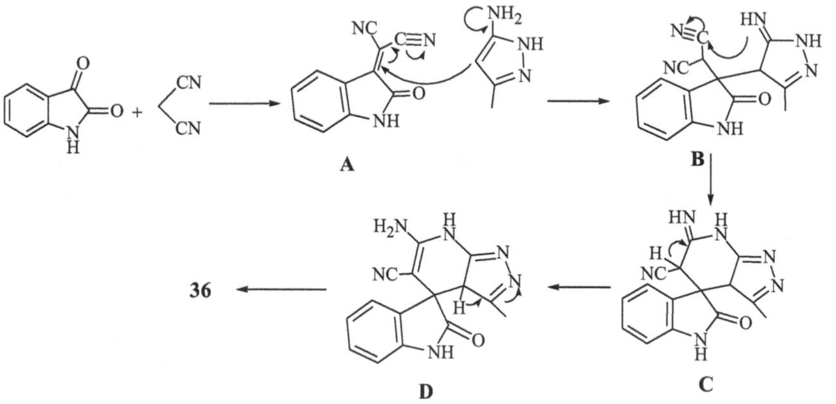

SCHEME 12.56 Proposed mechanism for synthesis of **36**.

12.5.5 Spiro[Oxindole-3,4′-Pyrano[3,2-c]Pyridine]Derivatives

Wang et al.[77] reported an efficient cyclization procedure for the synthesis of tetrahydrospiro[oxindole-3,4′-pyrano[3,2-*c*]pyridin] derivatives **37** via one-pot, three-component condensation of isatins, malononitrile, and (*E*)-3-arylidene-1-methylpiperidin-4-ones using piperidine as a base and ethanol as an environmentally benign solvent (Scheme 12.57).

The reaction proceeded through the activation of malononitrile by piperidine to produce a nucleophile, followed by a nucleophilic addition on the C-3 carbonyl group in isatin with the generation of α, β-unsaturated nitrile **A**. Michael adduct **B** is formed by the nucleophilic addtion of (*E*)-3-arylidene-1-methylpiperidin-4-one on intermediate **A**. The resulting intermediate undergoes consequent intramolecular cyclization with simultaneous [1,3]-sigmatropic proton shift lead to the formation of the final spiro compound **37** (Scheme 12.58).

The *in vitro* antitumor activity of these compounds was evaluated in human cervical carcinoma cell line (Hela), human liver hepatocellular carcinoma cell line (HepG2), and human breast carcinoma cell line (MDA-MB-231). The data revealed that some compounds exhibited outstanding growth inhibitory activity against the tested subpanel tumor cell lines.

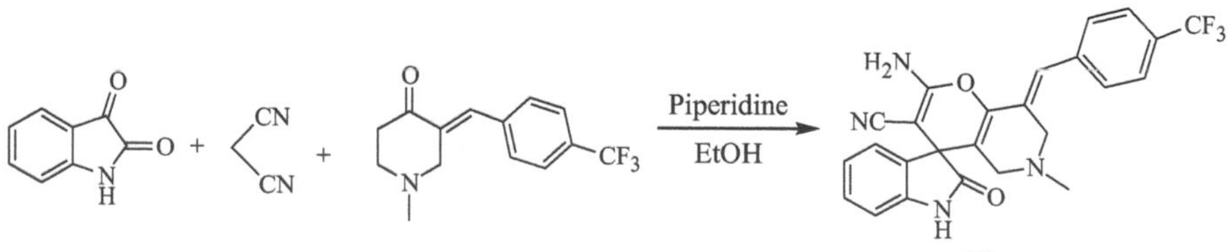

SCHEME 12.57 Synthesis of tetrahydrospiro[oxindole-3,4′-pyrano[3,2-*c*]pyridin] derivatives **37**.

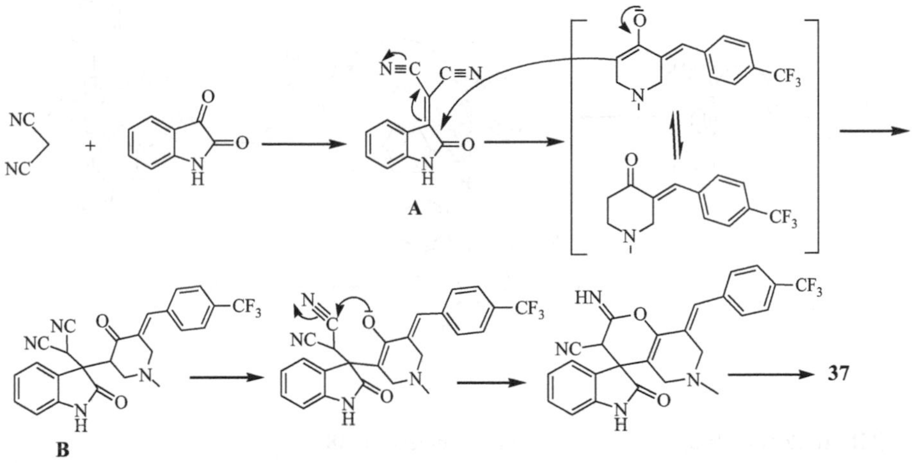

SCHEME 12.58 Proposed mechanism for synthesis of **37**.

12.5.6 Spiro[Pyridines/Pyridinones]

A proficient synthesis of diverse spiro dihydropyridines **38** via one-pot, four-component reactions of various cyclic ketones (isatin, acenaphthenequinone, ninhydrine, and indeno-[1,2-*b*]quinoxalinone), malononitrile, primary amines, and DMAD, using triethylamine as base catalyst, has been reported by Perumal et al.[78] (Scheme 12.59).

The mechanism for the synthesis of spiro dihydropyridines **38** involves, first, Knoevenagel condensation of isatin with malononitrile in the presence of Et_3N to give isatylidene malononitrile **A**. Primary amines add on to dimethyl acetylenedicarboxylate DMAD to give the zwitterionic intermediate **B**, which undergoes Michael addition with **A** to form **C**; the latter, through the migration of a hydrogen atom, furnishes **D**. The intramolecular addition of the amino group to the cyano triple bond provides **E**, which tautomerizes to give spiro dihydropyridines (Scheme 12.60).

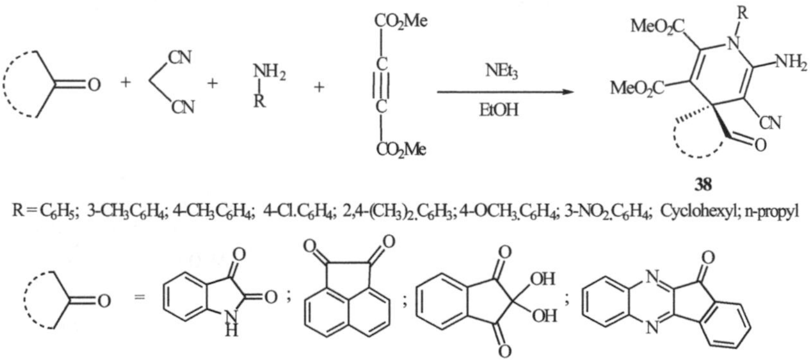

R = C_6H_5; 3-$CH_3C_6H_4$; 4-$CH_3C_6H_4$; 4-Cl.C_6H_4; 2,4-$(CH_3)_2.C_6H_3$; 4-$OCH_3.C_6H_4$; 3-$NO_2.C_6H_4$; Cyclohexyl; n-propyl

SCHEME 12.59 Synthesis of spiro[pyridines/pyridinones] **38**.

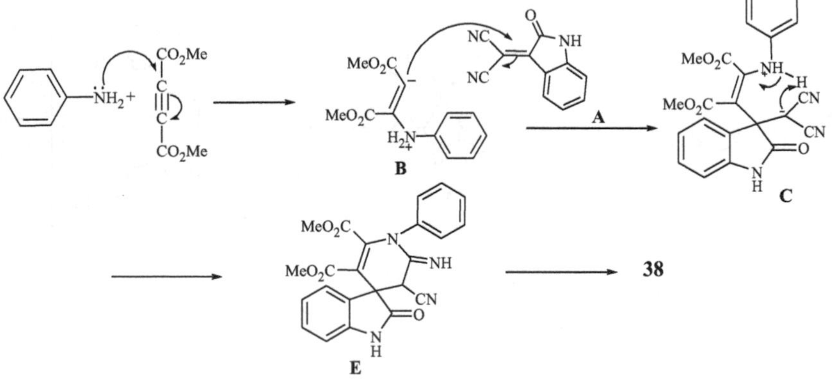

SCHEME 12.60 Proposed mechanism for synthesis of **38**.

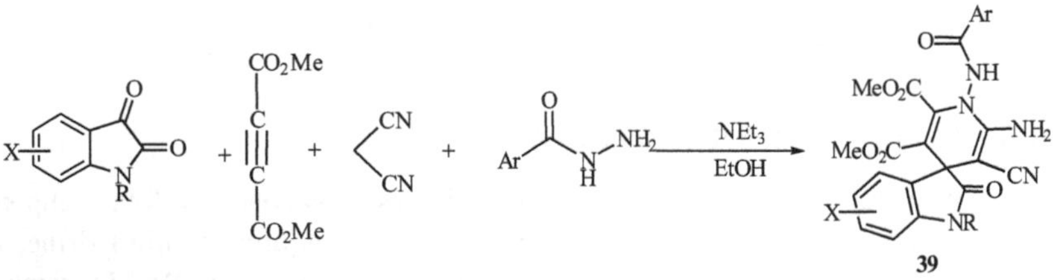

$Ar = C_6H_5$; $4\text{-}CH_3.C_6H_4$; C_5H_4N
$X = H$; Cl; CH_3
$R = CH_2.C_6H_4$; H

SCHEME 12.61 Synthesis of 1′-picolinamidospiro[indoline-3,4′-pyridines] **39**.

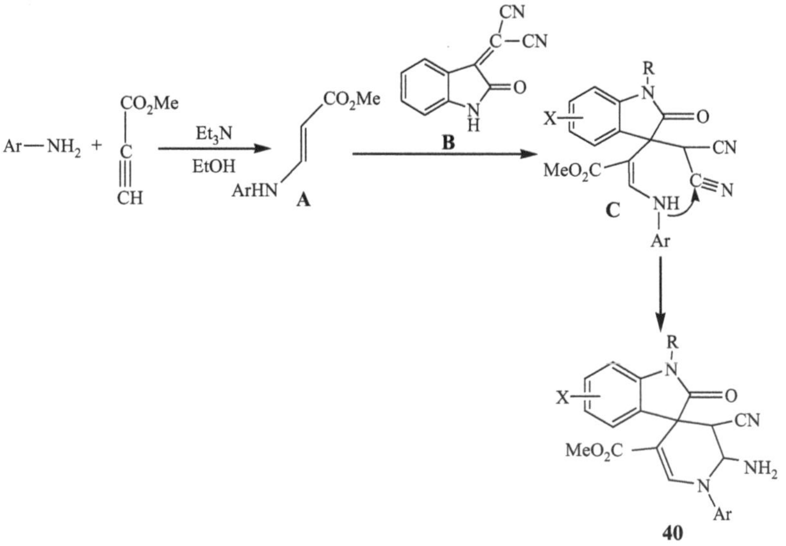

SCHEME 12.62 Synthesis of spiro dihydropyridines **40**.

Later, Yan et al.[79] studied this reaction using 2-picolinohydrazide instead of aniline to give the corresponding 1′-picolinamidospiro[indoline-3,4′-pyridines] **39** (Scheme 12.61).

Yan et al.[80] have also synthesized spiro dihydropyridines **40** by replacing dimethyl acetylenedicarboxylate by methyl propiolate. The first step is formation of a key intermediate β-enamino ester **A** from the addition of arylamine to methyl propiolate. The second step is a Michael addition of β-enamino ester intermediate **A** with isatinylidene derivative **B** synthesized by isatin and malononitrile to yield intermediate **C**. Further, the nucleophilic addition of the amino group to the C–N triple bond in intermediate **C** resulted in spiro compound **40** with the tautomerization of the imino group to an amino group (Scheme 12.62).

12.5.7 Spiro[Acenaphthylene-Isoquinoline]-Tricarbonitrile

A new type of catalytic cascade reaction for direct multicomponent transformation of acenaphthenequinone, cyclic ketones, and two molecules of malononitrile into

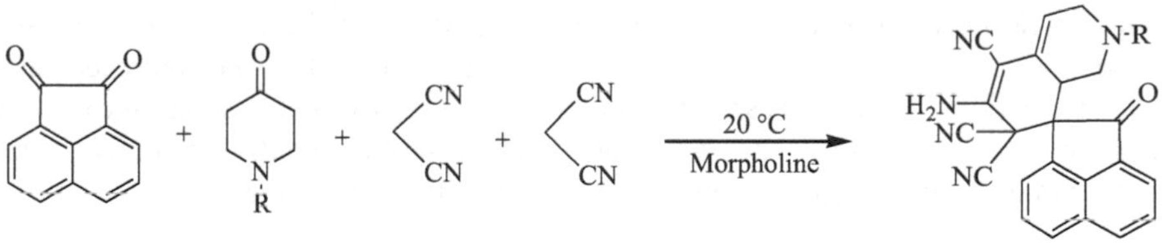

R = CH_3; C_2H_5; $CH_2C_6H_5$; $COCH_3$; $CO_2C_2H_5$

SCHEME 12.63 Synthesis of spiro[acenaphthylene-isoquinoline]-tricarbonitrile **41**.

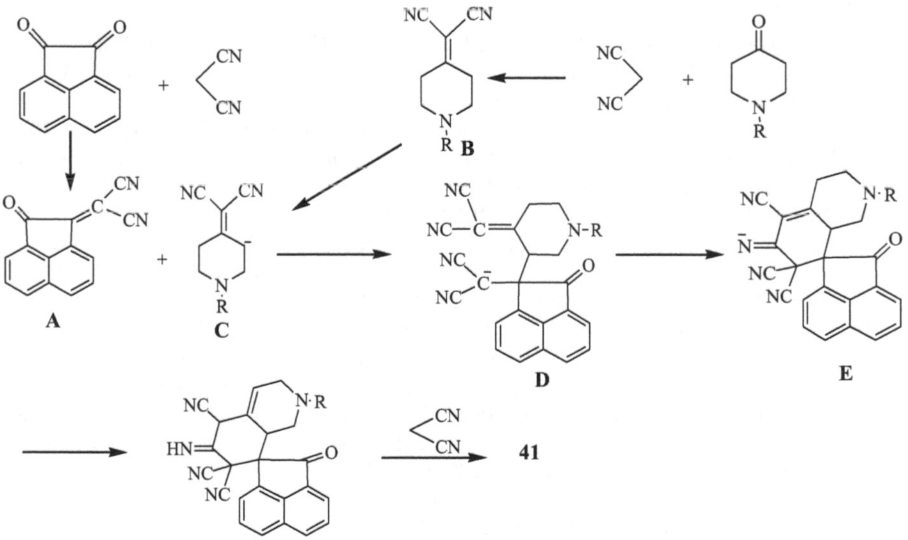

SCHEME 12.64 Proposed mechanism for synthesis of **41**.

spiro acenaphthylene pentacyclic systems **41** was studied by Elinson et al.[81] at ambient temperature in 5 min with 70%–95% yields. The application of this convenient multicomponent method is also beneficial from the viewpoint of diversity-oriented large-scale processes (Scheme 12.63).

The proposed mechanism involves reaction between the malononitrile and acenaphthenequinone to form a Knoevenagel adduct **A**. Simultaneous Knoevenagel condensation of cyclohexanone and malononitrile anion leads to Knoevenagel adduct **B**. Under basic conditions Knoevenagel adduct **B** forms anion **C**, which adds to the double bond of Knoevenagel adduct **A** with further cyclization into anion **C**. The latter reacts with malononitrile, forming spiro acenaphthylene with the regeneration of the malononitrile anion to continue the catalytic cycle (Scheme 12.64).

12.6 CONCLUSION

The present chapter has outlined reactions for the synthesis of fused and spiro pyridine derivatives. Several recent and convergent synthetic approaches to fused and spiro pyridine derivatives have been published. Many of these reports offer new modifications to existing methodologies, whereas others describe unprecedented transformations. Each of these new methods serves as a valuable addition to a field rich in chemical history. The vast majority of these medicinally persuasive compounds still require further exploration and applications in various fields.

ACKNOWLEDGMENT

R. S. thanks DST, New Delhi, for start-up grant YSS/2015/000972.

REFERENCES

1. Wender, P. A.; Verma, V. A.; Paxton, T. J.; Pillow T. H. *Acc. Chem. Res.* **2008**, *41*, 40.
2. Ahluwalia, V. K.; Varma, R. S. *Green solvents for organic synthesis*, Alpha Science International, Abingdon, UK, **2009**.
3. Zhu, J.; Bienaime, H. *Multicomponent reactions*, Wiley-VCH, Weinheim, Germany, **2005**.
4. Tietze, L. F.; Brasche, G.; Gericke, K. M. *Domino reactions in organic synthesis*, Wiley-VCH, Weinheim, Germany, **2006**.
5. Orru, R. V. A.; Greef, M. *Synthesis* **2003**, *10*, 1471–1499.
6. Ramon, D. J.; Yus, M. *Angew. Chem. Int. Ed.* **2005**, *44*, 1602.
7. Guillena, G.; Ramon, D. J.; Yus, M. *Tetrahedron—Asymmetry* **2007**, *18*, 693–700.
8. Padwa, A.; Bur, S. K. *Tetrahedron* **2007**, *63*, 5341–5378.
9. Voskressensky, L. G.; Festa, A. A.; Varlamov A. V. *Tetrahedron* **2014**, *70*, 551–572.
10. (a) Wasilke, J.-C.; Obrey, S. J.; Baker, R. T.; Bazan, G. C. *Chem. Rev.* **2005**, *105*, 1001–1020. (b) Nicolaou, K. C.; Montagnon, T.; Snyder, S. A. *Chem. Commun.* **2003**, *5*, 551–564. (c) Pellissier, H. *Tetrahedron* **2006**, *62*, 1619–1665. (d) Ruijter, E.; Scheffelaar, R.; Orru R. V. A. *Angew. Chem. Int. Ed.* **2011**, *50*, 6234–6246. (e) Cioc, R. C.; Ruijter, E.; Orru R. V. A. *Green Chem.* **2014**, *16*, 2958–2975.
11. Singh, M. S.; Chowdhury, S. *RSC Adv.* **2012**, *2*, 4547–4592.
12. Yanlong G. *Green Chem.* **2012**, *14*, 2091–2128.

13. (a) Jones, G. Comprehensive heterocyclic chemistry II, vol. 5 (eds.: Katritzky, A. R.; Rees, C. W.; Scriven, E. F. V.; McKillop, A.), Pergamon, Oxford, **1996**, 167–243. (b) Joule, J. A.; Mills, K. Heterocyclic chemistry, 4th ed., Blackwell Science, Cambridge. **2000**, 63–120.
14. Harrison, T. S.; Scott, L. J. *Drugs* **2005**, *65*, 2309–2336.
15. Deininger, M. W. N.; Druker, B. J. *Pharmacol. Rev.* **2003**, *55*, 401–423.
16. Kodimuthali, A.; Jabaris, S. S. L.; Pal, M. *J. Med. Chem.* **2008**, *51*, 5471–5489. (b) Price, D.; Chisholm, A.; Ryan, D.; Crockett, A.; Jones, R. *Prim. Care Resp. J.* **2010**, *19*, 342–351.
17. (a) Tu, S.; Jia, R.; Jiang, B.; Zhang, J.; Zhang, Y.; Yao C.; Ji, S. *Tetrahedron* **2007**, *63*, 381–388. (b) Krohnke, F. *Synthesis* **1976**, *1976(1)*, 1–24. (c) MacGillivray, L. R.; Diamente, P. R.; Reid, J. L.; Ripmeester, J. A. *Chem. Commun.* **2000**, 359–360.
18. Jayasinghe, L.; Jayasooriya, C. P.; Hara; Fujimoto, N. *Tetrahedron Lett.* **2003**, *44*, 8769–8771.
19. Boger, D. L.; Yasuda, M.; Mitscher, L. A.; Drake, S. D.; Kitos, P. A. Thompson S. C. *J. Med. Chem.* **1987**, *30(10)*, 1918–1928.
20. Egmond, S. V.; Hoedemaker ,C.; Sinclair, R. *Int. J. Dermatol.* **2006**, *46(3)*, 8318–8319.
21. Holdcroft, C. *Nurse practitioner* **1993**, *18(11)*, 13–14.
22. (a) Freeman, F. *Chem. Rev.* **1969**, *69*, 591–624. (b) Fatiadi, A. J. *Synthesis* **1978**, 165. (c) Freeman, F. *Chem. Rev.* **1980**, *80*, 329–350.
23. Saikia, A. *Synlett* **2004**, 2247–2248.
24. Girgis, A. S.; Mishriky, N.; Ellithey, M.; Hosnia, H. M.; Farag, H. *Bioorg. Med. Chem.* **2007**, *15*, 2403–2413.
25. (a) Wen, Y.-G.; Ghang, D.-W.; Xie, H.-Z.; Wang, X.-P.; Ni, X.-J.; Zhang, M.; Qui, C.; Liu, X.; Li, F.-F.; Li, X.; Luo, F.-T. *Psychopharmcol. Clin.* **2013**, *28*, 134–141. (b) Deeks, E. D.; Keating, G. M. *CNS Drugs* **2010**, *24*, 65–84.
26. Maharani, S.; Kumar, R. R. *Tetrahedron Lett.* **2015**, *56*, 179–181.
27. Pagadala, R.; Maddila, S.; Moodley, V.; van Zyl, W. E.; Jonnalagadda S. B. *Tetrahedron Lett.* **2014**, *55*, 4006–4010.
28. (a) Kolokythas, G.; Pouli, N.; Marakos, P.; Pratsinis, H.; Kletsas, D. *Eur. J. Med. Chem.* **2006**, *41*, 71–79. (b) Azuine, M. A.; Tokuda, H.; Takayasu, J., Enjyo, F., Mukainaka, T.; Konoshima, T.; Nishino, H.; Kapadia, G. J. *Pharmacol. Res.* **2004**, *49*, 161–169. (c) Srivastava, S. K.; Tripathi, R. P.; Ramachandran, R. *J. Biol. Chem.* **2005**, *280*, 30273–30281.
29. Olyaei, A.; Vaziri, M.; Razeghi, R. *Tetrahedron Lett.* **2013**, *54*, 1963–1966.
30. Mishra, S.; Ghosh, R. *Synth. Commun.* **2012**, *42*, 2229–2244.
31. Gan, H.-F.; Cao, W.; Fang, Z.; Li, X.; Tang, S.-G.; Guo K. *Chinese Chem. Lett.* **2014**, *25*, 1357–1362.
32. Damavandi, S. *Synth. React. Inorg. Met. Org. Chem.* **2011**, *41*, 1274–1277.
33. Mohammadzadeh, I.; Sheibani, H. *Chinese Chem. Lett.* **2012**, *23*, 1327–1330.
34. Elinson, M. N.; Gorbunov, S. V.; Vereshchagin, A. N.; Nasybullin, R. F.; Goloveshkin, A. S., Bushmarinov, I. S.; Egorov M. P. *Tetrahedron* **2014**, *70*, 8559–8563
35. Borah, P.; Naidu, P. S.; Bhuyan, P. J. *Tetrahedron Lett.* **2012**, *53*, 5034–5037.
36. (a) Bi, X. H.; Dong, D. W.; Li, Y.; Liu, Q.; Zhang, Q. *J. Org. Chem.* **2005**, *70*, 10886–10889. (b) Brown, M. J.; Carter, P. S., Fenwick, A. E.; Fosberry, A. P.; West, A.; Witty, D. R. *Bioorg. Med. Chem. Lett.* **2002**, *12*, 3171–3174.
37. Wen, L.; Ji, C.; Li, Y.; Li, M. *J. Comb. Chem.* **2009**, *11*, 799–805.
38. (a) Jardosh, H. H.; Patel, M. P. *Eur. J. Med. Chem.* **2013**, *65*, 348–359. (b) Starcevic, K.; Pesic, D.; Toplak, A.; Landek, G.; Alihodzic, S.; Herreros, E.; Ferrer, S.; Spaventi, R.; Peric, M.. *Eur. J. Med. Chem. Mar.* **2012**, *49*, 365–378. (c) Kathrotiya, H. G.; Patel, M. P. *Eur. J. Med. Chem.* **2013**, *63*, 675–684. (d) Muruganantham, N.; Sivakumar, R.;

Anbalagan, N.; Gunasekaran, V.; Leonard, J. T. *Biol. Pharm. Bull.* **2004**, *27*, 1683–1687. (e) Zieba, A.; Sochanik, A.; Szurko, A.; Rams, M.; Mrozek, A.; Cmoch, P. *Eur. J. Med.Chem.* **2010**, *45*, 4733–4739.
39. Pannala, M.; Kher, S.; Wilson, N.; Gaudette, J.; Sircar, L.; Zhang, S.; Bakhirev, A.; Yang, G.; Yuen, P.; Gorcsan, F.; Sakurai, N.; Barbosa, M.; Cheng J. F. *Bioorg. Med. Chem. Lett.* **2007**, *17*, 5978–5982.
40. Wan, Y.; Yuan, R.; Zhang, F.-R.; Pang, L.-L.; Ma, R.; Yue, C.-H.; Lin, W.; Yin, W.; Bo, R.-C.; Wu, H. *Synth. Commun.* **2011**, *41*, 2997–3015.
41. Sangani, C. B.; Makawana, J. A.; Duan, Y.-T.; Yin, Y.; Teraiya, S. B.; Thumar, N. J.; Zhu, H.-L. *Bioorg. Med. Chem. Lett.* **2014**, *24*, 4472–4476.
42. (a) Wenkert, E.; Liu L. H.; Johnston, D. B. R. *J. Org. Chem.* **1965**, *30*, 722–728. (b) Polniaszek, R. P.; Stevens, R. V. *J. Org. Chem.* **1986**, *51*, 3023.
43. (a) Herle′, B.; Wanner, M. J.; van Maarseveen J. H.; Hiemstra, H. *J. Org. Chem.* **2011**, *76*, 8907–8912. (b) Mergott, D. J.; Zuend, S. J.; Jacobsen, E. N. *Org. Lett.* **2008**, *10*, 745–748.
44. (a) Lim, C. W.; Tissot, O.; Mattison, A.; Hooper, M. W.; Brown, J. M.; Cowley, A. R.; Hulmes, D. I.; Blacker, A. *Org. Process Res. Dev.* **2003**, *7*, 379–384. (b) Durola, F., Sauvage, J. P.; Wenger, O. S. *Chem. Commun.* **2006**, *2006*, 171–173. (c) Sweetman, B. A.; Muller, B. H.; Guiry, P. J. *Tetrahedron Lett.* **2005**, *46*, 4643–4646.
45. Feng, X.; Wang, J.-J.; Xun, Z.; Zhang, J.-J.; Huang Z.-B.; Shi. D.-Q. *Chem. Commun.* **2015**, *51*, 1528–1531.
46. Feng, X.; Wang, J.-J.; Xun, Z.; Huang, Z.-B.; Shi. D.-Q. *J. Org. Chem.* **2015**, *80*, 1025–1033.
47. Schmidt A. W.; Knölker, H. J.; Reddy, K. R. *Chem. Rev.* **2002**, *102*, 3193–3328.
48. Indumathi, T.; Ahamed, V. S. J.; Moon, S.-S.; Fronczek, F. R.; Prasad, K. J. R. *Eur. J. Med. Chem.* **2011**, *46*, 5580–5590.
49. Siddiqui, I. R.; Rai, P.; Rahila; Srivastava A.; Srivastava, Anushree; Srivastava, Arjita; Srivastava, Anjali. *New J. Chem.* **2013**, *37*, 3798–3804.
50. Ziarani, G. M.; Nasab, N. H.; Rahimifard, M.; Soorki, A. A. *J. Saudi Chem. Soc.* **2015**, *19(6)*, 676–681.
51. Bhattacharyya, P.; Paul S.; Das A. R. *RSC Adv.* **2013**, *3*, 3203–3208.
52. (a) Aoki, S.; Wei, H.; Matsui, K.; Rachmat, R.; Kobayashi, M. *Bioorg. Med. Chem.* **2003**, *11*, 1969–1973. (b) Larghi, E. L.; Obrist, B. V.; Kaufman, T. S. *Tetrahedron* **2008**, *64*, 5236–5245.
53. (a) Rudys, S.; Ríos-Luci, C.; Pérez-Roth, E.; Cikotiene, I.; Padrón. J. M. *Bioorg. Med. Chem. Lett.* **2010**, *20*, 1504–1506. (b) Johns, B. A.; Weatherhead, J. G.; Allen, S. H.; Thompson, J. B.; Garvey, E. P.; Foster, S. A.; Jeffrey, J. L.; Miller, W. H. *Bioorg. Med. Chem. Lett.* **2009**, *19*, 1802–1806. (c) Li, Y.; Liang, J.; Siu, T.; Hu, E.; Rossi, M. A.; Barnett, S. F.; Jones, D. D.-; Jones, R. E.; Robinson, R. G.; Leander, K.; Huber, H. E.; Mittal, S.; Cosford, N.; Prasit, P. *Bioorg. Med. Chem. Lett.* **2009**, *19*, 834–836. (d) Ghotekar, B. K.; Ghagare, M. G.; Toche, R. B.; Jachak, M. N. *Monatsh. Chem.* **2010**, *141*, 169–175.
54. Mukhopadhyay, C.; Das, P.; Butcher, R. J. *Organic Lett.* **2011**, *13*, 4664–4667.
55. Cao, C.-P.; Lin, W.; Hu, M.-H.; Huang Z.-B.; Shi, D.-Q. *Chem. Commun.* **2013**, *49*, 6983–6985.
56. (a) Jordan, A. D.; Vaidya, A. H.; Rosenthal, D. I.; Dubinsky, B.; Kordik, C. P.; Sanfilippo, P. J.; Wu, W.-N.; Reitz, A. B. *Bioorg. Med. Chem. Lett.* **2002**, *12*, 2381–2386. (b) Reitz, A. B.; Gauthier, D. A.; Ho, W.; Maryanoff, B. E. *Tetrahedron* **2000**, *56*, 8809–8812. (c) Alajaro, M.; Vidal, A.; Tovar, F. *Tetrahedron Lett.* **2000**, *41*, 7029–7032.
57. Yan, C. G.; Wang, Q. F.; Song, X. K.; Sun, J. *J. Org. Chem.* **2009**, *74*, 710–718.
58. Mullan, M. J.; Paris, D.; Bakshi, P. WO 2008070875, **2008**; Chem. Abstr. **2008**, *149*, 45259.

59. Vadivelan, S.; Sinha, B. N.; Irudayam, S. J.; Jagarlapudi, S. A. *J. Chem. Inf. Model.* **2007**, *47*, 1526–1535.
60. Park, H.; Hwang, K. Y.; Oh, K. H.; Kim, Y. H.; Lee, J. Y.; Kim, K. *Bioorg. Med. Chem.* **2008**, *16*, 284–292.
61. Shi, F.; Li, C.; Xia, M.; Miao, K.; Zhao, Y.; Tu, S.; Zheng, W.; Zhang, G.; Ma, N. *Bioorg. Med. Chem. Lett.* **2009**, *19*, 5565–5568.
62. Altug, C.; Burnett, A. K.; Caner, E.; Durust, Y.; Elliott, M. C.; Roger, P. J.; Glanville; G. C.; Westwell, A. D. *Tetrahedron* **2011**, *67*, 9522–9528.
63. Altug, C.; Caner. E. *C. R. Chimie* **2013**, *16*, 217–221.
64. Pagadala, R.; Kommidi, D. R.; Rana, S.; Maddila, S.; Moodley, B.; Koorbanally, N. A.; Jonnalagadda S. B. *RSC Advances* **2015**, *5*, 5627–5632.
65. Huang, S.; Lin, R.; Yu, Y.; Lu, Y.; Connolly, P. J.; Chiu, G.; Li, S.; Emanuel, S. L.; Middleton, S. A. *Bioorg. Med. Chem. Lett.* **2007**, *17*, 1243–1245.
66. Saggar, S. A.; Sisko, J. T.; Tucker, T. J.; Tynebor, R. M.; Su, D. S.; Anthony, N. J. U. S. Patent Appl. **2007**, *021*, 442, 2007.
67. Zare, L.; Mahmoodi, N. O.; Yahyazadeh, A.; Mamaghani, M. *Synth. Commun.* **2011**, *41*, 2323–2330.
68. Yan, S.-J.; Niu, Y.-F.; Huang, R.; Lin, J. *Synlett* **2009**, 2821–2824.
69. Cheng, D.-C.; Croft, L.; Abdi, M.; Lightfoot, A.; Gallagher, T. *Org. Lett.* **2007**, *9*, 5175–5178.
70. Wen, L.-R.; Jiang, C.-Y.; Li, M.; Wang, L.-J. *Tetrahedron* **2011**, *67*, 293–302.
71. Fan, Y.-F.; Zhang, W.-W.; Shao, X.-S.; Xu, Z.-P.; Xu, X.-Y.; Li Z. *Chinese Chem. Lett.* **2015**, *26*, 1–5.
72. (a) Ali, M. A.; Ismail, R.; Choon, T. S.; Yoon, Y. K.; Wei, A. C.; Pandian, S.; Kumar, R. S.; Osman, H.; Manogaran, E. *Bioorg. Med. Chem. Lett.* **2010**, *20*, 7064–7066. (b) Ito, Y.; Takuma, K.; Mizoguchi, H.; Nagai, T.; Yamada, K. *J. Pharmacol. Exp. Ther.* **2007**, *320,* 819–827. (c) Pradhan, R.; Patra, M.; Behera, A. K.; Mishra, B. K.; Behera, R. K. *Tetrahedron* **2006**, *62*, 779–828.
73. Zhu, S.-L.; Zhao, K.; Su, X.-M.; Ji S.-J. *Synth. Commun.* **2009**, *39*, 1355–1366.
74. Ghahremanzadeh, R.; Rashid, Z.; Zarnani, A.-H.; Naeimi, H. *Ultrasonics Sonochem.* **2014**, *21*, 1451–1460.
75. Zhang, L.-J.; Yan, C.-G. *Tetrahedron* **2013**, *69*, 4915–4921.
76. Dandia, A.; Laxkar, A. K.; Singh, R. *Tetrahedron Lett.* **2012**, *53*, 3012–3017.
77. Wang, D.-C.; Fan, C.; Xie, Y.-M.; Yao, S.; Song, H. *Arabian J. Chem.* **2015**. In press.
78. Kiruthika, S. E.; Lakshmi, N. V.; Banu, B. R.; Perumal P. T. *Tetrahedron Lett.* **2011**, *52*, 6508–6511.
79. Wang, C.; Jiang Y.-H.; Yan C.-G. *J. Org. Chem.* **2014**, *10*, 2671–2676.
80. Zhang, L.-J.; Wu, Q.; Sun J.; Yan C.-G. *J. Org. Chem.* **2013**, *9*, 846–851.
81. Elinson, M. N.; Ilovaisky, A. I.; Merkulova, V. M.; Barba, F; Batanerob, B. *Tetrahedron* **2013**, *68*, 5833–5837.

13 Synthesis of 3,4-Dihydropyrimidin-2(1*H*)-Ones/Thiones and Polyhydroquinolines via Multicomponent Reactions

*Thomas J. Laughlin, Joshua S. Yoo, and Ram S. Mohan**

CONTENTS

13.1 INTRODUCTION

Multicomponent reactions (MCRs) are particularly attractive from a green chemistry perspective because they reduce the number of synthetic steps and eliminate waste products. Herein we report two examples of the use of an MCR for the synthesis of heterocycles with biological activities.

13.2 SYNTHESIS OF 3,4-DIHYDROPYRIMIDIN-2(1*H*)-ONES/THIONES

The synthesis of dihydropyrimidinones and dihydropyrimidine thiones has attracted interest because of their biological activities. They have been reported to show antibacterial and cytotoxic properties,[1] calcium-channel blocking activities,[2] and antimitotic properties.[3] Although the first synthesis of dihydropyrimidinones was reported in 1893 by Biginelli,[4] it is only since the 1990s that several new methods for their synthesis have been reported. The Biginelli reaction is remarkable not only for its

* Corresponding author: rmohan@iwu.edu.

simplicity (the reaction can be carried out by heating a mixture of an aldehyde, urea, and ethyl acetoacetate though yields are somewhat low) but also for the fact that it represents one of the early examples of an MCR.[5] Kappe and coworkers have extensively studied the Biginelli reaction and reported several elegant mechanistic studies on this reaction.[6] Several catalysts have been reported for the Biginelli reaction including Lewis acids and bases. A few representative examples of Lewis acid/acidic catalysts used for the Biginelli reaction are listed here: Mesoporous aluminosilicate,[7] PTSA,[8] tetra-Butyl ammonium bromide,[9] CaF_2,[10] $LaCl_3$,[11] $ZrCl_4$ and $ZrOCl_2$,[12] HBF_4,[13] $Y(NO_3)_3 \cdot 6H^2O$,[14] CuI,[15] $PhB(OH)_2$,[16] $RuCl_3$,[17] $SnCl_2 \cdot 2H_2O$,[18] $CuCl_2 \cdot 2H_2O$,[19] $Bi(OTf_3)_3$,[20] $Cu(OTf)_2$,[21] H_3BO_3,[22] VCl_3,[23] LiBr,[24] $Yb(OTf)_3$,[25] $InBr_3$,[26] $LiClO_4$,[27] $InCl_3$,[28] $La(OTf)_3$,[29] $FeCl_3 \cdot 6H_2O$,[30] Montmorillonite KSF,[31] and $BF_3 \cdot Et_2O/CuCl$.[32]

The Biginelli reaction has also been catalyzed under basic conditions, but there are fewer such reports. Some such catalysts include $(NH_4)_2CO_3$,[33] Ph_3P,[34] and *t*-BuOK.[35] Organocatalysts such as L-proline[36] and pyrazolidine dihydrochloride[37] have also been used to catalyze the Biginelli reaction. A few examples of the asymmetric Biginelli reaction have been reported as well.[38] Although a handful of the reported catalysts are environmentally friendly, many others are either toxic or corrosive, and difficult to handle. Our continued interest in developing environmentally friendly synthetic methodologies prompted us to investigate the utility of iron(III) tosylate, $Fe(OTs)_3 \cdot 6H_2O$ as a catalyst for the Biginelli reaction. Iron(III) tosylate is an inexpensive, commercially available, and easy to handle solid[39] that has only been recently utilized as a catalyst in organic synthesis.[40]

The results of this study are summarized in Table 13.1. As can be seen from this table, iron(III) tosylate (5.0 mol%) is an efficient catalyst for the synthesis of dihydropyrimidinones and dihydropyrimidine thiones from a wide range of aldehydes. Although the reaction worked with lower catalyst loading, the best results were obtained with the use of 5.0 mol% catalyst. Both isopropanol and octane worked as solvents. Although octane is less environmentally friendly than isopropanol, it was easily recovered using a rotary evaporator or by decantation, and recycled. Both procedures avoid an aqueous workup and hence large aqueous waste streams were avoided, thus adding to the green aspect of the methodology. Additionally, we have previously shown that iron(III) tosylate is an efficient catalyst for the deprotection of acetals.[40d] Hence, we attempted the Biginelli reaction with a range of acetals and found promising results (entries 7a–d and 8d). To the best of our knowledge, this is the first example of the use of acetals in the Biginelli reaction. Since acetals are frequently used to protect aldehydes, this method allows the direct synthesis of useful heterocycles from acetals without the need for an additional deprotection step.

The observation that a solution of iron(III) tosylate in water is acidic (pH ~ 2) suggests that the *p*-TsOH might be an active catalyst in this reaction, though the role of Fe^{3+} as a Lewis acid cannot be ruled out. The use of $Fe(OTs)_3$ is still preferable to *p*-TsOH because the latter compound is highly toxic and poses a health hazard.[39]

In summary, a new method for the Biginelli reaction that works with both aldehydes and acetals has been developed using an inexpensive, commercially available catalyst, iron(III) tosylate.

TABLE 13.1
Iron(III) Tosylate Catalyzed Synthesis of 3,4-Dihydropyrimidin-2(1*H*)-Ones/Thiones via the Biginelli Reaction

$Fe(OTs)_3{\cdot}6H_2O$ (5.0 mol%)
isopropanol or octane
reflux

X = O or S

Entry[a]	Aldehyde[b]	X	Solvent	Temp (°C)[c]	t (h)[d]	Yield (%)[e]
1a[41]	*p*-$CH_3OC_6H_4CHO$	O	Isopropanol	70	28	90
1b[41]	*p*-$CH_3OC_6H_4CHO$	O	Octane	125	17	65
1c[41]	*p*-$CH_3OC_6H_4CHO$	S	Isopropanol	70	15	83
1d[41]	*p*-$CH_3OC_6H_4CHO$	S	Octane	125	3.5	82
2[41]	*p*-HOC_6H_4CHO	O	Isopropanol	70	43	62
3[42]	CHO	O	Isopropanol	70	40.25	80
4a[41]	*p*-$CH_3C_6H_4CHO$	O	Isopropanol	70	23	67
4b[41]	*p*-$CH_3C_6H_4CHO$	S	Octane	125	2.5	73
5[43]	*m*-$CH_3OC_6H_4CHO$	O	Isopropanol	70	8	78
6[44]	*p*-BrC_6H_4CHO	O	Isopropanol	70	24.25	86
7a[44]	*p*-$BrC_6H_4CH(OEt)_2$	O	Isopropanol	70	15.5	74
7b[44]	*m*-$BrC_6H_4CH(OEt)_2$	O	Isopropanol	70	22.75	77
7c[44]	*o*-$BrC_6H_4CH(OEt)_2$	O	Isopropanol	70	41	84
7d[44]	*o*-BrC_6H_4	O	Isopropanol	70	27.5	79
8a[41]	*p*-ClC_6H_4CHO	O	Isopropanol	70	19.5	90
8b[41]	*p*-ClC_6H_4CHO	O	Octane	125	2.5	83
8c[41]	*p*-ClC_6H_4CHO	S	Octane	125	2.5	77
8d[41]	*p*-$ClC_6H_4CH(OMe)_2$	O	Isopropanol	70	21	65
9[45]	2,4-$Cl_2C_6H_3CHO$	O	Isopropanol	70	26.75	77

[a] Superscript against entry number refers to literature reference for product.
[b] All starting materials are commercially available and were used as received.
[c] Reaction mixtures were heated using a temperature-controlled hot plate.
[d] The progress of the reaction was followed by TLC.
[e] Refers to yield of the isolated product that was determined to be ≥98% pure by ^1H and ^{13}C NMR. All products have been reported previously and were characterized by ^1H and ^{13}C NMR spectroscopy, and by comparing their melting points to literature values.

Two representative procedures are given here:

- *Reaction in isopropanol (entry 1c).* A mixture of *p*-anisaldehyde (2.00 g, 1.79 mL, 14.7 mmol), thiourea (3.35 g, 44.1 mmol, 3.0 equiv.) and ethyl acetoacetate (5.73 g, 5.56 mL, 44.0 mmol, 3.0 equiv.) in isopropanol (20 mL) was stirred at room temperature as iron(III) tosylate (0.497 g, 0.734 mmol, 5.0 mol%) was added. The reaction mixture was heated at 70 °C (temperature-controlled hot plate). After 15 h, the mixture was cooled to room temperature. The solvent was then removed on a rotary evaporator, and the residue was recrystallized (ethanol) to yield 5.48 g of an off-white solid. This solid was then triturated using methanol/water (1:1) to yield 3.75 g (83%) of a light, off-white solid. ^1H NMR (d_6 DMSO, 270 MHz) δ 1.1 (t, 3H, J = 7.2 Hz), 2.3 (s, 3H), 3.7 (s, 3H), 4.0 (q, 2H, J = 7.2 Hz), 5.1 (d, 1H, J = 3.7 Hz), 6.9 (d, 2H, J = 8.9 Hz), 7.1 (d, 2H, J = 8.91 Hz), 9.6 (s, 1H), 10.3 (s, 1H); ^{13}C NMR (d_6 DMSO, 67.5 MHz) (13 peaks) δ 14.0, 17.2, 53.5, 55.1, 59.6, 101.0, 113.9, 127.6, 135.7, 144.8, 158.8, 165.2, 174.0.

- *Reaction in octane (entry 1d).* A mixture of *p*-anisaldehyde (0.501 g, 0.447 mL, 3.68 mmol), thiourea (0.839 g, 11.02 mmol, 3.0 equiv.) and ethyl acetoacetate (1.43 g, 1.39 mL, 11.02 mmol, 3.0 equiv.) in octane (10 mL) was stirred at room temperature as iron(III) tosylate (0.125 g, 0.185 mmol, 5.0 mol%) was added. The reaction mixture was heated at 125 °C (temperature-controlled hot plate). After 3.5 h, the mixture was cooled to room temperature. The solvent was then decanted off, and the residue was recrystallized (ethanol) to yield 0.9075 g (82%) of a solid with a yellow tint.

13.3 SYNTHESIS OF POLYHYDROQUINOLINES VIA A MODIFIED HANTZSCH REACTION

The synthesis of 1,4-dihydropyridine derivatives such as polyhydroquinolines (PHQs) is of considerable interest due to their biological properties. Their derivatives, such as nifedipine, are calcium channel antagonists.[46] Polyhydroquinolines bearing a tetraazole moiety have been shown to possess antimicrobial activity.[47] 2,4-Disubstituted polyhydroquinolines also show promise as antihyperglycemic and lipid modulating agents.[48] Polyhydroquinolines have also been shown to reduce cellular tau levels and therefore hold promise in the search for therapies for Alzheimer's disease.[49] Many PHQ derivatives also show activity against multidrug resistance proteins.[50] Hence, there has been considerable interest in methods for their synthesis. The most common route to polyhydroquinolines **5** involves a modified Hantzsch reaction[51] involving the multicomponent coupling of an aldehyde **1** with a β-ketoester (such as ethyl acetoacetate) **2**, a diketone (such as dimedone) **3**, and ammonium acetate **4** (Scheme 13.1).

Some representative examples of catalysts used for this coupling include I_2,[52] metal triflates,[53] ceric ammonium nitrate,[54] Baker's yeast,[55] organocatalysts,[56] zeolites,[57] Ph_3P,[58] $GaCl_3$,[59] thiourea dioxide,[60] thiazolium ion,[61] sulphamic acid,[62]

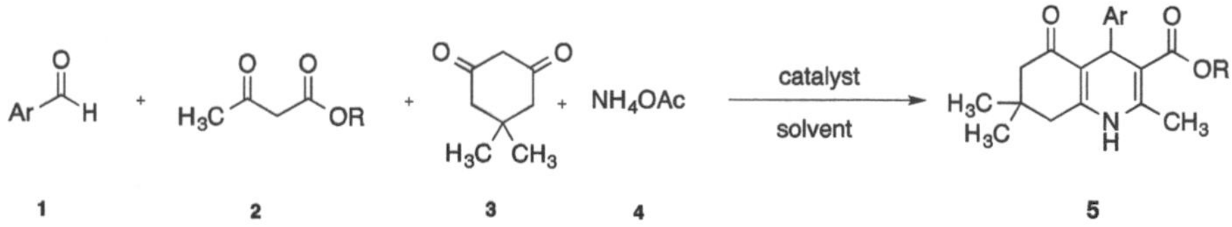

Ar–CHO (**1**) + H_3C–CO–CH_2–CO–OR (**2**) + dimedone (**3**) + NH_4OAc (**4**) → (catalyst, solvent) → **5**

SCHEME 13.1 Synthesis of polyhydroquinolines.

$Al_2(SO_4)_3$,[63] La_2O_3,[64] and nanoparticles.[65] Syntheses of PHQs have also been carried out in various ionic liquids as solvents.[66] Their syntheses have also been reported under catalyst-free conditions, but the methods suffer from longer reaction times.[67] Solvent and catalyst-free methods involving grinding the reaction mixtures have also been reported and are certainly attractive from a green chemistry perspective. But these methods require the use of solvent for product isolation and are not always amenable to scale up.[68] PHQs have also been synthesized under microwave irradiation conditions.[69] The most elegant synthesis of PHQs to date is their enantioselective synthesis using a phosphoric acid derived organocatalyst.[70]

Given our continued interest in bismuth(III) salts as catalysts due to their remarkably nontoxic nature (many bismuth salts are even less toxic than NaCl) and ease of handling, we explored the utility of bismuth bromide, $BiBr_3$ as a catalyst for the synthesis of PHQs.[71] Herein we report an efficient synthesis of a variety of polyhydroquinolines using bismuth(III) bromide as a catalyst in ethanol as the solvent (Table 13.2). The reactions are fast, and product is isolated by filtration of the reaction mixture. Hence, no aqueous waste stream is generated. In the absence of $BiBr_3$, the reactions took a considerably longer time and lower yields were obtained. For example, *p-N,N*-dimethylaminobenzaldehyde (Table 13.2, entry 2) yielded 61% product (t = 23 h), and *p*-chlorobenzaldehyde (Table

TABLE 13.2
Synthesis of Polyhydroquinolines Catalyzed by $BiBr_3$ (2.0 mol%)[72]

Ar–CHO + H_3C–CO–CH_2–CO–OR + dimedone + NH_4OAc → ($BiBr_3$ (2 mol%), CH_3CH_2OH, rt) polyhydroquinoline

Entry	Ar	R	Time (h)[a]	Yield (%)[b,c]
1	*p*-$CH_3C_6H_4$	CH_2CH_3	1.5	86[73]
2	*p*-$NMe_2C_6H_4$	CH_2CH_3	2.5	87[8a]
3	*p*-$CH_3OC_6H_4$	CH_2CH_3	3	89[28]
4	*p*-ClC_6H_4	CH_2CH_3	2.5	93[20f]
5	*p*-$NO_2C_6H_4$	CH_2CH_3	1.5	87[20f]
6	*m*-$CH_3OC_6H_4$	CH_2CH_3	1.5	89[74]
7	*m*-$NO_2C_6H_4$	CH_2CH_3	1.5	79[20f]
8	*m*-$CH_3C_6H_4$	CH_2CH_3	2	85[75]
9	2-Naphthyl	CH_2CH_3	2.5	83[25]
10	*p*-biphenyl	CH_2CH_3	3	88[25]
11	2-Furyl	CH_2CH_3	1	87[28]
12	*p*-$CH_3C_6H_4$	CH_3	1.5	90[20f]

[a] Reaction progress was monitored by TLC.

[b] Refers to yield of isolated, purified product.

[c] All products have been previously reported and were identified by comparing their melting point and spectral data to those reported in the literature. Superscript against yield refers to literature reference for product. ^1H and ^{13}C-NMR are provided in the SI section.

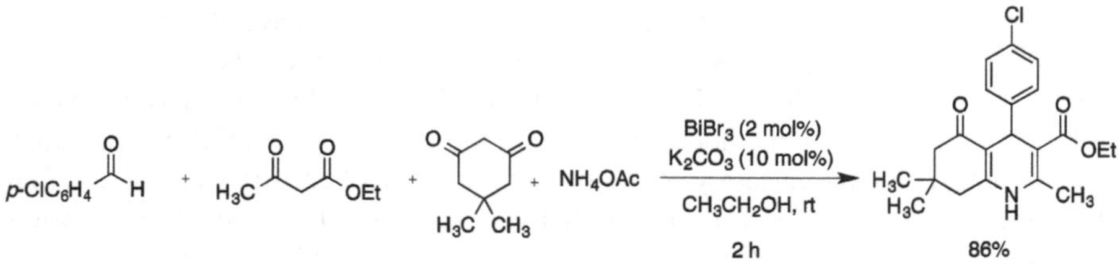

SCHEME 13.2 Synthesis of polyhydroquinolines using $BiBr_3$.

13.2, entry 4) yielded 65% product (t = 7 h) under catalyst-free conditions. We found that the reaction was also efficiently catalyzed by bismuth triflate (2.0 mol%). For example, *p*-tolualdehyde (Table 13.2, entry 1) yielded 86% product (t = 2 h), but owing to the higher cost of bismuth triflate ($225/25 g from Acros Organics) relative to bismuth bromide ($25/25 g) and its increased moisture sensitivity, we chose bismuth bromide. Ytterbium triflate, $Yb(OTf)_3$ (5.0 mol%), has been reported as a catalyst[8a] for the formation of polyhydroquinolines but a higher catalyst loading (5 mol%) and its high cost detract from its utility ($395/25 g for the anhydrous salt, and $136/25 g for the monohydrate). Although we used anhydrous ethanol as the reaction solvent, one cannot rule out hydrolysis of $BiBr_3$ by adventitious water to generate HBr, which could act as a Bronsted acid catalyst. In order to get some insight into the role of $BiBr_3$ in the reaction, we carried out the reaction (Scheme 13.2) of *p*-chlorobenzaldehyde in the presence of $BiBr_3$ (2.0 mol%) and solid potassium carbonate (10 mol%). The pH of this reaction mixture was found to be ~6–7, and yet product was obtained in 86% yield, suggesting that $BiBr_3$ is acting as a Lewis acid and not just as a source of HBr.

The mild reaction conditions with $BiBr_3$ allow for easy scale up as well. In summary, a mild and efficient method for the multicomponent synthesis of polyhydroquinolines using a relatively nontoxic and noncorrosive catalyst is reported.

ACKNOWLEDGMENTS

This material is based on work supported by the National Science Foundation under CHE-1229133, which funded the purchase of 400 MHz NMR spectrometer. We also acknowledge funding from The American Chemical Society–Petroleum Research Fund (ACS PRF# 51036-UR1).

REFERENCES

1. (a) For a comprehensive review, see Kappe, O. C. *Eur. J. Med. Chem.* **2000**, *35*, 1043. (b) Deshmukh, M. B.; Salunkhe, S. M.; Patil, D. R.; Anbhule, P. V. *Eur. J. Med. Chem.* **2009**, *44*, 2651. (c) Azizian, J.; Mohammadi, M. K.; Firuzi, O.; Mirza, B.; Miri, R. *Chem. Biol. Drug. Des.* **2010**, *75*, 375.
2. (a) Singh, K.; Arora, D.; Poremsky, E.; Lowery, J.; Moreland, R. S. *Eur. J. Med. Chem.* **2009**, *44*, 1997. (b) Kappe, C. O. *Molecules* **1998**, *3*, 1. (c) Rovnyak, G. C.; Atwal, K. S.; Hedberg, A.; Kimball, S. D.; Moreland, S.; Gougoutas, J. Z.; O'Reilly, B. C.; Schwartz, J.; Malley, M. F. *J. Med. Chem.* **1992**, *35*, 3254. (d) Jauk, B.; Pernat, T.; Kappe, C. O. *Molecules* **2000**, *5*, 227.

3. Mayer, T. U.; Kapoor, T. M.; Haggarty, S. J.; King, R. W.; Schreiber, S. L.; Mitchison, T. J. *Science* **1999**, *286*, 971.
4. Biginelli, P. *Gazz. Chim. Ital.* **1893**, *23*, 360.
5. (a) Tour, B. B.; Hall, D. G. *Chem. Rev.* **2009**, *109*, 4439. (b) Tietze, L. F. *Chem. Rev.* **1996**, *96*, 115. (c) Dömling, A.; Ugi, I. *Angew. Chem. Int. Ed. Eng.* **1993**, *32*, 563.
6. Kappe, C. O. *Acc. Chem. Res.* **2000**, *33*, 879.
7. Shobha, D.; Chari, M. A.; Mano, A.; Selvan, S. T.; Mukkanti, K.; Vinu, A. *Tetrahedron* **2009**, *65*, 10608.
8. Matache, M.; Dobrota, C.; Bogdan, N. D.; Dumitru, I.; Ruta, L. L.; Paraschivescu, C. C.; Farcasanu, I. C.; Baciu, I.; Funeriu, D. P. *Tetrahedron* **2009**, *65*, 5949.
9. Ahmed, B.; Khan, R. A.; Habibullah; Keshari, M. *Tetrahedron Lett.* **2009**, *50*, 2889.
10. Chitra, S.; Pandiarajan, K. *Tetrahedron Lett.* **2009**, *50*, 2222.
11. Khabazzadeh, H.; Saidi, K.; Sheibani, H. *Bioorg. Med. Chem. Lett.* **2008**, *18*, 278.
12. Rodríguez-Domínguez, J. C.; Bernardi, D.; Kirsch, G. *Tetrahedron Lett.* **2007**, *48*, 5777.
13. Chen, W.; Qin, S.; Jin, J. *Catal. Commun.* **2007**, *8*, 123.
14. Nandurkar, N. S.; Bhanushali, M. J.; Bhor, M. D.; Bhanage, B. M. *J. Mol. Catal. A: Chem.* **2007**, *271*, 14.
15. Kalita, H. R.; Phukan, P. *Catal. Commun.* **2007**, *8*, 179.
16. Debache, A.; Boumoud, B.; Amimour, M.; Belfaitah, A.; Rhouati, S.; Carboni, B. *Tetrahedron Lett.* **2006**, *47*, 5697.
17. De, S. K.; Gibbs, R. A. *Synthesis* **2005**, *11*, 1748.
18. Russowsky, D.; Lopes, F. A.; da Silva, V. S. S.; Canto, K. F. S.; D'Oca, M. G. M.; Godoi, M. N. *J. Braz. Chem. Soc.* **2004**, *15*, 165.
19. Gohain, M.; Prajapati, D.; Sandhu, J. S. *Synlett* **2004**, *2*, 235.
20. Varala, R.; Alam, M. M.; Adapa, S. R. *Synlett* **2003**, *1*, 67.
21. Paraskar, A. S.; Dewkar, G. K.; Sudalai, A. *Tetrahedron Lett.* **2003**, *44*, 3305.
22. Tu, S.; Fang, F.; Miao, C.; Jiang, H.; Feng, Y.; Shi, D.; Wang, X. *Tetrahedron Lett.* **2003**, *44*, 6153.
23. Sabitha, G.; Reddy, G. S. K. K.; Reddy, K. B.; Yadav, J. S. *Tetrahedron Lett.* **2003**, *44*, 6497.
24. Maiti, G.; Kundu, P.; Guin, C. *Tetrahedron Lett.* **2003**, *44*, 2757.
25. Dondoni, A.; Massi, A.; Sabbatini, S. *Tetrahedron Lett.* **2002**, *43*, 5913.
26. Fu, N-Y.; Yuan, Y-F.; Cao, Z.; Wang, S-W.; Wang, J-T.; Peppe, C. *Tetrahedron* **2002**, *58*, 4801.
27. Yadav, J. S.; Reddy, B. V. S.; Srinivas, R.; Venugopal, C.; Ramalingam, T. *Synthesis.* **2001**, *9*, 1341.
28. Ranu, B. C.; Hajra, A.; Jana, U. *J. Org. Chem.* **2000**, *65*, 6270.
29. Ma, Y.; Qian, C.; Wang, L.; Yang, M. *J. Org. Chem.* **2000**, *65*, 3864.
30. Lu, J.; Ma, H. *Synlett* **2000**, 63.
31. Bigi, F.; Carloni, S.; Frullanti, B.; Maggi, R.; Sartori, G. *Tetrahedron Lett.* **1999**, *40*, 3465.
32. Hu, E. H.; Sidler, D. R.; Dolling, U-H. *J. Org. Chem.* **1998**, *63*, 3454.
33. Tamaddon, F.; Ramzi, Z.; Jafari, A. A. *Tetrahedron Lett.* **2010**, *51*, 1187.
34. Debache, A.; Amimour, M.; Belfaitah, A.; Rhouati, S.; Carboni, B. *Tetrahedron Lett.* **2008**, *49*, 6119.
35. Shen, Z-L.; Xu, X-P.; Ji, S-J. *J. Org. Chem.* **2010**, *75*, 1162.
36. Pandey, J.; Anand, N.; Tripathi, R. P. *Tetrahedron* **2009**, *65*, 9350.
37. Suzuki, I.; Iwata, Y.; Takeda, K. *Tetrahedron Lett.* **2008**, *49*, 3238.
38. (a) Jiang, J.; Yu, J.; Sun, X-X.; Rao, Q-Q.; Gong, L-Z. *Angew. Chem. Int. Ed.* **2008**, *47*, 2458. (b) Li, N.; Chen, X-H.; Song, J.; Luo, S-W.; Fan, W.; Gong, L-Z. *J. Am. Chem Soc.* **2009**, *131*, 15301. (c) Chen, X-H.; Xu, X-Y.; Liu, H.; Cun, L-F.; Gong, L-Z. *J. Am. Chem. Soc.* **2006**, *128*, 14802.

39. According to the material safety data sheets (MSDSs), iron(III) tosylate is safer to use than *p*-toluenesulfonic acid. Iron(III) tosylate has a HMIS (hazardous material identification system) rating of 2 (moderate hazard), while *p*-toluenesulfonic acid has a HMIS rating of 3 (serious hazard).
40. For examples of use of $Fe(OTs)_3$ as a catalyst in organic synthesis, see (a) Spafford, M. J.; Anderson, E. D.; Lacey, J. R.; Palma, A. C.; Mohan, R. S. *Tetrahedron Lett.* **2007**, *48*, 8665. (b) Mansilla, H.; Afonso, M. M. *Synth. Commun.* **2008**, *38*, 2607. (c) Bothwell, J. M.; Angeles, V. V.; Carolan, J. P.; Olson, M. E.; Mohan, R. S. *Tetrahedron Lett.* **2010**, *51*, 1056. (d) Carolan, J. P.; Chiodo, M. V.; Lazzara, P. R.; Olson, M. E.; Mohan, R. S. *Tetrahedron Lett.* **2010**, *51*, 3969. (e) Bockman, M. R.; Angeles, V. V.; Martino, J. M.; Vagadia, P. P.; Mohan, R. S. *Tetrahedron Lett.* **2011**, *52*, 6939. (f) Baldwin, N. J.; Nord, A. N.; O'Donnell, B. D.; Mohan, R. S. *Tetrahedron Lett.* **2012**, *53*, 6946.
41. Fu, N-Y.; Yuan, Y-F.; Cao, Z.; Wang, S-W.; Wang, J-T.; Peppe, C. *Tetrahedron* **2002**, *58*, 4801.
42. Ghosh, R.; Maiti, S.; Chakraborty, A. *J. Mol. Cat.* **2004**, *217*, 47.
43. Badaoui, H.; Bazi, F.; Tamani, S.; Boulaajaj, S. *Synthetic Commun.* **2005**, *35*, 2561.
44. Pasunooti, K. K.; Chai, H.; Jensen, C. N.; Gorityala, B. K.; Wang, S.; Liu, X-W. *Tetrahedron Lett.* **2011**, *52*, 80.
45. Karade, H. N.; Sathe, M.; Kaushik, M. P. *Molecules* **2007**, *12*, 1341.
46. (a) Rosenberger, L.; Triggle, D. J. in *Calcium and drug action*, G. B. Weiss, Ed., Plenum Press, New York, 1978, p. 1. (b) van Zweiten, P. A.; E. Schönbaum, Eds., *Prog. Pharmacol.* **1978**, *2(1)*, 39. (c) Fleckenstein, A. *Annu. Rev. Pharmacol. Toxicol.* **1977**, *17*, 149.
47. Ladani, N. K.; Mungra, D. C.; Patel, M. P.; Patel R. G. *Chinese Chem. Lett.* **2011**, *22*, 1407.
48. Kumar, A.; Sharma, S.; Tripathi, V.D.; Maurya, R.A.; Srivastava, S. P.; Bhatia, G.; Tamrakar, A. K.; Srivastava, A. K. *Bioorga. Med. Chem.* **2010**, *18*, 4138.
49. Evans, C. G.; Jinwal, U. K.; Makley, L. N.; Dickey, C. A.; Gestwicki, J. E. *Chem. Commun.* **2011**, *47*, 529.
50. Miri, R.; Mehdipour, A. *Bioorg. Med. Chem.* **2008**, *16*, 8329.
51. Hantzsch, A. *Justus Liebigs Annalen der Chemie* **1882**, *215*, 1.
52. Ko, S.; Sastry, M. N. V.; Lin, C.; Yao, C-F. *Tetrahedron Lett.* **2005**, *46*, 5771.
53. (a) Yb(OTf)3: Wang, L-M.; Sheng, J.; Zhang, L.; Han, J-W.; Fan, Z-Y.; Tian, H.; Qian, C-T. *Tetrahedron* **2005**, *61*, 1539. (b) Sc(OTf)3: Donelson, J. L.; Gibbs, R. A.; De, S. K. *J. Mol. Cat. A: Chemical* **2006**, *256*, 309.
54. Ko, S.; Yao, C-F. *Tetrahedron* **2006**, *62*, 7293.
55. Kumar, A.; Maurya, R. A. *Tetrahedron Lett.* **2007**, *48*, 3887.
56. Kumar, A.; Maurya, R. A. *Tetrahedron* **2007**, *63*, 1946.
57. Gadekar, L. S.; Katkar, S. S.; Mane, S. R.; Arbad, B. R.; Lande, M. K. *Bull. Korean Chem. Soc.* **2009**, *30*, 2532.
58. Debache, A.; Ghalem, W.; Boulcina, R.; Belfaitah, A.; Rhouati, S.; Carboni, B. *Tetrahedron Lett.* **2009**, *50*, 5248.
59. Patil, D. R.; Deshmukh, M. B.; Salunkhe, S. M.; Salunkhe, D. K.; Kolekar, G. B.; Anbhule, P. V. *Der Pharma Chemica* **2010**, *2*, 342.
60. Kumar, N.; Verma, S.; Jain, S. L. *Chemistry Lett.* **2012**, *41*, 920.
61. Fatma, S.; Ankit, P.; Singh, M.; Singh, S. B.; Singh, J. *Synthetic Commun.* **2014**, *44*, 1810.
62. Lambat, T.; Deo, S.; Deshmukh, T. *J. Chem. Pharm. Res.* **2014**, *6*, 888.
63. Kulkarni, P. *J. Chil. Chem. Soc.* **2014**, *59*, 2319.
64. Tekale, S. U.; Pagore, V. P.; Kauthale, S. S.; Pawar, R. P. *Chinese Chem. Lett.* **2014**, *25*, 1149.

65. (a) Ranjbar-Karimi, R.; Hashemi-Uderji, S.; Bazmandegan-Shamili, A. *Chinese J. Chem.* **2011**, *29,* 1624. (b) Shirini, F.; Atghia, S. V.; Khoshdel, M. A. *Iranian J. Catal.* **2011**, *1*, 93. (c) Saikia, L.; Dutta, D.; Dutta, D. K. *Catal. Commun.* **2012**, *19*, 1. (d) Vahdat, S. M.; Chekin, F.; Hatami, M.; Khavarpour, M.; Baghery, S.; Roshan-Kouhi, Z. *Chinese J. Chem.* **2013**, *34*, 758. (e) Kiasat, A. R.; Almasi, H.; Saghanezhad, S. J. *Rev. Roum. Chim.* **2014**, *59*, 61. (f) Nasr-Esfahani, M.; Hoseini, S. J.; Montazerozohori, M.; Mehrabi, R.; Nasrabadi, H. *J. Mol. Catal. A: Chemical* **2014**, *382,* 99.
66. (a) Nirmal, J. P.; Dadhaniya, P. V.; Patel, M. P.; Patel, R. G. *Indian J. Chem. B* **2010**, *49B*, 587. (b) Rajendran, A.; Karthikeyan, C.; Rajathi, K. *Int. J. ChemTech Res.* **2011**, *3*, 810. (c) Raghuvanshi, D. S.; Singh, K. N. *Indian J. Chem. B* **2013**, *52B*, 1218. (d) Rostamnia, S.; Hassankhani, A. *Synlett* **2014**, *25*, 2753.
67. Undale, K. A.; Shaikh, T. S.; Gaikwad, D. S.; Pore, D. M. *Comptes Rendus Chimie* **2011**, *14*, 511.
68. (a) Arumugam, P.; Perumal, P. T. *Indian J. Chem. B* **2008**, *47B*, 1084. (b) Kumar, S.; Sharma, P.; Kapoor, K. K.; Hundal, M. S. *Tetrahedron* **2008**, *64*, 536.
69. (a) Zhang, X-L.; Sheng, S-R.; Liu, X-L.; Liu, X-L. *ARKIVOC* **2007**, *xiii*, 79. (b) Das, S.; Santra, S.; Roy, A.; Urinda, S.; Majee, A.; Hajra, A. *Green Chem. Lett. Rev.* **2012**, *5*, 97. (c) Joshi, V. M.; Pawar, R. P. *Eur. Chem. Bull.* **2013**, *2*, 679.
70. Evans, C. G.; Gestwicki, J. E. *Org. Lett.* **2009**, *11*, 2957.
71. For a review on the utility of bismuth compounds in organic synthesis, see (a) Ollevier, T. *Org. Biomol. Chem.* **2013**, *11*, 2740. (b) Krabbe, S. W.; Bothwell, J. M.; Mohan, R. S. *Chem. Soc. Rev.* **2011**, *40*, 4649. For some recent examples of the use of bismuth(III) compounds in organic synthesis, see (c) Coca, A.; Feinn, L.; Dudley, J. *Synthetic Commun.* **2015**, *45*, 1023. (d) Sabitha, G.; Shankaraiah, K.; Sindhu, K.; Latha, B. M. *Synthesis* **2015**, *47*, 124. (e) Sheng, J.; Wu, J. *Org. Biomol. Chem.* **2014**, *12*, 7629. (f) Moskalenko, A. I.; Boev, V. I. *Russian. J. Org. Chem.* **2014**, *50*, 1117. (g) Nitsch, D.; Bach, T. *J. Org. Chem.* **2014**, *79*, 6372. (h) Murai, M.; Origuchi, K.; Takai, K. *Org. Lett.* **2014**, *16*, 3828. (i) Rasmussen, M. R.; Marqvorsen, M. H. S.; Kristensen, S. K.; Jensen, H. H. *J. Org. Chem.* **2014**, *79*, 11011. (j) Schneider, A. E.; Beisel, T.; Shemet, A.; Manolikakes, G. *Org. Biomol. Chem.* **2014**, *12*, 2356.
72. Representative procedure: A homogeneous mixture of *p*-tolualdehyde (0.5000 g, 4.16 mmol), dimedone (0.6417 g, 4.58 mmol, 1.1 equiv.), ethyl acetoacetate (0.5968 g, 0.58 mL, 4.58 mmol, 1.1 equiv.), and ammonium acetate (0.3528 g, 4.58 mmol, 1.1 equiv.) was stirred in anhydrous ethanol (10.0 mL) at room temperature as BiBr3 (0.0373 g, 2.0 mol%) was added. The reaction progress was monitored by TLC (2,4-DNP stain). After 1.5 h, the reaction mixture was poured onto 20 g of ice and the resulting yellow precipitate was collected via suction filtration. The crude product was recrystallized using anhydrous ethanol (approximately 30 mL) to yield 1.2694 g (86% yield) of a slightly off-white, powdery compound. Mpt: 258 °C–259 °C (Lit: 258°C–259°C).
73. Heravi, M. M.; Saeedi, M.; Karimi, N.; Zakeri, M.; Beheshtiha, Y. S.; Davoodnia, A. *Synthetic Commun.* **2010**, *40*, 523.
74. Maleki, B.; Tayebee, R.; Kermanian, M.; Ashrafi, S. S. *J. Mex. Chem. Soc.* **2013**, *57*, 290.
75. Ghasemzadeh, M. A.; Safaei-Ghomi, H. *J. Chem. Res.* **2014**, *38*, 313.

Index

Page numbers followed by f and t indicate figures and tables, respectively.

C

D

G

H

I

J

K

N

O

P

Q

R

S

T

U

V

W

X

Y

Z

Milton Keynes UK
Ingram Content Group UK Ltd.
UKHW021825071024
449327UK00021B/1435